Structures and Their Analysis

Maurice Bernard Fuchs

Structures and Their Analysis

Maurice Bernard Fuchs
School of Mechanical Engineering
Tel Aviv University
Tel Aviv
Israel

ISBN 978-3-319-80958-8 ISBN 978-3-319-31081-7 (eBook)
DOI 10.1007/978-3-319-31081-7

Softcover reprint of the hardcover 1st edition 2016

Printed on acid-free paper

This Springer imprint is published by Springer Nature
The registered company is Springer International Publishing AG Switzerland

For
Rinna

Preface

There are two main methods for analyzing a structure: the *force method* and the *displacement method*. Some will argue the former is passé. We do not follow that argument. The force method may have lost its prominence as a structural analysis tool, but it is still conducive to the way engineers think about a structure. Forces and equilibrium are intuitive from infancy, for example, when we tried out our first solo steps. There is also the early childhood sense of triumph when an eighth wooden block is positioned correctly on a stack of seven, or the feeling of disappointment when the last cube makes the tower tumble.

This innate feeling of forces and equilibrium has remained with us. We usually think of structures in terms of forces, although in practice our computers will perform the analysis by way of displacements. This duality is reflected in this treatise.

My first book on structures was a text written by André Paduart of the Free University of Brussels (ULB). It was called M, N, T, R, δ. The title started out with the internal loads: M for bending Moment, N for Normal force, T for shear force (effort Tranchant—the text was in French), R for Reactions I guess, and lastly δ for displacements. At that time (1963) displacements were indeed a nuisance and you computed them only when you absolutely needed them. In those days the analysis method was the force method.

Things have changed. Nowadays the analysis method of choice is indisputably the displacement method, where the basic variables are the displacements of the structure. Once these have been computed the internal forces and reactions easily follow suit. The shift in paradigm was heralded by the advent of tremendous, readily available computing power. Arguably, a basic tenet in the classical days of structural analysis was never to have more than three linear equations with three unknowns at any time, no matter how complex the structure. This led to some ingenious modeling and clever assumptions in order to avoid the three-equations obstacle.

The displacement method, on the other hand, albeit deceptively simple and relatively boring, because it involves tedious matrix calculations, requires the solution of many linear equations with as many unknowns. With the finite-element method, an offshoot of the displacement method, it is not unusual to solve tens of thousands of linear equations when, for instance, considering the analysis of an aircraft wing or a cable-stayed bridge, a task perfectly suited to computerized calculations. Moreover, the writing of the equations, what we call assembly, is also automated in the displacement method; again something computers do well.

This is also how we present these two aspects of analysis: performing an analysis by the force method is a human endeavor. It can be applied to small and medium-sized structures. It requires an understanding of the behavior of the structure. Its results, the internal loads, can be understood and related to and, as we shall see, even qualitatively predicted in some cases. The displacement method, on the other hand, is for computers. It is algorithmic and can easily be programmed, it manipulates vast arrays of numbers, and does not require too deep an understanding of the process.

This book is intended for a graduate level of instruction although undergraduate material is embedded in the text. Like most structural courses it assumes that the reader has some basic knowledge of the field. Courses in Statics and Solid Mechanics (Strength of Materials) are prerequisites, and essentials of these courses are the first topics of this book.

Part I: Preliminaries

We start with principles of equilibrium (Chap. 1) from which we need to postulate the existence of internal loads. This is followed immediately by a probably too lengthy chapter (Chap. 2) on how to compute and display these internal forces by means of the *nsm*-diagrams. We next discuss the principle of virtual work (Chap. 3) without having introduced Hooke's law. This is very important from a didactic viewpoint. It shows that the principle, in its basic form, is a mathematical concept and not really connected to a structure (a structure is something that deforms under loading). We also show that for the principle to work we need to define quantities that turn out to be deformations (Chap. 4). This way the principle is not some hat trick but simply integration by parts, a mathematical tool. In the next chapter (Chap. 5) we introduce elasticity, that is, the linear relation between the internal forces and the deformations. When external loads are applied to a structure, it deforms so as to engender internal forces which balance the applied ones. We discuss the Timoshenko beam and its relation to the Bernoulli–Euler beam used in engineering beam theory, a graduate course topic. Finally, in the last chapter of Part I (Chap. 6) we introduce a technique for calculating a displacement or a rotation at any location along a structure. The method is closely connected to virtual work and is often taught in introductory courses to structures.

Part II: The Structures

The first chapter in Part II (Chap. 7) describes the types of structures dealt with in this book: beams, plane frames and, to a lesser extent, plane trusses. Such structures are composed of line elements fully or partially connected to one another or to the ground, as described in this chapter. The following chapter (Chap. 8) is an introduction to and a comparison of the two main methods of analysis. This is done by analyzing a stylized structure, a rigid beam on several elastic supports, by both the force method and the displacement method. The last chapter of Part II (Chap. 9) is usually taught to architectural students but unfortunately less so to engineering students (See, for instance, D.L. Schodek and M. Bechthold, *Structures*, 6th edition, Pearson Prentice Hall 2008.) We consider this material extremely important. It often gives a hint of the expected results before actually performing the analysis.

Part III: Flexibility

The flexibility matrix (Chap. 10) lies at the heart of the force method. The chapter starts with the Betti–Maxwell reciprocal theorems, the notion of degree of freedom, leading to the definition of influence or flexibility coefficients, which are the constituents of the flexibility matrix. Several illustrated examples close the chapter, but not before discussing why normal deformations in the presence of bending deformation are usually neglected. Redundancy is treated in the following chapter (Chap. 11). The force method is applicable to redundant structures only. Determining the degree of redundancy is therefore the first step of an analysis by the force method. We also consider redundancy in the context of safety and stiffness of structures. The next chapter is the core of the force method (Chap. 12). We learn the important notions of releasing a structure and writing the compatibility equations. Thereafter, the task is reduced to the analysis of a statically determinate structure where all the applied loads are known. The chapter closes with several illustrated problems. This is followed by a chapter dealing with the effect of heat on structures (Chap. 13). It includes prestressing and manufacturing deformations or misalignments. The chapter emphasizes that all this is applicable to redundant structures only. Determinate structures without applied forces are guaranteed to be free of internal forces.

Part IV: Stiffness

We introduce the stiffness method by way of trusses (Chap. 14). We find the analysis of trusses by the force method too tedious. The stiffness method, which is essentially a numerical method to be solved algorithmically, is ideally suited to trusses as an analysis tool. The following chapter (Chap. 15) introduces the stiffness matrix of an element. This potent concept allows us to analyze with the same ease any structure if we know the stiffness matrices of all its elements. An intermediate chapter (Chap. 16) essentially indicates a systematic way (congruent transformations) to obtain the element stiffness matrix in oblique coordinates. The following chapter (Chap. 17) is the core of the displacement (stiffness) method. We show the meaning of the equilibrium equations which compute the unknown nodal displacements. Next, we indicate the concept of 'assembly', an efficient algorithm to write the equilibrium equations, essentially, the system stiffness matrix. Everything we have done until now assumed that the structure was submitted to point loads applied to the nodes. This excludes distributed forces and temperature effects (applied strain). The final chapter of Part IV (Chap. 18) addresses this subject. By means of fixed-end reactions the problem is circumvented in a most elegant manner. This concludes the analysis methods presented in this text.

Part V: Four Additional Topics

The next part of the book touches on a few aspects of structural analysis and design which complement well the subject matter of this text.

A nice theorem which may be of practical use is presented next (Chap. 19). It states that a roller support of a continuous beam is optimally positioned (from a stiffness perspective) if the slope of the beam over the support is horizontal. A simple numerical technique to find the optimal location is also indicated.

A redundant truss cannot, in general, be fully stressed. In the following chapter (Chap. 20) we show a class of redundant trusses which can be fully stressed.

The next chapter (Chap. 21) indicates a way to cast the analysis equations of a frame in the form of truss equations by means of unimodal normal, shear, and moment elements. Consequently, a unified approach to structural analysis is possible by considering any frame as a generalized truss.

In the penultimate chapter of this book (Chap. 22) we present a very elegant formula for the internal forces in trusses explicitly in terms of the stiffnesses of the bars. Unfortunately, the formula is plagued by the so-called curse of dimensionality.

Part VI: Epilogue

Finally, in the concluding chapter (Chap. 23) we summarize the essentials of all that was said by means of the simplest structure of them all: an assembly of linear springs.

One of our guidelines when writing the book was to keep everything as simple as possible so as to convey the basic ideas without unnecessary burdens. We use plane structures, straight-linear uniform elements, uniformly distributed forces, beams with lateral forces only (which precludes normal internal forces in beams). Using planar structures ignores torsional effects, which is a drawback. Students will have ample leisure to include all this when taking a course in finite elements or when performing a finite-element analysis. We have also avoided talking of energy-related concepts (Castigliano, Clapeyron, etc.). Instead, we emphasize Virtual Work. The latter has a large enough gangplank to take all these methods on board. This was also the reason for avoiding, as much as possible, analyzing trusses manually. In our view trusses are computer territory. All this made it possible to make space for understanding structural behavior which should after all be the main objective of a book on structures.

I have read and browsed through many books on structures. A few stand out. KH Gerstle's *Basic Structural Analysis*, Prentice-Hall, 1974, is probably my first choice. I have benefitted a lot from Schodek's *Structures*, mentioned earlier. Then there is, of course, SO Asplund (*Structural Mechanics*: Classical and Matrix Methods, Prentice-Hall, 1966), an amazing book. I have seen it referred to as the bible of modern structural theory. I fully agree with this description.

I am indebted to Professor Z. Hashin (TAU) and to the late Professors M.A. Brull (TAU) and A. Libai (Technion), who in their own way have been instrumental in making my career in structures possible. I am grateful to my graduate and PhD students who were an inspiration. Many thanks go to Tim Love of Cambridge University, a rare altruist, who was so helpful during a 2-months' summer stay at the Engineering Department. Thanks to Miriam Hercberg for her help in checking the manuscript and to Guy Shiber for verifying and correcting the equations, calculations, and figures.

Finally, I wish to express my gratitude to the Faculty of Engineering at Tel Aviv University, a haven of serenity and genuine academic spirit, for providing the material conditions to do research, support students, finance for sabbaticals, and participation in scientific meetings and conferences.

December 2015

Acknowledgments

Figure 19.5 reprinted from Computers & Structures, M.B. Fuchs and M.A. Brull, A New Strain Energy Theorem and Its Use in the Optimum Design of Continuous Beams, Vol. 10, pp. 647-657, 1979, with permission from Elsevier.

Figure 20.2, 20.3 and the figure on the cover, reprinted from AIAA Journal, M.B. Fuchs and L.P. Felton, On a Class of Fully Stressed Trusses, Vol. 12, No. 11, pp.1597-1599, 1974, with permission from AIAA.

Figures 21.1, 21.2, 21.3, 21.4, Problem 21.5 reprinted from Computers & Structures, M.B. Fuchs, Unimodal Formulation of the Analysis and Design of Framed Structures, Vol. 63, No. 4, pp. 739-747, 1997, with permission from Elsevier.

Contents

Part I Preliminaries

About the Author

Maurice Bernard Fuchs was brought up in Antwerp, Belgium, and graduated from the Ecole Polytechnique of the Free University of Brussels (ULB) in civil and aeronautical engineering, specializing in structures. He earned an ScD in Technology at the Faculty of Aeronautical Engineering at the Israel Institute of Technology (Technion) and joined the department of Solid Mechanics, Materials and Structures of Tel Aviv University in 1977, where he has been a faculty member ever since. He was the first head of the then newly established School of Mechanical Engineering and is currently Professor Emeritus at that school.

The author has taught basic and advanced undergraduate courses in structural analysis and aircraft analysis and design, and graduate courses in finite-element analysis and structural optimization. He has also taught structures courses to architectural students at the Faculty of Architecture at the Technion and at the School of Architecture of Tel Aviv University. The author was a visiting assistant professor at the School of Engineering at UCLA, a visiting scholar at the Engineering Department of Cambridge University, and a visiting professor at the Joseph Fourier Université de Grenoble. He is a founding member of the International Society of Structural and Multidisciplinary Optimization (ISSMO) and was a regular contributor to its conferences and meetings. His main research field is the design of optimal structures on which he has published over 50 papers.

The author has two married daughters living in Israel. Sossy Fuchs-Aroch, a clinical psychologist, raises her family in Binyamina, a sleepy village some 70 km north of Tel Aviv, not far from a beautiful Mediterranean beach. Dr. Judith Weiss, a postdoctoral fellow at the Hebrew University, specializes in Christian Kabbala and other mystic stuff. She lives in Jerusalem with her family in a calm oasis in an otherwise bustling capital city.

Part I
Preliminaries

Most students taking a first course in Structures will already have passed courses in Statics and Solid Mechanics, (Strength of Materials). Statics is concerned with the fundamental relations between forces applied to elements, and Solid Mechanics deals with forces inside the solid elements and their related deformations.

In Structural Theory we are applying statics and solid mechanics theories to *structures*. It is, therefore, only natural to begin a book on Structures with the basics of Statics and Solid Mechanics. Part I of this book, the Preliminaries, recaps these subjects, naturally emphasizing the structural aspects involved.

We recall the essentials of statics in Chap. 1, and discuss the ubiquitous internal forces in Chap. 2. Chapter 3, shows how we tailor structural theory to produce the virtual work principle, and in Chap. 4 we describe the ensuing deformations. In Chap. 5, we remind the reader that the ranges of loads which will be applied to structures are such that the material is linear elastic. The final chapter of Part I, Chap. 6, deals with a technique for computing displacements and rotations, which we will use extensively in the sequel. It is part of preliminaries because it is usually taught as Castigliano's theorem in Solid Mechanics courses.

It is important to emphasize that these preliminaries include the three pillars of structural theory, Equilibrium, Deformation and Elasticity. The other three chapters either complete the picture or are included as preparation for things to come.

Chapter 1
Equilibrium

This chapter deals with forces and couples and recalls the basic laws of static equilibrium. We emphasize the important concept of *equivalent forces* and show that a system of forces is in static equilibrium if it is statically equivalent to zero. In other words, a system of forces is in static equilibrium if it is equivalent to no forces at all.

We also postulate the existence of internal forces, better known as stresses, without which the principles of equilibrium do not make sense. There are a myriad of internal forces which come in pairs; no-one has ever seen them, but these forces are essential for equilibrium.

1.1 In the Beginning

In the beginning there was equilibrium.

Consider the cantilever beam subjected to two forces P and Q shown in Fig. 1.1a. Under the loading the beam will bend and eventually come to rest in a shape close to that in Fig. 1.1b. Well-designed beams do not bend that much. In fact, except for a few cases, such as the wings of a wide-body aircraft, you will hardly notice the bending. In diagrams it is customary to grossly exaggerate the displacements of the structures, for the sake of clarity.

Since the beam is now at rest it *must* have managed to cancel the applied loads by providing equal and opposite forces, symbolically indicated by black arrows. The displaced form of the beam is called the *equilibrium configuration* or *equilibrium shape* or also the *displaced shape* of the structure.

When the beam was straight we did not have equilibrium. On the contrary, the beam sagged and moved until it reached its resting configuration. We must conclude that by deforming, the beam somehow created the black-arrow forces P and Q. These equilibrating forces result from internal forces which were generated by the deformations.

M.B. Fuchs, *Structures and Their Analysis*,
DOI 10.1007/978-3-319-31081-7_1

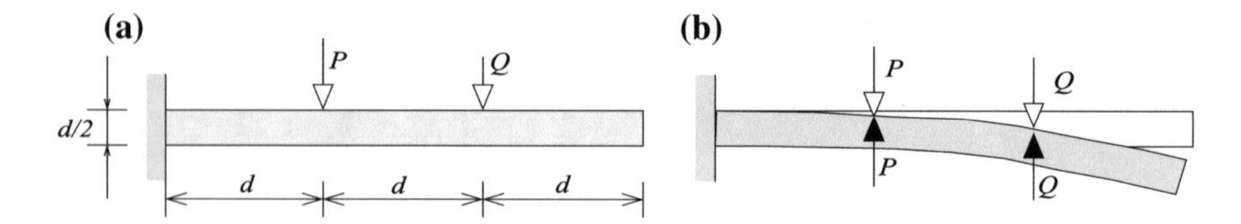

Fig. 1.1 Equilibrium by deforming

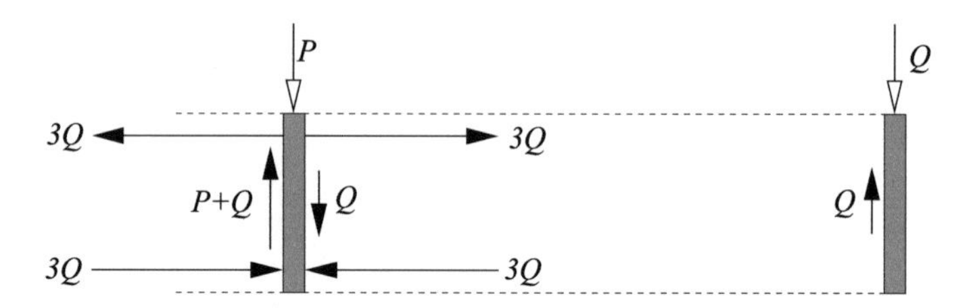

Fig. 1.2 Annulling the applied forces

When we isolate a very thin disc of the beam (a lamella in the parlance of Asplund[1]) directly under the force Q (Fig. 1.2) we will learn that the external force Q is balanced by an equal and opposite internal force (called shear force) Q applied tangentially to the left facet of the lamella. The lamella is infinitely thin so the shear force Q can be as close to the applied force Q as necessary.

The equilibrium of the lamella under P, in the same figure, is a little more interesting. We will see that on the right facet we have an internal down-shear force Q and on the left facet an internal shear force $P + Q$ in the up direction. Together they balance the external force P. Here also, the lamella is infinitely thin and the 3 forces $P + Q$, P and Q can be as close to collinear as required.

But this is not the end of the story. In addition, the lamella experiences compressive forces in the lower region and tensional forces in the upper region. Assuming that the beam is of length $3d$ and has a rectangular cross-section of height $d/2$, the magnitude of the internal compression and tension forces is $3Q$. Internal forces appear not only under the applied loads. In fact, the structure is literally awash with internal forces; and all to balance P and Q.

How did all this come about? Well, a structure creates the internal forces it needs for equilibrium by deforming. There exists a beautiful theorem regarding uniqueness of solution which states that there is only one such equilibrium configuration of a structure for a given set of applied forces. In other words, there is no other way for the structure to equilibrate P and Q save the one schematically shown in Fig. 1.1b.

Frankly, if we had a choice we would ignore the horizontal forces. They apparently do not contribute in canceling the applied P and Q, and if the beam is made of concrete (without steel reinforcements), for instance, the tensile forces along the upper chord

---

[1] S.O. Asplund, Structural Mechanics: Classical and Matrix Methods, Prentice-Hall, Englewood Cliffs, N.J., 1966 (A leading book on contemporary structural analysis).

do not bode well, because concrete has little resistance to tension and the beam may crack along the upper fibers. Nevertheless, this is the *only* way the beam can perform the task of equilibrating P and Q.

You will have noted that we wrote the equilibrium equations for the undeformed beam, Fig. 1.1a, where we do not have equilibrium, instead of writing them for the deformed beam, Fig. 1.1b, which is in equilibrium. This is not an oversight; it has to do with the assumption of *small displacements*. Writing the equilibrium equations in the deformed configuration is called a *second-order theory* and is an unnecessary headache at this stage.

Determining the internal forces is what structural analysis is all about

and for the rest of this book we will mostly be doing just that.

Over the years many techniques have been developed to compute the internal forces. They all boil down to just two methods: (1) *the force or flexibility method* which is suitable for manual calculations, and helps in understanding the structural response; and (2) *the displacement or stiffness method* which is to be programmed for computer use. Chances are that if you will be doing structural analysis while sipping your coffee in the morning you will be thinking in terms of forces. The displacement approach is however the method of choice for all industrial applications. It easily tackles high-rise buildings, long-span bridges and complex aerospace structures. These two methods will figure prominently in this text. But first a review of principles of equilibrium.

1.2 Resultant Force and Resultant Moment

Forces are modeled as vectors. The *vector-sum* of any two forces $P_i + P_j$ follows the parallelogram rule (Fig. 1.3a).

The *moment* with respect to O of a force P_i (Fig. 1.3b) is the algebraic product $P_i\ r_i$

$$M_O(P_i) \equiv P_i\ r_i \qquad \text{Positive counter-clockwise} \tag{1.1}$$

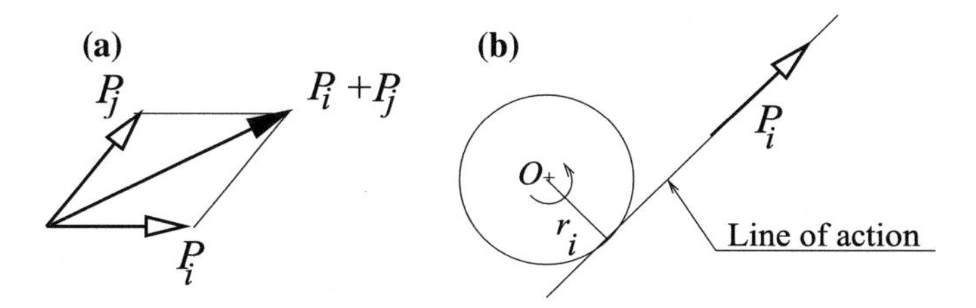

Fig. 1.3 Vector sum and moment with respect to O

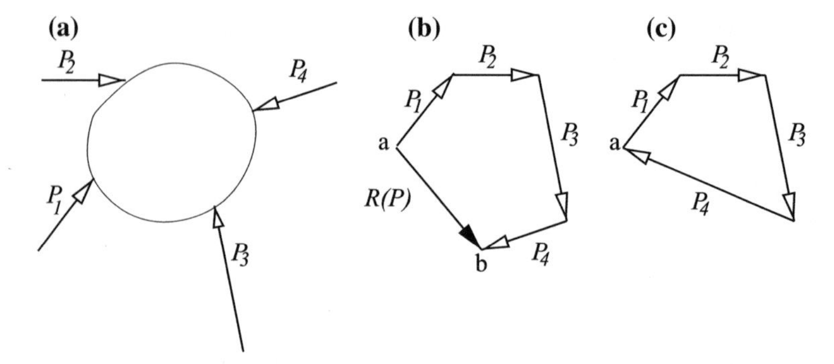

Fig. 1.4 Resultant force. **b** Resultant. **c** Zero resultant

where r_i is the distance from O of the line of action of P_i. If we imagine a circle of radius r_i centered at O the moment of the force is said to be positive if the force tends to rotate the circle counter-clockwise.

In Fig. 1.4a we show a set or a system of forces P_1–P_4. The system of forces is represented by $\boldsymbol{P}$. The *resultant* $R(\boldsymbol{P})$ of a set of forces is the vector-sum of all the forces[2]:

$$R(\boldsymbol{P}) \equiv \sum_i P_i \qquad M_O(\boldsymbol{P}) \equiv \sum_i M_O(P_i) \tag{1.2}$$

A vector-sum can be visualized by connecting the forces (vectors) head to tail in any sequence.

The *resultant* is then the vector joining the tail (a) of the first force to the head (b) of the last one as shown in Fig. 1.4b. The resultant is independent of the actual location of the forces; only their direction and magnitude matters. The resultant is therefore a free vector; its position is not determined. When the forces form a closed polygon (Fig. 1.4c) the resultant of the forces is zero.

The *resultant moment* with respect to O of a set of forces P_i, represented by $M_O(\boldsymbol{P})$, is the algebraic sum of the individual moments with respect to O (1.2).

1.3 Equilibrium of a Solid

A solid body is said to be in equilibrium under applied forces $\boldsymbol{P}$ if the resultant of the forces and the resultant moment of the forces relative to an arbitrary point O are both zero

$$\begin{aligned} R(\boldsymbol{P}) &= 0 \quad \text{translational equilibrium} \\ M_O(\boldsymbol{P}) &= 0 \quad \text{rotational equilibrium} \end{aligned} \tag{1.3}$$

[2]Here and throughout this text we deal with plane structures and forces.

Equilibrium implies that if the body is initially at rest it will remain so under forces in equilibrium. When forces with a zero resultant are applied to a body, the resultant body will not translate. Similarly when forces with a zero resultant moment are applied to a body, the latter will not rotate. The solid can be rigid as a rock or flexible as a bridge or an airplane wing, but not a mechanism.[3] The conditions for equilibrium (1.3) do not apply, as is, to mechanisms.

Note, flexible bodies, that is structures, will of course experience some very small deformations due to forces in equilibrium, but after settling in the equilibrium configuration the body will remain at rest.

1.3.1 Three Scalar Conditions

The rotational equilibrium condition is a scalar equation. The condition for translational equilibrium is a vector equation. Now, the resultant can be decomposed into its components R_x and R_y, usually in an orthogonal framework x, y. A zero resultant is equivalent to zero values of its scalar components $R_x = 0$ and $R_y = 0$. In other words, conditions (1.3) can be replaced by three scalar equations

$$\begin{aligned} R_x(\boldsymbol{P}) &= 0 \quad \text{x-translational equilibrium} \\ R_y(\boldsymbol{P}) &= 0 \quad \text{y-translational equilibrium} \\ M_O(\boldsymbol{P}) &= 0 \quad \text{rotational equilibrium} \end{aligned}$$

The beam in Fig. 1.5a, for example, is in equilibrium. There is no need to check x-equilibrium because unless specified to the contrary, in this text beams will not have longitudinal external forces. Vertical translational equilibrium is satisfied, starting from the left extremity ($0.7\,Q - Q + 0.3\,Q = 0$) and so is rotational equilibrium ($M_c = -0.7\,Q \times L - Q \times 0.7\,L = 0$). We could have used $M_b(Q) = 0$, that is, $M_b = -0.7\,Q \times 0.3\,L + 0.3\,Q \times 0.7\,L = 0$ or the moment relative to any other point on this page.

The frame in Fig. 1.5b has forces P at locations a, b, c, d. Clearly, the horizontal forces at a, c are in equilibrium and so are the vertical forces at b, d. For the rotational equation we check the moment with respect to d. The forces at a, d pass through d and their moments with respect to d are of course zero. We are left with the forces at b, c: $M_d = P \times 0.5L - P \times 0.5L = 0$. (Counter-clockwise is positive.) The frame is indeed in equilibrium.

[3] A mechanism is a body whose shape can significantly be changed by very small forces, for example a pair of scissors.

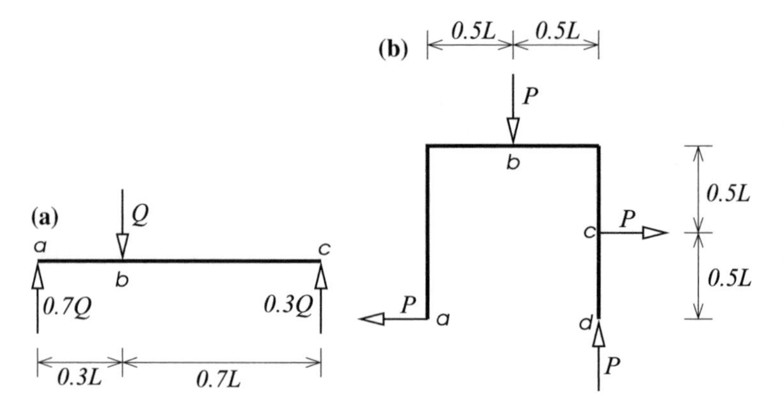

Fig. 1.5 Line segments in equilibrium

1.4 Statically Equivalent Forces

The concept of static equivalence of two systems of forces is extremely important for the equilibrium of solids. Consider two sets of forces $\boldsymbol{P}$ and $\boldsymbol{Q}$, which may have a different number of forces. The two sets are said to be statically equivalent if they have the same resultant and the same resultant moment with respect to some point O

$$R(\boldsymbol{P}) = R(\boldsymbol{Q}) \quad \text{and} \quad M_O(\boldsymbol{P}) = M_O(\boldsymbol{Q}) \tag{1.4}$$

Why do we call them statically equivalent? Because they have the same effect on the static equilibrium equations. Indeed, let a solid be in equilibrium under a set of forces $\boldsymbol{F}$ and $\boldsymbol{P}$, that is, $R(\boldsymbol{F}) + R(\boldsymbol{P}) = 0$ and $M_O(\boldsymbol{F}) + M_O(\boldsymbol{P}) = 0$. Clearly, we can replace system $\boldsymbol{P}$ by $\boldsymbol{Q}$ and maintain equilibrium of the solid. This is the equivalence.

Statically equivalent forces can be replaced by one another *when writing equilibrium equations*. We will see that they may be very different from one another and may have very different effects when applied to a structure, but from a statics viewpoint, that is, when writing equilibrium equations, they are nevertheless equivalent. Further on are some typical systems of forces which are equivalent when writing equilibrium equations.

1.4.1 Sliding Force

A force Q is statically equivalent to the same force Q positioned anywhere along the line of action of the force as in Fig. 1.6a. Indeed, the force sliding along its line of action has, of course, at any location on the line a same resultant $R(Q) = Q$ and a same moment $M_O(Q)$ with respect to some point O.

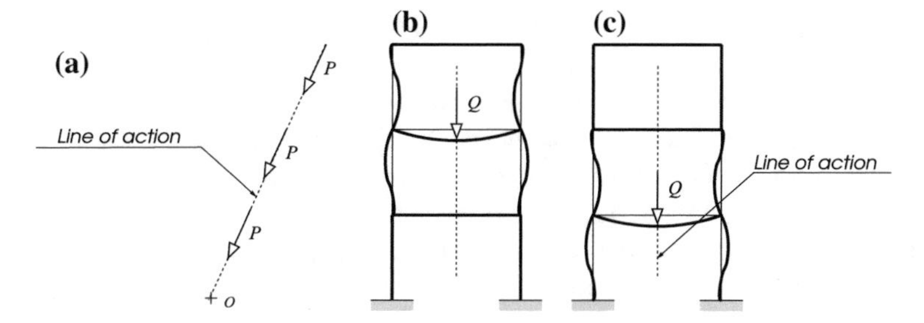

Fig. 1.6 Sliding forces on frame

We cannot fail to note that the equivalence is only valid for writing the equilibrium equations. The actual behavior of the structure depends on the exact location of the force. A piano (Q) on the second floor of the building in Fig. 1.6b is statically equivalent to the same piano on the first floor Fig. 1.6c but clearly the musical instrument will have a different effect on the structure depending on the floor it is located.

1.4.2 Couples

This is perhaps a good opportunity to introduce the *couple of forces* concept or simply the *couple*. Two equal and opposite parallel forces, as in Fig. 1.7a, form a couple. The resultant of a couple is obviously zero but a couple has a resultant moment.

What makes the couple so special is that the moment of the couple, that is, the sum of the moments of the two forces, with respect to some point O is independent of O. The couple will have the same moment with respect to O, O', O'' or any other point.

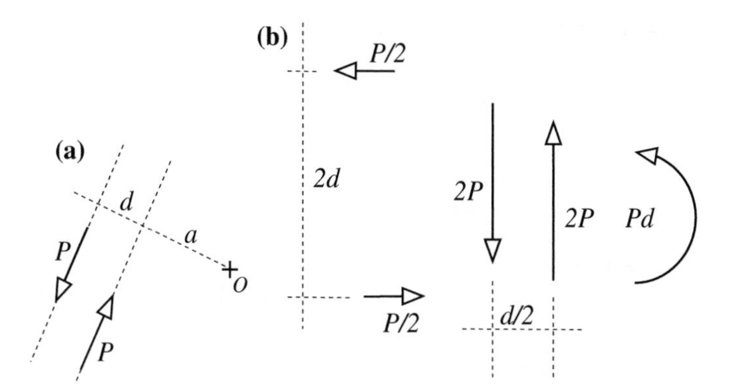

Fig. 1.7 Equivalent systems of forces. a Couple. b Equivalent couples

Indeed, the moment of the two forces in Fig. 1.7a with respect to O is $M_O = (a + d)P - aP = Pd$ (positive counter-clockwise). Since a does not appear in the expression of the moment, the moment is the same irrespective of point O.

Many seasoned engineers will denote the 'couple', $C \equiv Pd$, that is the set of two equal and opposite parallel forces, the 'moment' $C \equiv Pd$, because this is the net effect the couple has on the equilibrium equations. So instead of applying a 'couple' Pd they apply a 'moment' Pd. This is often confusing for the student of structures and we will try to avoid this shortcut. For us, a couple is and will remain a couple of two forces which apply a moment to the system.

Note, as far as equilibrium goes, the couple is a free entity. Wherever the 2 forces are, they will have the same effect on the equilibrium equations.

Consider the couples in Fig. 1.7b. Forces $P/2$ with $2d$ or $2P$ at $d/2$ or the point couple Pd are statically equivalent. They have a zero resultant and their moment is Pd (positive, that is, counterclockwise).

When the two forces are very large and very close to one another the product of the magnitude and the distance can be finite. We have then a *concentrated couple* or *point couple*. The two forces can obviously not be drawn apart in any significant manner. The concentrated couple is therefore represented by a crescent (an icon reminiscent of the rotation a couple imparts to the solid) signified by its sign (the sense of the arrow) and its magnitude (the value of the couple). It is not superfluous to remind the reader that this circular arc does not physically exist. It is an icon. It merely indicates the presence of a localized, or point couple of forces as in Fig. 1.7b where the couple equals Pd.

The three couples in Fig. 1.7b are *statically* equivalent. If one of them acts on a solid in equilibrium you can replace it by any of the two others and you will still have equilibrium. You can move it to other locations on the solid and maintain equilibrium. The structure will however react (deform) differently depending on where it is applied.

1.4.3 Reduction to One Force and One Couple

Any set of forces is statically equivalent to a force and a couple.

Let $\boldsymbol{P}$ be a set of forces which has a resultant $R(\boldsymbol{P})$ and a moment $M_O(\boldsymbol{P})$ with respect to some point O. A simple statically equivalent system is a force F equal to the resultant, $F = R(\boldsymbol{P})$, positioned at O (or better, whose line of action passes through O) and a couple $C = M_O(\boldsymbol{P})$. The couple can be anywhere. Indeed, the resultant of system (F, C) is F, that is, $R(F, C) = R(\boldsymbol{P})$ and the resultant moment with respect to O of system (F, C) is C, that is, $M_O(F, C) = M_O(\boldsymbol{P})$. Consequently, system (F, C) is statically equivalent to system $\boldsymbol{P}$.

A force and a couple can be further reduced to a force only.

Since O is an arbitrary point there are many such force-couple systems. There exists a point (rather a line) for which the forces $\boldsymbol{P}$ reduce to a force without a couple. In Fig. 1.8a, we

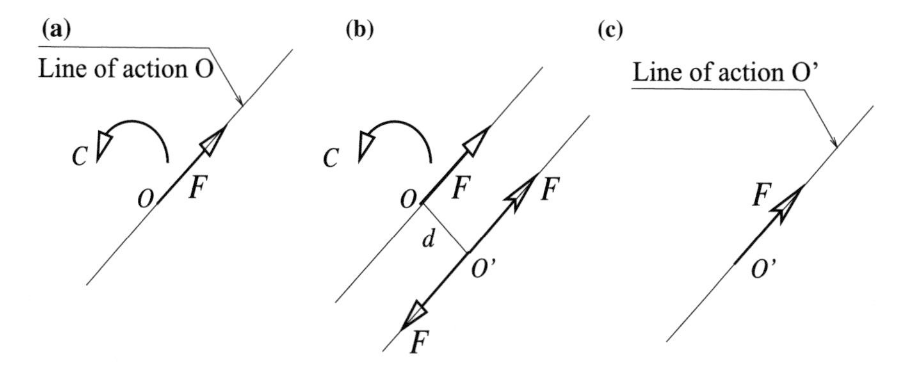

Fig. 1.8 Reduction to a single force

have the reduced system F through O and a couple C which can be anywhere. In Fig. 1.8b, we have added at O' (a distance $d = C/F$ apart from the line of action through O) equal and opposite forces F (these two forces are statically equivalent to nothing). Systems (a) and (b) are obviously equivalent. Since the couples C and Fd (F at O and F at O' pointing down) cancel out, system (a) is also equivalent to a single force F at O' (Fig. 1.8c).

Changing a force into a force and a couple.

We can also go the other way. Force F in Fig. 1.8c passing to O' is of course statically equivalent to the force F passing through O plus the couple C in Fig. 1.8a.

Sometimes the crescent is used for forces some distance apart. For instance, the couple represented by the crescent in Fig. 1.7b can be concentrated but also represents two distinct parallel forces and can be located anywhere on the structure under consideration.

1.4.4 Uniformly Distributed Forces

In Fig. 1.9 we show that uniformly (that is, constant) distributed forces q over a length d (Fig. 1.9a) are statically equivalent to the total load qd concentrated in the middle of the segment (Fig. 1.9b).

Indeed, we can imagine gradually moving the forces in a symmetric fashion to the middle. Coming from both sides the required additional couples cancel out. So all we

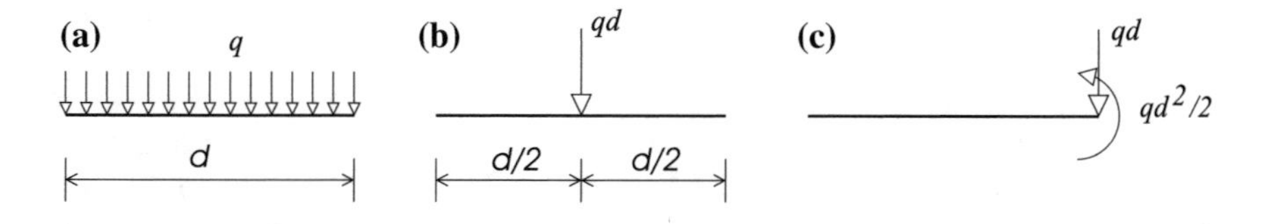

Fig. 1.9 Uniformly distributed forces

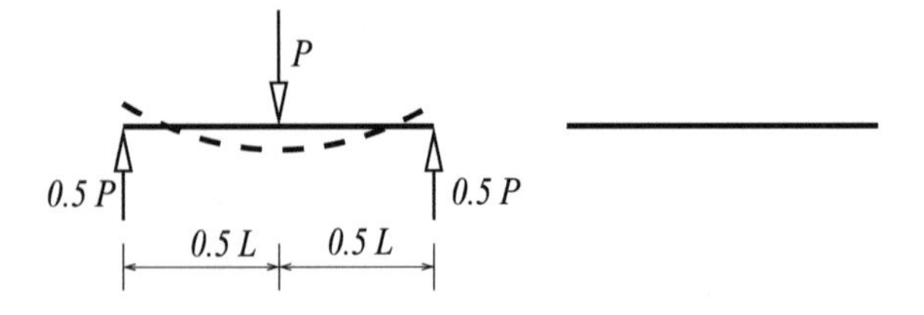

Fig. 1.10 Systems statically equivalent to zero

are left with at the end of the process is the total force in the middle of d. Next, moving qd to the right extremity requires a counter-clockwise couple $qd \times (d/2) = qd^2/2$.

That the three systems in the second line of Fig. 1.9 are statically equivalent is obvious, but it puts the notion of statical equivalence in the correct perspective, because

statically equivalent forces are not *structurally* equivalent.

Far from it.

The beam on the left in Fig. 1.10, for instance, is submitted to forces statically equivalent to zero.

The dashed line is a typical deformed shape of the beam. The same beam to the right is also subjected to forces statically equivalent to zero. The difference between the two cases is rather obvious.

1.4.5 Three Conditions Out of Five

One can show that $M_O = 0$, $M_{O'} = 0$ and $M_{O''} = 0$ are three independent equations of equilibrium subject to the restriction that the distinct arbitrary points O, O' and O'' are not collinear. It is therefore perfectly admissible to replace a translational equilibrium equation or both translational equations by rotational equations about two other points, O' and O''.

In practice we often work with a pool of five equations, two translational and three rotational

$$R_x(\boldsymbol{P}) = 0 \qquad R_y(\boldsymbol{P}) = 0 \qquad M_O(\boldsymbol{P}) = 0 \qquad M_{O'}(\boldsymbol{P}) = 0 \qquad M_{O''}(\boldsymbol{P}) = 0 \tag{1.5}$$

Any three equations of this set are linearly independent and can be used to establish equilibrium. Taking a rotational equilibrium equation instead of a translational one is common practice because it very often facilitates calculations.

To check the equilibrium of the frame in Fig. 1.11, we could have used $M_a = 0$ ($M_a = -3P \times 0.5L + 2.5P \times L + P \times L - 2P \times L$), $M_b = 0$ (starting from a: $M_b = 2P \times L - 3P \times 0.5L + 2.5PL - 2P \times 2L + P \times L = 0$) and $M_d = 0$ ($M_d = -2P \times L - 2P \times L + 3P \times 0.5L + 2.5PL = 0$), but not $M_a = M_b = M_e = 0$ because point e is on line ab.

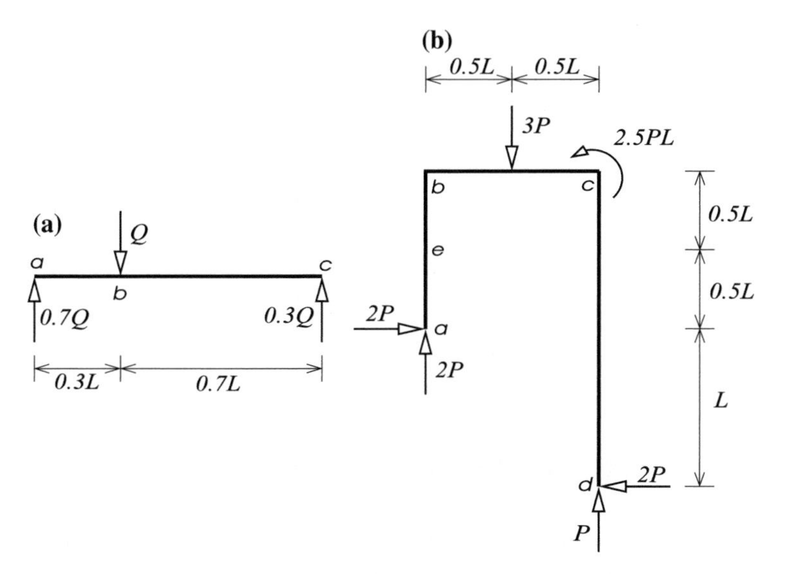

Fig. 1.11 Three conditions out of five

Earlier, for the beam in Fig. 1.5a we used the vertical equilibrium condition $R_y = 0$ and a rotational condition. But we could have written two rotational conditions such as $M_a = 0$ and $M_c = 0$ or $M_b = 0$ and $M_c = 0$ to check the equilibrium.

1.5 Illustrated Examples

1.5.1 Moments with Respect to 3 Collinear Points

Show that if O, O', O'' are collinear $M_O = M_{O''} = 0$ means that $M_{O'} = 0$ is therefore not a new equation.

We assume that the system of forces was reduced to a single force F. If $M_0(F) = 0$ the force is either zero or it passes through O. If in addition $M_{O'} = 0$ the force is either zero or the line of action of the force passes also through O', that is, the line of action of the force overlaps segment OO'. If point O'' is on the line joining O to O' then obviously $M_{O''} = 0$. So this is not new information.

However, if O, O', O'' were not collinear then $M_O = M_{O'} = M_{O''} = 0$ means that the force is either zero or it is aligned with OO', $O'O''$ and $O''O$. The second alternative is impossible and therefore $M_O = M_{O'} = M_{O''} = 0$ means $F = 0$, that is the system of forces reduces to nothing, in other words, we have equilibrium.

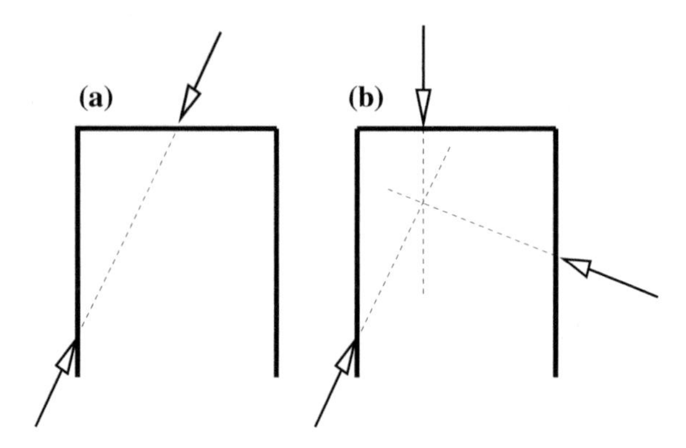

Fig. 1.12 Two-force and three-force bodies

1.5.2 *Auto-Equilibrated 2-Force and 3-Force Systems*

The study of bodies subjected to auto-equilibrated systems of two forces (Fig. 1.12a) or three forces (Fig. 1.12b) can take advantage of intrinsic properties of such systems.

1. Show that two forces in equilibrium are necessarily equal and opposite (zero resultant) and they must have a common line of action (zero moment).
   Indeed if the lines of action were apart they would create a couple and hence a non-zero moment.
2. Show that when in the presence of an auto-equilibrated system of three forces (excluding the case of parallel forces), the forces must intersect at a common point.
   The condition for zero moment requires that the forces intersect at a common point. Indeed, in the opposite case we cannot have equilibrium. Two forces (P_1, P_2) will always intersect at some point O and let us assume that the line of action of the third force (P_3) does not pass through the intersection point. The resultant moment with respect to O of the system of forces is $M_O(\boldsymbol{P}) = M_O(P_1) + M_O(P_2) + M_O(P_3)$. Now the first two terms are obviously zero and the third term is different from zero hence $M_O(\boldsymbol{P}) = M_O(P_3) \neq 0$. Therefore, all three forces must intersect at a common point.

1.6 The Internal Forces

1.6.1 *Free-Body Diagrams*

Consider the free-body[4] diagram in Fig. 1.13a of a beam subject to a set of three auto-equilibrated forces. It is easy to see that the resultant and resultant moment of the forces are zero.

[4]A free body is a free-floating structure or part of a structure.

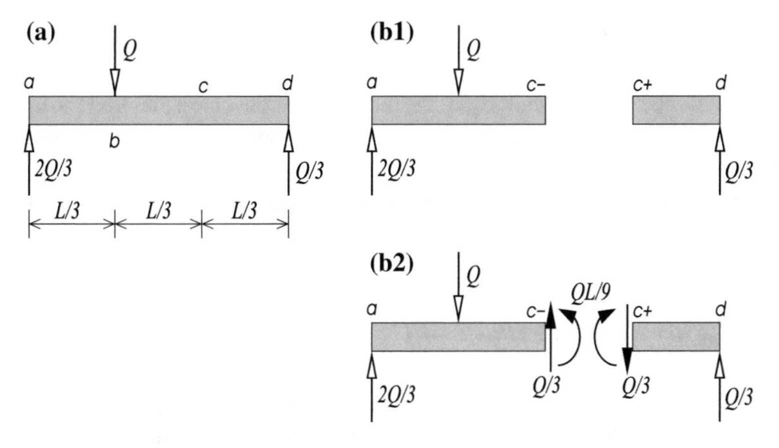

Fig. 1.13 Interal forces

We now use a common technique to, figuratively, expose the internal forces in a structure. We perform an *imaginary cut* at some arbitrary section c (at $2L/3$ from the left in our case) thus creating two segments ac^- and c^+d (Fig. 1.13b1).

Note: We assign a positive direction x for the axis of the beam. For horizontal beams this will always run from left to right. When performing a cut we have thus a facet to the left and a facet to the right. The left facet will be called the minus facet and labeled with a − sign. The facet to the right is therefore the plus facet and labeled with a + sign.[5]

Clearly when a structure is in equilibrium, every part of it is also in equilibrium. The conditions of equilibrium must thus apply individually to the free bodies ac^- and c^+d.

However, segment c^+d is manifestly not in equilibrium (we expect c^+d to take-off under the action of force $Q/3$). The only way to explain this paradox is to assume that there are other forces at work. But where could they be? Along the upper and bottom surfaces of the segment we see no other forces and neither are there forces at end-section d, except for $Q/3$. The only place where there could be additional forces is along facet c^+, that is, inside the beam. Since nobody can look inside the material, we can assume that they are there, and indeed, we hypothesize the existence of *internal* forces at facet c^+ which will equilibrate c^+d (black arrows in Fig. 1.13b2).

In a first instance we need a down-force $Q/3$ to get a zero resultant in the y-direction. But this is not enough. We have now an anti-clockwise couple acting on the segment $(Q/3)(L/3) = QL/9$. So we also need to hypothesize the existence of additional forces which are statically equivalent to a clockwise couple of magnitude $QL/9$ applied to facet c^+. We don't know the nature of the internal forces at facet c^+ but they must add up to respectively a down-force $Q/3$ and a clockwise couple

[5]We differ here from the convention used in continuum mechanics where the positive facet is the one with its outer normal in the direction of x.

$QL/9$. These are *internal forces* applied by facet c^- to facet c^+. Now segment c^+d in equilibrium.

We could repeat the procedure for segment ac^- of the beam and find the up-force $Q/3$ and the counterclockwise couple $QL/9$ that should be applied to facet c^- in order to have ac^- in equilibrium.

A shorter procedure is to note that internal forces come in equal and opposite pairs, by virtue of Newton's third Law.

1.6.2 Equivalent System of Internal Forces

You will have noted that we did not delve into the exact nature of these internal forces or stresses. A detailed description of the stresses is the concern of such disciplines as Solid Mechanics, Strength of Materials and Theory of Elasticity. In structural theory we reduce the plethora of forces (in fact stresses) acting on both sides of a section to their simplest mechanical equivalent: their resultant force and couple.

Consider a beam with a virtual cut at some cross-section b and let us assume that the stresses that are applied to facet b^- by its neighbor to the right look as depicted in Fig. 1.14a. As far as equilibrium is concerned any set of forces is statically equivalent to the resultant of the forces passing through some point and a corresponding couple, for instance, force P at O and couple C in Fig. 1.14b. In other words, P and C are statically equivalent to the stresses in Fig. 1.14a.

But one can rightfully argue that the stresses are also equivalent to force P passing through other points such as O' with a different couple as in Fig. 1.14c. (Note, for the two systems to be equivalent they need to have the same moment with respect to O, that is, $C' - Pd = C$ or $C' = C + Pd$.)

In structural mechanics, the point at which the resultant of the internal forces is positioned is invariably the centroid G or *center of area* of the cross-section. For cross-sections with double symmetry the center of area is in the middle of the cross-section. In Fig. 1.15a we have an example of a beam with a symmetric I-cross-section where the lower flange is wider than the upper one and the center of area is therefore closer to the lower fibers. Please bear in mind that since every internal force has a counterpart on the opposite facet, the resultant force and couple appear on both facets with opposite directions.

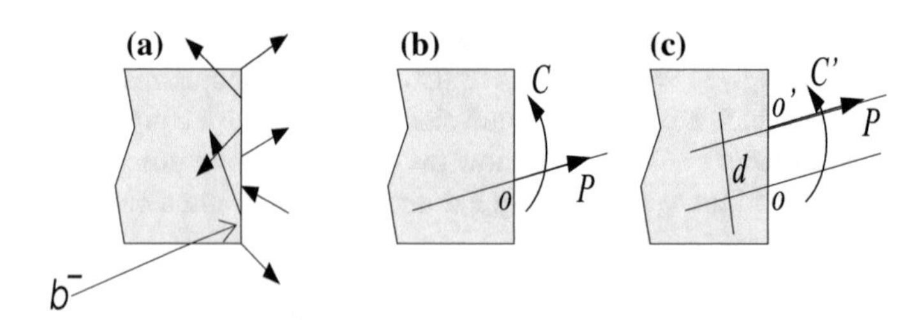

Fig. 1.14 Statically equivalent internal forces

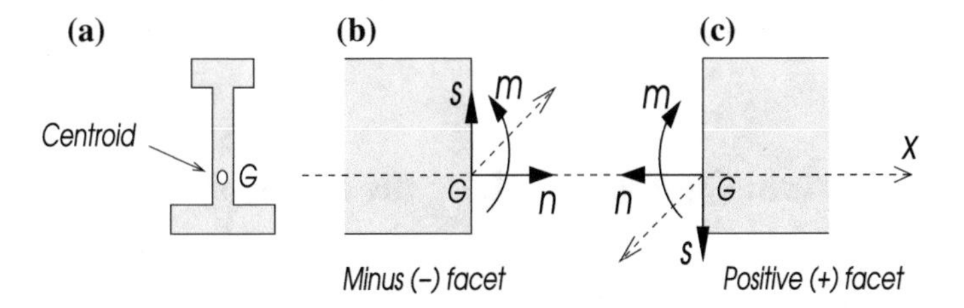

Fig. 1.15 Definition of nsm

1.6.3 Normal Force, Shear Force and Bending Moment

The line joining the centroids is the centroidal axis or simply axis of the element. For our purposes this line will be assumed straight or moderately curved with a positive direction x defined on the axis.

The resultant of the stresses (dashed vector in Fig. 1.15b) is positioned at the centroid of the facet. With the resultant at the centroïd G the related couple is called the bending moment m. We now take the axial n and lateral s components of the resultant (Fig. 1.15b).

- n the axial component of the resultant is the *normal or axial force*,
- s the lateral component of the resultant is the *shear force* and
- m the couple is the *bending moment* or sometimes simply the *moment*.

Clearly, n, s and m always appear as opposite pairs, one on each facet of the section (Fig. 1.15b, c). When we consider the bending moment m, for instance, we are referring simultaneously to both couples. The same is valid for the normal and shear forces.

The positive directions for the internal forces on $-$ and $+$ facets of an imaginary cut are as shown in Fig. 1.15b, c.

Depending on the context we will often denote the internal forces by the 3×1 vector

$$\boldsymbol{n}(x) \equiv \{n(x)\; s(x)\; m(x)\}^T \qquad \text{abbreviated to} \qquad nsm(x) \tag{1.6}$$

A positive $n(x)$ denotes tension at x, a positive $m(x)$ will cause tensile stresses in the bottom fibers and compression stresses in the upper fibers at x. A positive $s(x)$ represents tangential stresses pointing upwards on the left facet and downwards on the right facet[6] at x.

[6]For the reconstruction of the actual internal stresses from the stress resultants $nsm(x)$ the reader is referred to treatises on Strength of Materials or Solid Mechanics.

Determining the values of the internal forces $\boldsymbol{n}(x)$ or $n(x)$, $s(x)$ and $m(x)$ along x for all the elements of a structure is what structural analysis is mostly all about.

1.7 Equilibrium Conditions Along the Element

It is important to note that no matter how many point forces and couples may be applied to an element, the latter can always be construed as a sequence of segments subjected to distributed loads interspersed by point loads (which take up no space).

Indeed, consider a typical element a^+e^- which was removed from a structure (Fig. 1.16). The element has a point force at b, a couple at c and distributed forces along de. We perform virtual cuts at b_L (a little to the left of b) and b_R (a little to the right of b). We will see in the sequel that b_L is in fact b_L^- and similarly b_R is b_R^+. The segment can therefore be decomposed into parts $a^+b_L^-$, $b_R^+c_L^-$, $c_R^+d^-$ and d^+e where c_L and c_R are sections just in front and behind the point couple at c.

As already mentioned, all segments are subjected to distributed loads (segment $a^+b_L^-$, for example, has a distributed load of zero magnitude).

1.7.1 *Equilibrium of a Lamella dx of a Prismatic Element*

We consider the equilibrium of a very thin slice dx of a straight element, a lamella, (see segment d^+e^- in Fig. 1.16b. The lamella is subjected to general distributed loads and to (what were internal but are now) external loads applied to the end facets. (We will recall that the equilibrium equations of a free body involve all the external loads applied to it, but no internal forces.)

The distributed loads are axially distributed forces $p(x)$, laterally distributed forces $q(x)$ and distributed couples $c(x)$ shown to the right in Fig. 1.17a. Distributed couples are seldom used in classical analysis but we will keep them for the time being for the sake of generality and clarity. The equations are better balanced when $c(x)$ is included. This is an instance where the general case is easier to tackle than the specific one. As with the internal forces the distributed loads can be referred to

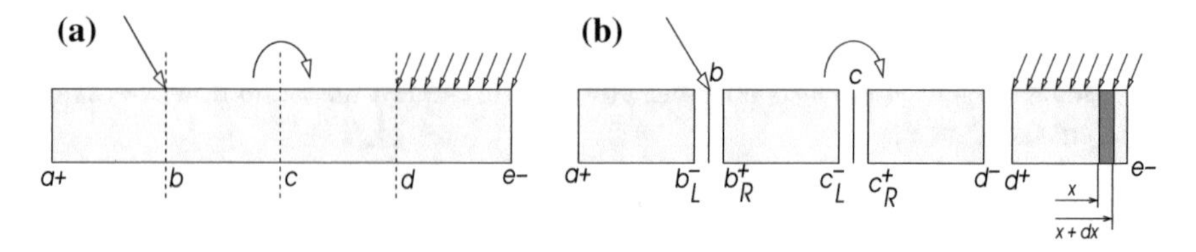

Fig. 1.16 Segments with distributed forces

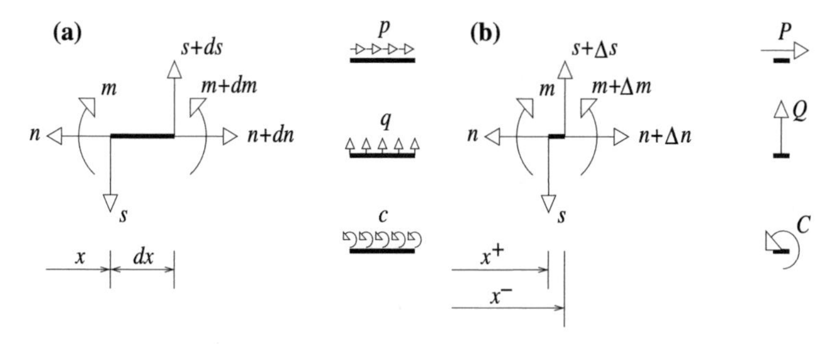

Fig. 1.17 Equilibrium: distributed and concentrated loads. **a** Distributed loads on dx. **b** Concentrated loads at x

as vector $\boldsymbol{p}(x)$ of components $p(x)$, $q(x)$ and $c(x)$ or also by the shortened form $pqc(x)$.

$$\boldsymbol{p}(x) \equiv \{p(x)\, q(x)\, c(x)\}^T \qquad \text{abbreviated to} \qquad pqc(x) \tag{1.7}$$

The boundaries of the lamella are sections x and $x + \mathrm{d}x$ (x runs from left to right). When we consider the outer normals of these sections we notice that x is a negative facet (its outer normal points in the negative x direction) and $x + dx$ is a positive facet (the outer normal points in the positive x direction). Accordingly, the internal forces are positive as depicted in Fig. 1.17a.

Equilibrium of the lamella requires that the resultant of the axial forces is zero $-n + (n + \mathrm{d}n) + p\mathrm{d}x = \mathrm{d}n + p\mathrm{d}x = 0$, the resultant of the lateral forces is zero $-s + (s + \mathrm{d}s) + q\mathrm{d}x = \mathrm{d}s + q\mathrm{d}x = 0$ and that the moment of all the forces with respect to, say, point $x + \mathrm{d}x$ on the x–axis are also zero, $-m + s\mathrm{d}x + (m + \mathrm{d}m) + c\mathrm{d}x - q\mathrm{d}x(\mathrm{d}x/2) = s\mathrm{d}x + \mathrm{d}m + c\mathrm{d}x + (-q/2)\,\mathrm{d}x^2 = 0$.

The last term on the left-hand side of this equation is infinitely smaller than the other terms and can be discarded.[7] Dividing throughout by $\mathrm{d}x$ the 3 equilibrium equations become

$$\begin{aligned} \frac{dn}{dx} + p &= 0 \\ \frac{ds}{dx} + q &= 0 \\ s + \frac{dm}{dx} + c &= 0 \end{aligned} \tag{1.8}$$

In this book we will be assuming that the distributed loads are *uniform*, that is, *constant* on any given segment of the element.

When the distributed loads are zero, the equilibrium equations become $dn/dx = 0$, $ds/dx = 0$, that is, the normal and shear forces are constant, and

[7] If $\mathrm{d}x$ is, for instance, of the order of 10^{-100} then the last term is of the order of 10^{-200}, that is, the last term is 10^{-100} smaller than the other terms.

$dm/dx = -s$. The shear force being constant makes the moment linear in x on that segment. If the shear force happens to be zero, the bending moment is constant. We should remember this when drawing the nsm-diagrams.

On a segment *without* distributed loads, n and s are constant and m is linear.

If the distributed loads are constant on a segment, it is clear from (1.8) that the normal and shear forces are linear and the moment is parabolic. By limiting the continuous loads to constant distributions we avoid lengthy calculations without loosing the essence of the method. We have tried to follow this didactic principle throughout this text. So,

On a segment *with constant* distributed loads, n and s are linear and m is parabolic.

The basic equilibrium equations of a lamella of the element (1.8) can be written in vector form

$$\frac{\mathrm{d}}{\mathrm{d}x}\begin{Bmatrix} n \\ s \\ m \end{Bmatrix} = -\begin{Bmatrix} p \\ q \\ c \end{Bmatrix} - \begin{Bmatrix} 0 \\ 0 \\ s \end{Bmatrix} \tag{1.9}$$

or also

$$\frac{\mathrm{d}\boldsymbol{n}}{\mathrm{d}x} = -\boldsymbol{p} - \begin{Bmatrix} 0 \\ 0 \\ s \end{Bmatrix} \tag{1.10}$$

where we recall that $\boldsymbol{n} \equiv \{n(x)\ s(x)\ m(x)\}^T$ and $\boldsymbol{p} \equiv \{p(x)\ q(x)\ c((x)\}^T$.

1.7.2 Equilibrium at a Concentrated Load

Distributed loads induce *gradual changes* in the internal forces. Concentrated or point loads, on the other hand, cause a sudden change or *jump* in the value of the corresponding internal force.

In Fig. 1.17b we have a case where forces P, Q and a couple C are applied at point x. The internal forces immediately to the left of x (that is, at x when approached from the left) are termed $(nsm)_{x_L}$ and those immediately to the right are $(nsm)_{x_R}$. The forces applied at x are thus as depicted in Fig. 1.17b: internal loads on both sides of x and applied point loads at x.

The conditions of equilibrium require that the sum of the forces in the axial and lateral directions are zero and that the sum of the couples is zero

$$\begin{aligned} \Delta n + P &= 0 \\ \Delta s + Q &= 0 \\ \Delta m + C &= 0 \end{aligned} \tag{1.11}$$

with $\Delta n = n_R - n_L$, $\Delta s = s_R - s_L$ and $\Delta m = m_R - m_L$. With concentrated loads we have discontinuities (sudden variations) in the normal force, the shear force and the moment. In the absence of distributed couples c, $dm/dx = -s$ and thus $\Delta(dm/dx) = -\Delta s = Q$. Consequently,

A lateral point force causes a *sudden change in slope* of the moment distribution.

Here $\Delta(dm/dx) = (dm/dx)_R - (dm/dx)_L$.

The shear force and the bending moment are related quantities since the former is the derivative of the latter. We say that the shear force and the bending moment are coupled. The axial force is uncoupled from the shear force and the bending moment in straight elements (in curved elements this may not always be true).

1.8 Summing Up

Equilibrium, in conjunction with deformation and elasticity, forms the basis of the behavior of structures and our understanding of their response to applied loads. We have seen in this chapter how the laws of equilibrium, arguable the most important ingredient of the three, intervene in the context of structures composed of line elements.

These laws are based on the necessary assumption of the existence of internal forces and the reduction of all the internal forces to statically equivalent $n(x)$, $s(x)$ and $m(x)$, three functions of the running coordinate x.

Chapter 2
The $nsm(x)$ Diagrams

The structures in this book are composed of beams and bars (and cables). These are long and relatively slender line elements which have a longitudinal axis passing through the centers of area of the cross-sections. The coordinate along the axis is x.

On both sides of virtual sections perpendicular to the axis we find equal and opposite internal forces: normal forces n acting along the axis, shear forces s perpendicular to the axis (along the faces of the section), and equal and opposite couples, the bending moments m. Calculating these internal forces throughout the structure is the main purpose of structural analysis.

This chapter shows how to compute and present the internal forces along any line element if the external (applied) forces are all known. The internal forces are functions of x, leading to the normal force distribution $n(x)$, the shear force distribution $s(x)$ and the bending moment distribution $m(x)$ or diagrams, also called $nsm(x)$ or $\mathbf{n}(x)$ for short.

These distributions are drawn along the elements in a traditional manner, and we will show how to compute and draw these distributions or diagrams.

2.1 Line Elements

The essence of the analysis of a structure is the computation of its internal forces: the normal force $n(x)$, the shear force $s(x)$ and last, but by far not least, the bending moment $m(x)$ at any section x along the structure. We will refer to these quantities also in their vector form $\boldsymbol{n}(x) = \{n(x)\ s(x)\ m(x)\}^T$ and also symbolically as $nsm(x)$.

The element is represented by the element axis x which is assumed to follow a line on the center of mass of the cross-sections. The line need not be straight; it can be moderately curved or kinked or composed of rigidly connected segments. The line must be open; no closed loops are allowed at this point. To begin, we consider lines with one starting point and one end point (Fig. 2.1).

M.B. Fuchs, *Structures and Their Analysis*,
DOI 10.1007/978-3-319-31081-7_2

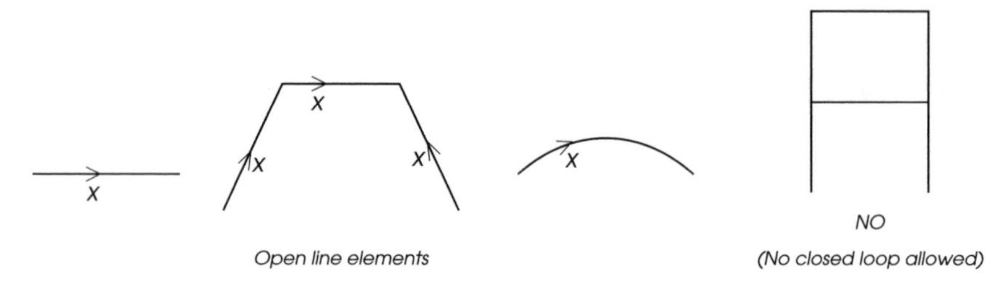

Fig. 2.1 Line elements

The 'floating' line elements will be subjected to concentrated or point forces and couples, and to uniform distributed forces but not to distributed couples. These loads will always be self-equilibrating.

2.2 Positive, Negative *nsm*

Structural analysis is based on the premise that if you could pry open a structural element by a cut perpendicular to the axis, you would discover, on both sides of the section, equal and opposite internal forces. The facet to the left is called the negative or $-$ facet and the facet to the right is the positive or $+$ facet.[1] The plethora of forces on the facets can be reduced to statically equivalent equal and opposite normal forces n and shear forces s positioned at the center of area of the cross-section and corresponding equal and opposite couples, called bending moments or simply moments m.

When the nsm are as depicted in the first row of Fig. 2.2 they are said to be positive. If they have opposite sense (second row) they are said to be negative.[2]

Physically, a positive normal force n causes tension at that location, a positive moment sets the lower fibers in tension and the upper fibers in compression, and positive shear would deform a segment as depicted in Fig. 2.3. Assuming that in the depicted beam segment the shear is positive, the shear forces at two neighboring sections a and b will look as shown (everything there is positive shear).

Having computed the internal forces along x, they are presented in the form of a normal (force) diagram $n(x)$, a shear (force) diagram $s(x)$ and a (bending) moment diagram $m(x)$. Computing the internal forces and displaying them are two separate albeit related activities.

The computational aspects of $nsm(x)$ are practically universal. We will present some essential computational techniques although, in due time, you will certainly

---

[1] Here we deliberately digress from Solid Mechanics convention where 'positive' depends on the direction of the outside normal and the left facet is therefore the positive facet. It would be too confusing in this text.

[2] Do not attach too much importance to positive or negative here. Different authors, industries and countries have different signs for these forces.

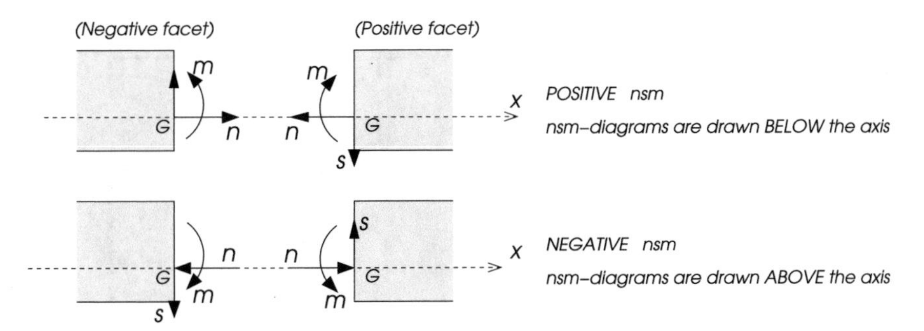

Fig. 2.2 *nsm* sign convention

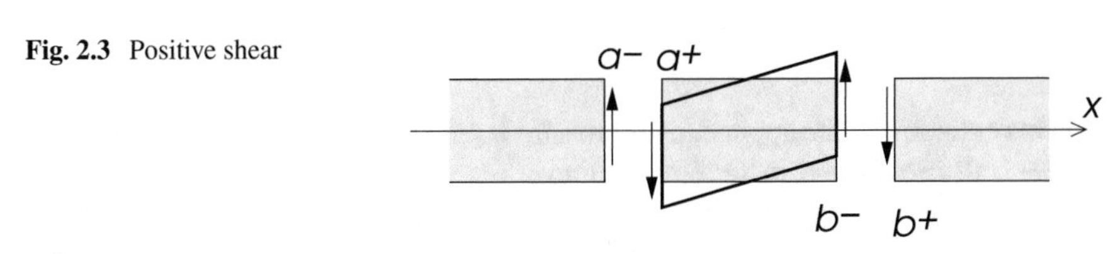

Fig. 2.3 Positive shear

develop your own personal methods. As long as they produce correct results they are valid techniques.

Displaying the diagrams is another matter. In a diagram of a function $f(x)$ for instance, the coordinate x usually runs from left to right and, positive $f(x)$ are usually depicted above the x axis. This is not so for the internal forces. It varies from country to country and often from discipline to discipline. Now, whether you draw a positive moment above or below the line depends on the specific environment you are working in. Since there are two alternatives for every diagram, and three diagrams in total, there are eight ways to depict the $nsm(x)$ diagrams.

In this text we have opted for one out of these eight possibilities.

POSITIVE *nsm* (upper row in Fig. 2.2) are drawn BELOW the axis.

If you find yourself in an environment or browsing though a textbook with a different protocol for the diagrams, you will no doubt adapt to that specific presentation.

2.3 Calculating *nsm* at a Section

To compute the internal forces *nsm* in an element, we need to know all the external loads applied to that element. These must of course be in equilibrium. (If you attempt to compute the internal forces in a floating structure subjected to forces which are not auto-equilibrated you will obtain nonsensical results.)

The frame represented by line $a..f$ in Fig. 2.4, for instance, is subjected to loads in equilibrium. (When properly supported this structure is called a *portal frame*.)

Fig. 2.4 Frame-like line

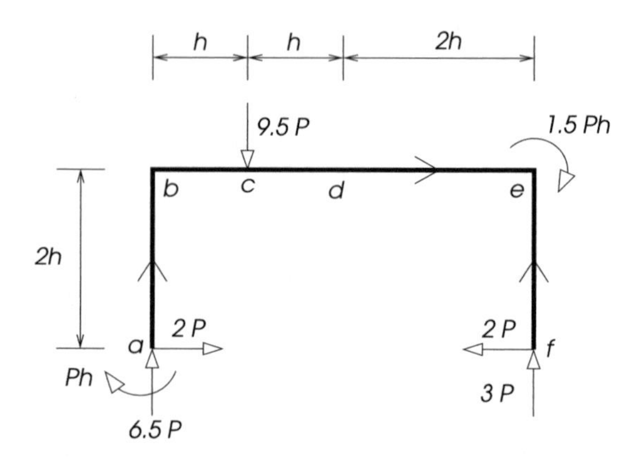

You may check that the applied loads are auto-equilibrated, in other words, the loads are statically equivalent to zero (to nothing).

To compute the internal forces nsm_d at mid-span of segment be, for instance, we perform a virtual cut at d thus creating two segments ad^- and d^+f. Enforcing equilibrium on ad^- will give the internal forces nsm_{d-}. Alternatively, requiring equilibrium of d^+f will produce nsm_{d+}. (The positive sense of x along the elements is indicated by the arrows.) This is classical structural mechanics. However, we would like to advocate one specific technique to calculate the internal forces. It is based on considering the left segment (ad^-) and calculating loads at d^- which are statically equivalent to the external loads on segment ad^-.

- **Equivalent loads at d^-**: In Fig. 2.5a we consider segment ad^- of the frame including the external load applied to the segment. We compute new external loads at d^- which are statically equivalent to the applied loads (Fig. 2.5b). This is done in two steps.

  In the first step we move the applied loads to d^-, including couple $-Ph$ (assuming positive couple counterclockwise).

  In the second we add to couple $-Ph$ the moments with respect to d^- of the applied forces, that is $2P \times 2h - 6.5P \times 2h + 9.5P \times h = 0.5Ph$, which gives the equivalent loads shown in Fig. 2.5b.[3]

- **nsm$_{d-}$**: Clearly, to equilibrate ad^- with loads at d^- all we need to do is apply loads equal and opposite to the equivalent loads (black arrows in Fig. 2.5c). These are the nsm_{d-}.
- **nsm$_{d+}$**: The nsm_{d+} are equal and opposite to nsm_{d-} (black arrows in Fig. 2.5d). It turns out that the nsm_{d+} (Fig. 2.5d) are identical to the equivalent loads (Fig. 2.5b).
- **nsm$_d$**: This notation refers simultaneously to the internal forces on both facets of section d.

[3]The equivalent couple can, in principle, be anywhere on the segment and not specifically at d, but since the bending moment is at d we also place the couple of the equivalent system there.

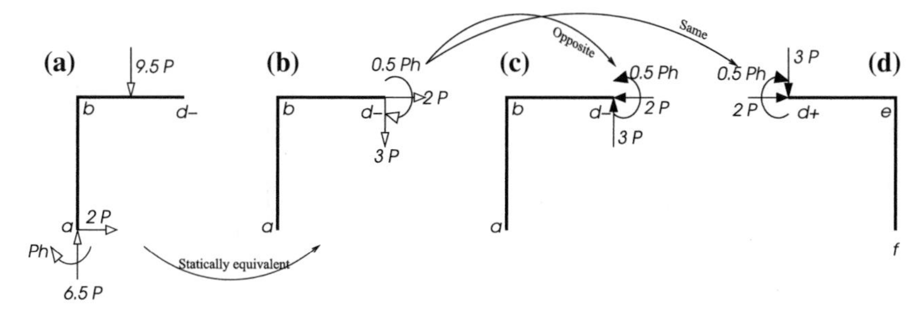

Fig. 2.5 Calculating nsm at a fictitious section. **a** Loads on segment abd^{-}. **b** Equivalent loads at d^{-}. **c** nsm at d^{-}. **d** nsm at d^{+} (Below the axis in nsm diagrams)

In conclusion, if you want to determine nsm at some section d take the segment from the left end until d and compute the equivalent loads at d. These loads are the nsm_{d+}, that is, the internal forces applied to the $+$ facet of d. An important fringe benefit of computing the internal forces at d^{+} is that the shear force s_{d+} points in the direction of where the shear is drawn, as we shall see in the sequel.

2.4 The nsm Diagrams

The values of the internal forces are needed not only at one section but usually everywhere along the structure. Repeating the above procedure for all the points along the structure may seem a daunting task but, as we shall see, shortcuts exist.

The internal forces are displayed in three separate diagrams: the normal, shear and moment *diagrams or distributions*. The diagrams are in fact graphs $n(x)$, $s(x)$ and $m(x)$ where the abscissa x is a line which follows the layout of the elements, and the ordinates, which are always perpendicular to the x-line, carry the respective nsm values.

2.4.1 Basic Rules

The first two *rules* for drawing the graph of the nsm functions, as for any function $f(x)$, are to define the direction of the x axis and the direction of the ordinate axis.

1. First we define the sense of the axis of the element. The running variable is denoted by x. If the element is horizontal or almost horizontal, it will run from left to right. If the element is vertical (or almost vertical), it will usually run from bottom to top, unless otherwise specified. (Remember, no axis sense, no diagrams!)

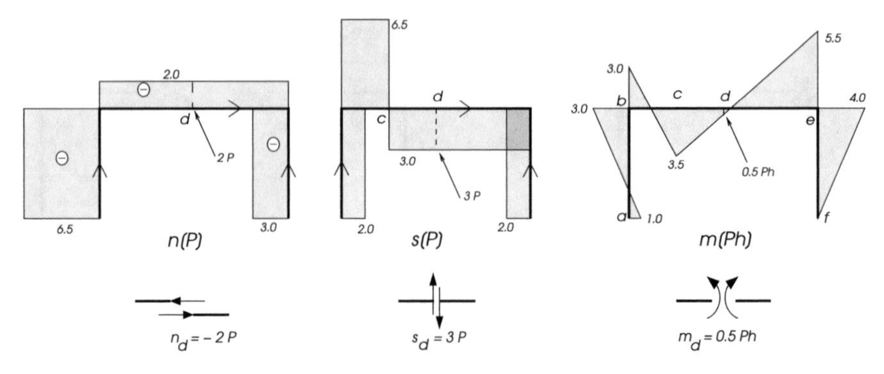

Fig. 2.6 Example of *nsm* diagrams

2. The positive ordinate axes of the $nsm(x)$ functions point *downwards*. This is akin to drawing a function $f(x)$ with the abscissa x running horizontally from left to right and the ordinate f pointing down. In other words, positive *nsm* (see Fig. 2.2) are drawn below the x axis. Negative *nsm* are drawn above the x axis. If the axis of the element is curved, as for an arch, the normal force is the component of the internal force (locally) tangent to the axis and the shear force is the component of the internal forces (locally) perpendicular to the axis. The ordinates of the respective *nsm* functions are also drawn perpendicular to the (curved) element axis.

In Fig. 2.6, we show the *nsm* diagrams for the frame in Fig. 2.4. We note that the diagrams indeed conform with the values nsm_d mid-way on beam *be* (the dashed line in the diagrams). The normal force $n_d = -2P$ being negative appears above the axis, the shear force is positive and is therefore drawn below the axis, and the bending moment is positive and is also drawn below the axis. The units of the diagrams appear in square brackets below each graph.

You will have noted that we have added a '−' character in the boxes of the normal force. This is because the conventions for the sign of the normal force are poorly standardized and the difference between tension and compression normal force can be a matter of survival of the structure (in compression slender members may buckle). To avoid any misunderstandings it is good practice to add a '−' for compression members and a '+' for tension members. Even this may be ambiguous, so you may add any accepted notation to make sure that the difference between compression and tension is clear, such as C, T for instance.

2.4.2 Practical Guidelines for Positioning the s and m Diagrams

The rules for positioning the *nsm* diagrams are rather clear. Positive *nsm* are below the axis. But you still have to remember what positive means.

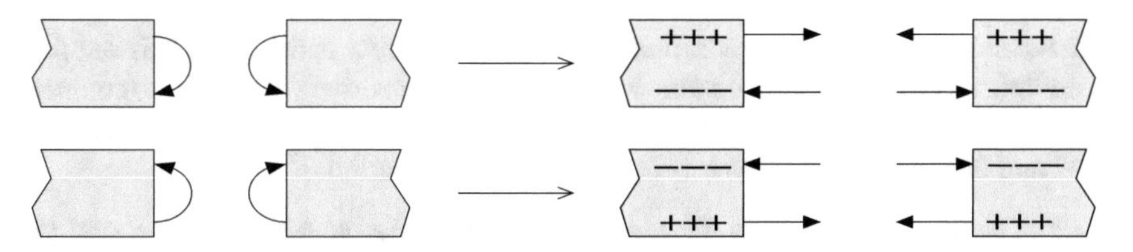

Fig. 2.7 Conventions for drawing the bending moment

So here are two simple guidelines for the m and s diagrams respectively:

- The *bending moment* is drawn on the side of the fibers in *tension*.

Consider the *bending moment* in Fig. 2.7. We remember that the curved arrows represent couples, the simplest representation of which are two parallel forces. One set of equal and opposite forces pulls on the facets and sets the material behind it in tension. The other set of forces pushes, causing compression. So in bending, one part of the cross-section is in tension and the other in compression. *The moment is drawn on the part of the beam where the tension is.* In the first row of Fig. 2.7 the upper fibers of the beam are in tension (+) and the lower part in compression (−). The moment will be drawn on top of the beam along the fibers in tension. In the second row of Fig. 2.7 the lower fibers are in tension and the diagram will appear below the axis.

In Fig. 2.6 at section d the fibers in tension are in the bottom part. The bending moment at d is therefore below the line. (See also Fig. 2.5c.)

- The shear diagram is drawn in the direction of the shear force on the + facet

(the facet to the right). For the *shear force*, look at the direction of the shear applied to the *far*-facet. It points to where the shear should be drawn. If the *shear force to the right goes up*, the shear is drawn *above* the line. If it points down, the shear is below the line.

In Fig. 2.5 the shear force at (d) on the + facet points down. The shear diagram at d is therefore below the line.

2.4.3 Relations Between the s and m Diagrams

The equilibrium equations of a lamella of a straight (or slightly curved) beam are

$$\frac{\mathrm{d}s}{\mathrm{d}x} + q = 0 \qquad \frac{\mathrm{d}m}{\mathrm{d}x} + s = 0 \tag{2.1}$$

We recall that, assuming a coordinate system $xy\theta$ where x is the beam axis running from left to right, y points up and θ is the rotation, positive from x to y (counter-clockwise), the nsm internal forces on the $-$ facet of a section are in the direction of $xy\theta$ and the distributed force q is positive in the y direction.

1. *The shear force is* minus *the* tan *of the slope of the moment diagram, and the distributed force q is* minus *the* tan *of the slope of the shear diagram.* Indeed, $\mathrm{d}m/\mathrm{d}x + s = 0$ gives $\mathrm{d}m/\mathrm{d}x = -s$ (and similarly $\mathrm{d}s/\mathrm{d}x = -q$).

   The slope of the bending moment distribution $\mathrm{d}m/\mathrm{d}x = -s$ thus equals minus the value of the shear force at that point. To make any use of this relation, we need to remember the positive directions of the ordinates of the s and m diagrams, and that the slope is the negative of the shear. This is not easy. It is useful to keep one typical example in mind and use it for relating the slope of the moment to the shear.

This example is the classical beam with a force in the middle of the span supported by end-forces (Fig. 2.8). The corresponding shear and moment diagrams encapsulate the entire idea. A *shear above* the line (left part) has the *moment sloping down* and a shear below the line corresponds to a moment sloping up. This is the correspondence. Remember this and you will be safe.

We have intentionally avoided using the words 'positive' and 'negative'. It would only confuse things, and a picture is worth a thousand words.

2. *A segment with a constant distributed lateral force has a linear shear function and a parabolic moment function.* The shear force is linear because it has a constant slope. Taking the derivative of the second equation in (2.1) gives $\mathrm{d}^2m/\mathrm{d}x^2 + \mathrm{d}s/\mathrm{d}x = 0$ and introducing the first equation of (2.1) yields $\mathrm{d}^2m/\mathrm{d}x^2 = q$. Integrating this relation twice produces the algebraic equation of a parabola.
3. *The moment is locally an extremum (either maximum or minimum) where the shear force is zero.* Indeed, a local extremum of a function corresponds to a zero

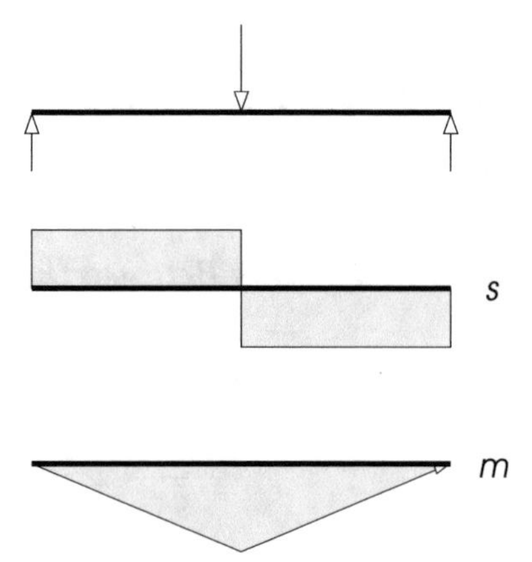

Fig. 2.8 Relation between s and the slope of m

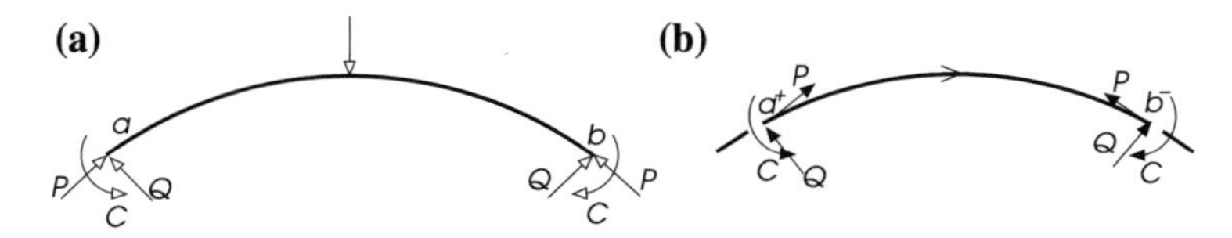

Fig. 2.9 End loads and end internal forces

slope of the function, and the slope of the moment function $\mathrm{d}m/\mathrm{d}x = -s$ is zero when $s = 0$.

4. *A segment without distributed forces has a constant shear force and a linear moment function.* This is self-evident. If the shear happens to be zero over a segment the moment is constant along that segment.
5. *The difference in the values of the bending moment between two sections along a segment is equal to the area under the shear diagram between these two sections.* Multiplying $\mathrm{d}m/\mathrm{d}x + s = 0$ by $\mathrm{d}x$ gives $\mathrm{d}m + s\,\mathrm{d}x = 0$. When we integrate this differential equation over the interval $[x_a, x_b]$ with $(x_b > x_a)$ we obtain $m_b - m_a + \int_{x_a}^{x_b} s\,\mathrm{d}x = 0$ ($\mathrm{d}m$ is a total differential). Consequently

$$m_b = m_a - \int_{x_a}^{x_b} s\,\mathrm{d}x \tag{2.2}$$

In other words, if you know m_a, the integral of the shear function gives you m_b.
6. *The internal forces on a section very close to an extremity are identical to the applied loads at that extremity.* This is obvious but not always clear to the novice, who may be tempted to draw a moment diagram with a non-zero bending moment at an extremity without an applied couple (*creatio ex nihilo*).
In the symmetric arch, symmetrically loaded, shown in Fig. 2.9a, the external loads at extremities a and b are forces P and Q and a couple C. The forces are conveniently expressed in the tangential and lateral directions. The internal loads on facet a_R^+ (a_R is a section to the right and infinitely close to a) are of course identical to the external loads at a (see Fig. 2.9b) and the internal loads on facet b_L^- (b_L is a section to the left and infinitely close to b) are identical to the external loads at b. The internal forces are thus $\boldsymbol{n}_a = \{-P \ \ -Q \ \ -C\}^T$ and $\boldsymbol{n}_b = \{-P \ Q \ \ -C\}^T$.

2.5 Typical Diagrams

The readers are reminded that, as a rule, straight *beams* will not have external axial forces and, therefore, also no normal internal forces. So, for beams we consider only *sm*-diagrams.

Beam with Point Force Along its Span

In Fig. 2.10 we have an example of a beam with basic auto-equilibrated forces: a point force along the span equilibrated by end forces.

The beam is composed of two unloaded segments $a_R b_L$ and $b_R c_L$, both starting immediately after a point force (subscript 'R', to the right) and ending immediately in front of a point force (subscript 'L', to the left). Note,

we NEVER make a virtual section exactly at a point-load.

We can make a virtual cut either in front or after a point load, but one cannot split a load in the same way as one cannot split a point. In Fig. 2.10 for instance, force Q is at b, and sections b_L and b_R are respectively in front and after b. Recall that the default sense of a horizontal axis is from left to right.

Since the segments are not loaded (without distributed forces), along every segment the shear is constant and the bending moment linear.

For the *shear diagram* we need *one* value along $a_R b_L$ and *one* value along $b_R c_L$. There are many ways to come by these values. We will in due course use most of them.

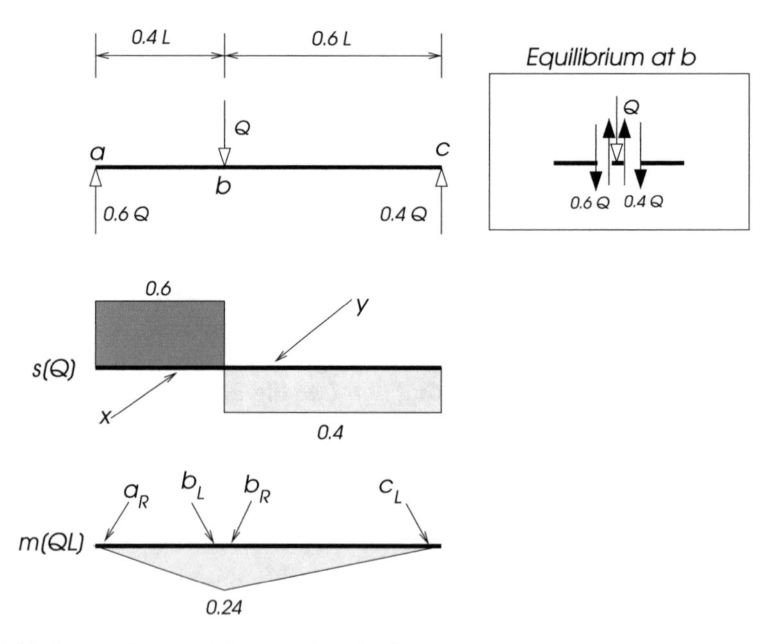

Fig. 2.10 Shear and moment diagrams for point force

(a) For the shear at some point x on the first segment (see Fig. 2.10) we collect all the forces along ax^- (only one up force $0.6Q$). We have thus a shear of $0.6Q$ (up) at x^+. When the shear goes *up on the right facet* it is drawn *above the axis*, hence the box on ab is above the axis (see Fig. 2.10).
(We could have used the property that the shear force on the + facet near the left extremity is the applied 'shear', that is, the component perpendicular to the axis of the force, at that extremity.)

(b) For the shear at y on $b_R c_L$ we consider a section at y (Fig. 2.10), sweep over all the forces on segment ay^-, and their algebraic sum is the shear at y^+. In our case we find a down force of $0.4Q$ from where it follows that the shear is drawn below the axis.

For the *moment diagram* we remember that the moment is linear on $a_R b_L$ and on $b_R c_L$. To position a line we need its ordinates at two coordinates. For the bending moment along the first segment, we use points a_R (right after $0.6Q$ at a) and b_L (right in front of force Q at b), and for the bending moment along the second segment, we use b_R (right after force Q at b) and c_L (right in front the force at c).

You will have noticed the change in slope of the bending moment at b. It follows the sudden change in the value of the shear at that section.

Finally, in the box to the right of the figure you see the shear forces which equilibrate Q at b: forces $0.6Q$ at b_L and $0.4Q$ at b_R.

Note. We would be satisfied with these two forces but the beam could not achieve this feat without bending moments along its entire span, as evidenced by the m diagram in Fig. 2.10. These unwanted internal couples cause flexion of the beam and are sometimes the cause of beam failure.

Beam with Point Couple Along Span

The counterclockwise point couple QL along the span of the beam in Fig. 2.11 is equilibrated by the clockwise couple composed of the equal and opposite forces Q at its extremities.

Interestingly, the shear force is constant along the entire span of the beam (Fig. 2.11). It is easy to see that

a shear diagram is insensitive to point couples.

If we replace the point couple QL icon by what it represents, two equal and opposite very large forces very close to one another, such that they create a couple QL, a sweep through the couple may cause a blip in the shear diagrams (Fig. 2.12). But since the forces are infinitesimally close to one another it is inconsequential.

For the moment diagrams, we note that on both segments the moments are linear (no distributed forces) and have the same slope (same shear) but there is a discontinuity in the moment diagram at b. Clearly, close to the free ends (a_R and c_L), the bending moments are zero because there are no applied couples at the extremities. From Fig. 2.11 we conclude that $m_{b_L} = 0.4QL$, and immediately after the couple we have $m_{b_R} = -0.6QL$, hence the moment diagram.

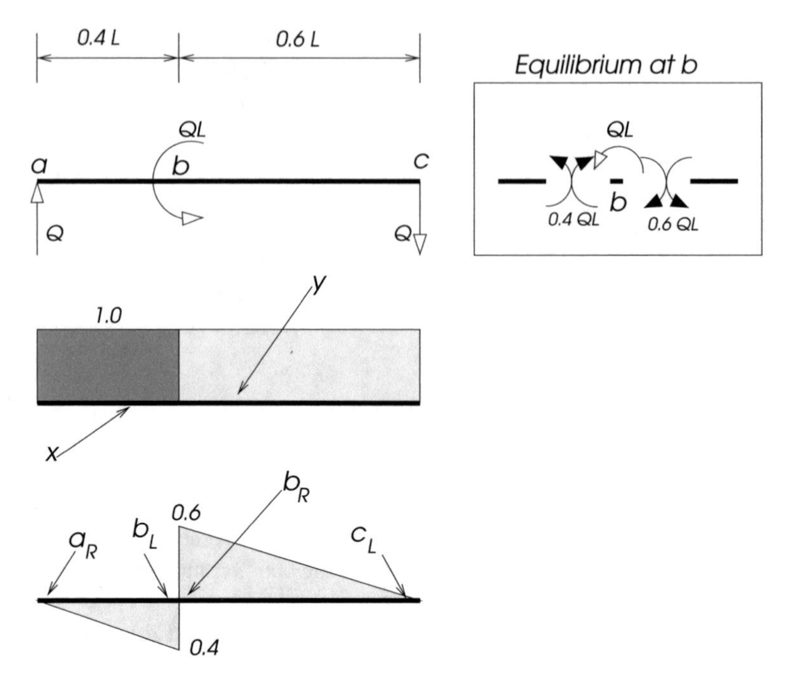

Fig. 2.11 Shear and moment diagram for point couple

Fig. 2.12 Point couple

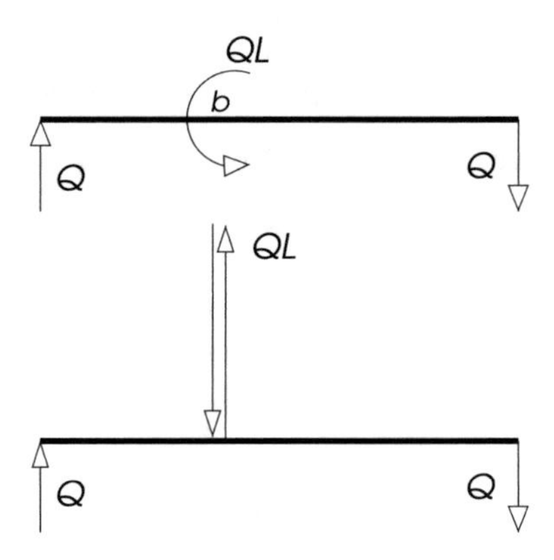

Note, m_{b_L} could also have been obtained from the relation $m_{b_L} - m_a = -\int_{x_a}^{x_b} s \, \mathrm{d}x = -0.4L \times -Q = 0.4QL$, that is, the area of the shear diagram over the interval (shaded area in shear diagram of Fig. 2.11). The relation is not valid for m_{b_R} because of the point couple of which the shear diagram is not aware.

Finally, the point counterclockwise couple QL at b is equilibrated by the clockwise bending moments $0.4QL$ and $0.6QL$ on both sides of b (see box in Fig. 2.11).

Beam with Uniform Distributed Forces Along its Span

The beam in Fig. 2.13 is subjected to uniformly (constant) distributed forces Q/L equilibrated by symmetric forces $0.5Q$ at the extremities. Consequently, the shear force is linear and the bending moment parabolic. For the shear force we need two values along the span. For parabolic bending moment distributions, we will require the end values, extremum values along the span (if applicable) and any other values necessary to draw a decent free-hand approximation of the distribution.

Now, at the extremities the shear forces and bending moments are equal to the applied forces and couples. This is enough to draw the shear diagram.

We notice that the shear is zero at mid-span (point b). The bending moment will have there an extremum which we must evaluate. An easy way is to use $m_b - m_a = -\int_{x_a}^{x_b} s \, \mathrm{d}x = -0.5L \times (-0.5Q)/2 = 0.125QL$ ($m_a = 0$), that is, the area of the shaded triangle.

Note. The example deals with a symmetric beam under symmetric loading and still the shear diagram is anti-symmetric. This stems from the fact that the shear diagram is biased. It shows the shear on + facets at any virtual section.

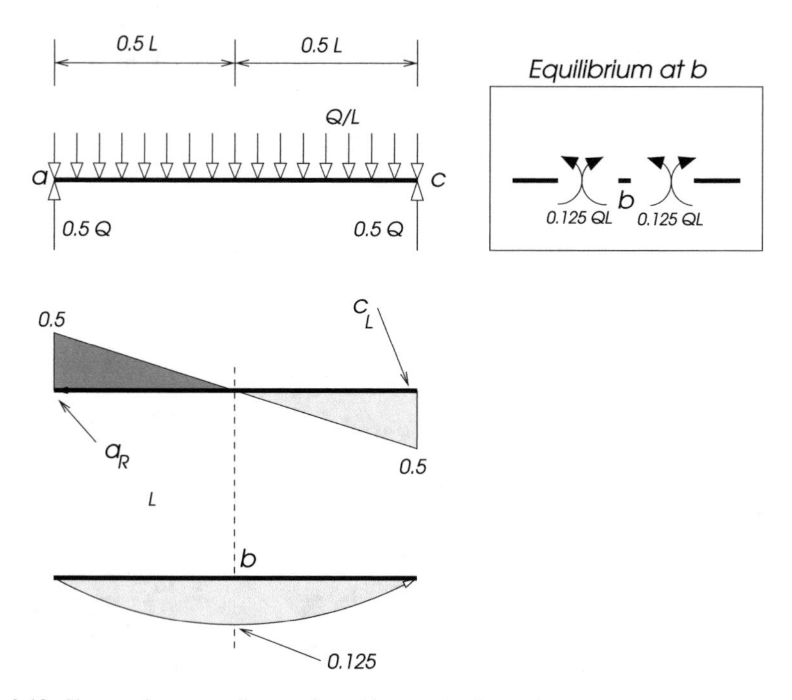

Fig. 2.13 Shear and moment diagram for uniformly distributed forces

2.6 Frames

2.6.1 Point Force on a Symmetric Frame with Inclined Columns

Frame *abcde* in Fig. 2.14a is symmetric with respect to the vertical line passing through *c* and is also loaded in a symmetrical fashion by self-equilibrating forces. So each internal force *nsm* in segment *ce* is a mirror image of a corresponding *nsm* in segment *ac*. This symmetry includes the *n* and *m* diagrams but not the *s* diagram, as we shall see.

First, we define the positive x-axis on the elements of the frame (see Fig. 2.14b). In this case we keep the same sense of x for the entire frame.

Segments *ab*, *bc*, *cd* and *de* (taken on intentionally inbetween forces and corners) are without distributed forces and consequently the $n(x)$ and $s(x)$ distributions are constant on every segment, and $m(x)$ is linear. For the *sm* distributions along leg *ab*, we use conveniently the axial and lateral projections of $0.5Q$, respectively $0.433Q$ and $0.25Q$, shown in Fig. 2.14b. The *nsm* distributions are shown in Fig. 2.15.

The structure and external loading are symmetric and so are the shear forces inside the elements (see Fig. 2.16c). However, the shear *diagram* is not (Fig. 2.15). As you

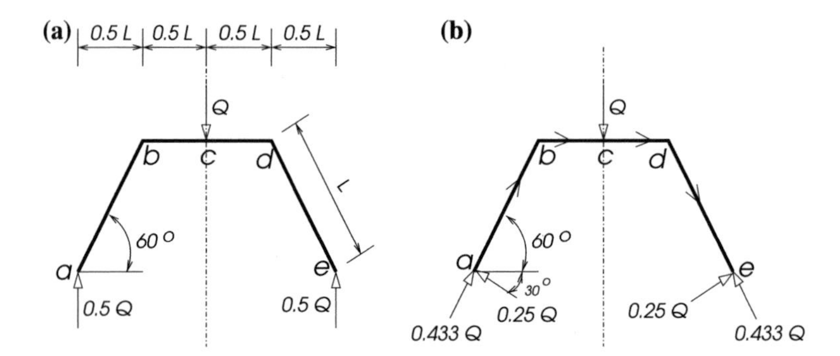

Fig. 2.14 Symmetric frame with point forces

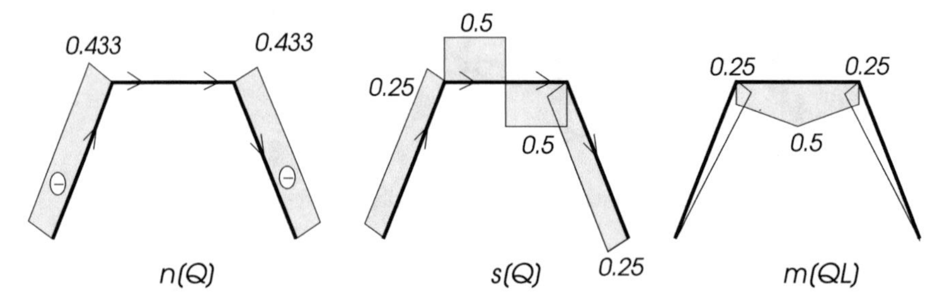

Fig. 2.15 *nsm* distributions in a symmetric frame

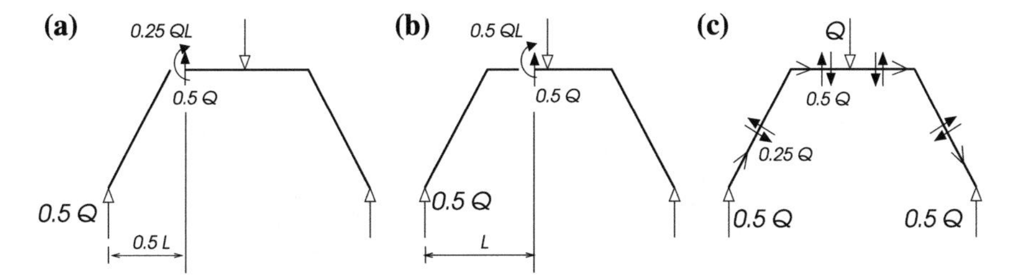

Fig. 2.16 **a**, **b** Calculating nsm; **c** symmetry of s

probably realize, this is because for the shear diagram we are always looking at the + facets. For a correct mirror image we need to compare the internal shear forces on the + facet of the left half of the structure with those at the − facet of the right half of the structure.

2.6.2 Discontinuities at Point Loads

In our view, structures are essentially subjected to zones of distributed forces with, here and there, concentrated forces and couples at the zone boundaries. Forces cause discontinuities in the n and s diagrams, and couples create discontinuities in m diagrams, as was established earlier when considering equilibrium. This was observed, for instance, in the s diagram of Fig. 2.10 and in the m diagram of Fig. 2.11.

Note, since the shear is a measure of the slope of the bending moment ($\mathrm{d}m/\mathrm{d}x + s = 0$), a lateral *force* also causes a sudden *change in slope of the bending moment* (see cross-section b in the bending moment diagram in Fig. 2.10).

On the other hand, a concentrated couple does not affect the shear diagram(!). In Fig. 2.11 there is a concentrated couple at b and the moment has a jump there. The shear force, however, is oblivious of the presence of the couple, and passes through that point as if there were nothing there. The slope of the bending moment is thus also not affected by the couple (the slope is $0.4QL/0.4L = Q$ on ab and the same $-(-0.6QL)/0.6L = Q$ on bc).

We would like to show one example of a discontinuity in the normal distribution. This kind of discontinuity can occur in this text only in frames as, for instance, the two-legged structure shown in Fig. 2.17a.

We will, in this case, divide the structure into *nodes* (where point loads are applied or at corners) and *elements* (segments in between nodes). This leads to the exploded view depicted in Fig. 2.17b. In the figure, the white arrows are external forces and the black arrows are internal forces. Starting from a, we impose sequentially vertical equilibrium at the nodes and at the elements. (We postpone rotational equilibrium for clarity's sake.)

Assuming positive forces upwards, the $0.75P$ force at a is balanced by $-0.75P$ on facet a_R^-, which causes $0.75P$ on facet a_R^+ of element ab. This applies an equal and

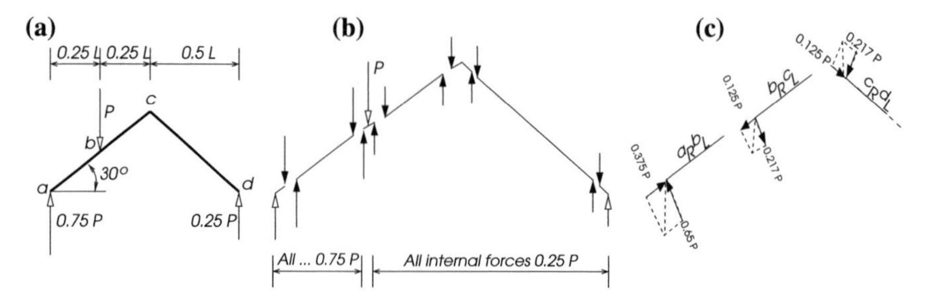

Fig. 2.17 Two-legged frame

opposite force on facet b_L^- of node b. Node b has now $0.75P$ and $-P$ (the external force) so we need to assume that facet b_R^- has $0.25P$ and so on until extremity c. (Sections b_L and b_R are infinitely close to b.) As emphasized in the drawing, all the internal forces from a to b_L are of magnitude $0.75P$ and the internal forces on the remainder of the structure have magnitude $0.25P$.

For the nsm diagrams, we recall that we are considering the three segments in between the point forces, a_Rb_L, b_Rc_L and c_Rd_L. In preparation of the n and s, diagrams we take conveniently the normal and lateral components of the internal forces at the left end of the three segments (see Fig. 2.17b).

For the m diagram we need to compute the bending moments at b and c only. (The bending moments are zero at a and d and are continuous at b and c.)

We leave it to the reader to show that rotational equilibrium of segment a_Rb_L requires a counterclockwise couple $0.75P \times 0.25L = 0.1875PL$ at b_L^- (see Fig. 2.18). Similarly, rotational equilibrium of segment c_Rd_L requires a clockwise couple $0.25P \times 0.5L = 0.125PL$ at c_R^+. The results in Fig. 2.18 show a discontinuity both in the n and s diagrams at b. The m diagram is continuous but has a slope discontinuity at b.

The following Table 2.1 is a summary.

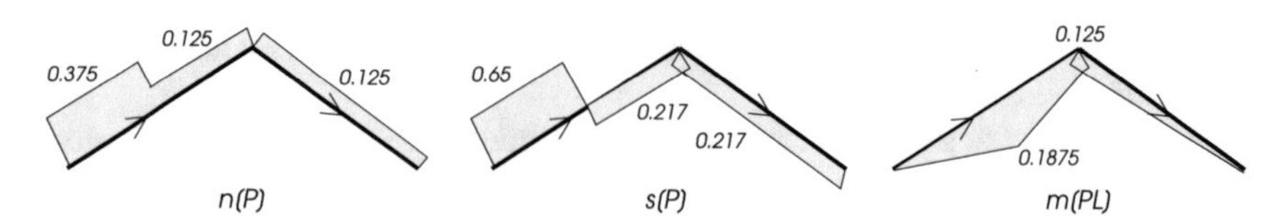

Fig. 2.18 nsm in the two-legged frame

Table 2.1 nsm discontinuities

Diagram	n	s	m
Axial force	Jump	–	–
Lateral force	–	Jump	Sudden change in slope
Couple	–	– (!)	Jump

2.7 Inverse Analysis

In broad outline, an inverse problem is when you are given the answer and you are asked for the question. Interestingly, such problems appear very often in engineering practice and research, such as when you are given the vibration profile of a structure and you want to learn something regarding the excitation.

Inverse problems are sometimes easier to come by than direct ones. For example, the standard equations of the displacement method (to be seen later on) are $\boldsymbol{K}\boldsymbol{u} = \boldsymbol{p}$ where $\boldsymbol{K}$ is a square matrix, called stiffness matrix, $\boldsymbol{u}$ is the nodal displacements vector and $\boldsymbol{p}$ is the vector of applied loads at the nodes. The direct approach is to compute the displacement, given the loads. This involves the solution of the linear system $\boldsymbol{K}\boldsymbol{u} = \boldsymbol{p}$.

The corresponding inverse problem would be, knowing the displacements $\bar{\boldsymbol{u}}$, what are the applied loads? This problem is much easier to solve. Simply multiplying $\boldsymbol{K}$ by $\bar{\boldsymbol{u}}$ yields $\boldsymbol{p}$

$$\boldsymbol{K}\bar{\boldsymbol{u}} \to \boldsymbol{p} \tag{2.3}$$

With this knowledge we realize that it can be much easier to determine the loads applied to a line element when we are given the bending moment distribution (inverse problem) than the other way around (direct approach). The rationale is clear. Ignoring the discontinuities, given $m(x)$ the first derivative produces the shear $s(x) = -\mathrm{d}m/\mathrm{d}x$ and the second derivative the distributed forces $q(x) = \mathrm{d}^2m/\mathrm{d}x^2$. In the direct approach we are dealing with integrals.

The purpose here is mainly didactic. We are, after all, learning to understand the workings of the $nsm(x)$ distributions.

Example

Consider the bending moment distribution shown in the box in Fig. 2.19. On the left half $m(x)$ is linear and on the right half $m(x)$ is a parabola starting at $-2QL$ at b with a zero value and a zero slope at c. What are the applied loads?

- Segment ab has no distributed loads but couples $5QL$ and $4QL$ at a and b respectively. This entails two equal and opposite forces $18Q$ for rotational equilibrium $5QL - 18Q \times L/2 + 4QL = 0$ (Fig. 2.19a′).
- Due to the parabolic $m(x)$ segment bc has distributed forces of magnitude, say q. The applied loads are a couple $2QL$ at b and a force $-qL/2$ at mid-span, statically equivalent to q over $L/2$ (Fig. 2.19a″). This entails a force $qL/2$ at b for vertical equilibrium. (Note, the shear at c is zero because the bending moment has there a horizontal slope.) Rotational equilibrium requires $2QL - qL/2 \times L/4 = 0$ from where we find $q = 16Q/L$.

Hence the shear distribution in Fig. 2.19b and the applied loads (the purpose of this exercise) in Fig. 2.19c.

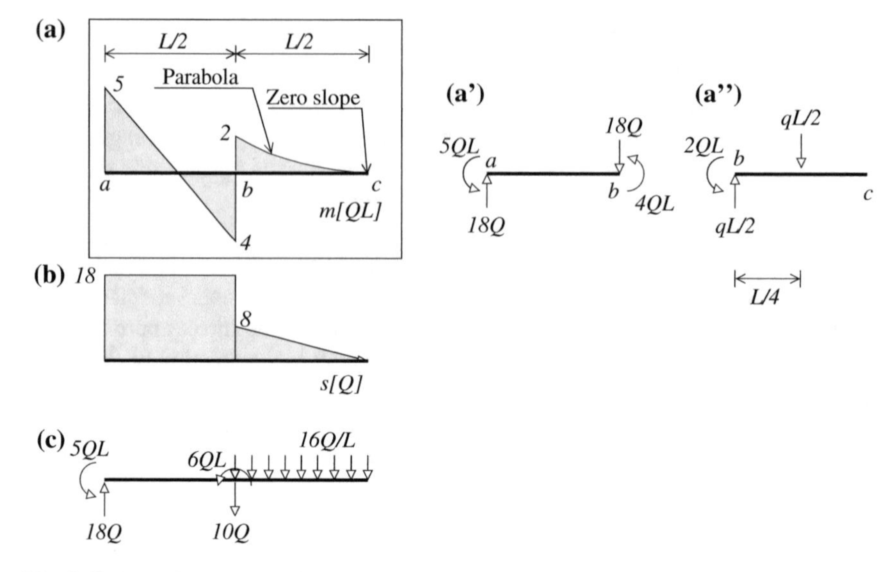

Fig. 2.19 Beam inverse analysis

2.8 Notes

We elaborate here on some notions related to the above examples, not necessarily in their order of importance.

2.8.1 *You Cannot Split a Point*

You will have noticed that we perform sections in front or after point loads or corners. It is not recommended to perform a section right at a corner or at a load because you cannot split a point. Simply stay away from these problematic cross-sections and be either in front or behind them. Needless to add that, however localized they may be, loads are in fact continuous forces of high magnitude distributed over a very short distance. Whatever goes on in the beam right under these forces or at corners, for that matter, is a complicated affair and belongs more to the theory of elasticity and continuum mechanics, so we had better steer clear of these regions in structural analysis. We should either be in front or behind the troublesome zones.

2.8.2 *What Is in a Name?*

At corners of frames without applied forces, such as b in Fig. 2.6, the 'physical' internal forces at b_L and b_R are the same (and opposite), only their components

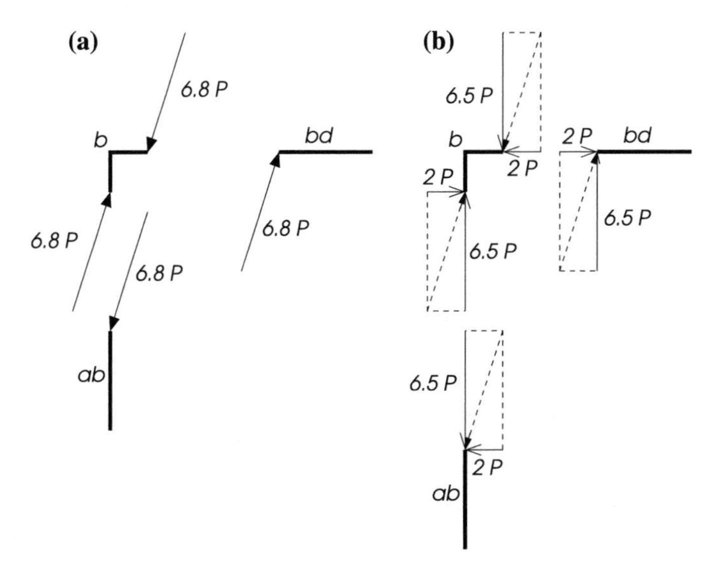

Fig. 2.20 *s* and *n* interchange at a corner

along the elements connected to the corner (node) change their names. This is clearly visualized in Fig. 2.20a. The resultant of the internal forces (stresses) acting on both sides of sections b_L and b_R is $\sqrt{2^2 + 6.5^2}P = 6.8P$.

The horizontal component of this force is a *normal* internal force $2P$ in beam *bd* and a *shear* force $2P$ in column *ab* (Fig. 2.20b). Similarly, the vertical component of the force is a *shear* force $6.5P$ in beam *bd* and a *normal* force $6.5P$ in column *ab*.

2.8.3 Why Tensile Side for Bending Moments?

Beams are primarily bending elements and as such have tension along the bottom fibers and compression along the upper fibers when the bending moment is positive. The opposite is true for segments with negative bending.

When you design beams made of reinforced concrete it is very important to know where the tensile stresses are. Concrete is an artificial composite material made of an aggregate of cement, stone, sand and water. It is a rather easy material to come by and, similar to rock, it behaves very well in compression but not in tension. Mountain ranges have peaks which measure in kilometers, their bases incur very heavy compressive forces (self weight) and still they are perfectly stable, if not for the tectonic forces. Arguably, the heaviest man-made structure, if it can be called a structure, the Great Pyramid of Cheops in Upper Egypt may have lost its erstwhile glitter but its base still holds the punishing pressure of over two million huge blocks of limestone. The same prevails for concrete.

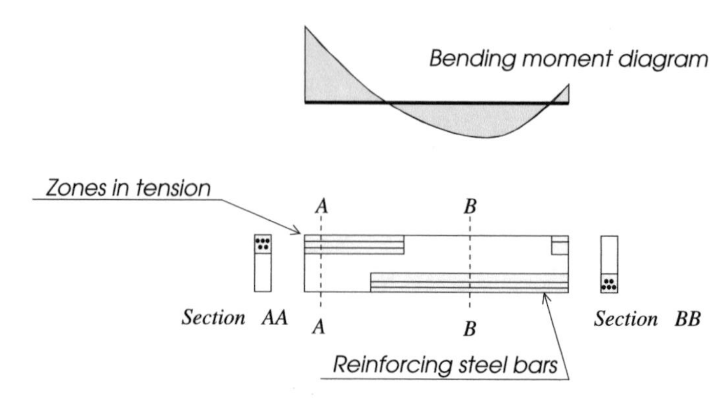

Fig. 2.21 Reinforced concrete example

Like rock, concrete is brittle in tension. Pull on a concrete bar and chances are that it will crack in no time. This is why we use reinforced concrete in our structures. Steel is very effective in tension and steel bars are embedded in the concrete in regions where tension is expected. Consequently, when designing a beam made of reinforced concrete we will position the reinforcing bars near the fibers in tension, that is, *on the side where we draw the bending moment.*

The parabolic bending moment diagram in Fig. 2.21 is typical for beams under distributed uniform forces. We notice that to the left of the point of zero bending the upper fibers are in tension while the lower ones are in compression and to the right of that point the lower fibers have tensile stresses. The diagram is telling us to position the reinforcing bars in the correct zone. Starting from the left, the steel reinforcing bars should hug the upper surface of the beam cross-section and then move to the underside tracking the tensile tresses. (The drawing is of course very schematic.)

It has been argued that in countries where concrete beams are the rule the diagrams are on the tensile side, where concrete is poised to crack, whereas in regions where the beam are usually made of steel, the bending moment is on the compressive side. Steel beams have more to fear from compression, which may cause local buckling and wrinkling of the flanges, than from tensile stresses.

2.8.4 How Are the Applied Forces Annulled?

We should say a word regarding the direct equilibrium of the external forces. Here, we need opposite internal forces applied exactly at the external forces to annul them. (Note, the internal forces will be read off the nsm diagrams in Fig. 2.6.) Consider a thin lamella at the base of column ab (node a in Fig. 2.22). On one side it has two applied forces and an applied couple (white arrows) and on the other side, inside the material, the internal forces (black arrows). This is trivial.

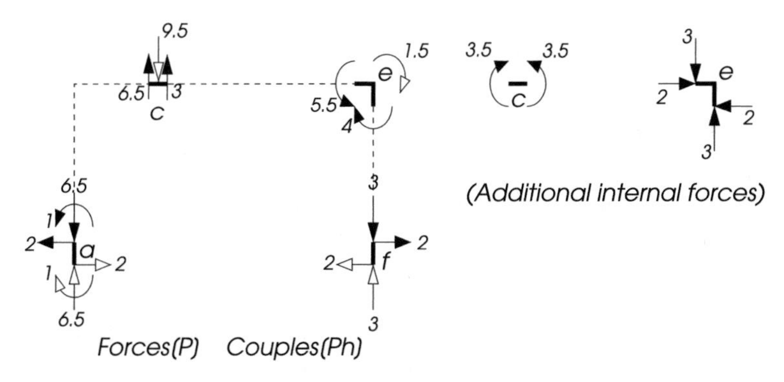

Fig. 2.22 Equilibrating the applied loads in a frame

Section c is a little more interesting. Taking the values off the shear force diagram in Fig. 2.6, we find that the $9.5P$ force is annulled by the sum of the shear forces on both sides of the thin lamella beneath the force.

At corner e, we have a similar situation where the applied $1.5Ph$ couple is annulled by the bending moments 5.5 and $4.0Ph$.

Finally, at f the shear and normal forces annul directly the two applied forces.

2.8.5 Noise on the Way to Equilibrium

At node c and corner e, we can get a glimpse of how structures behave (see the additional drawings in Fig. 2.22. At e, for instance, the applied couple is indeed balanced but by a difference of two couples, each of which is, however, about three times larger than the applied one. We would have preferred smaller bending moments to do the job, but this is the only way the structure is capable of annulling the external forces. In fact, most of the internal forces are unwanted guests. All we want is to annul the applied forces, and the rest, as far as we are concerned, is superfluous. But the frame in Fig. 2.6 cannot do a better job than that.

Finally:

How do structures produce the internal forces? By deforming.

But this is for later.

2.9 Illustrated Examples

2.9.1 *Frame with Point Loads*

Again consider frame $abcdef$ in Fig. 2.4. We will now attempt to determine the $nsm(x)$ shown in Fig. 2.6 (in a frame we have all three types of internal forces). You will have noticed that the positive axes are bottom-up for columns ab, fe and left-right for beam be. It is subjected to point forces and point couples in self-equilibrium. Segments ab, bc, ce, ef are without loads and cover the entire length of the frame. Consequently, the shear force on every segment is constant and the bending moment is linear. Thus, we need one value per segment for the shear force and two for the linear bending moment (usually chosen close to the extremities). It is by now clear that we compute the internal forces on the + facet of every virtual cut.

In Fig. 2.23, we have computed the internal forces on the + facet at a_R, b_L (just in front of corner b), b_R (right after corner b), c_L (right in front of the point load at c), c_R (right after the point load) and e_L. Note, when passing from e^- to e^+ the normal force 2.0 becomes a shear force, and the shear force 3.0 is now a normal force. The bending moment drops from 5.5 to 4.0 because of the concentrated couple at the corner.

These results are universal. They are true all over the world. The way we are displaying the results, however (see Fig. 2.6), is region-dependent. In this text we use the results in Fig. 2.23 to draw the diagrams in Fig. 2.6 as follows.

- For the *shear diagram* we recall that the shear is drawn in the physical direction of the shear force on the + facet (right facet) of the section. In the first instance, we need to indicate the begin and end directions by means of arrows. The arrows

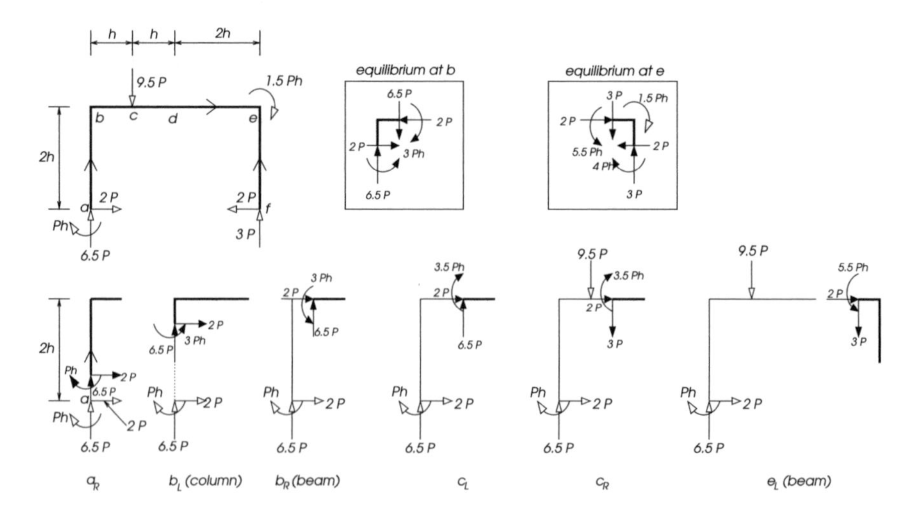

Fig. 2.23 Calculating nms in frame

on the columns were chosen in the up direction and the arrow on the beam is, of course, pointing right. For column ab starting from a, we have a shear force on the + facet of any section pointing inside the frame, consequently the constant shear is drawn on the inside. On segment bc, we have at b_R^+ a shear force $6.5P$ pointing up, consequently the shear diagram is there above the line. On segment ce, on the other hand, at c_R^+ we have a down force $3P$ and the shear diagram is there under the line (down) with magnitude $3P$. Finally, the direction of column fe was chosen from f to e as indicated by the arrow. Starting from the left, that is f, we note that the shear force on a + facet points inwards and thus the constant shear diagram is there above the line (or inside the frame).

- For the *moment diagram* we remember that it is composed of straight lines which follow the tensile side of the beam. For instance, on ab the moment at a (Ph) causes tension in the fibers inside the frame (see sketch 'a_R') and at b $(3.0Ph)$ the fibers on the outside of the frame are tensile (see 'b_L'). Hence the moment diagram on segment ab in Fig. 2.6 is a straight line running from inside the frame at a to outside the frame at b.
  From b to c we go from tension above the axis $(3.0Ph)$ to tension below the axis $(3.5Ph)$. On ce the moment diagram starts with tension inside at c $(3.5Ph)$ to tension outside at e $(5.5Ph)$.
- Finally, the *normal forces diagram* is drawn below the axis for tensional (that is positive) forces and above the line for compressive forces. Here the stresses are uniform over the section and are either in tension or in compression. 'Above' and 'below' have the same meaning as in the shear diagram. (We define the axis direction and, if it is horizontal, the beginning should be at our left.) The representation of the normal forces in the normal diagram in Fig. 2.6 is self-evident. Incidently, the normal forces in this example are all compressive.

It is important to keep in mind that of the three diagrams the most important by far is the moment diagram.

2.9.2 Distributed *Forces Between Point Loads*

The beam in Fig. 2.24 is subjected by downwards uniformly distributed forces over the left half of the span, two clockwise point couples and two forces at the extremities. One can check that the loading is auto-equilibrated, that is, the applied forces and couples are statically equivalent to zero but, they induce internal forces which we plan to determine.

We already know that along ac the shear is linear and the bending moment parabolic, and along ce the shear is constant (we recall that shear diagrams are not affected by point couples), and the bending moment has a constant slope (because the shear is constant), but is discontinuous due to the presence of point couples, as can be seen in the shear and moment diagrams of Fig. 2.24. As always the beam and matching sm diagrams are drawn one beneath the other.

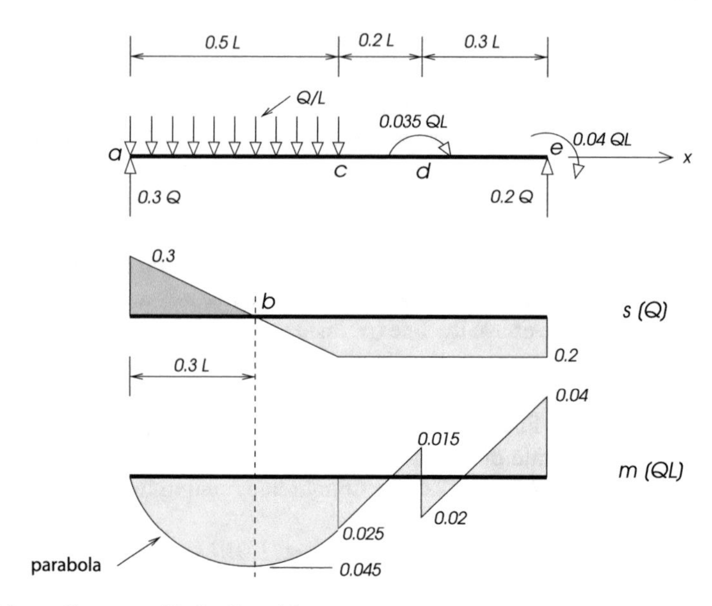

Fig. 2.24 *sm* diagrams with distributed forces

What is left to compute are the extreme values of the diagrams. In the present case, we will follow a different route for computing the internal forces. We divide the beam into nodes where point loads are applied (a, d and e) and segments with constant distributed forces. This requires another node c where the magnitude of the distributed forces changes from Q/L to zero. The segments are thus $a_R c$, cd_L and $d_R e_L$ (see the exploded view in Fig. 2.25).

We now determine the internal forces at the + facets by using the equilibrium conditions, starting with node aa_R^- then cascading to $a_R^+ c^-$ and so on until e_L^+. The individual sketches are shown in Fig. 2.25 and are self-explanatory. For every sketch, the known forces are denoted by white arrows and the ones that are computed are shown with black arrows. (There are at most two black arrows in every diagram, and we have only two equilibrium equations to compute them.)

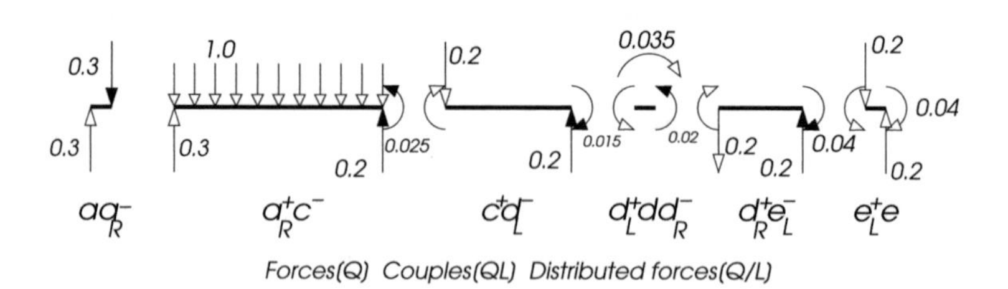

Fig. 2.25 Segments end loads

At extremity a, the external force $0.3Q$ is balanced by the shear force. The latter is calculated, hence the arrow is black. For segment $a_R^+c^-$ the same shear force as at a^+ is now known and is white, while the shear and moment at c^- are computed (black), etc. (At node d we have ignored the shear forces in the diagram for reasons of clarity.)

Note, when we reach e all the forces are known (white). There is nothing you can do regarding the equilibrium of e only hope that it works out well. This is a verification (necessary but not sufficient) that everything has been computed accurately. If e is in equilibrium, there is a good chance that everything is in order. If e is not in equilibrium, something is wrong. We can now draw the sm distributions.

We can draw the diagrams for every segment separately and glue them together or do it in one stretch from left to right. The results are in Fig. 2.24. We draw the m diagram below the s diagram to follow the correlation between loads, shear and moment. Moving information from one diagram to the other is both a verification and an aid in establishing the distributions.

For the bending moment, we have determined the end-values on the interval ac, and since the shear force passes through zero in the interval we also evaluate the moment there. Indeed, $\mathrm{d}m/\mathrm{d}x = -s$, and when $s = 0$, the derivative of the moment is zero, in other words, the moment is at an extremum.

We can easily determine the coordinate of x_b where the shear is zero. The right-angled triangles built on segments ab and bc are similar and consequently $x_b/0.3 = (0.5L - x_b)/0.2$ or $x_b = 0.3L$. This is also the point of extremum (largest) bending moment.

There are several ways to compute the moment at b. One of them is by integrating $s\mathrm{d}x = -\mathrm{d}m$, or, $\int_a^b s\mathrm{d}x = -\int_a^b \mathrm{d}m = m_a - m_b$. In our case $m_a = 0$ and therefore $m_b = -\int_a^b s\mathrm{d}x$, or, m_b is equal to the area of the triangle built on segment ab, that is, $m_b = -(0.3L \times -0.3Q)/2 = 0.045QL$ (the shaded area in the s-diagram of Fig. 2.24). As for the sign, you can verify that s above the line is algebraically negative, so is the integral. The moment is thus positive and therefore drawn below the line.

We must admit that juggling with the signs and coordinate axes is an error-prone activity. It is good practice to use the area method to determine the unsigned value of the bending moment and then rely on your engineering judgment as to where the moment should be drawn.

Note that the slope of the bending moment at c depends on s_c only ($\mathrm{d}m/\mathrm{d}x = -s$). The slope is thus the same on the parabola at c^- and on the straight line at c^+. In other words, the straight line which is the bending moment along cd is the same as the slope of the parabola at c.

This should be enough to draw an adequate representation of the bending moment diagram.

2.9.3 Inverse Analysis of a Frame

It is often much easier to reconstruct the applied external loads from the $nsm(x)$ distributions than the other way around. The following example is a simple illustration. It is also a good exercise for checking whether you have read the nsm diagrams correctly.

Given the nsm distributions for the frame in Fig. 2.26 what would the applied loads be? (Notice the arrows in the sm distributions.) In a first instance, the ns on segments ab, bc and cd are piecewise constant; consequently there are no distributed applied forces on these segments. Next we look for discontinuities at the limits of the segments from which we deduce the presence of point loads.

At extremities a and d the normal and shear values *are* the applied loads there. The bending moments are zero, so there are no applied couples at the extremities (Fig. 2.27b). At node b, we note discontinuities in both the n and s distributions. The sketch in Fig. 2.27a is one way to compute the external forces causing the discontinuities. The black arrows are internal forces read from the nsm diagrams at b. The white arrows are computed external forces which maintain equilibrium, written conveniently in the axial, lateral and rotational directions. (We also show in Fig. 2.27a that corner c is perfectly balanced by the internal forces so there are no external loads there.)

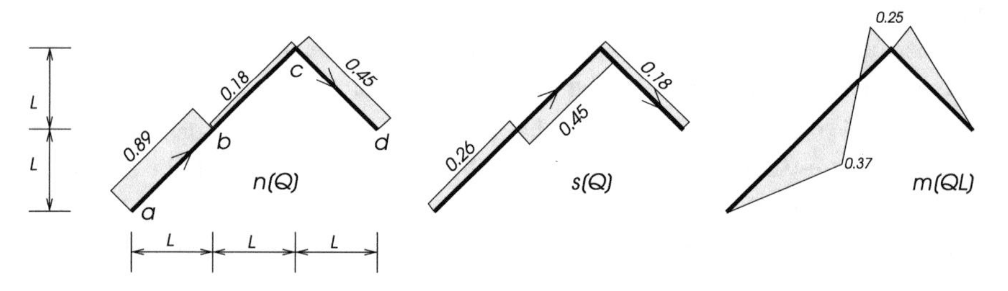

Fig. 2.26 Inverse analysis: the nsm distribution

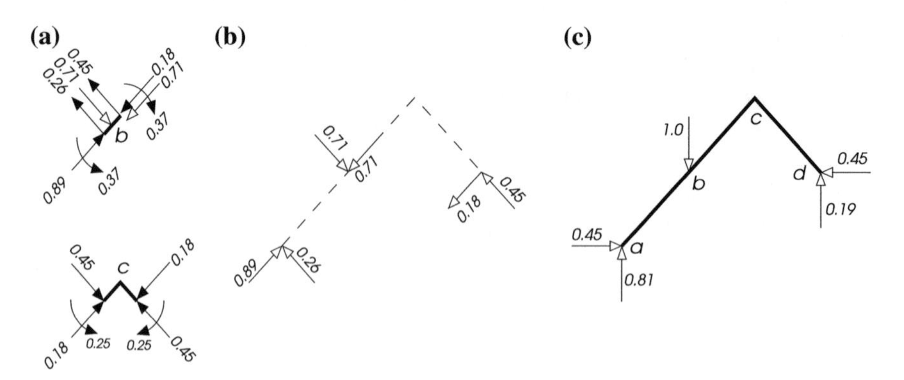

Fig. 2.27 Inverse analysis: calculating the loads

The results are shown in Fig. 2.27b. When we study the displacement method, such components of the forces will be called 'local' whereas the ones in Fig. 2.27c will be 'global'.

Finally, you were right to note that the s diagram is superfluous. It can be obtained from the m diagram ($s = -\mathrm{d}m/\mathrm{d}x$).

2.9.4 Kinked Bar

Consider the kinked bar[4] in Fig. 2.28a subjected to co-linear forces F at a and c. The element can be considered as having been straight of length $2L$ and having received a kink in its middle.

When the bar was straight ($h = 0$) the internal forces reduced to a compressive axial forces $n(x) = -F$ (no shear, no bending moments). What happened when the element was bent by a kink at b? First and foremost we have now internal bending moments, which is not a good omen.

Indeed, in Fig. 2.28b we have depicted half the structure, ab as a free body. (Since the structure and the loading are symmetric, so is the solution. The forces on the other half are the mirror image.) At a we have F. Somewhere else there must be an equal and opposite force, if not, the body under consideration would accelerate to the right. Since the half-element is devoid of any additional external forces, we must assume that there is an equal and opposite internal force F at the right end b of the (now) free body. This force is inside the structure, and there we can surmise anything plausible. So far translational equilibrium is satisfied. But there is now an anti-clockwise couple Fh (the two forces F at a distance h). We need a clockwise couple Fh to maintain rotational balance and this can only be the bending moment inside the element at b. The free body is now in equilibrium.

To draw the diagrams of the internal forces, it is convenient to decompose the end forces in axial and lateral components as in Fig. 2.28c. We can now draw the internal forces diagrams. Since there are no distributed forces, and the element is straight, the normal and shear distributions are constant. We will note that $n(x) = -F\cos\alpha$ and $s(x) = F\sin\alpha$ (α is the slope of the bar) are both drawn below the axis. The bending moment is linear, joining the value at a (zero) to Fh at b (tension above). This covers the left part of the diagrams. We note, as expected that the normal internal forces n are reduced with α and the shear s and bending moments m are increased.

The right part is their mirror images, but for the shear diagram. We are reminded that the shear diagram of a symmetric problem[5] is anti-symmetric! The actual shear forces are of course symmetric (Fig. 2.29a), but because the x-coordinate runs from left to right on both sides of the axis of symmetry, the shear diagram is anti-symmetric (Fig. 2.29b).

---

[4]We usually reserve the term 'bar' for a straight element under axial loads and 'beam' for an element in bending. The kinked bar is something which was a bar and acquired characteristics of a beam.

[5]A symmetric problem is a symmetric structure with symmetric loads.

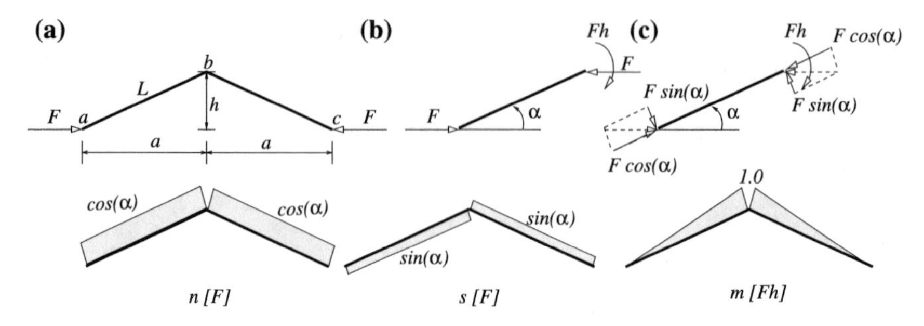

Fig. 2.28 nsm of kinked bar under compressive forces

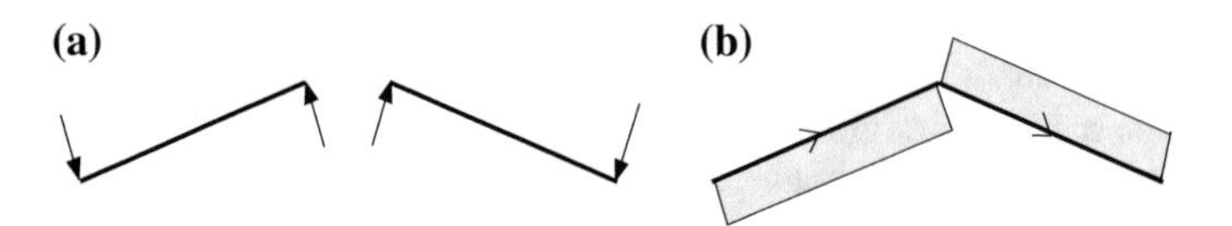

Fig. 2.29 Symmetric shear forces; anti-symmetric $s(x)$. **a** Shear forces (symm.). **b** Shear diagram (anti-symm.)

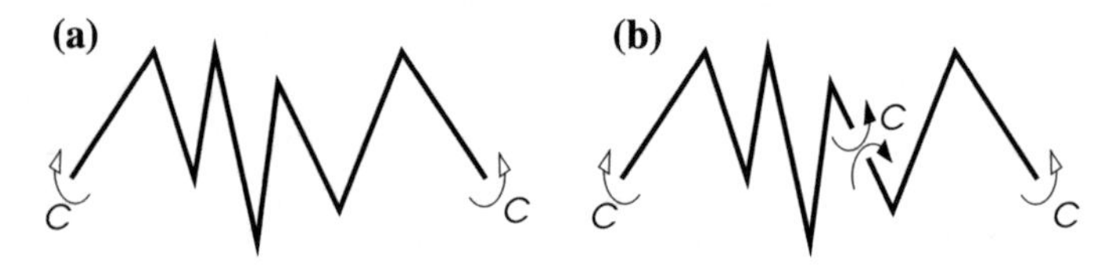

Fig. 2.30 Zigzag frame

2.9.5 Broken Line with End-Couples

We shall conclude on a light note. The zigzag (rigid) frame in Fig. 2.30a is subjected to equal and opposite couples C at its extremities. What are the nsm?

Take a section anywhere and you will note that all you need to enforce equilibrium is a positive bending moment of magnitude C (Fig. 2.30b). Therefore, $n = s = 0$ and $m = C$ (tension below the line) everywhere.

2.10 Summing Up

We have concluded a somewhat lengthy description of how to compute the internal forces in line segments, and more so of how to depict them in $nsm(x)$ diagrams. You will have noticed that it is all an exercise in equilibrium.

Typical cases were shown and some examples solved. The nsm diagrams permeate the field of structures, and will appear repeatedly throughout this and other texts on structural analysis and design.

Learning is a repeating process, done in ever widening circles. The reader will have many opportunities to either return to this chapter or to improve her/his skills by studying and drawing other $nsm(x)$ diagrams which will appear further on.

Chapter 3
Virtual Work

Virtual Work is a concept which will help us to develop most of the properties and methods pertaining to the theory of structures. In parallel with 'equilibrium systems' between external and internal loads (Chap. 1), we independently construct 'compatible systems' composed of deformations which result from displacements. It is shown that the virtual work of the external loads through the displacements is equal to the virtual work of the internal forces through the deformations. This is the principle of virtual work.

The concept is illustrated with simple examples of discrete particles, and is further developed for a beam segment under continuous and point loads undergoing an arbitrary displacement.

3.1 What Is Virtual Work?

We will introduce Virtual Work by means of a rather simple model. Consider a system of, say, $N = 4$ collinear particles, a, b, c and d (such as four billiard balls on a very smooth horizontal table) subjected to collinear external forces P_a to P_d (first row in Fig. 3.1) and $M = 3$ internal (interaction) forces between adjacent particles (second row in the same figure). We assume that forces and everything else happen along the line of particles.

The interaction forces n_{ab} to n_{cd} are internal forces that every particle exerts on its immediate neighbors. By virtue of the action–reaction principle, if particle c, for instance, applies n_{cd} to particle d, then the latter applies an opposite force n_{cd} to particle c. The particles need not have any mechanical properties but they help in visualizing the interaction forces.

We now assume that the particles are in equilibrium under the action of the external and internal forces. The equilibrium equations for the N particles $P_a + n_{ab} = 0$, $P_b + n_{bc} - n_{ab} = 0$, $P_c + n_{cd} - n_{bc} = 0$ and $P_d - n_{cd} = 0$, can conveniently be written in matrix form:

M.B. Fuchs, *Structures and Their Analysis*,
DOI 10.1007/978-3-319-31081-7_3

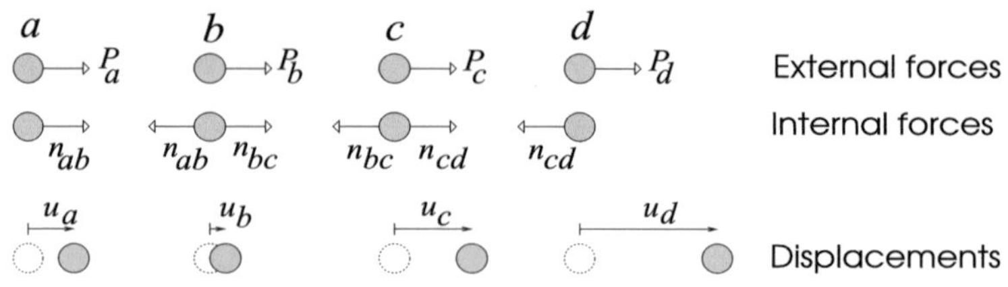

Fig. 3.1 Virtual work: particles (1D)

$$\begin{bmatrix} -1 & 0 & 0 \\ 1 & -1 & 0 \\ 0 & 1 & -1 \\ 0 & 0 & 1 \end{bmatrix} \begin{Bmatrix} n_{ab} \\ n_{bc} \\ n_{cd} \end{Bmatrix} = \begin{Bmatrix} P_a \\ P_b \\ P_c \\ P_d \end{Bmatrix} \quad \text{or} \quad \boldsymbol{Q}\,\boldsymbol{n} = \boldsymbol{p} \tag{3.1}$$

where the $N \times M$ coefficients matrix $\boldsymbol{Q}$ is the equilibrium matrix, $\boldsymbol{n}$ is the M-vector of internal forces and $\boldsymbol{p}$ is the N-vector of external forces.

Next, we ask a bystander who knows nothing about the forces, to displace all the particles by arbitrary distances u_a to u_d (third row in Fig. 3.1). Of course, there is no relation whatsoever between the forces and the displacements, and the latter are certainly not due to the forces.

We define the *virtual work of a force* as the algebraic product of the force times the distance by which it has been displaced.[1] It is called *work* because in physics, work and, for that matter, also energy, have the dimensions of *force* $\times$ *length*, such as Nm (Newton meter.)

For example, the virtual work of force P_c is $P_c u_c$ and the virtual work of the force n_{cd} (the one applied to d) is $-n_{cd} u_d$, which is negative because the force and the displacement have opposite senses.

We now sum up the work of all the external forces in the figure $P_a u_a + P_b u_b + P_c u_c + P_d u_d$ or $\boldsymbol{p}^T \boldsymbol{u}$. This is the virtual work of the external forces or evw (external virtual work.) Introducing (3.1) into the expression of evw, we have $\boldsymbol{p}^T \boldsymbol{u} = (\boldsymbol{Q}\boldsymbol{n})^T \boldsymbol{u} = \boldsymbol{n}^T \boldsymbol{Q}^T \boldsymbol{u} = \boldsymbol{n}^T (\boldsymbol{Q}^T \boldsymbol{u})$. The term in parenthesis is an M-vector $(M \times N \times N \times 1)$ which we call $\boldsymbol{e}$:

$$\begin{Bmatrix} e_{ab} \\ e_{bc} \\ e_{cd} \end{Bmatrix} = \begin{bmatrix} -1 & 1 & 0 & 0 \\ 0 & -1 & 1 & 0 \\ 0 & 0 & -1 & 1 \end{bmatrix} \begin{Bmatrix} u_a \\ u_b \\ u_c \\ u_d \end{Bmatrix} \quad \text{or} \quad \boldsymbol{e} = \boldsymbol{Q}^T \boldsymbol{u} \tag{3.2}$$

It is easy to see that the components $e_{ab}, e_{bc}, \ldots,$ of $\boldsymbol{e}$ are the elongations or lengthenings of the distances ab, bc, ... The nodal displacements $\boldsymbol{u}$ and the deformations $\boldsymbol{e}$ are called a compatible system.

[1] In what follows we will not construe anything from the fact that this product was called work. We could have called it 'virtual melody' and discuss the virtual melody of a force.

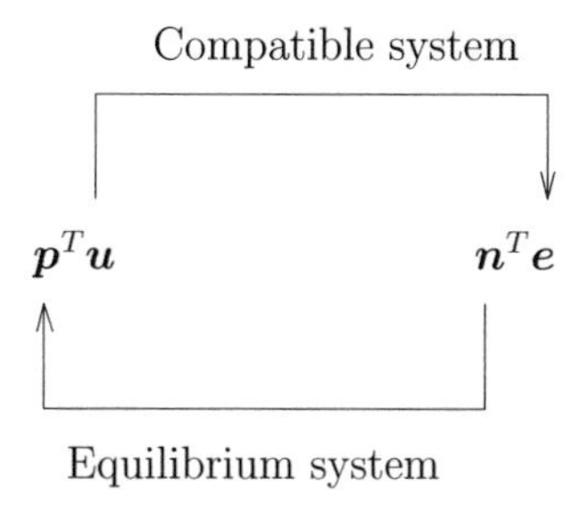

Fig. 3.2 Principle of virtual work

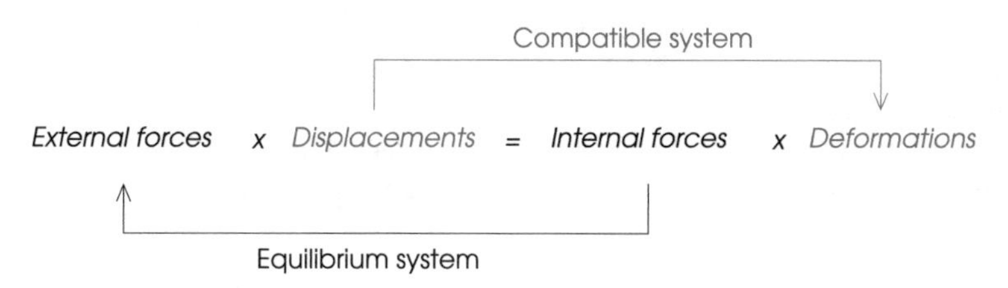

Fig. 3.3 General case

In consequence, for a system in equilibrium $\boldsymbol{Q}\,\boldsymbol{n} = \boldsymbol{p}$ and an independent compatible system $\boldsymbol{e} = \boldsymbol{Q}^T\boldsymbol{u}$, we have the *principle of virtual work*: $\boldsymbol{p}^T\boldsymbol{u} = \boldsymbol{n}^T\boldsymbol{e}$ also shown in Fig. 3.2. To the left we have the external virtual work (evw) and the term to the right is called the internal virtual work,[2] ivw. This is, of course, a scalar equation.

This is a basic theorem and it lies at the heart of the entire field of structures.

The direction of the arrows, from $\boldsymbol{n}$ to $\boldsymbol{p}$ and from $\boldsymbol{u}$ to $\boldsymbol{e}$ in Fig. 3.2, and likewise in Fig. 3.3, indicates the natural way to build equilibrium and compatible systems.

We start by arbitrarily assigning *any* internal forces and, by pre-multiplying these forces by $\boldsymbol{Q}$, we obtain the external forces in equilibrium with the internal ones, $\boldsymbol{Q}\,\boldsymbol{n} \rightarrow \boldsymbol{p}$. For a compatible system, we assign arbitrary displacements at the nodes, and pre-multiplying them by $\boldsymbol{Q}^T$ gives the deformations compatible with the displacements $\boldsymbol{Q}^T\,\boldsymbol{u} \rightarrow \boldsymbol{e}$.

Next, we will show how the principle of virtual work applies to particles in a plane (a forerunner of the truss) and finally to a continuous frame-type element.

Note Regarding the Equilibrium System. Clearly, for all this to work out nicely the external forces must be auto-equilibrated. Indeed, summing the equilibrium equations in (3.1) gives $0 = P_a + P_b + P_c + P_d$. We will not find a set of internal forces that can equilibrate a system of external forces which are not auto-equilibrated.

[2] On the right-hand side we have in fact minus the work of the internal forces, however it is commonly called the internal virtual work.

3.2 Virtual Work for Particles in a Plane

Consider the five particles *abcde* arranged in a truss-like configuration in Fig. 3.4a. The ten degrees of freedom at the particles are numbered so as to facilitate the equations in what follows. $N = 10$ external forces $\boldsymbol{p}$ can be applied along the degrees of freedom. We assume that $M = 7$ internal interaction forces between the particles act along the lines shown in the figure. To build an equilibrium system, we arbitrarily assign M internal forces $\boldsymbol{n}$, and compute the applied forces to be in equilibrium with these internal forces ($\boldsymbol{Q}\boldsymbol{n} \rightarrow \boldsymbol{p}$). For instance, the equilibrium equations at degrees of freedom 1 and 2 (see Fig. 3.4b) are respectively: $-n_1 - c\,n_4 + c\,n_5 + n_2 + p_1 = 0$ and $sn_4 + sn_5 + p_2 = 0$ where $s = \sin(60°)$ and $c = \cos(60°)$. Likewise, the equilibrium equations for degrees of freedom 3 and 4 (Fig. 3.4c) are $-n_7 - c\,n_5 + c\,n_6 + p_3 = 0$ and $-s\,n_5 - s\,n_6 + p_4 = 0$.

This can be repeated for the remaining six degrees of freedom, yielding the equilibrium system with an $N \times M$ equilibrium matrix $\boldsymbol{Q}$, the first four lines of which are shown:

$$\begin{bmatrix} 1 & -1 & 0 & c & -c & 0 & 0 \\ 0 & 0 & 0 & -s & -s & 0 & 0 \\ 0 & 0 & 0 & 0 & c & -c & 1 \\ 0 & 0 & 0 & 0 & s & s & 0 \\ \dots & & & & & & \end{bmatrix} \begin{Bmatrix} n_1 \\ n_2 \\ n_3 \\ n_4 \\ n_5 \\ n_6 \\ n_7 \end{Bmatrix} = \begin{Bmatrix} p_1 \\ p_2 \\ p_3 \\ p_4 \\ \dots \end{Bmatrix} \quad \text{or} \quad \boldsymbol{Q}\,\boldsymbol{n} = \boldsymbol{p} \tag{3.3}$$

We will now create a compatible system. We take as arbitrary nodal displacements N-vector $\boldsymbol{u}$ from which we deduce the M-vector $\boldsymbol{e} = \boldsymbol{Q}^T\,\boldsymbol{u}$. Clearly, the *evw* $\boldsymbol{p}^T\boldsymbol{u}$ equals the *ivw* $\boldsymbol{n}^T\boldsymbol{e}$.

Physical Interpretation of *e*

The question now is, what are the components of $\boldsymbol{e}$? We can show that they are the lengthenings (elongations) of the lines joining the respective nodes. For instance, the fifth row of $\boldsymbol{e} = \boldsymbol{Q}^T\,\boldsymbol{u}$ is $e_5 = \boldsymbol{q}_5^T\boldsymbol{u}$ where $\boldsymbol{q}_5$ is the fifth column of $\boldsymbol{Q}$. Note, the

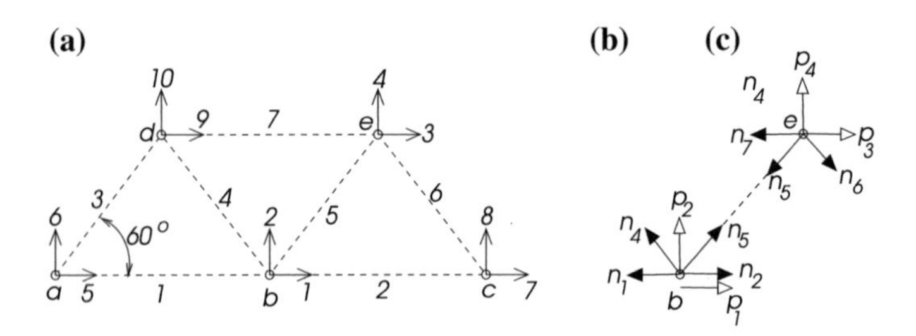

Fig. 3.4 Virtual work: particles (2D)

last three entries of $\boldsymbol{q}_5$ (not shown in (3.3)) are all zero. Consequently, $e_5 = \boldsymbol{q}_5^T \boldsymbol{u} = -c\,u_1 - s\,u_2 + c\,u_3 + s\,u_4 = -(c\,u_1 + s\,u_2) + (c\,u_3 + s\,u_4)$. The term in the second parenthesis is the component of the displacements of node e along be, and the term in the first parenthesis is the component of the displacements of node b along be. With the minus sign in front of that parenthesis, it is clear that entry e_5 is indeed the change in length of distance be.

3.3 Virtual Work for a Frame Element

We will now repeat all this for a more general case of a frame element. The prismatically shaped element ab depicted in Fig. 3.5 was removed from a structure with all the forces acting on it. It is therefore subjected to an auto-equilibrated system of external loads and internal forces. External fores are applied to the boundaries of the element and include all the forces that we can actually see. (Body forces such as gravity forces are here included in the distributed forces.) The external forces are distributed axial forces $p(x)$, distributed lateral forces $q(x)$ and distributed couples $c(x)$ (we include the couples temporarily for the sake of clarity). Segment ab was part of a longer element. Consequently, on sections a and b, where the segment was 'cut' from the mother element, the internal forces are now 'external', that is, nsm_a and nsm_b. The internal forces are hidden inside the element and are, as we recall, the axial forces $n(x)$, shear forces $s(x)$ and bending moments $m(x)$ (there are no concentrated applied forces at this point).

In Fig. 3.5, the distributed forces are symbolically shown; however, the internal forces, being internal, do not appear except at the two extremities where the internal forces become external. The positive direction of these forces at extremity b (whose outer normal points in the $+x$ sense) are shown in the figure. At section a (outer normal in the $-x$ sense) the corresponding forces and couple are in the opposite sense.

Independently of the distributed and the internal forces, we impose on the element arbitrary displacements of Bernoulli type, that is, plane cross-sections displace as rigid plane lamellas. The displacements of the element can therefore be described by the displacements of these 'rigid' discs. The center of the area of every plane cross-section at x is given translations $u(x)$ and $v(x)$ and rotations $\theta(x)$, keeping in mind that we have small displacements (see positive directions in the figure). It is important to emphasize that *these displacements have*, at this point, *no relation to any forces*.

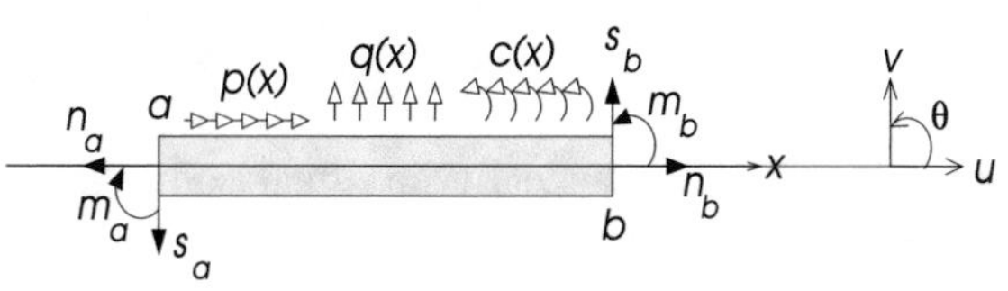

Fig. 3.5 Virtual work: element

We assume that the functions are continuous enough to handle the derivatives in the sequel. Beyond this the element may or may not have any mechanical properties. What follows is therefore independent of whatever the beam is made of.

In the ensuing equations we will be using the following 3×1 vector functions: the internal forces $\boldsymbol{n} \equiv \boldsymbol{n}(x) \equiv \{n(x)\ s(x)\ m(x)\}^T$, the external distributed forces $\boldsymbol{p} \equiv \boldsymbol{p}(x) \equiv \{p(x)\ q(x)\ c(x)\}^T$, and the displacements of the centers of area $\boldsymbol{u} \equiv \boldsymbol{u}(x) \equiv \{u(x)\ v(x)\ \theta(x)\}^T$, with $\boldsymbol{u}_D$, for instance, meaning $\boldsymbol{u}(x_D)$ or $\boldsymbol{u}(x)$ evaluated at x_D.

The virtual work expression for prismatic elements (1D) is based on integration by parts. (For 2D elements, such as plates, we would use the theorem of divergence.) Consider the product of the internal forces and the displacements $\boldsymbol{n}^T\boldsymbol{u} \equiv (nu + sv + m\theta)$. Such a product does not make much physical sense, but it becomes helpful when we take its derivative with respect to x and then integrate. The derivative is

$$\frac{\mathrm{d}}{\mathrm{d}x}(\boldsymbol{n}^T\boldsymbol{u}) = \frac{\mathrm{d}\boldsymbol{n}^T}{\mathrm{d}x}\boldsymbol{u} + \boldsymbol{n}^T\frac{\mathrm{d}\boldsymbol{u}}{\mathrm{d}x} \tag{3.4}$$

We now integrate both sides between a and b and, noting that the left-hand side is the integration of a total differential, we obtain

$$(\boldsymbol{n}^T\boldsymbol{u})_b - (\boldsymbol{n}^T\boldsymbol{u})_a = \int_a^b \left(\frac{\mathrm{d}\boldsymbol{n}^T}{\mathrm{d}x}\boldsymbol{u} + \boldsymbol{n}^T\frac{\mathrm{d}\boldsymbol{u}}{\mathrm{d}x}\right)\mathrm{d}x \tag{3.5}$$

We recall that the internal and external forces are in equilibrium. In other words they obey the equilibrium equations

$$\frac{\mathrm{d}\boldsymbol{n}}{\mathrm{d}x} = -\boldsymbol{p} - \begin{Bmatrix} 0 \\ 0 \\ s \end{Bmatrix} \tag{3.6}$$

Introducing (3.6) in (3.5) yields

$$(\boldsymbol{n}^T\boldsymbol{u})_b - (\boldsymbol{n}^T\boldsymbol{u})_a = \int_a^b \left[\left(-\boldsymbol{p} - \begin{Bmatrix} 0 \\ 0 \\ s \end{Bmatrix}\right)^T \boldsymbol{u} + \boldsymbol{n}^T\frac{\mathrm{d}\boldsymbol{u}}{\mathrm{d}x}\right]\mathrm{d}x \tag{3.7}$$

Moving (minus) the virtual work of the distributed forces to the left-hand side we obtain

$$(\boldsymbol{n}^T\boldsymbol{u})_b - (\boldsymbol{n}^T\boldsymbol{u})_a + \int_a^b \boldsymbol{p}^T\boldsymbol{u}\ \mathrm{d}x = \int_a^b \left(-\begin{Bmatrix} 0 \\ 0 \\ s \end{Bmatrix}^T \boldsymbol{u} + \boldsymbol{n}^T\frac{\mathrm{d}\boldsymbol{u}}{\mathrm{d}x}\right)\mathrm{d}x \tag{3.8}$$

The left-hand side is evidently the virtual work of the external forces. Indeed, expanding $(\boldsymbol{n}^T\boldsymbol{u})_b$, for instance, gives $(nu+sv+m\theta)_b$, where every term has the dimensions of 'work' ($force \times displacement$). You will note that, in the last term, m carries by itself $force \times displacement$ and θ is non-dimensional.

Since the left-hand side is the work of the external forces, the right-hand side is necessarily (minus) the work of the internal forces, because the sum of the two is zero (principle of virtual work $evw + ivw = 0$). The right-hand side is therefore (minus) the internal virtual work.

Let us clean up the $-ivw$. Noting that

$$-\begin{Bmatrix} 0 \\ 0 \\ s \end{Bmatrix}^T \boldsymbol{u} = \boldsymbol{n}^T \begin{Bmatrix} 0 \\ -\theta \\ 0 \end{Bmatrix}$$

(they are both equal to $-s\ \theta$) Eq. (3.8) becomes

$$(\boldsymbol{n}^T\boldsymbol{u})_b - (\boldsymbol{n}^T\boldsymbol{u})_a + \int_a^b \boldsymbol{p}^T\boldsymbol{u}\ \mathrm{d}x = \int_a^b \boldsymbol{n}^T \begin{Bmatrix} \mathrm{d}u/\mathrm{d}x \\ \mathrm{d}v/\mathrm{d}x - \theta \\ \mathrm{d}\theta/\mathrm{d}x \end{Bmatrix} \mathrm{d}x \tag{3.9}$$

We shall see that we find on the *right-hand side* is in fact the definition of deformation $\boldsymbol{\epsilon}$

$$\boldsymbol{\epsilon} \equiv \begin{Bmatrix} \epsilon \\ \gamma \\ \kappa \end{Bmatrix} \equiv \begin{Bmatrix} \mathrm{d}u/\mathrm{d}x \\ \mathrm{d}v/\mathrm{d}x - \theta \\ \mathrm{d}\theta/\mathrm{d}x \end{Bmatrix} \tag{3.10}$$

Consequently, we have here a nice and compact expression for the virtual work of the internal forces. The principle of virtual work $evw = -ivw$ becomes

$$\underbrace{(\boldsymbol{n}^T\boldsymbol{u})_b - (\boldsymbol{n}^T\boldsymbol{u})_a + \int_a^b \boldsymbol{p}^T\boldsymbol{u}\ \mathrm{d}x}_{evw} = \underbrace{\int_a^b \boldsymbol{n}^T\boldsymbol{\epsilon}\ \mathrm{d}x}_{-ivw} \tag{3.11}$$

Note the simplicity of the internal virtual work expression[3]

$$ivw = \int_a^b \boldsymbol{n}^T\boldsymbol{\epsilon}\ \mathrm{d}x \equiv \int_a^b [n\ \epsilon + s\ \gamma + m\ \kappa]\ \mathrm{d}x \tag{3.12}$$

We will now see that the external forces $\boldsymbol{p}(x)$, together with the internal forces $\boldsymbol{n}(x)$, form an equilibrium set or *equilibrium system*, and the displacements $\boldsymbol{u}(x)$ with its related deformations $\boldsymbol{\epsilon}(x)$ are called a compatible set or *compatible system*.

[3]This is in fact minus the work of the internal forces but in line with probably most authors we call it simply ivw.

3.4 Equilibrium and Compatible Systems

When external forces acting on a body and internal forces in the body are auto-equilibrated they constitute an *equilibrium system*. Displacements of the body and the deformations which are computed on the basis of the displacements are called a *compatible system*.

The principle of virtual work states that the *sum* of the work of the external forces of the equilibrium system times the displacements of the compatible system and the work of the internal forces of the equilibrium system times the deformations of the compatible system is zero.

The force and displacements sets may be unrelated.

3.4.1 Equilibrium Systems

External forces are by definition in equilibrium. Internal forces are always in self-equilibrium because they are made of sets of equal and opposite forces. The two constitute an *equilibrium system* if they also equilibrate every particle in the body. For line elements under distributed forces, for instance, mutual equilibrium means that the internal forces $\boldsymbol{n}(x)$ in conjunction with the external forces $\boldsymbol{p}(x)$ equilibrate every lamella of the element.

A simple way to create an equilibrium system is to proceed in the reverse sense: imagine internal forces $\boldsymbol{n}(x)$ and determine the external forces through (3.13).

$$\boldsymbol{p} \leftarrow -\frac{\mathrm{d}\boldsymbol{n}}{\mathrm{d}x} - \begin{Bmatrix} 0 \\ 0 \\ s \end{Bmatrix} \tag{3.13}$$

Usually, we proceed the other way. We are given the external loads $\boldsymbol{p}(x)$, and by integration we compute internal forces.

3.4.2 Compatible Systems

Displacements $\boldsymbol{u}(x)$ and deformations (strains) $\epsilon(x)$ constitute a compatible system if the strains are computed, or can be computed, on the basis of the displacements by (3.14). Consequently, to create a compatible system imagine displacement functions $\boldsymbol{u}(x)$ and compute the corresponding strains $\epsilon(x)$

$$\epsilon \leftarrow -\frac{\mathrm{d}\boldsymbol{u}}{\mathrm{d}x} + \begin{Bmatrix} 0 \\ \theta \\ 0 \end{Bmatrix} \tag{3.14}$$

Fig. 3.6 Non-compatible deformations

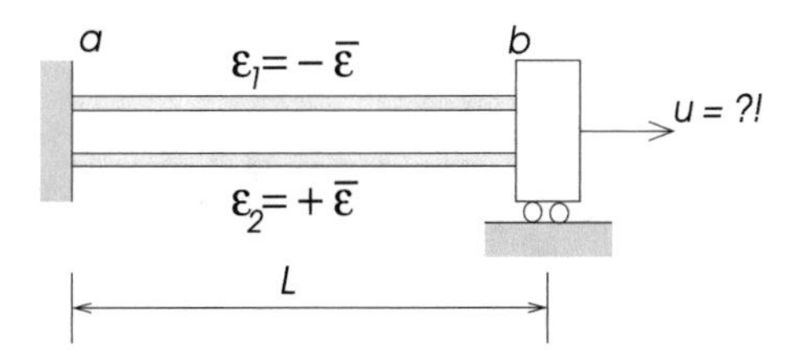

Why do we call the system compatible? Because in general, when you imagine deformations $\epsilon(x)$ they are usually not compatible with one another and no displacements could induce such deformations.

In Fig. 3.6, we consider two parallel bars of length L, built into the wall at a and attached at b to a rigid block that can only translate horizontally by u but cannot rotate. We now want to create a compatible system for that structure starting from the deformations. We assign to bar 1 a contraction, say $\epsilon_1 = -\bar{\epsilon}$ and to bar 2 an extensional strain, say $\epsilon_2 = \bar{\epsilon}$. But this is an impossible configuration. What would the displacement u be? The strains are not compatible and so neither is the system.

However, if we start from some displacement $\bar{u}$ the strains are $\epsilon_1 = \epsilon_2 = \bar{u}/L$. The system u, $\epsilon_1 = \epsilon_2 = \bar{u}/L$ is a compatible system.

In the sequel, the compatible systems will usually be real displacements and strains of real structures, which of course are compatible.

Note: Later we will show that the structure in Fig. 3.6 is redundant, and indeed, only for redundant structures will arbitrary strains be incompatible. For determinate designs any set of (small enough) strains is compatible, that is, there always exist corresponding displacements.

3.5 Point Loads

What if there are point (concentrated) forces and couples on the element? How does it affect the virtual work equation? Well, the ivw expression is unchanged, because no matter how many point loads there are on the element the sum of the lengths of the segments between the loads constitute the entire length of the element. We simply write a separate integral $\int \boldsymbol{n}^T \boldsymbol{\epsilon}\, \mathrm{d}x$ for every segment.

In the evw term we add the virtual work of the point loads. For instance, if $\boldsymbol{P} \equiv \{P \ Q \ C\}^T$ are point loads which are given corresponding virtual displacements $\boldsymbol{u} \equiv \{u \ v \ \theta\}^T$ the virtual work performed is $\boldsymbol{P}^T \boldsymbol{u}$. It is almost intuitive. We will develop the expression in a simple case.

In Fig. 3.7a we have an element ad with concentrated forces $\boldsymbol{P} \equiv \{P \ Q \ C\}^T$ at b and uniform distributed loads $\boldsymbol{p}(x) \equiv \{p(x) \ q(x) \ c(x)\}^T$ along the entire span and

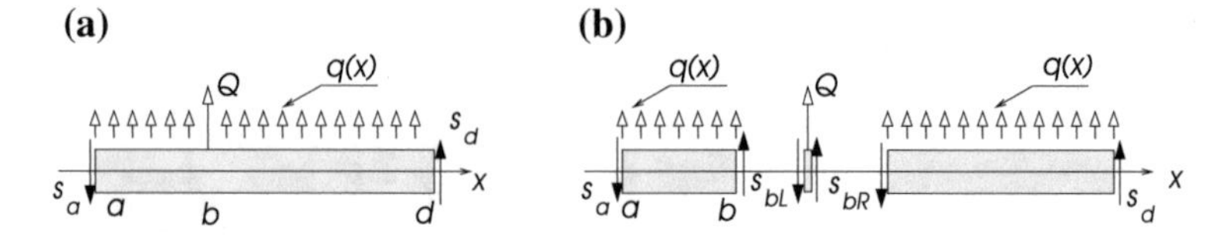

Fig. 3.7 Virtual work: with point loads

the end-forces $\boldsymbol{n}_a \equiv \{n_a\ s_a\ m_a\}^T$ and $\boldsymbol{n}_d \equiv \{n_d\ s_d\ m_d\}^T$. The latter two are internal forces which became external at the cuts.[4]

We consider the element to be subjected basically to distributed loads interspersed here and there with point loads. There may be very many point loads, but the entire span of the beam is, for practical purposes, subjected to distributed forces only ('no loads' can be construed as distributed loads of zero magnitude).

We now decompose the element into ab_L, where b_L is a section a little to the left of the point loads, and b_Rd, where b_R is a section a little to the right of the point loads. You will notice that we have left out the lamella (of zero thickness) where Q is applied (Q represents $\boldsymbol{P}$).

For each segment we can write a virtual work equation

$$(\boldsymbol{n}^T\boldsymbol{u})_{b_L} - (\boldsymbol{n}^T\boldsymbol{u})_a + \int_a^{b_L} \boldsymbol{p}(x)^T\boldsymbol{u}(x)\ \mathrm{d}x = \int_a^{b_L} \boldsymbol{n}(x)^T\boldsymbol{\epsilon}(x)\ \mathrm{d}x$$

$$(\boldsymbol{n}^T\boldsymbol{u})_d - (\boldsymbol{n}^T\boldsymbol{u})_{b_R} + \int_{b_R}^{d} \boldsymbol{p}(x)^T\boldsymbol{u}(x)\ \mathrm{d}x = \int_{b_R}^{d} \boldsymbol{n}(x)^T\boldsymbol{\epsilon}(x)\ \mathrm{d}x$$

Summing the two equations and noting that $\int_a^{b_L} + \int_{b_R}^d$ is simply $\int_a^d$, we have

$$(\boldsymbol{n}^T\boldsymbol{u})_d - (\boldsymbol{n}^T\boldsymbol{u})_a + \underbrace{[(\boldsymbol{n}^T\boldsymbol{u})_{b_L} - (\boldsymbol{n}^T\boldsymbol{u})_{b_R}]} + \int_a^d \boldsymbol{p}(x)^T\boldsymbol{u}(x)\ \mathrm{d}x = \int_a^d \boldsymbol{n}(x)^T\boldsymbol{\epsilon}(x)\ \mathrm{d}x \tag{3.15}$$

There is still the under-braced term to clean up. The displacements at b are continuous, $\boldsymbol{u}_{b_L} = \boldsymbol{u}_{b_R} = \boldsymbol{u}_b$, therefore, $(\boldsymbol{n}^T\boldsymbol{u})_{b_L} - (\boldsymbol{n}^T\boldsymbol{u})_{b_R} = (\boldsymbol{n}_{b_L} - \boldsymbol{n}_{b_R})^T\boldsymbol{u}_b$.

Now consider the forces applied to lamella b on the right of Fig. 3.7b. We note that equilibrium dictates $-\boldsymbol{n}_{b_L} + \boldsymbol{P} + \boldsymbol{n}_{b_R} = 0$ or $\boldsymbol{n}_{b_L} - \boldsymbol{n}_{b_R} = \boldsymbol{P}$, which when introduced into (3.15) yields the final form of the virtual work equation with distributed and discrete applied loads

$$(\boldsymbol{n}^T\boldsymbol{u})_d - (\boldsymbol{n}^T\boldsymbol{u})_a + \underbrace{(\boldsymbol{P}^T\boldsymbol{u})_b} + \int_a^d \boldsymbol{p}(x)^T\boldsymbol{u}(x)\ \mathrm{d}x = \int_a^d \boldsymbol{n}(x)^T\boldsymbol{\epsilon}(x)\ \mathrm{d}x \tag{3.16}$$

[4]For clarity, the figure depicts only the lateral components, that is, s for the internal forces, and q and Q for the external forces.

The right-hand side is still the same form of (minus) the work of the internal forces. The new term on the left-hand side is, as expected, the work of the discrete forces at b, $(\boldsymbol{P}^T\boldsymbol{u})_b \equiv (Pu + Qv + C\theta)_b$.

In summary, for the external virtual work we write the work relation for all the external forces, distributed and discrete loads alike. The internal virtual work is the classical expression $\int \boldsymbol{n}(x)^T \boldsymbol{\epsilon}(x)\,\mathrm{d}x$ along the entire structure.

3.6 Notes

3.6.1 $C\theta$ and $(m\kappa dx)$ Are Virtual Work Expressions

We recall that a point couple C represents two equal and opposite very large forces, say P, a small distance δx apart such that $\lim_{\delta x\to 0} P\ \delta x = C$. In Fig. 3.8a, we depict two such forces applied to an element.

When the element is given a lateral displacement $v(x)$, not necessarily as a result of the forces, as shown in Fig. 3.8b, the forces perform virtual work $-Pv + P(v+\delta v) = P\ \delta v$. We now multiply this result by $\delta x/\delta x$ which yields $(P\ \delta x)\ \delta v/\delta x$. Taking the limit of that expression, we note that $\lim_{\delta x\to 0}(P\ \delta x) = C$ and $\lim_{\delta x\to 0} \delta v/\delta x = \mathrm{d}v/\mathrm{d}x \equiv \theta$, that is, the slope of the element at x. Consequently, the virtual work of the couple is $C\theta$.

Similarly, $m\kappa\,\mathrm{d}x$ is the virtual work of the bending moment m when the facets of the lamella rotate, due to the flexion of the element (here also the flexion is not necessarily related to the bending moment). In Fig. 3.9a, we have a lamella of width $\mathrm{d}x$ subjected to couples of forces P, a distance h apart, such that $m = Ph$. Let us assume that the lamella undergoes a virtual deformation of flexure due the (small) rotation of its facets by θ and $\theta + \mathrm{d}\theta$ (Fig. 3.8b).

We have chosen, without loss of generality, to position the center of rotation of the rigid facets (remember Bernoulli's assumption) at the point of application of the tension forces. Thus, only the compression forces perform work. The virtual work of the forces is thus $-P(h\theta) + P[h(\theta + \mathrm{d}\theta)] = (Ph)\mathrm{d}\theta$. (Note, $\tan\theta \cong \theta$ for small θ.) We multiply this result by $\mathrm{d}x/\mathrm{d}x$, and since by definition $\kappa \equiv \mathrm{d}\theta/\mathrm{d}x$, we find that $m\kappa\,\mathrm{d}x$ is indeed the virtual work of the bending moment due to the deformation of a lamella $\mathrm{d}x$.

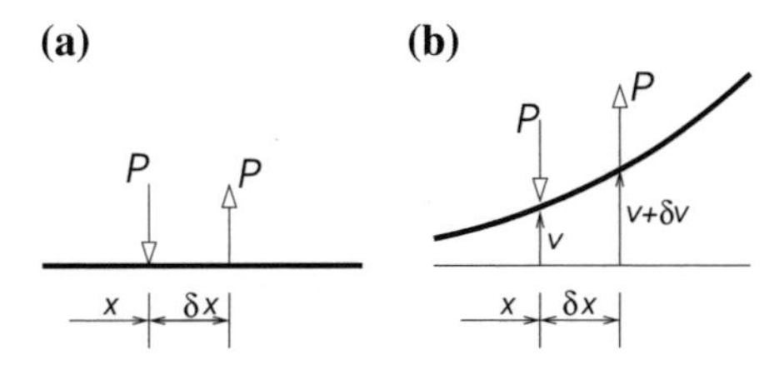

Fig. 3.8 External work: $C\theta$

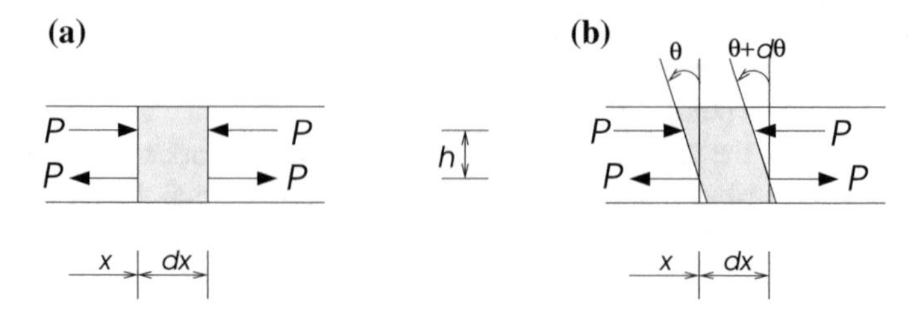

Fig. 3.9 Internal work: $m\kappa\,\mathrm{d}x$

3.7 Illustrated Examples

3.7.1 An Example with Axial Forces

Let us apply the virtual work principle on a very simple example. We construct an equilibrium system and then independently a compatible system, and then we show that they obey the virtual work principle.

In Fig. 3.10, we consider an element (a rod) of length L. The cross-sections are not given, and we know nothing regarding the material properties of the element. What follows is true regardless of whether the object is elastic or plastic, or whether one bar is made of wood and the other of titanium.

For the *equilibrium system*, we will assume that there are only axial distributed forces $p(x)$ and corresponding normal internal forces $n(x)$. For the two functions to constitute an equilibrium system they have to satisfy $p(x) = -\mathrm{d}n(x)/\mathrm{d}x$. An easy way to imagine such a system is to start with some internal normal forces, say, $n(x) = P(1 - 3(x/L))$ and take the derivative $(-3P/L)$ which we then equate to $-p(x)$. Note, P is a constant having the dimension of a force. The equilibrium system is given in the second column of Table 3.1 and visualized in Fig. 3.10.

To the left we have the internal forces from which we started. It turns out that the external forces which are in equilibrium with these internal forces are a constant distributed axial load $3P/L$ with two concentrated forces P and $2P$ at the end-sections. It is noteworthy that the external forces at the extremities are the internal

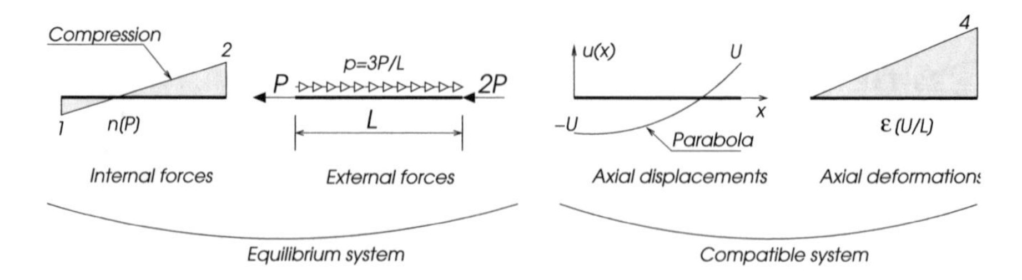

Fig. 3.10 Virtual work: graphical representation

Table 3.1 Virtual work: axial forces and displacements

Type	Equilibrium	Compatible
Ext. Forces—Displacements	$p(x) = 3P/L$	$u(x) = U(2(x/L)^2 - 1)$
Int. Forces—Deformations	$n(x) = P(1 - 3(x/L))$	$\epsilon(x) = 4(U/L)(x/L)$

forces at $x = 0$ and $x = L$ ($n(0) = P$ and $n(L) = -2P$), and these forces together with $p(x)$ constitute external forces in equilibrium.

To construct a *compatible system*, we imagine displacements, say,

$$u(x) = U(2(x/L)^2 - 1)$$

where U is a constant with dimensions of a displacement. Deformations compatible with these displacements are simply their derivative $\epsilon(x) = \mathrm{d}u(x)/\mathrm{d}x$ (see the third column in Table 3.1 and Fig. 3.10). The equilibrium and compatible systems are, of course, unrelated.

Now that we have constructed an equilibrium and a compatible system we can evaluate the external and internal virtual works

$$evw = (-2P)(U) - (P)(-U) + \int_0^L \left(\frac{3P}{L}\right)\left[U\left(2\left(\frac{x}{L}\right)^2 - 1\right)\right] \mathrm{d}x = -2PU$$

$$-ivw = \int_0^L P\left(-1 + 3\frac{x}{L}\right)\left(\frac{4U}{L}\frac{x}{L}\right), \mathrm{d}x = 2PU$$

and indeed $evw = ivw$.

3.7.2 Beam with Arc Shape Deformation

The straight beam ab in Fig. 3.11a is deformed into an arc of a circle of constant curvature $\kappa = 1/R$, where R is the radius of the circular segment. The intercept angle 2α is taken small such that the distance L between a and b remains constant. The displaced shape of the beam with the deformations κ are a *compatible system.*

We construct an *equilibrium system*, the external forces of which consist of a point couple C applied to a and two equal and opposite forces C/L at the extremities. These auto-equilibrated forces induce the internal bending moments shown in Fig. 3.11b. The external and internal forces shown in Fig. 3.11b are an equilibrium system. Clearly, the two systems are unrelated, not least because the compatible system is symmetric and the equilibrium system is not.

We now show that they obey the virtual work principle. The evw is the work of the external loads of the equilibrium system over the corresponding displacements of the compatible system. The forces do not work because points a and b were kept in place. The point couple, on the other hand, rotates by α. Consequently, $evw = C\alpha$.

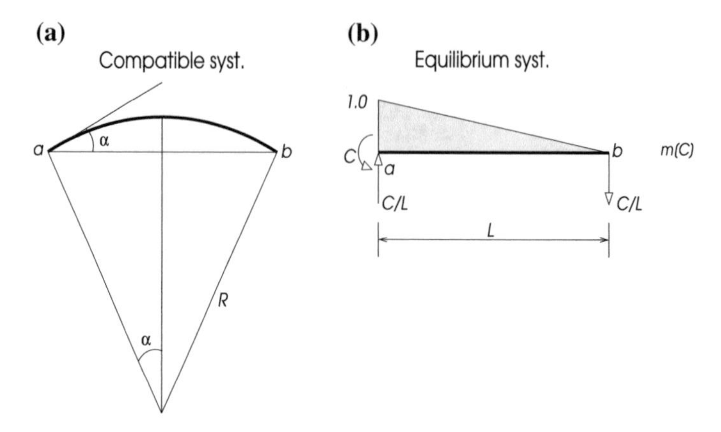

Fig. 3.11 Virtual work: beam element

The $ivw = \int_a^b m\kappa\,dx$ is in fact the area of the bending moment diagram times the curvature, or $ivw = (CL/2)(1/R)$. Rewriting this $ivw = C(0.5L/R) = C\sin\alpha$ and, because for small α we have $\sin\alpha \cong \alpha$, we find $ivw = C\alpha$. Consequently, $evw = ivw$.

3.8 Summing Up

Usually, after discussing equilibrium systems, we define the deformations resulting from displacement of an element, and then proceed to show that we can prove the principle of virtual work. This often leaves the reader with the uneasy feeling that some sort of trick has been played because it does not explain why there is such a principle at all.

We have followed other authors who define the deformations resulting from displacements so as to satisfy the principle of virtual work. This leads to the basic components of deformation: $\epsilon(x)$, $\gamma(x)$ and $\kappa(x)$ and their definition as a function of the displacements of the rigid 'discs': $u(x)$, $v(x)$ and $\theta(x)$. The principle of virtual work does not appear *ex-nihilo*; on the contrary. We are using deformations, conducive to the principle of virtual work.

Chapter 4
Deformations

Quite independently of what has been done till this point, in this chapter, we develop a Theory of Deformation, based on Bernoulli's assumption of rigid sections. We show that the deformations are exactly those suggested by virtual work: $\epsilon(x)$, $\gamma(x)$ and $\kappa(x)$ as a function of the displacements of the rigid 'discs', $u(x)$, $v(x)$ and $\theta(x)$. A physical interpretation is given: $\epsilon(x)$ is the extensional deformation (axial stain), $\gamma(x)$ is the shear deformation (shear strain) and $\kappa(x)$ is the bending deformation (flexural strain).

4.1 The Bernoulli Hypothesis of Plane Sections

Jacob Bernoulli, a Swiss-born scion of an illustrious family of mathematicians originally from Antwerp in Belgium, made an assumption which became one of the cornerstones of the theory of structures. First expressed in the seventeenth century, the assumption is used not only in beam theory but has also been extended to plates and shells. The conjecture is called the plane section hypothesis and states that

Cross-sections, initially plane, remain plane after deformation of the element.

Moreover, the shape of the cross-section remains undeformed.[1] In other words, if we assume that, inside the beam there is a plane surface perpendicular to the beam's axis before deformation, this surface will remain plane and undeformed after deformation of the element. Another way of visualizing the hypothesis is to consider an infinitesimally thin lamella before deformation of the element. The end surfaces of the lamella are plane surfaces. After deformation the lamella will assume a deformed shape but the end surfaces will remain plane and undeformed.

[1]Bernoulli may not have actually been the first or the only one to make the assumption, he may not even have enunciated the assumption in these words. However, the conjecture is credited to him.

M.B. Fuchs, *Structures and Their Analysis*,
DOI 10.1007/978-3-319-31081-7_4

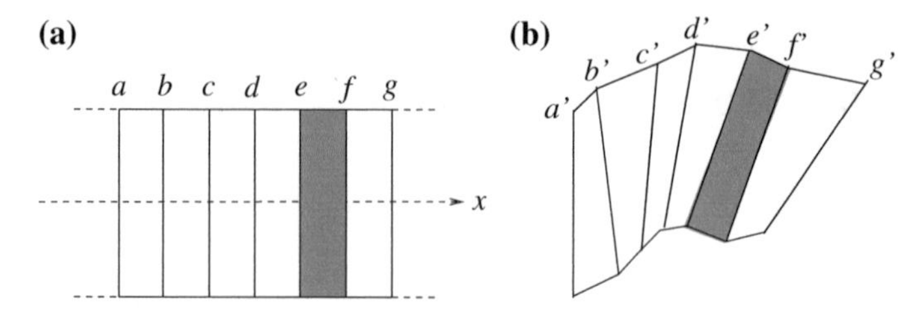

Fig. 4.1 Displacements and deformations

When we comply with the assumption of rigid cross-sections, the deformation of a lamella occurs as a result of the relative displacements of its end cross-sections. The deformation of the entire element is the juxtaposition of the deformed lamellae. In Fig. 4.1a, for instance, we have depicted an initially straight element composed of a series of lamellae. The lamella boundaries $a, b, \ldots$ are straight surfaces which will remain straight after deformation. The displacements of the cross-sectional surfaces $abc\ldots$ to $a'b'c'\ldots$ define the deformed shape (Fig. 4.1b).

It should be clear that displacements and deformations are two different, albeit related, concepts. On the basis of the Bernoulli assumption of rigid plane sections, we can visualize the two concepts by noting that displacements define the new positions of cross-sections of the element, whereas deformation depends on the relative positions of these cross-sections, that is, on whether some or more lamellae have deformed as a consequence of the displacements of adjacent cross-sections. The two end surfaces of lamella ef are displaced but the lamella is not deformed. It has merely undergone a rigid-body movement.

The importance of the Bernoulli assumption cannot be overemphasized. It is physically a good representation of the actual deformation of slender elements and, from a computational viewpoint, the hypothesis has made the structural problem manageable. Instead of following every single particle in the solid, we can track the positions of 'rigid' thin discs. As we shall see, the Bernoulli assumption allows us to replace the task of determining two functions of two variables $U(x, z)$ and $V(x, z)$ by three functions of one variable $u(x)$, $v(x)$ and $\theta(x)$ (4.7).

4.2 Study of Deformation

Consider two particles P and Q located at two neighboring positions in a solid. Let point P be at coordinates (x, z) and point Q at coordinates $(x + dx, z + dz)$ as illustrated to the left in Fig. 4.2. The x coordinate is the axis of the centers of area of the cross-sections and z is chosen perpendicular to the x axis. We use the x, z coordinates to position a particle within the element. Let ds be the length of PQ. Following Pythagoras, we have

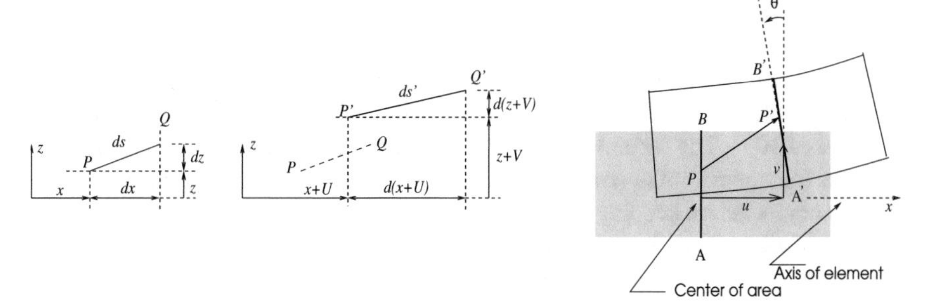

Fig. 4.2 Deformation

$$\mathrm{d}s^2 = \mathrm{d}x^2 + \mathrm{d}z^2 \tag{4.1}$$

Now, the solid experiences some displacements, and let $U = U(x, z)$ and $V = V(x, z)$ be the axial and lateral displacements of particle P. Likewise, the displacements of particle Q are $U + \mathrm{d}U$ and $V + \mathrm{d}V$. These displacements transform line PQ into line $P'Q'$. The square of the length of $P'Q'$ is

$$\mathrm{d}s'^2 = (\mathrm{d}x + \mathrm{d}U)^2 + (\mathrm{d}z + \mathrm{d}V)^2 \tag{4.2}$$

If $\mathrm{d}s' = \mathrm{d}s$ then the line has simply moved (translated and/or rotated) as a rigid body (rigid line). Deformation occurs if the two lengths are different. A measure of the deformation could then be the difference of the (square of the) lengths

$$(\mathrm{d}s')^2 - \mathrm{d}s^2 = (\mathrm{d}x)^2 + (\mathrm{d}U)^2 + 2\mathrm{d}x\mathrm{d}U + (\mathrm{d}z)^2 + (\mathrm{d}V)^2 + 2\mathrm{d}z\mathrm{d}V - (\mathrm{d}x)^2 - (\mathrm{d}z)^2 \tag{4.3}$$

In this equation $(\mathrm{d}x)^2$ and $(\mathrm{d}z)^2$ cancel out. Also, since U and V are functions of x and z, their total differentials can be written as

$$\begin{aligned} \mathrm{d}U &= \frac{\partial U}{\partial x}\mathrm{d}x + \frac{\partial U}{\partial z}\mathrm{d}z \\ \mathrm{d}V &= \frac{\partial V}{\partial x}\mathrm{d}x + \frac{\partial V}{\partial z}\mathrm{d}z \end{aligned} \tag{4.4}$$

where ∂ indicates partial derivatives. Introducing these values in (4.3) and grouping the terms by $(\mathrm{d}x)^2$, $(\mathrm{d}z)^2$ and $(\mathrm{d}x\mathrm{d}z)$ gives

$$\begin{aligned} (\mathrm{d}s')^2 - \mathrm{d}s^2 = &\left[2\frac{\partial U}{\partial x} + \left(\frac{\partial U}{\partial x}\right)^2 + \left(\frac{\partial V}{\partial x}\right)^2\right]\mathrm{d}x^2 + \left[2\frac{\partial V}{\partial z} + \left(\frac{\partial U}{\partial z}\right)^2 + \left(\frac{\partial V}{\partial z}\right)^2\right]\mathrm{d}z^2 \\ &+ \left[2\left(\frac{\partial V}{\partial x} + \frac{\partial U}{\partial z}\right) + 2\frac{\partial U}{\partial x}\frac{\partial U}{\partial z} + 2\frac{\partial V}{\partial x}\frac{\partial V}{\partial z}\right]\mathrm{d}x\mathrm{d}z \end{aligned} \tag{4.5}$$

In every bracket we notice one or more partial first-order derivatives, such as, $\partial V/\partial x$ in the company of products of partial derivatives, for instance, $\partial V/\partial x\, \partial V/\partial z$. The partial derivatives are non-dimensional since we divide lengths by lengths. Now assume that all the partial derivatives are much smaller then 1 in absolute value, say of the order of 10^{-3}. The second-order terms will be of the order of 10^{-6}. Consequently products (or squares) of derivatives will be much smaller than individual derivatives. We will neglect them keeping in mind that everything that follows is acceptable only when these derivatives are much smaller than 1.

After dropping the second-order terms, (4.5) becomes

$$\frac{(\mathrm{d}s')^2 - \mathrm{d}s^2}{2} = \frac{\partial U}{\partial x}\mathrm{d}x^2 + \frac{\partial V}{\partial z}\mathrm{d}z^2 + \left(\frac{\partial V}{\partial x} + \frac{\partial U}{\partial z}\right)\mathrm{d}x\mathrm{d}z \tag{4.6}$$

In its present form there is not much we can learn from (4.6). To write the deformation in a more tractable form *we introduce the Bernoulli assumption of plane sections.* In essence, this assumption states that when the element deforms, plane sections initially normal to the line of centroids of the element, will remain plane and undeformed.

Consider cross-section AB on the right of Fig. 4.2. After deformation it will have moved as a rigid body to position $A'B'$. Any point P within the cross-section will move along with it while retaining its relative position in the cross-section. The displacement of the section can be defined by the translation u and v of the center of area of the cross-section and by the rotation θ as shown in the figure. Note that these quantities are a function of x only: $u(x)$, $v(x)$ and $\theta(x)$. For small rotations we can calculate the displacement of any point P on the basis of $u(x)$, $v(x)$ and $\theta(x)$.

$$\begin{aligned} U(x, z) &= u(x) - z\theta(x) \\ V(x, z) &= v(x) \end{aligned} \tag{4.7}$$

Recall that $U(x, z)$ and $V(x, z)$ are the horizontal and lateral displacements of particles at x and z, whereas $u(x)$, $v(x)$ and $\theta(x)$ are the translations and rotation (with respect to the center of area) of the 'rigid' cross-sections at x. The partial derivatives of U and V are

$$\begin{aligned} \frac{\partial U}{\partial x} &= \frac{\mathrm{d}u}{\mathrm{d}x} - z\frac{\mathrm{d}\theta}{\mathrm{d}x} \\ \frac{\partial U}{\partial z} &= -\theta \\ \frac{\partial V}{\partial x} &= \frac{\mathrm{d}v}{\mathrm{d}x} \\ \frac{\partial V}{\partial z} &= 0 \end{aligned} \tag{4.8}$$

Introducing these results in (4.6) gives

$$\frac{(\mathrm{d}s')^2 - \mathrm{d}s^2}{2} = \left(\frac{\mathrm{d}u}{\mathrm{d}x} - z\frac{\mathrm{d}\theta}{\mathrm{d}x}\right)\mathrm{d}x^2 + \left(\frac{\mathrm{d}v}{\mathrm{d}x} - \theta\right)\mathrm{d}x\mathrm{d}z \tag{4.9}$$

This expression looks much better. Indeed, if the right-hand side of (4.9) is different from zero, the element has deformed locally at location x. And since dx and dz are arbitrary, if any one of the terms within the parentheses is different from zero, we have deformation.

You will have noted that, in the first term on the right-hand side, we have coordinate z which defines the position of point P within the section. Since we want to include all possible points of the cross-section, $\mathrm{d}u/\mathrm{d}x$ and $\mathrm{d}\theta/\mathrm{d}x$ can separately cause deformation. We will therefore define the following deformation quantities:

$$\begin{aligned} \epsilon(x) &\equiv \frac{\mathrm{d}u}{\mathrm{d}x} \\ \gamma(x) &\equiv \frac{\mathrm{d}v}{\mathrm{d}x} - \theta \\ \kappa(x) &\equiv \frac{\mathrm{d}\theta}{\mathrm{d}x} \end{aligned} \tag{4.10}$$

Introducing (4.10) in (4.9) yields the final expression for the deformation of ds at x

$$\frac{(\mathrm{d}s')^2 - \mathrm{d}s^2}{2} = [\epsilon(x) - z\ \kappa(x)]\ dx^2 + [\gamma(x)]\ dxdz \tag{4.11}$$

In other words,

We have deformation if either one of ϵ, γ and κ is different from zero.

These quantities rightfully define local deformation and are often also referred to as *generalized strains or simply strains*. Equation (4.10) define how the strains originate from the displacements and are called the *strain-displacement relations*. The strain-displacement relations allow us to compute the strain components at x if the displacements $u(x)$, $v(x)$, and $\theta(x)$ are known.

4.3 Geometric Interpretation of the Deformations

We will now show that the three generalized strains $\epsilon(x)$, $\gamma(x)$ and $\kappa(x)$ correspond to the three basic deformation modes: extensional, shear and bending deformation, shown in Fig. 4.3.

4.3.1 Extensional Deformation Mode $\epsilon(x)$

As shown in Fig. 4.3, when ϵ is different from zero the initial rectangle of base length dx transforms into a rectangle with a base length $\mathrm{d}x + \epsilon\ \mathrm{d}x$.

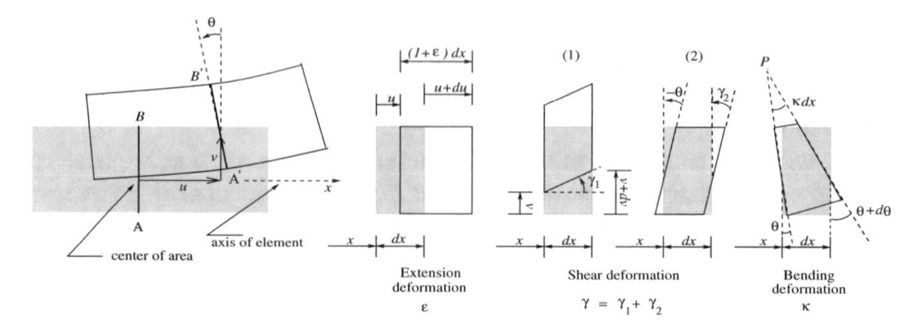

Fig. 4.3 Basic deformations of a lamella

4.3.2 *Shear Deformation Mode* $\gamma(x)$

The shear deformation γ has a little more flavor to it. In this mode the end sections remain parallel but slide past one another and make a parallelogram out of what was initially a rectangle. The left lower right-angled corner $\pi/2$ is reduced by an angle γ (positive anti-clockwise). It has two parts, $\gamma = \mathrm{d}v/\mathrm{d}x - \theta$ and, in Fig. 4.3, we notice that a parallelogram can indeed be created by two types of relative displacement of the end sections of the lamella ($\gamma = \gamma_1 + \gamma_2$.)

In case (1) deformation happens due to differential v displacements. Facet x moves up by an amount v and facet $x + dx$ moves up by $v + dv$. Consequently, $\tan \gamma_1 = \mathrm{d}v/dx$, and since the tangent of a small angle is equal to the angle (expressed in radians) we have $\gamma_1 = \mathrm{d}v/dx$.

There is an additional possibility for creating γ. In case (2) we have that both faces assume a $-\theta$ rotation. Angle $\pi/2$ is here reduced by $\gamma_2 = -\theta$. Combining (1) and (2) gives $\gamma = \mathrm{d}v/\mathrm{d}x - \theta$, which is exactly the definition of the shear deformation.

4.3.3 *Bending Deformation Mode* $\kappa(x)$

Last but certainly not least (bending usually being the most critical of the three modes of deformation), the bending deformation is a relative rotation of section $x + \mathrm{d}x$ with respect to x, the faces of the slice making an intercept angle κdx. (By this definition κ has the dimensions $1/length$.) Now, face x rotates counterclockwise by θ and face $x + dx$ rotates by $\theta + \mathrm{d}\theta$. Since the intercept angle is κdx, we have $\kappa dx = \mathrm{d}\theta$, and hence the definition of κ.

4.4 Zero Deformation—Rigid-Body Displacement

If $\mathrm{d}u/\mathrm{d}x = 0$ there is no axial strain. If $\mathrm{d}v/\mathrm{d}x = \theta$ there is no shear deformation and if $\mathrm{d}\theta/\mathrm{d}x = 0$ no bending deformation takes place. When all three are simultaneously zero, the lamella at this position x may have moved as a rigid body, but it has not deformed.

4.5 In Vector Notation

With the Bernoulli assumption in mind, the displacements of a prismatic element are reduced to the displacements and rotation of rigid cross-sections at x: $\boldsymbol{u}(x) = \{u(x)\ v(x)\ \theta(x)\}^T$. The deformations are characterized by three basic quantities $\boldsymbol{\epsilon}(x) = \{\epsilon(x)\ \gamma(x)\ \kappa(x)\}^T$. The deformations can be obtained from the displacements by the following relations:

$$\frac{\mathrm{d}}{\mathrm{d}x}\begin{Bmatrix} u \\ v \\ \theta \end{Bmatrix} = \begin{Bmatrix} \epsilon \\ \gamma \\ \kappa \end{Bmatrix} + \begin{Bmatrix} 0 \\ \theta \\ 0 \end{Bmatrix} \tag{4.12}$$

or also

$$\frac{\mathrm{d}\boldsymbol{u}}{\mathrm{d}x} = \boldsymbol{\epsilon} + \begin{Bmatrix} 0 \\ \theta \\ 0 \end{Bmatrix} \tag{4.13}$$

4.6 Summing Up

It is noteworthy that the strains which appear when studying deformation are exactly those needed to satisfy the principle of Virtual Work.

Chapter 5
Elasticity

A major player in the behavior of a structure is the material it is made off. The stone columns of Greek temples or the wrought iron elements of the Eiffel tower may perform similar tasks but they do it in a very different manner. In this chapter we introduce the linear elastic materials composing the structures described in this book.

5.1 Elastic Materials

Robert Hooke (1635–1703) noted that when you suspend weights from metal rods or apply forces to metal springs or wooden cantilever beams, as in Fig. 5.1a, the elongations of the rods, the contractions of the springs and the bending of the beams (δ) were proportional to the weights or forces P. When the weights or forces are removed the structures resumed their initial configurations.

This is not always the case. Granted, the deformations increase with weights or forces but, depending on the type of material and the range of the weights, the displacements δ (resulting from the deformations of the elements) may not be proportional to the loads, and the specimen may not resume its original shape when the loads decrease to zero. In Fig. 5.1b, we depict a stylized force-displacement curve for a more general case.

Initially, we are in the linear-elastic range, where P is proportional to δ and where the loading and unloading path are identical and follow a straight line. With larger forces and displacements we may enter a non-linear-elastic region but where the loading and unloading still follow the same path. Further on we enter a plastic range where the unloading follows a different route than the loading and when the force is totally removed we are left with a permanent deformation.

Non-linear elasticity and plasticity are bugbears in this basic theory of structures, where everything is kept neatly linear. We will therefore assume that the loads and deformations are relatively small (this also helps our assumption of small displacements), so as to stay snugly in the linear elastic range. With time Hooke's modest

M.B. Fuchs, *Structures and Their Analysis*,
DOI 10.1007/978-3-319-31081-7_5

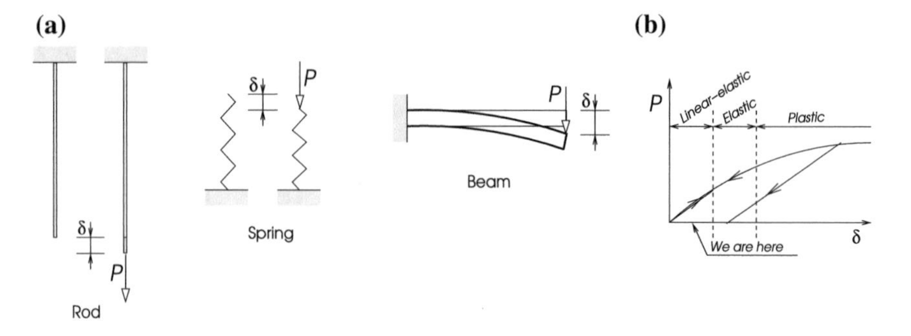

Fig. 5.1 Loads and deformation

beginnings blossomed into the theory of elasticity, a very general method for the analysis of elastic bodies.

If linear elasticity holds for the entire structure, then the structure is said to be linear elastic, or better, it is said to perform in the linear-elastic range. This will be the working assumption of our structures. They will all perform in the linear-elastic range and, after being unloaded, they will assume their original shape.

5.2 Hooke's Law at Location y, z on the Cross-Section

You will remember that we have slender elements whose cross-sections perpendicular to the x-axis are rigid (y, z) but in the (x, z) plane the material is elastic. It may elongate in the x direction and it may experience shear in the x, z plane (Fig. 5.2).

In the axial and shear modes Hooke's law is

$$\begin{Bmatrix} \sigma_{xx} \\ \tau_{xz} \end{Bmatrix} = \begin{bmatrix} E & 0 \\ 0 & G \end{bmatrix} \begin{Bmatrix} \epsilon_{xx} \\ \gamma_{xz} \end{Bmatrix} \tag{5.1}$$

where σ_{xx} is the axial stress in the x direction and τ_{xz} is the shear stress in the z direction on a facet perpendicular to the x direction. The axial and shear strains are defined by

$$\begin{Bmatrix} \epsilon_{xx} \\ \gamma_{xz} \end{Bmatrix} = \begin{bmatrix} \partial/\partial x & 0 \\ \partial/\partial z & \partial/\partial x \end{bmatrix} \begin{Bmatrix} U \\ V \end{Bmatrix} \tag{5.2}$$

Fig. 5.2 Coordinate axis

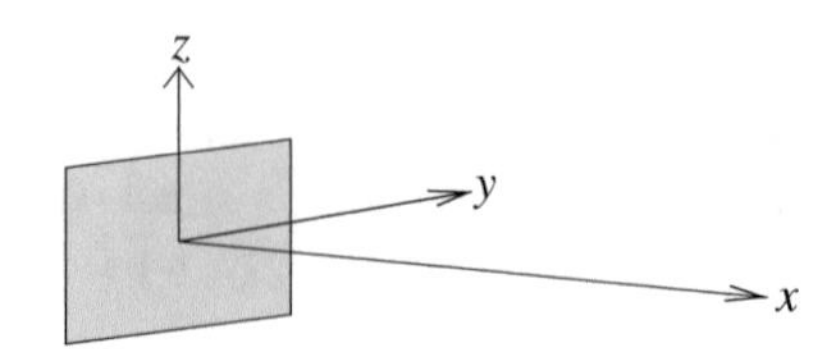

where U, V are the axial (in direction x) and lateral (in direction z) displacements assumed small, of point x, y, z.

The modulus of elasticity E or Young's modulus (Thomas Young, 1773–1829) and the shear modulus G express the stiffness of the material in the axial and shear modes, and are established experimentally. Since the strains are dimensionless, E and G have the same units as stresses, *force/area*. For instance, Young's modulus in MN/m^2 (Mega-Newtons are 10^6 Newtons) is 7 for rubber, 14,000 for wood, 70,000 for aluminum alloys and 210,000 for structural steel. The larger E the stiffer the material. In fact, we have to read these equations in the form that they are written.

When a material deforms by ϵ_{xx} it produces a stress $E\epsilon_{xx}$.

By deforming, the materials produce stresses which eventually balance the applied forces.

When we introduce the Bernoulli assumption of undeformable cross-sections, it follows that the displacements are limited to the pattern

$$\begin{Bmatrix} U(x,z) \\ V(x,z) \end{Bmatrix} = \begin{Bmatrix} u(x) - z\theta(x) \\ v(x) \end{Bmatrix} \tag{5.3}$$

where, as we recall, $u(x)$, $v(x)$ are the translation of the center of area of the rigid cross-section and θ is the anti-clockwise rotation of the cross-section about the z axis. Consequently, the strains (5.2) with (5.3) become

$$\begin{Bmatrix} \epsilon_{xx} \\ \gamma_{xz} \end{Bmatrix} = \begin{Bmatrix} \mathrm{d}u/\mathrm{d}x - z\mathrm{d}\theta/\mathrm{d}x \\ -\theta + \mathrm{d}v/\mathrm{d}x \end{Bmatrix} \equiv \begin{Bmatrix} \epsilon - z\kappa \\ \gamma \end{Bmatrix} \tag{5.4}$$

We can now introduce these strains in Hooke's law (5.1) to produce the stresses in the cross-section at coordinate z of the cross-section

$$\begin{Bmatrix} \sigma_{xx} \\ \sigma_{xz} \end{Bmatrix} = \begin{Bmatrix} E(\epsilon - z\kappa) \\ G\,\gamma \end{Bmatrix} \tag{5.5}$$

This is true locally at a point x, y, z in the element. What we seek is a global relation between the internal forces $\boldsymbol{n}$ and the deformations $\boldsymbol{\epsilon}$ at location x of the element.

5.3 The Timoshenko Beam

What follows is computed at a cross-section x. We recall that the internal forces $\boldsymbol{n}(x)$ are the resultants of the stresses applied to y, z at x. Indeed, a typical cross-section is subjected to infinitesimal forces $\sigma_{xx}\,\mathrm{d}A$ and $\tau_{xy}\,\mathrm{d}A$ where $\mathrm{d}A$ is an infinitesimal area of the cross-section. By definition, the stress resultants n and s are equivalent axial and shear forces positioned at the center of area of the cross-section and m is the corresponding couple

$$n = \int_A \sigma_{xx} \, \mathrm{d}A \qquad s = \int_A \sigma_{xy} \, \mathrm{d}A \qquad m = -\int_A \sigma_{xx} \, z \, \mathrm{d}A \tag{5.6}$$

The minus sign in front of m is there because, on a positive x facet, a positive moment is by definition in the counterclockwise sense whereas positive $\sigma_{xx}\mathrm{d}A$ and positive z give a negative (clockwise) infinitesimal moment.

Introducing Hooke's law (5.5), we obtain

$$n = \int_A E(\epsilon - z\kappa) \, \mathrm{d}A \qquad s = \int_A G\gamma \, \mathrm{d}A \qquad m = -\int_A E(\epsilon - z\kappa) \, z \, \mathrm{d}A \tag{5.7}$$

The integrals are in terms of y, z variables. Consequently, $\epsilon(x)$, $\kappa(x)$, $E(x)$ and $G(x)$ can be factored out

$$n = E\epsilon \int_A \mathrm{d}A - E\kappa \int_A z\mathrm{d}A \qquad s = G\gamma \int_A \mathrm{d}A \qquad m = -E\epsilon \int_A z\mathrm{d}A + E\kappa \int_A z^2\mathrm{d}A \tag{5.8}$$

The three integrals are geometric properties of the cross-section at x

$$A \equiv \int_A dA \qquad \left(\int_A z dA = 0\right) \qquad I \equiv \int_A z^2 dA \tag{5.9}$$

The first integral is the cross-sectional area A. The next is called the first moment of area with respect to axis y. Since the centroid is located on that axis, this integral is zero. Incidentally, this is the reason for having chosen the centroidal line as our reference line. Any other line would not have canceled the first moment of area. Finally, the last integral is the second moment of area with respect to the y axis. It is denoted by I and for structural engineers it is called the *moment of inertia*. The moment of inertia depends on the shape of the cross-section and measures how far the material is removed from the center of area.

As is observed in Fig. 5.3, although all cross-sections are built up of four $a \times a$ squares, the further the material is removed (in the vertical direction) from the center of the area, the larger is I and, as we shall see, the stiffer is the member in bending.

For a rectangular cross-section (base b and height h) the moment of inertia is $I = bh^3/12$. Due to the cubic exponent of h, upright rectangles (large h) are preferable to horizontal ones (small h), as shown in Fig. 5.3. By simply tilting the beam from a horizontal position to an upright one, the moment of inertia is improved sixteen-fold, and so is the bending stiffness. This is why structural engineers use beams with I-shaped cross-sections (the rightmost one in the figure) and also circular or box tubes. The latter are also very effective in torsion, closed cross-sections being much stiffer in torsion than open ones. (Beware of I sections for torsion.)

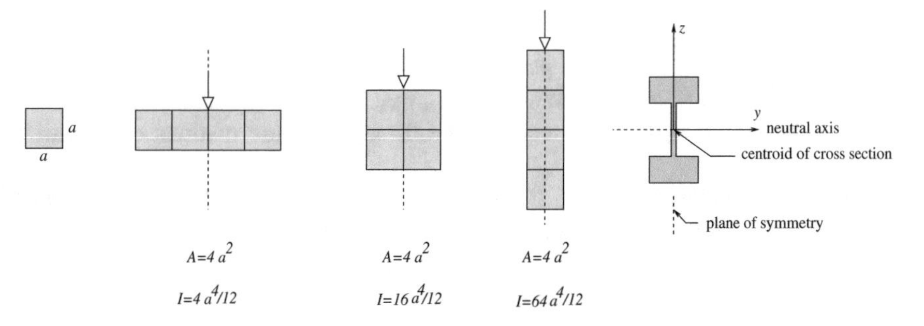

Fig. 5.3 Moment of inertia

Using (5.6)–(5.9), we thus obtain the final form of Hooke's Law for prismatic members

$$\begin{Bmatrix} n \\ s \\ m \end{Bmatrix} = \begin{bmatrix} EA & 0 & 0 \\ 0 & GA & 0 \\ 0 & 0 & EI \end{bmatrix} \begin{Bmatrix} \epsilon \\ \gamma \\ \kappa \end{Bmatrix} \tag{5.10}$$

or $\boldsymbol{n} = [\boldsymbol{EA}]\epsilon$ in matrix notation.

A beam element with the (5.10) stiffness relations, where we recall that the equilibrium equations are

$$\frac{\mathrm{d}}{\mathrm{d}x} \begin{Bmatrix} n \\ s \\ m \end{Bmatrix} + \begin{Bmatrix} 0 \\ 0 \\ s \end{Bmatrix} + \begin{Bmatrix} p \\ q \\ c \end{Bmatrix} = \begin{Bmatrix} 0 \\ 0 \\ 0 \end{Bmatrix} \tag{5.11}$$

and the deformations are given by

$$\begin{Bmatrix} \epsilon \\ \gamma \\ \kappa \end{Bmatrix} = \frac{\mathrm{d}}{\mathrm{d}x} \begin{Bmatrix} u \\ v \\ \theta \end{Bmatrix} - \begin{Bmatrix} 0 \\ \theta \\ 0 \end{Bmatrix} \tag{5.12}$$

is called a Timoshenko beam (Stephen P. Timoshenko, 1878–1972). Its peculiarity lies in the presence of the shear deformation mode γ.

These are remarkable equations in more than one sense. They are written at location x on the centroidal line and in terms of quantities which are functions of x only: the internal forces $\boldsymbol{n}$, the deformations ϵ and the stiffnesses **EA**. The elements of the diagonal matrix **EA**, i.e., *EA*, *GA* and *EI*, express the stiffness of the element at x, where E and G are the stiffnesses of the material (steel, concrete, wood, etc.) and A and I are the cross-sectional properties of the cross-section which govern the axial and shear stiffness (A) and the flexural stiffness (I). The terms *EA*, *GA* and *EI* are called the axial, shear and bending stiffnesses or rigidity of the element at x.

The uncoupled Eq. (5.10) are a generalized form of Hooke's law ($\sigma = E\epsilon$) and consequently **n** are often called *generalized stresses* or simply stresses and are the *generalized strains* or *strains*. (Whereas σ_{xx}, σ_{xz}, and ϵ_{xx}, γ_{xz}, refer to a particular point in the structure, $\boldsymbol{\sigma}$ and $\boldsymbol{\epsilon}$ capture the forces and deformations of an entire cross-section.)

The elastic constitutive law (5.10) in conjunction with the equilibrium conditions (5.11) and the deformation-displacement relations (5.12) constitute

the three pillars of structural analysis

and of solid mechanics, for that matter.

5.4 Euler–Bernoulli Beams

There exists a restricted description of the deformation of beams and it is the most widely used in structural analysis. This is the *Euler* description of deformation (Leonard Euler, 1707–1783) and it leads to Euler–Bernoulli beams. Euler's approach is based on the practical behavior of prismatic members. It turns out that the displacements due to the shear effect are much smaller than those due to flexion as depicted in Fig. 5.4. The initial square segments of the beam deform into parallelograms in shear mode and into trapezes in bending mode. The drawings are not to any scale, however practical experience indicates that, for standard (slender) elements, the displacements due to shear deformation are much smaller than those originating in flexion. Therefore, in Euler theory the shear strain is simply set to zero

$$\gamma = 0 \tag{5.13}$$

This leads to some interesting results. Setting $\gamma = 0$ entails (5.12)

$$\theta = \frac{dv}{dx} \tag{5.14}$$

that is, the angle that the Bernoulli section makes with the vertical is equal to the slope of the beam axis. In other words: plane sections, initially perpendicular to the beam axis,

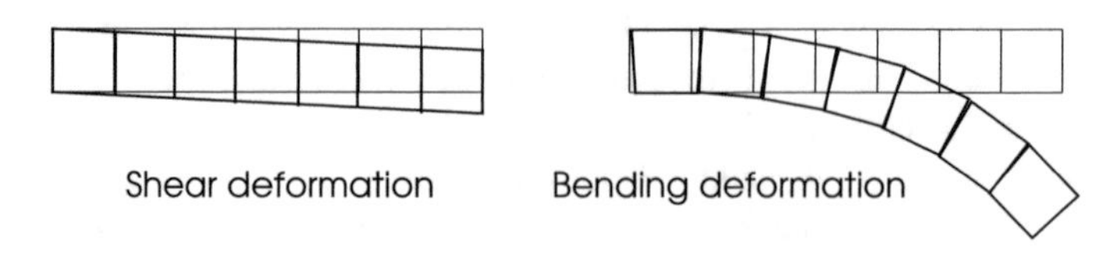

Fig. 5.4 Displacements due to shear and bending deformation

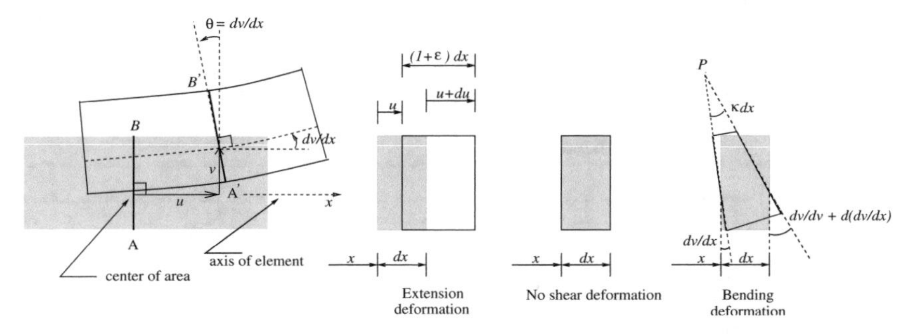

Fig. 5.5 Basic deformation modes of the Engineering Beam Theory

remain *plane and perpendicular* to the axis, after deformation (Euler–Bernoulli beams.)

The Euler–Bernoulli deformations of a lamella $\mathrm{d}x$ are depicted in Fig. 5.5. We note that the axial deformation pattern remains the same, that there is no shear deformation, and that the bending deformation can now be written in terms of $\mathrm{d}v/\mathrm{d}x$. Significantly,

$$\kappa \mathrm{d}x = \frac{\mathrm{d}}{\mathrm{d}x}(\mathrm{d}v/\mathrm{d}x)\,\mathrm{d}x = (\mathrm{d}^2 v/\mathrm{d}x^2)\,\mathrm{d}x$$

and as a result we can express the curvature in terms of the lateral displacements $\kappa = \mathrm{d}^2 v/\mathrm{d}x^2$. We note, that θ is no longer needed and that $\gamma = 0$.

As a result, for Euler–Bernoulli deformations, we have two displacement unknowns u and v and two deformation quantities ϵ and κ. We still have the shear force s but it has no deformations counterpart. The shear can be obtained from the bending moment by $s = -\mathrm{d}m/\mathrm{d}x$.

In the process, we have also lost one of the constitutive laws. Indeed, $s = GA\,\gamma$, with $s \neq 0$ and $\gamma = 0$ makes sense only if $GA \to \infty$. This is in line with our assumption that the member is very stiff in shear. However, the shear constitutive law is now utterly useless ($s = \infty \times 0$).

It is also customary to combine the shear and moment equilibrium equations. Assuming that there are no distributed couples

$$c = 0 \tag{5.15}$$

and there very seldom are, we take the derivative with respect to x of the moment equation, $\mathrm{d}^2 m/\mathrm{d}x^2 + \mathrm{d}s/\mathrm{d}x = 0$, in which we substitute $-q$ for $\mathrm{d}s/\mathrm{d}x$. The fundamental variables for the Euler–Bernoulli, also called, Engineering Beam Theory (EBT) are given in Table 5.1. These are n, ϵ and u for the extensional mode and m, κ and v for the flexural mode.

The governing equations of the EBT are thus the equilibrium equations ($\boldsymbol{L}_n \boldsymbol{n} + \boldsymbol{p} = \boldsymbol{0}$)

Table 5.1 Analysis variables and parameters in the EBT

Name	Alias	Rod	Beam
Internal forces	Generalized stresses	n	m
Deformations	Generalized strains	ϵ	κ
Displacements		u	v
Stiffnesses		EA	EI

Table 5.2 Engineering Beam Theory equations at x

Equilibrium equations	Hooke's Law	Strain-displacement relations
$dn/dx + p = 0$	$n = EA\ \epsilon$	$\epsilon = du/dx$
$d^2m/dx^2 - q = 0$	$m = EI\ \kappa$	$\kappa = d^2v/dx^2$

$$\begin{bmatrix} \mathrm{d}/\mathrm{d}x & 0 \\ 0 & \mathrm{d}^2/\mathrm{d}x^2 \end{bmatrix} \begin{Bmatrix} n \\ m \end{Bmatrix} + \begin{Bmatrix} p \\ -q \end{Bmatrix} = \begin{Bmatrix} 0 \\ 0 \end{Bmatrix} \quad \text{with} \quad s + \frac{\mathrm{d}m}{\mathrm{d}x} = 0 \tag{5.16}$$

Hooke's law ($\boldsymbol{n} = [\mathbf{EA}]\boldsymbol{\epsilon}$)

$$\begin{Bmatrix} n \\ m \end{Bmatrix} = \begin{bmatrix} EA & 0 \\ 0 & EI \end{bmatrix} \begin{Bmatrix} \epsilon \\ \kappa \end{Bmatrix} \tag{5.17}$$

and the strain-displacement relations ($\boldsymbol{\epsilon} = \boldsymbol{L}_{\epsilon}\boldsymbol{u}$)

$$\begin{Bmatrix} \epsilon \\ \kappa \end{Bmatrix} = \begin{bmatrix} \mathrm{d}/\mathrm{d}x & 0 \\ 0 & \mathrm{d}^2/\mathrm{d}x^2 \end{bmatrix} \begin{Bmatrix} u \\ v \end{Bmatrix} \tag{5.18}$$

We have now six equations in six unknowns n, m, u, v, ϵ and κ, as summarized in Tables 5.1 and 5.2.

An important benefit of the Euler–Bernoulli theory is that κ is now the curvature of the line $v(x)$. In the Timoshenko approach, κ was the difference in the rotational angles of two adjacent sections. This was not related to the curvature of the beam deformed axis $v(x)$. In principle, we could have $\kappa \neq 0$ for a straight line if there existed a $\mathrm{d}\theta/\mathrm{d}x$. Now, however, κ represents the curvature of the deformed beam axis.

5.4.1 *Virtual Work for Euler–Bernoulli Deformations*

In the following we will assume that there are no distributed couples, $c(x) = 0$, and we will be using Euler–Bernoulli theory for which there is no shear deformation $\gamma(x) = 0$. There are, of course, shear forces but the beams do not deform due to the shear. This is the usual engineering assumption and we will abide by it. For a beam segment ab under distributed loads the virtual work equation is here

$$(\boldsymbol{n}^T\boldsymbol{u})_{a^-} - (\boldsymbol{n}^T\boldsymbol{u})_{b^+} + \int_a^b \boldsymbol{q}(x)^T\boldsymbol{u}(x)\,\mathrm{d}x = \int_a^b \boldsymbol{n}(x)^T\boldsymbol{\epsilon}(x)\,\mathrm{d}x \tag{5.19}$$

When we expand (5.19) we obtain

$$(nu+sv+m\theta)_{a^-} - (nu+sv+m\theta)_{b^+} + \int_a^b (qu+pv)\mathrm{d}x = \int_a^b (n\epsilon+m\kappa)\mathrm{d}x \tag{5.20}$$

where, by definition, $\epsilon = \mathrm{d}u/\mathrm{d}x$ and $\kappa = \mathrm{d}^2v/\mathrm{d}x^2$ are the deformations of the displacement functions $u(x)$ and $v(x)$. The outer normal to end section a points in the negative x direction, hence the negative superscript (the opposite for b).

The external forces $\boldsymbol{p}(x)$ together with the internal forces $\boldsymbol{n}(x)$ form an *equilibrium set*, and the displacements $\boldsymbol{u}(x)$ with its deformations $\boldsymbol{\epsilon}(x)$ are a *compatible set*.

5.4.2 A Note on Curvature

Curvature can be defined as follows. Take three points A, B and C on a curve $v(x)$, and draw the circle which passes through these points (Fig. 5.6). (We have drawn an exaggerated lateral displacement in order to clarify the concept. For beams these displacements are in fact very small.) Now let points A and C approach point B. In the process, the circles passing through the three points have their radius modified and after some wobbling they settle on the limit circle as the three points merge. This limit circle does not collapse to a point but has a finite radius and is called the *osculatory circle* or the *circle of curvature* at point B on the curve. The radius R of this circle is the radius of curvature, and its inverse $\kappa = 1/R$ is by definition the curvature of the curve at the considered point. The more curved the line, the smaller is R and consequently the larger is κ. On a straight portion of the line, the radius of

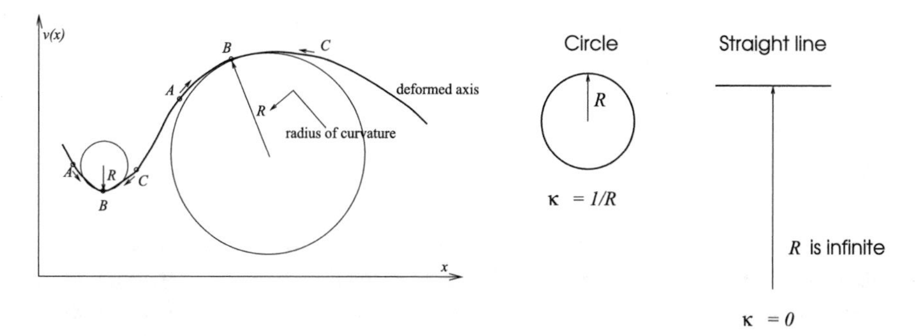

Fig. 5.6 Curvature of $v(x)$

curvature $R \to \infty$ and $\kappa = 0$. Also, the radius of curvature of a segment of a circle is the radius of the circle and consequently its curvature is (of course) constant.

It can be shown that the curvature of a curve is

$$\text{Curvature} = \frac{\mathrm{d}^2 v}{\mathrm{d}x^2} \frac{1}{[1 + (\mathrm{d}v/\mathrm{d}x)^2]^{3/2}} \tag{5.21}$$

For small displacements $(dv/dx \ll 1)$, $(dv/dx)^2$ is negligible compared to 1 and consequently the fraction on the right-hand side tends to unity. For κ, we thus recover the definition of the curvature

$$\kappa = \frac{\mathrm{d}^2 v}{\mathrm{d}x^2} \tag{5.22}$$

This is the model that we will employ in this book. It is also the theory that most structural engineers use.

5.4.3 *Rods and Beams*

A furtive glance at the structural equations reveals that the analysis equations of a prismatic straight element are uncoupled. It is as if, inside the element, we have a *rod* or strut which behaves in an *extensional* mode with variables n, ϵ, u under loading p and extensional stiffness EA

$$\frac{\mathrm{d}n}{\mathrm{d}x} + p = 0 \qquad n = EA\,\epsilon \qquad \epsilon = \frac{\mathrm{d}u}{\mathrm{d}x} \qquad \text{(Rod)} \tag{5.23}$$

and a *beam* with *flexural* behavior with variables m, κ, v under loading q and flexural stiffness EI

$$\frac{\mathrm{d}^2 m}{\mathrm{d}x^2} - q = 0 \qquad m = EI\,\kappa \qquad \kappa = \frac{\mathrm{d}^2 v}{\mathrm{d}x^2} \qquad \text{(Beam)} \tag{5.24}$$

with $s = -\mathrm{d}m/\mathrm{d}x$. This has important consequences for the design and analysis of structures in the linear range.

5.5 Summing Up

In this chapter we have emphasized the linear elastic nature of the materials the structures discussed in this book are made of. Recalling the rigid cross-section assumption of beam theory, we have shown that this leads to Timoshenko and Euler–Bernoulli type beams. The latter are used in standard structural analysis and also in this volume.

Chapter 6
The Unit-Load Method

In structural analysis, displacements are not always given the importance they deserve and are often even considered a nuisance. In truth, displacements lie at the heart of how structures sustain loads. In some sense they are the engine that drives the structural response. The unit-load method is a technique that will help us to quantify displacements and rotations of the equilibrium configuration, that is, the shape of the structure after it has managed to equilibrate the applied loads. We will introduce the method with the aid of beams, and show how the unit-load method computes a displacement or a rotation at some point along the beam.

6.1 The Equilibrium Configuration

Consider the simply supported beam ab in Fig. 6.1a1. As soon as we apply a couple C to the left extremity the beam will curve up and, after some undulation (a trial and error process), it will come to rest in a shape $v(x)$ similar to what is drawn in the figure. This is the *equilibrium configuration*, that is, the shape for which the beam has restored equilibrium.

We have ample evidence that when the couple was applied there was no equilibrium because the beam started to move. After it assumed the equilibrium configuration the structure was again at rest, hence every part of the beam was again in equilibrium and in particular the applied point couple was balanced by an equal and opposite point couple. We will not try to retrace the transient phases, that is, the shapes through which the beam passed until it found the correct configuration. Instead, we will concentrate on the equilibrium state. What is so special about this configuration?

The internal forces, generated by all possible displaced shapes but one, will not equilibrate the applied loads. The beam will be busy looking for that special shape until equilibrium is reached. The beam is then at rest in the equilibrium configuration. Fortunately, the procedure is quite fast.

M.B. Fuchs, *Structures and Their Analysis*,
DOI 10.1007/978-3-319-31081-7_6

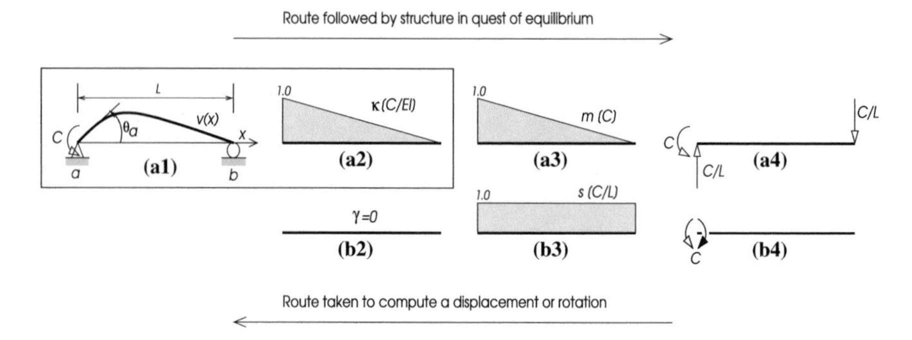

Fig. 6.1 Simple beam with end-couple

Following Fig. 6.1 from left to right, the beam assumes a displaced shape $v(x)$ (see (a1)), the most important aspect of which is the distribution of its curvatures $\kappa = \mathrm{d}^2v/\mathrm{d}x^2$ (see (a2)), which is the second derivative of the displacements. Indeed, these curvatures create the bending moments (see (a3)) by means of Hooke's law $m(x) = EI(x)\kappa(x)$, where, we remember, E is the elastic modulus of the material, I is the moment of inertia of the cross-section. The product $EI(x)$ is the bending stiffness along the beam. We usually assume a constant bending stiffness.

As with all Euler–Bernoulli beams the shear deformation $\gamma(x)$ is negligible and for all practical purposes it is set to zero (see (b2)). But it is there, and it engenders the shear forces which we know must be the derivative of the bending moment if we want to have equilibrium (see (b3)). Eventually the applied couple is balanced by the internal moment at extremity a (see (b4)).

Can we calculate the slope θ_b of the beam at b or the sagging of the beam v_d at, say, mid-span? The answer is yes, if we know the distribution of the curvatures $\kappa(x)$ shown in Fig. 6.1a2. To get the curvatures we follow the opposite route. By analysis of the beam we compute the reactions (see (a4)), find the bending moments $m(x)$ (see (a3)), and by Hooke's law also the curvature distribution $\kappa(x)$ (see (a2)).

We can now compute any slope or displacement by the unit-load method.

6.1.1 The Unit-Load Method for Beams

We have reproduced the equilibrium configuration $v(x)$ and the curvature distribution $\kappa(x)$ of the simply supported beam ab in the rectangle of Fig. 6.2(1). The exact displaced shape is of course unknown, it is only an approximate sketch of $v(x)$. You will note that what we have in the frame is a compatible system, called a real compatible system, because it is composed of the real displacements and curvatures of the beam, and the curvatures are indeed the second derivatives of the displacement function.

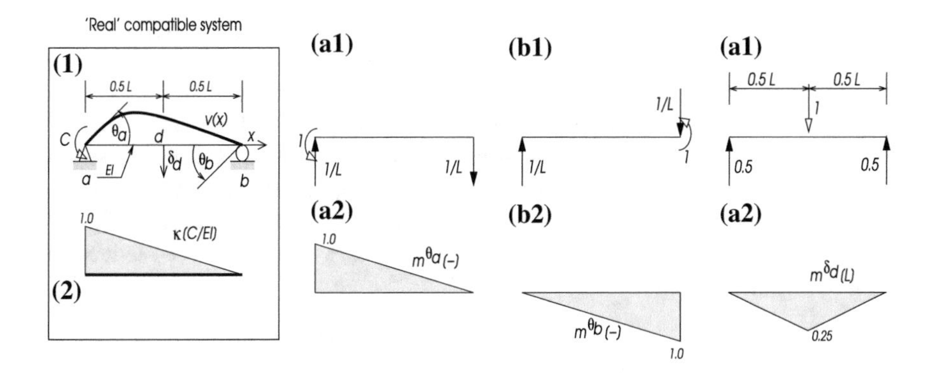

Fig. 6.2 Computing displacements by *vw*

We now want to compute the slope at the extremities of the beam and the lateral displacement at mid-span.

θ_a To compute the slope at the left extremity by the unit-load method we construct a virtual equilibrium system in the following manner (see Fig. 6.2a1, a2).

Draw a copy of the beam and apply a unit couple (a couple of magnitude 1, without dimensions) at a in the direction in which you want to compute the rotation. Then apply forces where the reactions occur (and only there) in the structure (that is, a and b in our case) such that the unit couple and the forces are auto-equilibrated. The result is in Fig. 6.2a1. Clearly the two forces form a clockwise couple $(1/L) \times L = 1$ which together with the unit couple are equivalent to zero, that is, in equilibrium. Let us call these loads symbolically $\boldsymbol{f}^{\theta_a}$ because they were built to compute θ_a.

Next draw the bending moment for these forces as shown in Fig. 6.2a2 and call them m^{θ_a}. These are non-dimensional moments since they originate from a non-dimensional couple. The external forces $\boldsymbol{f}^{\theta_a}$ and the bending moment m^{θ_a} are a virtual equilibrium system; virtual because everything was invented for the sole purpose of computing θ_a.

We are now ready to write a virtual work equation with the real compatible system and the virtual equilibrium equation. The external virtual work consists of the work of the couple and the two forces through the corresponding displacements/rotations taken from $v(x)$, that is, $evw = (1) \times \theta_a + (1/L) \times v(0) - (1/L) \times v(L)$. But $v(0) = v(L) = 0$, therefore $evw = \theta_a$. So the virtual work equation ($evw = ivw$) becomes

$$\theta_a = \int_0^L \kappa\, m^{\theta_a}\, \mathrm{d}x = \frac{1}{3}\frac{CL}{EI} \text{ (non-dimensional—in radians)}$$

Note that κ and m^{θ_a} are the distributions (2) and (a2) in Fig. 6.2.

θ_b For the slope at extremity b you apply a unit couple at b and equilibrate with forces at a and b (were the rigid reactions are in the actual beam) as shown in (b1). This auto-equilibrated system of external forces is called $\boldsymbol{f}^{\theta_b}$. Next we determine the moment distribution m^{θ_b} in equilibrium with these forces, given in (b2). The pair $\boldsymbol{f}^{\theta_b}$ and m^{θ_b} is the virtual equilibrium system devised to compute θ_b. We can now write a virtual work equation with the real compatible system (1) and (2) and the virtual equilibrium system (b1) and (b2) taken from Fig. 6.2. The external virtual work $(1) \times \theta_b + (1/L) \times v(0) - (1/L) \times v(L)$ reduces to $evw = \theta_b$, therefore

$$\theta_b = \int_0^L \kappa\, m^{\theta_b}\, \mathrm{d}x = -\frac{1}{6}\frac{CL}{EI}$$

The negative sign stems from κ and m^{θ_b} distributions (2) and (a2) in Fig. 6.2 being of opposite signs. The negative sign tells us that the slope is in the opposite direction of the one chosen for the unit couple.

δ_d For the deflection at mid-span, Fig. 6.2(1), we apply a unit force (a force of magnitude 1, without dimension) at d, and equilibrate with forces applied at a and b as shown in (d1). Next we determine the related bending moment m^{δ_b} (see Fig. 6.2d1, d2). Note, the moments are the product of non-dimensional forces by distances, hence their units are lengths (L). Consequently, the unit-load method gives

$$\delta_d = \int_0^L \kappa\, m^{\delta_d}\, \mathrm{d}x = -\frac{1}{16}\frac{CL^2}{EI} = -0.0625\frac{CL^2}{EI}$$

The negative sign indicates that the displacement is in the opposite sense of the applied unit force, that is, the beam displaces as expected in the up-direction at d.

6.1.2 Redundant Beams

The unit-load method for computing a displacement or rotation is particularly interesting, some will say intriguing, when the structure is redundant. The reader may not have studied redundant structures yet, for our purposes however, it is enough to note that in redundant structures there are usually more reactions than necessary. This gives the user a choice in selecting an equilibrium system.

The beam in Fig. 6.3(1) is an example of such a redundant structure. Supports b and c are superfluous (redundant). They are perfectly welcome, however the beam could support the applied force P without them. Indeed, remove these two supports and you obtain a sound cantilever.

Intuitively, under the force P the beam will assume an equilibrium position as shown in the figure. This partially explains the type of curvature distribution given in Fig. 6.3(2). Even if it at this stage it does not entirely make sense to you, let us accept that after a lengthy analysis we know the curvature distribution.

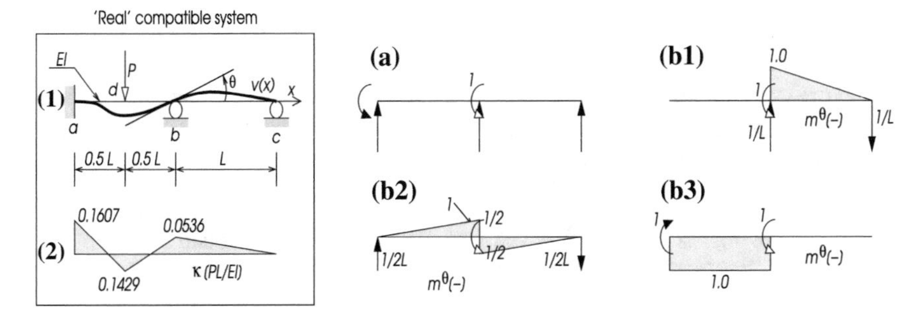

Fig. 6.3 Unit-load method for redundant beams

We now intend to compute the slope θ of the beam at b by the unit-load method. The displacements $v(x)$ and the deformations $\kappa(x)$ of the beam under load P given in Fig. 6.3(1, 2) constitute a compatible system, albeit 'real', but a compatible system all the same.

Next, we apply a unit couple at b, and equilibrate the couple by forces applied to the reactions as shown in Fig. 6.3a. (Note, the diagram does not represent a beam, and the forces are not reactions. We are in the realm of the study of equilibrium.) The wall at a prevents rotation and vertical displacements, the supports at b and c prevent vertical displacements, hence the four possible loads (not reactions) shown in the figure. With these four loads we must equilibrate the couple of magnitude 1 applied at b. There are many choices, three of which are given in Fig. 6.3b1–b3. We show the choice of forces and the ensuing virtual moment distributions on the same diagram.

In Fig. 6.3b2, for instance, the unit couple is balanced by an opposite couple formed by two equal and opposite forces of magnitude $1/2L$ at a and c ($1/2L \times 2L = 1$). In case you are wondering where the moment distribution m^θ comes from, note that the shear is constant and, therefore, the moments are linear. The moments are zero at the extremities, and to complete the diagram compute the moments at b_L and b_R.

It is noteworthy that all three diagrams will yield the same θ_b; but the easiest integral to calculate is by using case (b3). We find $\theta_b = 0.0179\,\mathrm{PL}^2/\mathrm{EI}$ measured in radians.

6.2 The General Case

The unit-load method is a numerical technique for computing a displacement or a rotation at some point along a structure *if you know* the deformations. The most general representation of structures are frames, and the deformation of their equilibrium configuration $\boldsymbol{\epsilon}(x) \equiv \{\epsilon(x) \;\; \kappa(x)\}^T$ consists of axial $\epsilon(x)$ and bending $\kappa(x)$

deformations where x is the running coordinate along the structure. Recall that we assume that the deformation due to shear γ is zero.

Having determined the distribution of the deformations, the unit-load method will allow us to calculate the displacement or the rotation at any location x by writing an appropriate virtual work equation.

Knowing $\epsilon(x)$ and $\kappa(x)$ we can calculate δ or θ at any location x .

In this sense the unit-load method is almost a *post factum* technique, but the method has its importance.

6.2.1 Computing a Displacement

Let us assume that we have analyzed a structure and have determined the distribution of the deformations $\boldsymbol{\epsilon}(x) \equiv \{\epsilon(x)\ \kappa(x)\}^T$. They are usually obtained from the internal forces $\boldsymbol{n}(x)$ by means of Hooke's law.

We now want to compute the displacement δ at some point of the structure in the direction of δ. For this purpose, we will write a virtual work equation with the following compatible and equilibrium systems:

1. The *compatible system* is the real compatible system of the structure that we have just analyzed. It is composed of the displacements $\boldsymbol{u}(x) \equiv \{u(x)\ v(x)\}^T$ and the deformations $\boldsymbol{\epsilon}(x) \equiv \{\epsilon(x)\ \kappa(x)\}^T$ of the structure. The displacements are as yet unknown but there are displacements. The deformations, on the other hand, must be known.
2. The *equilibrium system* is constructed from scratch for the sole purpose of computimg δ. We consider an outline of the structure without the supports and apply a unit force (a force equal to 1, without dimension) at the location and in the direction of δ. Then we equilibrate the unit load with loads at the location of the supports *and at the supports only*, such that the whole structure is in equilibrium. It is important to emphasize that if the structure is fixed at a support, we can apply a force and a couple there. If it is only hinged at the support, we can apply a force but not a couple. In other words, we apply a unit force to the floating structure where we want to compute the displacement (in the real structure), and balance the force with loads at the supports, thus creating a auto-equilibrated system of loads.
   For redundant structures there exist many ways to balance the unit force. We may choose any that are to our liking, but remember to put loads only at the locations of the supports.

3. We then proceed to *compute the internal forces distributions* $\boldsymbol{n}^{\delta}(x) \equiv \{n^{\delta}(x)\ m^{\delta}(x)\}^T$ for this auto-equilibrated loading. The superscript δ reminds us that the internal forces are not the internal forces of the real structure under the real loads, but the internal forces of an image of a floating structure under auto-equilibrated loads with a unit force in the direction of δ.
   The virtual equilibrium system is the unit force 1^{δ}, the loads at the locations of the supports and the internal forces $\boldsymbol{n}^{\delta}$.
4. We can now *write the unit-load formula* which is the virtual work equation with the compatible and equilibrium systems $(\delta)\ (1^{\delta}) = \int_{\text{Struct}} \boldsymbol{\epsilon}^T(x)\boldsymbol{n}^{\delta}(x)\mathrm{d}x$. On the left-hand side, we have the work of the real displacements times the forces of the virtual equilibrium system. The latter are the unit load and the loads applied where the rigid supports are. At the rigid supports the displacements are of course zero. Consequently, all we are left with for the virtual work of the external forces is $(\delta)\ (1^{\delta})$, that is, δ.

This is then the unit-load equation which directly gives the displacement we want to determine.

$$\delta = \int_{\text{Struct}} \boldsymbol{\epsilon}^T(x)\boldsymbol{n}^{\delta}(x)\mathrm{d}x = \int_{\text{Struct}} [\epsilon(x)\ n^{\delta}(x) + \kappa(x)\ m^{\delta}(x)]\mathrm{d}x \qquad (6.1)$$

6.2.2 Computing a Rotation

There is nothing much to add to compute a rotation θ at some point along the structure. Instead of applying a unit force, we apply a *unit concentrated couple* (the value of the couple is 1, without dimension) at the same point and in the sense of θ on the free floating structure, and balance the couple with loads at the locations of the supports of the real structure.

We then proceed to calculate the distributions of the internal forces of this auto-equilibrated system of forces. These will now be called $\boldsymbol{n}^{\theta}$ because they were built to compute the angle θ. The virtual equilibrium system is the unit couple 1^{θ}, the loads at the locations of the supports and the internal forces $\boldsymbol{n}^{\theta}$.

We write a work equation with the real compatible system and the virtual equilibrium system $(\theta)\ (1^{\theta}) = \int_{\text{Struct}} \boldsymbol{\epsilon}^T(x)\boldsymbol{n}^{\theta}(x)\mathrm{d}x$. On the left-hand side, we notice that only the unit couple moves through the rotation θ, whereas the other loads are applied along the supports and do not move. Consequently, we obtain the unit-load expression (6.1), this time for a rotation

$$\theta = \int_{\text{Struct}} \boldsymbol{\epsilon}^T(x)\boldsymbol{n}^{\theta}(x)\mathrm{d}x = \int_{\text{Struct}} [\epsilon(x)\ n^{\theta}(x) + \kappa(x)\ m^{\theta}(x)]\mathrm{d}x$$

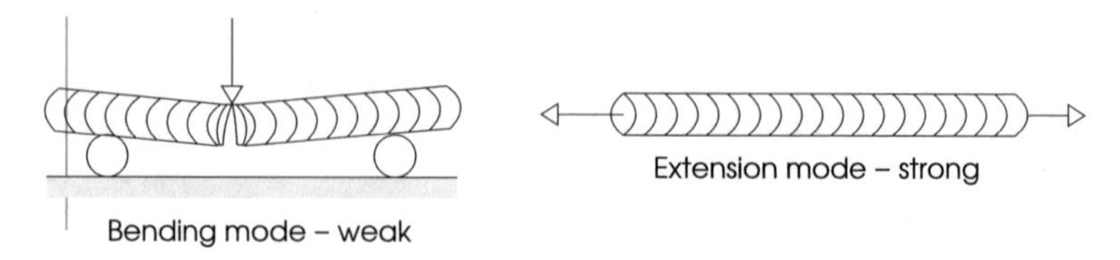

Fig. 6.4 Bending versus extensional deformations

6.2.3 Frames, Beams and Trusses: Numerical Aspects

When we observe the right-hand side of the unit-load relation we notice that a displacement is the effect of normal extensions $\epsilon(x)\ n^{\delta}(x)$ and of flexural strains $\kappa(x)\ m^{\delta}(x)$.

Frames. Frames have both normal and bending deformations. However, it so happens that the flexural effect is usually by far the dominant factor

$$\int_{\text{Struct}} \kappa(x)\ m^{\delta}(x)\mathrm{d}x \gg \int_{\text{Struct}} \epsilon(x)\ n^{\delta}(x)\mathrm{d}x$$

One way to look at this is to remember that bending is a relatively flexible mode of deformation and extension is a relatively stiff mode of deformation. If we want to break a bundle of spaghetti, we never pull on its extremities (axial mode) but break it by bending. Or, if you need short pieces of wood for a bonfire, we would not pull on a branch of a tree, but rather try to produce the pieces by breaking the branch by bending it (Fig. 6.4). Indeed, bending produces larger stresses which originate from larger strains.

By a same reasoning, the part of the displacements due to bending deformations is much larger than that due to extensional deformations and, in general, we will neglect the axial term when a flexural term exists, because the latter overshadows axial deformations. Consequently, in practice the unit-load formula is

$$\delta, \theta = \int_{\text{Struct}} \kappa(x)\ m^{\delta,\theta}(x)\ \mathrm{d}x \qquad \text{for beams and frames} \tag{6.2}$$

It is valid for calculating both a displacement and a rotation.

Beams. For beams there are, by definition, no normal effects $\int_{\text{Struct}} \epsilon(x)\ n^{\delta}(x)\mathrm{d}x = 0$ and, consequently, we will be using only the $\kappa(x)\ m^{\delta}(x)$ part in the unit-load method.

Trusses. The elements composing a truss, on the other hand, experience little bending (or no bending at all if it is an ideal truss) and, consequently, we have only axial effects.

Nothing is large or small but by comparison.

Axial effects are small compared to bending effects, but when we have only axial effects they cannot be ignored. The unit-load method is here $\delta = \int_{\text{Struct}} \boldsymbol{\epsilon}^T(x)\boldsymbol{n}^\delta(x)\mathrm{d}x$. Now, since a truss is composed of elements, this expression becomes $\delta = \sum_i \int_{\text{Struct}} \boldsymbol{\epsilon}_i^T(x)\boldsymbol{n}_i^\delta(x)\mathrm{d}x$ where the summation index i runs over all the elements of the truss. We assume that the elements of the truss are straight and uniform, which means that the deformations $\epsilon_i(x)$ and forces $n_i(x)$ in any element i are constant. We obtain the expression for the unit-load method in the case of trusses

$$\delta = \sum_i \epsilon_i \, n_i^\delta \, L_i \qquad \text{for trusses} \tag{6.3}$$

where L_i is the length of element i.

6.2.4 Formula for Calculating the Integrals

From what we have seen so far, the unit-load method involves the evaluation of integrals of products of two functions $\int_x \kappa(x)\; m^{\delta,\theta}(x)\; \mathrm{d}x$. But these in turn can often be broken down, at least in the examples that we will treat, into the sum of integrals over intervals j in which the functions are linear in x

$$\int_x \kappa(x)\; m^{\delta,\theta}(x)\; \mathrm{d}x = \sum_j \int_{a_j}^{b_j} g_j(x)\; h_j(x)\; \mathrm{d}x \tag{6.4}$$

where a_j and b_j are the boundaries of interval j, and the sum is taken over all the intervals into which the integral was broken up. For two linear functions (see Fig. 6.5), the results of the integration are (subscript j has been dropped)

$$\int_a^b g(x)\; h(x)\; \mathrm{d}x = \frac{L_{ab}}{3}\;(g_a h_a + g_b h_b) + \frac{L_{ab}}{6}\;(g_a h_b + g_b h_a) \tag{6.5}$$

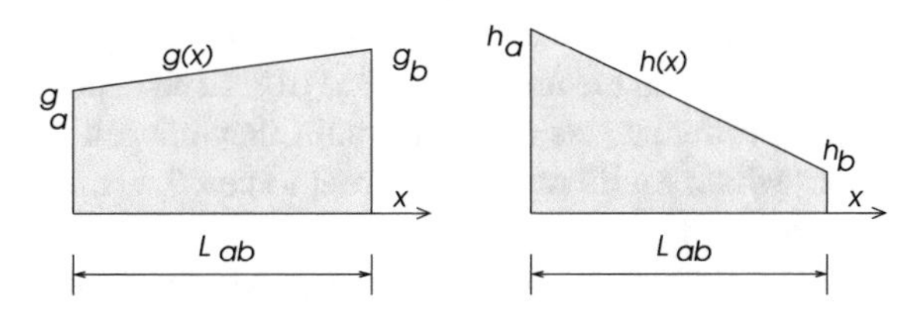

Fig. 6.5 Formula

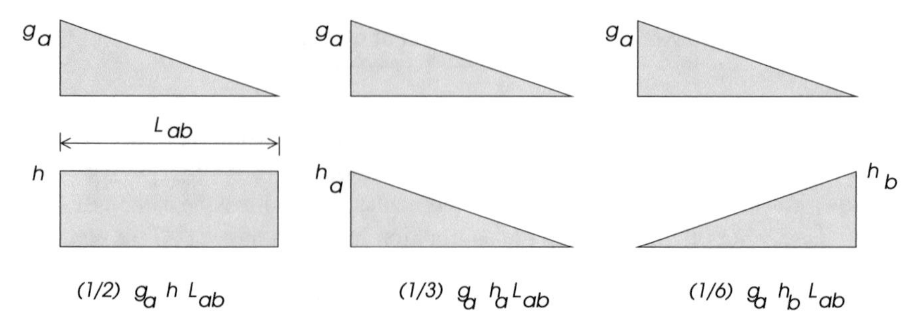

Fig. 6.6 Special cases

The values of g and h are algebraic and can, individually, have positive or negative values. We also give some special cases (Fig. 6.6).

We can of course use any other method for the integration. We can find tables with integrals of more intricate functions in many textbooks.

6.3 Trusses

Trusses, similar to any other structure, equilibrate the applied loads by moving into *the* equilibrium configuration (there is always only one such configuration).

We remember that what sets trusses in a class apart is that the slender members have no significant bending stiffness $I/AL^2 \to 0$, and that the external forces (no couples are allowed) are applied at the junctions only. If the geometry has a triangulated nature, one can assume that every truss element i has only a normal force n_i and an axial extension/contraction ϵ_i. The latter two are related by Hooke's law for axial members $n_i = (EA)\epsilon_i$ where EA is the axial stiffness per unit length, which is assumed constant for the entire element. We note, as a corollary, that the normal forces and the axial extensions have the same values but in different units.

How does a truss react to forces applied to its nodes? The sequence of events is

$$\boldsymbol{u} \to \boldsymbol{\epsilon} \to \boldsymbol{n} \tag{6.6}$$

where $\boldsymbol{u}$ is the vector of displacements of the open degrees of freedom of the truss, $\boldsymbol{\epsilon}$ is the vector of strains of the members of the truss and $\boldsymbol{n}$ is the vector of the normal forces in the members of the truss. The nodes try various displacements thus causing different sets of strains, until they happen on strains which create internal forces which equilibrate the external forces. The truss has settled in the equilibrium configuration.

Consider truss abc in Fig. 6.7a with a vertical force P applied at the free node b. The two elements have axial stiffness EA and the height of the truss is h. Clearly, in order to equilibrate the applied force the truss should try to create a tensile force P

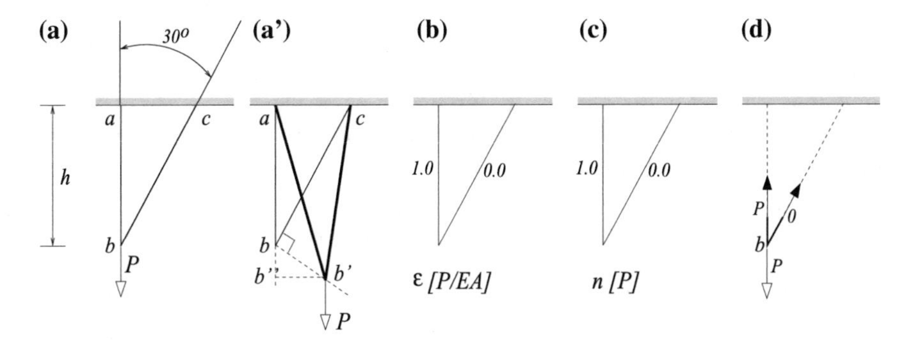

Fig. 6.7 Simple truss

in bar ab and a zero force in bar bc as shown in the equilibrium of forces (Fig. 6.7d). It will do so by extending its bars accordingly, that is, $\epsilon_{ab} = P/EA$ and $\epsilon_{bc} = 0$ (Fig. 6.7c). Indeed, applying Hooke's law, that is, multiplying the axial strains by the axial stiffnesses, yields the necessary internal forces.

It is noteworthy that $\epsilon_{bc} = 0$ means that bar bc is not supposed to elongate. This is only possible if bar bc swings like a rigid element about c thus bringing node b in position b' (Fig. 6.7a'). For very small displacements the trajectory bb' which is an arc of a circle can be assimilated to its tangent. Hence bb' is perpendicular to bc. The elongations of bar ab is bb''.

To compute the displacement of the free node in truss abc of Fig. 6.7, we use the real deformations of its members, and we construct appropriate equilibrium systems. Since we do not know the direction of the displacements, we are going to compute the $x - y$ components of the displacements. The compatible system was copied from Fig. 6.7 and appears in the frame in Fig. 6.8a. Note that the axial strains are dimensionless (EA is a force and so is P) which is in accordance with the definition of ϵ $[length/length]$.

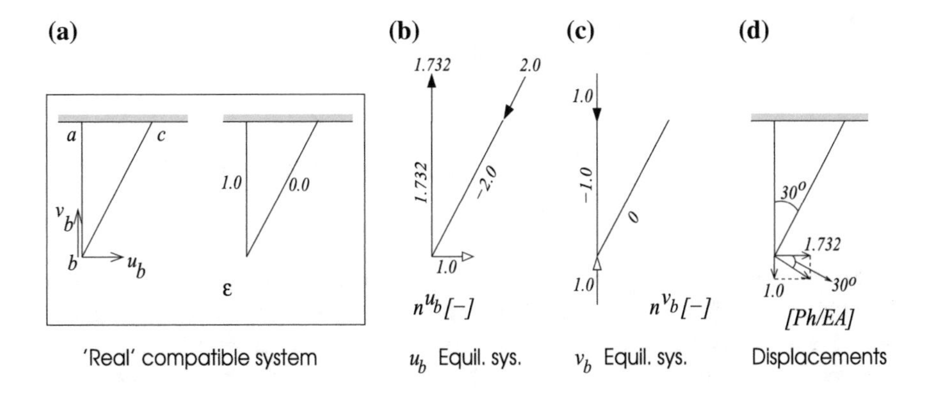

Fig. 6.8 Using vw for a simple truss

To compute the horizontal component u_b of the displacement, we construct an equilibrium system the external forces of which are a unit horizontal force at b and equilibrating forces at the supports. In Fig. 6.8b we give the external forces and the corresponding internal forces $\boldsymbol{n}^{u_b}$. Together they constitute an equilibrium system. We now write the unit-load equation

$$u_b = \epsilon_{ab}\, n_{ab}^{u_b}\, L_{ab} + \epsilon_{bc}\, n_{bc}^{u_b}\, L_{bc}$$

Introducing the appropriate values ($L_{ab} = h$, $L_{bc} = 1.155\ h$) we obtain $u_b = 1.732\ Ph/EA$.

For the vertical displacement v_b, we apply a unit vertical force at b and equilibrating forces at the supports. The internal forces are trivial, $\boldsymbol{n}^{v_b} = \{-1.0\ \ 0.0\}^T$ (see Fig. 6.8c for the equilibrium system), and the vertical displacement is here $v_b = \epsilon_{ab}\, n_{ab}^{v_b}\, L_{ab} + \epsilon_{bc}\, n_{bc}^{v_b}\, L_{bc} = -Ph/EA$. The negative sign indicates that the displacement is downwards (the unit force was taken up).

One will appreciate that b moves perpendicular to bar bc as was expected from the fact that the bar does not elongate (Fig. 6.8d). Indeed, $\arctan(1/1.732) = \pi/6°$ or 30°.

6.4 Illustrated Examples

The unit-load method hinges on the selection of an adequate equilibrium system. When we apply a unit load to a statically determinate system, there will be only one possible set of equilibrium loads applied to the reactions and corresponding distribution of internal forces. This is what determinacy is all about.

When you apply a unit load to a redundant structure, you will find that there are many possible equilibrium loads at the reaction points and corresponding distributions of internal forces. All give the same result (the sought after displacement or rotation) but some will be numerically more efficient than others. This makes the redundant cases arguably more interesting.

We will start with determinate examples.

6.4.1 A Kinked Bar or Shifting from Normal to Bending Mode

In this example we will apply the unit-load method to a bent bar and, in doing so, learn something regarding the interplay between normal forces and bending moments, which is a recurring phenomenon in structures. It will give us an indication of the relative importance of the normal and bending terms in the unit-load equation.

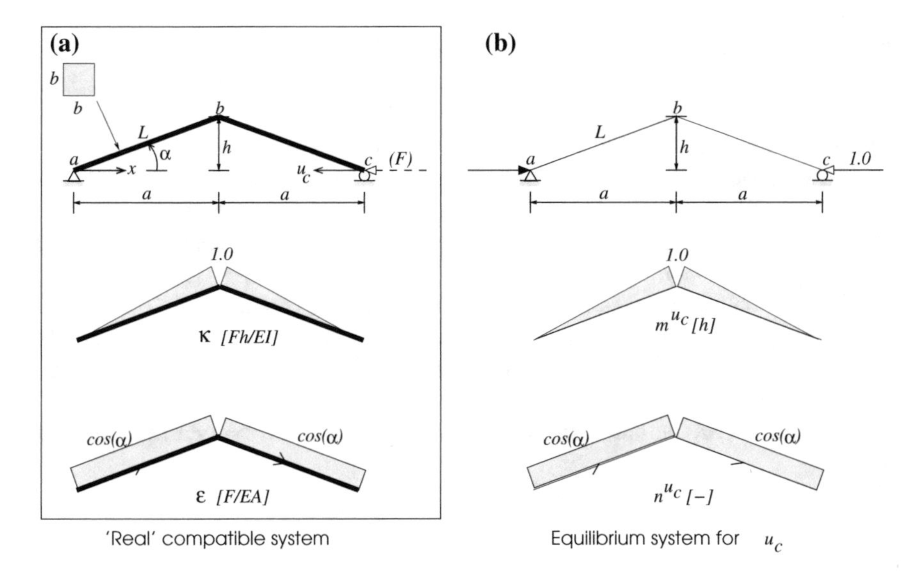

Fig. 6.9 Kinked bar

Consider the kinked bar[1] in Fig. 6.9a, hinged at a and supported by a roller at c, subjected to a compressive force F at c. The element can be considered as having been straight of length $2L$ and having received a kink in its middle. We want to compute the loss in longitudinal rigidity due to the eccentricity h/L. We note the symmetry of the problem (both the structure and, as we shall see, the loading are symmetric.)

We will determine the contraction of the bar by computing the horizontal displacement u_c. When the bar was straight ($h = 0$), the internal forces reduced to compressive axial forces $n(x) = -F$ (no shear, no moments), and so were the strains $\epsilon(x) = -F/EA$ (axial deformations only.) Note that the actual axial strains $\epsilon(x)$ and the horizontal displacements $u(x)$ of the beam constitute a *compatible system* (real).

To compute u_c we apply a unit force at c (positive to the left) which produces internal forces $n^{u_c} = -1$ (forces n devised to compute u_c). The force at c, the reaction at a and the internal forces are an *equilibrium system*. The unit force method gives $u_c = (-1)\ (-F/EA)(2L)$ which is akin to the contraction

$$\delta_o = (2L/EA)\ F \tag{6.7}$$

What happens when the element is bent? First and foremost, we have now internal bending moments, which makes it more flexible longitudinally.

We will now use the unit-load method to compute the contraction δ of chord ac (originally $2L$). The same case was analyzed earlier in the text and the nsm diagrams

[1] We usually reserve the term 'bar' for a straight element under axial loads and 'beam' for an element in bending. The kinked bar is an element which was a bar and acquired characteristics of a beam.

were given. Dividing the moments by EI gives the curvature κ, and dividing the normal forces by EA produces the normal extensions ϵ. These are shown in Fig. 6.9a. The equilibrium configuration of the kinked bar under F, not shown in the figure (we do not need to know the exact shape), and the deformations κ and ϵ constitute a (real) compatible system. The real deformations thus are $\epsilon(x) = -F\cos(\alpha)/EA$ and $\kappa(x) = Fhx/(EIL)$, where x is the coordinate along the chord. (We remind the reader that although there is now shear, the shear deformations are very small and assumed zero ($\gamma = 0$)).

We now apply a horizontal force of magnitude 1 at c to compute u_c. The internal forces in equilibrium with the unit force and the reaction are the same as those due to F, divided by the scalar F and multiplied by their respective rigidity, that is, for ab: $n^{u_c}(x) = -\cos(\alpha)$, and $m^{u_c}(x) = hx/L$.

Taking into account both extensional and flexural effects, the unit-load formula gives

$$u_c = 2\int_a^b (n^{u_c}(x)\epsilon(x) + m^{u_c}(x)\kappa(x))\,\mathrm{d}x$$

$$\text{or} \quad \delta = LF\left(\frac{\cos^2\alpha}{EA} + \frac{1}{3}\frac{h^2}{EI}\right) \tag{6.8}$$

By factoring out $1/EA$ from the *right-hand side* and dividing throughout by LF/EA (which incidently is δ_o), with $h = L\sin\alpha$, we get

$$\frac{\delta(\alpha)}{\delta_o} = \cos^2\alpha + \frac{1}{3}\frac{AL^2}{I}\sin^2\alpha \tag{6.9}$$

The first term on the *right-hand side* is the contribution of the axial deformations, and the second term originates from the bending curvatures. For a straight bar $\alpha = 0$ we have of course $\delta(0)/\delta_o = 1$ or $\delta(0) = \delta_o$. As α increases, the influence of the axial part decreases and we get increasingly bending deformations (the sine term). This term is boosted by the $(1/12)(AL^2/I)$ factor.

This non-dimensional ratio

$$AL^2/I \tag{6.10}$$

will often appear in frame elements. It describes the interplay between the axial stiffness EA and the bending stiffness EI, and it depends also on the square of the length. So when AL^2/I is large, that is, small bending rigidity as compared to the axial rigidity, this factor will increase the influence of the bending deformations. On the other hand, relatively large bending stiffness will reduce the influence of the flexural deformations. Note the presence of L^2 in the numerator. Longer elements will be more sensitive to bending deformations.[2]

[2]Longer straight elements in compression will be prone to buckling (instability) but that is not dealt with here.

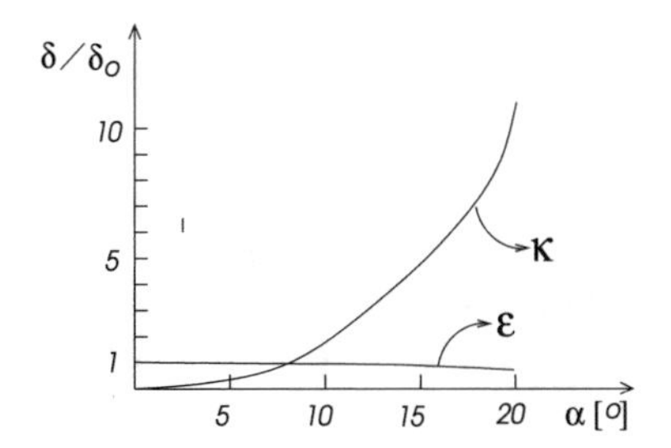

Fig. 6.10 Bending and normal deformations in kinked bar

To appreciate how bending deformations enter the game, let us assume that the cross-section of the bar is a square of side $b = 0.1L$. The height is 1/10th of the total length, which makes it a rather slender element with a poor cross-section to take bending. (Elements which bend should have larger heights than widths.)

The cross-sectional area is $A = b^2$ and the second moment of area is $I = b^4/12$. Consequently, (6.9) becomes $\delta_b/\delta_o = \cos^2\alpha + 400\ \sin^2\alpha$. In Fig. 6.10, we depict the influence of the axial component $\cos^2\alpha$, and the bending component $400\ \sin^2\alpha$ on the longitudinal flexibility of the bent bar as a function of the slope α in the range 0°–20°. The numbers are normalized to the axial flexibility of a straight bar. It will be noticed that, as soon as there is the slightest kink, bending takes over and the flexibility increases very significantly.

6.4.2 *Simple Beam with Distributed Forces*

Consider the simply supported beam in Fig. 6.11a0, of uniform bending stiffness EI, subjected to a uniformly distributed force $q = Q/L$ as indicated. The general shape of the equilibrium configuration is also indicated. We, of course, do not know the exact function $v(x)$.

Let us assume that we have performed an analysis of the structure and now we want to compute the displacement v_b and slope θ_b at $x = 2L/3$. To start with we need the actual (real) deformations $\kappa(x)$. We have already determined the parabolic moment distribution function $m(x)$ shown in Fig. 6.11a1, and the deformations $\kappa(x) = (1/EI)\ m(x)$ are the same function in a different scale.

To compute v_b, we apply a unit vertical force along v_b and in the sense of v_b (see Fig. 6.11b0), and indicate (black arrows) where we can apply forces to equilibrate the unit force. The structure is statically determinate, so there is only one configuration of auto-equilibrated forces and the corresponding distribution of bending moments, as shown in Fig. 6.11b1. Similarly, for the slope at b we apply a unit couple at b, and indicate where we can apply equilibrating forces (black arrows) (Fig. 6.11c0). The only possible forces and moment distribution (only one) are given in Fig. 6.11c1.

As a consequence, the unit-load formula (6.13) gives

$$v_b = \int_{\text{Beam}} \kappa(x)\, m^{v_b}(x)\, \mathrm{d}x; \qquad \theta_b = \int_{\text{Beam}} \kappa(x)\, m^{\theta_b}(x)\, \mathrm{d}x \tag{6.11}$$

In the absence of a formula for the integration of a product of a linear and a parabolic function one can perform the integration ad-hoc.

Setting $\lambda = x/L$ and noting that the curvature at mid-span is $\kappa(\lambda = 1/2) = 1/8(QL/EI)$, the parabolic curvature function can be written as

$$\kappa(x) = (1/2)\lambda(1 - \lambda)(QL/EI)$$

for an abscissa starting at a and an ordinate pointing down. Indeed, the polynomial is quadratic (a parabola); it is zero at $\lambda = 0$ and $\lambda = 1$, and at $\lambda = L/2$ its value is indeed $1/8(QL/EI)$. The equations for the virtual moments will be written in two parts, one for segment ab and the other for bd.

The moments for the displacement at b are, for segments ab and bd, $m^{v_b}_{(ab)} = (\lambda/3)L$ and $m^{v_b}_{(bd)} = (2/3)(1 - \lambda)L$, and the moments for the rotation at b are $m^{\theta_b}_{(ab)} = \lambda$ and $m^{\theta_b}_{(bd)} = (\lambda - 1)$. Allowing for $\mathrm{d}x = L\ \mathrm{d}\lambda$, the values of the displacement and rotation are respectively

$$v_b = (1/6)\left[\int_0^{2/3} \lambda^2(1 - \lambda)\ \mathrm{d}\lambda + 2\int_{2/3}^{1} \lambda(1 - \lambda)^2\ \mathrm{d}\lambda\right] QL^4/EI = 0.0113\, QL^3/EI$$

$$\theta_b = (1/2)\left[\int_0^{2/3} \lambda^2(1 - \lambda)\ \mathrm{d}\lambda - \int_{2/3}^{1} \lambda(1 - \lambda)^2\ \mathrm{d}\lambda\right] QL^3/EI = 0.02\ QL^2/EI$$

6.4.3 A Simple Frame

This example emphasizes that computing displacements in frames has nothing intriguing beyond the fact that in frames we neglect the deformations due to axial

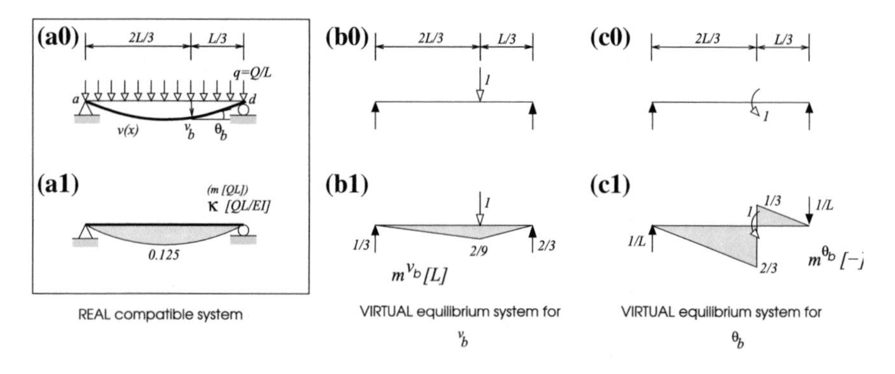

Fig. 6.11 Simple beam with distributed forces

elongations (they are significantly smaller than those due to bending) and, of course, the shear deformations, because we work under the Euler–Bernoulli bending assumption, which, by definition, has no shear deformation. Dismissing the axial deformations is the same as assuming that the axial stiffness of the elements is infinite. Since in addition we assume small displacements, the chord (distance between extremities) of such elements remains constant even when the element bends (slightly). More on this in the example.

The statically determinate frame $abcd$ in Fig. 6.12a is subjected to a horizontal force P at node c. Our aim is to compute the displacements of the dofs (degrees of freedom) at the nodes which will serve as a guideline for drawing a rough draft of the equilibrium configuration. Our method is of course the unit-load method, and for this we need to know the curvature distribution $\kappa(x)$ along the frame. For the time being, the curvatures originate from the bending moments $\kappa = m/EI$ (in the sequel, temperature effects will also cause curvature), so we need first to analyze the structure to determine $m(x)$. A full analysis produces $nsm(x)$ but if all we need are displacements the bending moments suffice.

Analysis

The reactions and moment distribution due to the external load are indicated in Fig. 6.12b. These are easy to come by. The applied force is equilibrated by an equal and opposite horizontal reaction at a. (Note, d is a horizontal roller which cannot impede the horizontal movement of the frame.) But now the two horizontal forces constitute a clockwise couple $P(2h) = 2Ph$. This couple is then equilibrated by two equal and opposite vertical forces $2P$ at a and d which together constitute an anticlockwise couple $(2P)h = 2Ph$. Vertical equilibrium is also satisfied and the reactions are therefore correct. The moment distribution is linear on every leg of the frame (no distributed forces). So we need, for instance, the end values to draw the lines. It is clear that the moments at a, c and d are zero and that the moment at b is $2Ph$, tension intrados. To establish this, consider the horizontal force P at a times

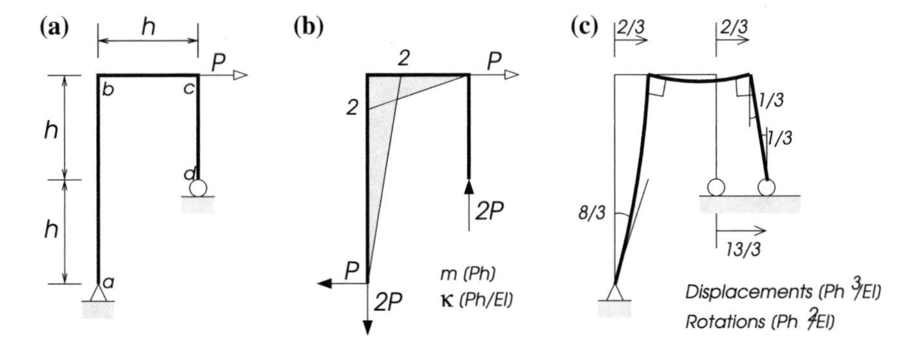

Fig. 6.12 Simple frame

the distance $2h$ to b on element ab, or the vertical force $2P$ at d times the distance h to b on element bcd (Fig. 6.12b). Note, the distance, as with every moment, is measured perpendicular to the line of action of the force.

This is all we need to draw the bending moment diagram (Fig. 6.12b). The units of the diagram are in Ph. We remind the reader that since the bending stiffness EI is uniform over the entire frame, the diagram can also serve as the diagram for the curvature in units Ph/EI.

Sample Displacements

All this was analysis. We will now determine displacements and rotations at the nodes of the frame. Since we assume $EA \to \infty$ for all the members of the structure, the unit load formula $\delta = \int_x [(m/EI)m^\delta + (n/EA)n^\delta]\mathrm{d}x$ reduces to $\delta = \int_x (m/EI)m^\delta \mathrm{d}x$. Because elements ab and cd do not elongate, we have immediately $v_b = v_c = 0$ and since bc does not elongate, forcibly $u_b = u_c$.

θ_a: In Fig. 6.13, we give the equilibrium systems for computing a sample of displacements and rotations, starting with θ_a, the only non-zero dof at a. To the free-floating frame, we apply a unit couple (white arrow) at a (chosen anti-clockwise), and equilibrate the couple with forces at the supports. The only possibility is the pair of vertical forces of magnitude $1/h$ applied at a and d (black arrows). Note, the forces are measured in $1/h$ because the moment is dimensionless, and indeed, $1/h \times h = 1$. The resulting moment diagram is named m^{θ_a} because it was built to compute θ_a. On every leg of the frame the moment is linear. Regarding cd you see immediately that the moment is zero along the entire span. For ab and bc, we note that to draw a straight line you need two points. The moment equals 1 (extrados) at a and 0 at c, so all we need is the value of the moment at b (it is the same on ba and bc) which is easy to compute.

This defines m^{θ_a}. The value of the rotation is $\theta_a = \int_a^b \kappa\, m^{\theta_a}\mathrm{d}x + \int_b^c \kappa\, m^{\theta_a}\mathrm{d}x$, the third integral being zero since both κ and m^{θ_a} are zero along cd. The functions being linear, we can use formula (6.5). Here, $g(x)$ stands for $\kappa(x)$ and $h(x)$ for m^{θ_a}.

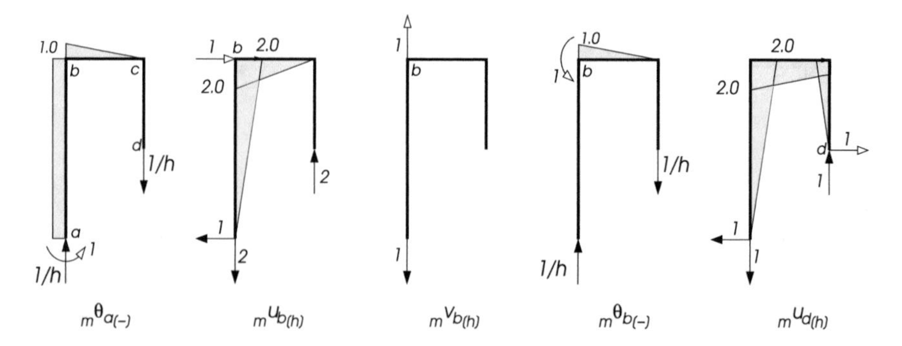

Fig. 6.13 Computing displacements

Since κ and m^{θ_a} appear on both sides of the coordinate axis we need to define the positive sense. We chose intrados as positive.

Consequently, $g_a = 0,$ $g_b = 2Ph/EI$, $g_c = 0$ and $h_a = -1$, $h_b = -1$, $h_c = 0$, which with $L_{ab} = 2h$, $L_{bc} = h$ yields $\theta_a = -(6/3 + 2/3)\ Ph^2/EI = -8/3\ Ph^2/EI$.

Note the values of g_a, g_b, g_c and $g_d = 0$ are taken from the real curvature and remain the same throughout all the computations.

u_b: For the horizontal displacement at b, we apply a unit force (white arrow) at b to the free floating frame, and equilibrate the force with forces (black arrows) applied to the locations of the supports (and only there). Disregarding the units, the forces are identical to those due to the external loads (Fig. 6.12a), and so are the moments. Note, the forces are dimensionless, consequently the moments which are $force \times length$ are measured in $length$, that is, h. The value of the displacement is $u_b = \int_a^b \kappa\ m^{u_b}\mathrm{d}x + \int_b^c \kappa\ m^{u_b}\mathrm{d}x$, the third integral being zero since $\kappa = 0$ along cd. The functions are linear and we can again use Fig. 6.6. We note, $h_a = 0$, $h_b = 2h$, $h_c = 0$, that is, $u_b = (8/3 + 4/3)\ Ph^3/EI = 4\ Ph^3/EI$.

We have already established that the vertical displacement v_b must be zero because element ab is infinitely stiff axially. The unit-load method numerically corroborates this physical requirement. We apply a unit vertical force to the free-floating structure at b, which can only be equilibrated by an equal and opposite vertical force at a. (Remember that we are allowed to apply forces only at the supports.) Now these two unit forces give no bending moments $m^{v_b} = 0$, only a unit axial tension force in ab. As a result $v_b = 0$.

θ_b: The rotation θ_b is computed with the help of the moment function m^{θ_b} due to a unit couple at b. The equilibrating forces (black forces at the supports) form an opposite unit couple shown in Fig. 6.13. The resulting moment diagram is obvious. The moment is zero along the two columns. For bc the moment varies linearly between the known end-values 1 and 0. For the computation of the rotation we have $h_b = -1$, $h_c = 0$ which yields $\theta_b = -(2/3)\ Ph^2/EI$ (radians). We would like to draw attention to the negative sign of h_b. If we use for intrados a positive value then extrados values are negative.

u_d: For the horizontal displacement at d, the moment diagram is the last drawing in Fig. 6.13. Using $h_a = 0$, $h_b = 2$ and $h_c = 1$, we obtain $u_d = (13/3)\ Ph^3/EI$. And to complete the picture, we find $\theta_c = \theta_d = (1/3)\ Ph^2/EI$ (note, cd remains a straight line because nothing bends it; there are no bending moments along cd). This can be computed directly through (6.4), but also from the geometry of the displacements. Indeed, $u_d = u_c + L \tan \theta_c$ which for small displacements is $u_d = u_c + L\ \theta_c$ or also $\theta_c = (u_d - u_c)/L$.

Equilibrium Configuration

On the basis of the movements of the nodes, we can sketch the equilibrium configuration of the frame (Fig. 6.12c). Bear in mind that this displaced configuration is the *only* way for the frame to equilibrate the horizontal applied force P.

Fig. 6.14 Doing it wrong

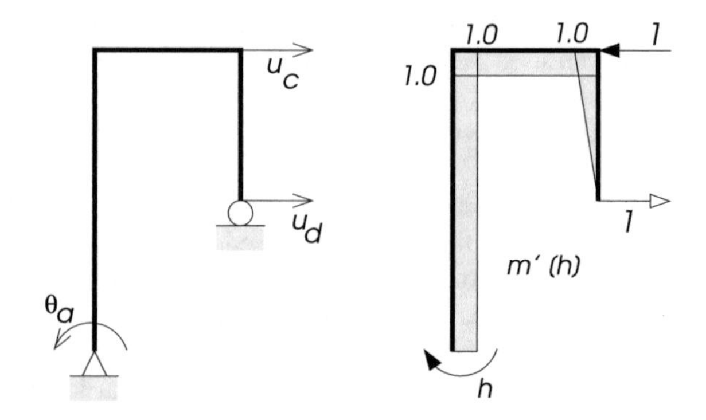

6.4.4 *Ineffective Equilibrium Systems*

If for computing u_d we decide to balance the unit force applied to d by an opposite force at c and a counter-clockwise couple at a, as in Fig. 6.14, you will encounter problems.

The applied loads at a, c and d with the moment diagram m' constitute a valid equilibrium system, which can be combined with the real deformations and displacements in a virtual work equation. Indeed, the virtual work equation will now yield $-h\theta_a - u_c + u_d = \int_x \kappa\ m'\mathrm{d}x$ which, with $h_a = 1$, $h_b = 1$ and $h_c = 1$, yields $-h\theta_a - u_c + u_d = 3\ Ph^3/EI$. There is nothing wrong with this equation. We can substitute for θ_a, u_c and u_d, and verify that the left-hand side is also $3\ Ph^3/EI$. However, for computing u_d the equation is useless, because of the presence of two unknown quantities θ_a and u_c.

The 'trick' of the unit-load method is to react the unit load by loads positioned were the supports are in the real structure. In the virtual work expression of the external loads only the unit load will then appear. In this 'wrong' example we have a couple h at extremity a which is free to rotate; so the θ_a term will appear in the work equation, and we have a unit force at node c which can move horizontally, hence the term in u_c.

6.5 Statically Redundant Examples

We have not yet formally introduced the notion of redundancy. Let us only mention that, for our purposes, the number of possible loads at the locations of supports exceeds two in the case of beams and three for frames. We shall see that this entails a choice in determining the external forces in equilibrium with the unit load.

6.5.1 The Propped Cantilever

The propped cantilever of uniform stiffness EI, fixed on one side and simply supported on the other, is subjected to a concentrated down force Q applied at mid-span as shown in Fig. 6.15(1). The structure is redundant of order $R = 1$. Indeed, remove the roller (one reaction), and we have a cantilever which is known to be statically determinate. The force method will enable us to determine the internal forces and deformations of this redundant structure, however, at present we are given the curvature distribution (Fig. 6.15(2)).

Note that the generalized constitutive law of a beam element in bending is $m = EI\ \kappa$ and, since EI is constant, the curvature distribution $\kappa(x)$ in Fig. 6.15(2) is also the distribution of the bending moments $m(x)$ in a different scale (moments in QL and curvatures in QL/EI).

Computing a Displacement

The 'real' compatible system is taken from the problem at hand (Fig. 6.15(1, 2)).

The equilibrium system is virtual. It is constructed in the following way. To calculate v_b, we apply a unit force at b in the direction of v_b, and we equilibrate the force by means of loads applied at the reactions points of the structure (black arrows in Fig. 6.15b0, that is, C_a, F_a and F_d).

The virtual work equation becomes

$$C_a\theta_a + F_a v_a + (1)\delta_b + F_d v_d = \int_x m^{v_b}(x)\ \kappa(x)\ \mathrm{d}x \tag{6.12}$$

Here δ_b is considered positive in the sense of the applied unit load. If we look at the equilibrium configuration in Fig. 6.15(1), we will notice that C_a, F_a and F_b do not

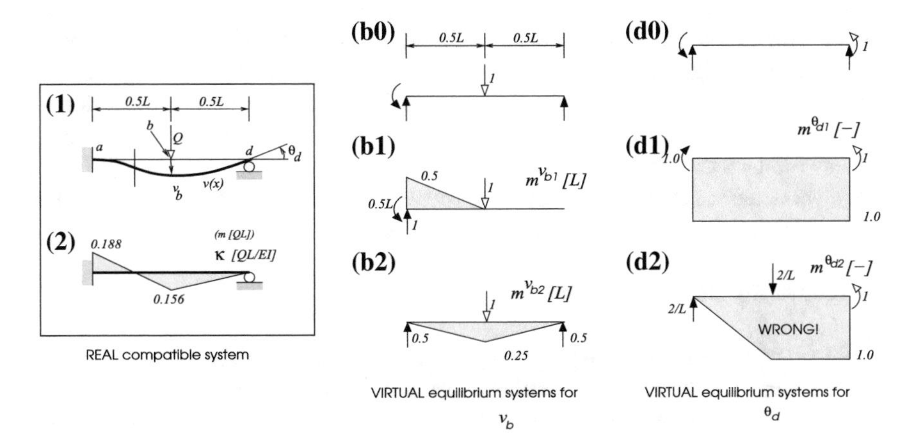

Fig. 6.15 Redundant beam

work because the corresponding displacements θ_a, v_a and v_b are null. Consequently, (6.12) reduces to what is called Mohr's formula

$$v_b = \int_x m^{v_b}(x)\ \kappa(x)\ \mathrm{d}x \tag{6.13}$$

where δ_i is the displacement at degree of freedom i due to a unit force at i, $m^i(x)$ is *any* moment distribution in equilibrium with the unit force at i and $\kappa(x)$ is, we will recall, the real deformations of the structure.

Now, the structure is redundant and there are many combinations of reactions which can equilibrate the unit force at mid-span. Two are shown in Fig. 6.15b1, 2. The corresponding moment distributions $m^{v_{b1}}$ and $m^{v_{b2}}$ are also indicated. We call any of these distributions m^{v_b} because whatever follows is true for both distributions. (Note, a unit force means a force without dimension. In this nomenclature, forces are dimensionless and couples and the moment distribution have the dimension of a length.)

If you are mystified by (6.13) you are in good company. It is counterintuitive to expect any equilibrium moment distribution $m^i(x)$ when inserted under the integration sign to produce the same v_i. But you can check the result: $\int_0^L m^{v_{b1}}\ \kappa(x)\ \mathrm{d}x = \int_0^L m^{v_{b2}}\ \kappa(x)\ \mathrm{d}x$. In time you will come to terms with this relation. It all has to do with finding enough anchor points to fix a deformed curve in the plane.

Numerically it is easier to use the integral with $m^{v_{b1}}$ than with $m^{v_{b2}}$. Using $m^{v_{b1}}$ with (6.4), we have $L_{ab} = 0.5L$, $h_a = 0.5$, $h_b = 0$ in units L and $g_a = 0.188$, $g_b = -0.156$ in units QL/EI and consequently $v_b = 0.00917\ QL^3/EI$.

Computing a Rotation

To compute the slope θ_d of the beam at extremity d, the unit-load method calls for applying a unit couple at d and finding loads at the reactions points which equilibrate the unit couple (see C_a, F_a and F_b in Fig. 6.15d0). Since we have three loads for two equilibrium equations there are many such equilibrium systems, one very simple one is given in Fig. 6.15d1 with $C_a = 1$ and $F_a = F_b = 0$. With an applied unit couple, the moment diagram is dimensionless and the reactions have a dimension of $1/length$. When we now write the virtual work equation with the real deformations and any moment distribution in equilibrium with a unit couple at i (rotation θ_d in our case), we obtain essentially the same generic equation (6.13) where the term on the left side is now a rotation: the slope of the section at d.

Ineffective Equilibrium System

In Fig. 6.15d2, we have an example of an ineffective equilibrium system. We can imagine very well that the counterclockwise unit couple at d is equilibrated by the clockwise couple of forces of magnitude $2/L$ at a distance $L/2$ ($2/L \times L/2 = 1$). The related moment distribution is $m^{\theta_{d2}}$. For this equilibrium system we can also write a virtual work equation

$$- 2/L \times v_b + \theta_d = \int_x m^{\theta_{d2}}(x)\ \kappa(x)\ \mathrm{d}x \tag{6.14}$$

This equation is correct but not very useful in computing θ_d. Indeed, there is 'noise' on the *left-hand side* of the equation in the form of v_b, a quantity we do not necessarily know. The method goes wrong because we have used a force which was not at a zero displacement point (force $2/L$ at b). In the unit-load method the unit load must be equilibrated with forces applied at the reactions.

6.5.2 A Redundant Truss

When we consider first principles the unit-load method has the form

$$\delta_i = \int_x \boldsymbol{\epsilon}(x)^T \boldsymbol{n}^i(x)\ \mathrm{d}x = \int_x (\epsilon(x)\ n^i(x) + \gamma(x)\ s^i(x) + \kappa(x)\ m^i(x))\ \mathrm{d}x \tag{6.15}$$

where the *right-hand side* makes allowance for all the deformations.

By definition, however, a truss has either no or negligible bending moments and shears. As a result, (6.15) reduces to

$$\delta_i = \int_x \epsilon(x)\ n^i(x)\ \mathrm{d}x \tag{6.16}$$

where ϵ are the real axial strains in the elements and $n^i(x)$ are normal forces at any location x in equilibrium with a unit force applied along the nodal degree of freedom i. Note, the degree of freedom must be attached to a node, and δ_i is the component of the nodal displacement along the degree of freedom i.

Now, a truss is composed of (slender) elements, so (6.16) is also

$$\delta_i = \sum_j \int_{x_j} \epsilon_j(x_j)\ n^i_j(x_j)\ \mathrm{d}x_j \tag{6.17}$$

where x_j is the local coordinate of element j, and j runs over all the members of the truss.

The normal force is constant in a truss $n^i_j(x_j) = n^i_j$ and, if the actual axial strain is constant, $\epsilon_j(x_j) = \epsilon_j$, then (6.17) reduces to the familiar form

$$\delta_i = \sum_j \epsilon_j\ n^i_j\ L_j \quad \text{(or also} \quad = \sum_j e_j\ n^i_j \ = \boldsymbol{e}^T \boldsymbol{n}^i) \tag{6.18}$$

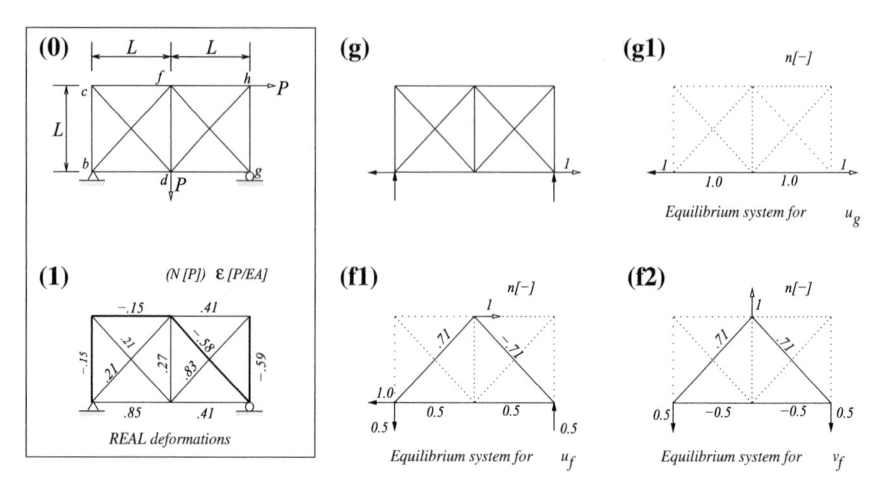

Fig. 6.16 Redundant truss

where L_j is the length of element j, $e_j = \epsilon_j L_j$ is the elongation of bar j and $\boldsymbol{e}$ and $\boldsymbol{n}^i$ are vectors of components e_j and n^i_j respectively.

Consider, for instance, the truss in Fig. 6.16(0) with constant axial stiffness EA of all its elements. This structure of redundancy $R = 2$ is analyzed elsewhere; the normal forces are indicated in Fig. 6.16(1).

We are now asked to compute the nodal displacements at g and f. With the unit-load method, we are in need of the actual deformations ϵ_j of the truss. This is easy. Following Hooke's law, the axial strains are simply the normal forces divided by EA.

Now, node g can only move horizontally, so we apply a unit horizontal force (the white arrow) at g and construct an equilibrium system. The possible external forces are shown by black arrows in Fig. 6.16g. Incidently, there is only one possible solution for the black arrows. Such structural arrangements are sometimes viewed as externally determinate and internally redundant, because there are many internal forces arrangements to create equilibrium, for instance, the one given in Fig. 6.16g1. In the examples, we have marked by thick lines the selected distribution of virtual internal forces in equilibrium with the external virtual loads. The other bars have zero virtual forces. Consequently, for the horizontal displacement u_g, (6.18) gives

$$u_g = \epsilon_{bd}\, n^{u_g}_{bd} L_{bd} + \epsilon_{dg}\, n^{u_g}_{dg} L_{dg} = \left(0.85 \frac{P}{EA} \times 1.0 \times L\right) + \left(0.41 \frac{P}{EA} \times 1.0 \times L\right)$$
$$= 1.26 \frac{PL}{EA} \tag{6.19}$$

Node f has two degrees of freedom (it is 'free' to move on the plane), we will therefore establish separately the component of the displacement in the horizontal u_f and in the vertical v_f directions. There are a multitude of possible equilibrium systems.

The ones selected are shown respectively in Fig. 6.16f1 (horizontal component of the displacement) and in Fig. 6.16f2 for the vertical component of the displacement.

The unit load is the white arrow and the equilibrating forces have black arrowheads.

Applying (6.18) yields the components of the displacement at f

$$\begin{aligned} u_f &= \epsilon_{bd}\, n_{bd}^{u_f} L_{bd} + \epsilon_{dg}\, n_{dg}^{u_f} L_{dg} + \epsilon_{bf}\, n_{bf}^{u_f} L_{bf} + \epsilon_{fg}\, n_{fg}^{u_f} L_{fg} \\ &= \left(\tfrac{0.85P}{EA} \times 0.5\right) L + \left(\tfrac{0.41P}{EA} \times 0.5\right) L + \left(\tfrac{0.21P}{EA} \times 0.71\right) 1.41L \\ &\qquad + \left(\tfrac{-0.58P}{EA} \times -0.71\right) 1.41L \\ v_f &= \epsilon_{bd}\, n_{bd}^{v_f} L_{bd} + \epsilon_{dg}\, n_{dg}^{v_f} L_{dg} + \epsilon_{bf}\, n_{bf}^{v_f} L_{bf} + \epsilon_{fg}\, n_{fg}^{v_f} L_{fg} \\ &= \left(\tfrac{0.85P}{EA} \times -0.5\right) L + \left(\tfrac{0.41P}{EA} \times -0.5\right) L + \left(\tfrac{0.21P}{EA} \times 0.71\right) 1.41L \\ &\qquad + \left(\tfrac{-0.58P}{EA} \times 0.71\right) 1.41L \end{aligned} \tag{6.20}$$

or

$$u_f = 1.42\frac{PL}{EA} \qquad v_f = -1.0\frac{PL}{EA} \tag{6.21}$$

Since the vertical unit load at f points upwards, the negative sign of v_f means that the node actually moves down to reach its equilibrium position.

6.5.3 A Portal Frame

In Fig. 6.17(0), we have a portal frame with both columns built-in subjected to a side load P at node c. This $R = 3$ structure is still off limits for us, so we have to go back to the future and bring the bending moment distribution along (Fig. 6.17(1)), which for a uniform EI is also the distribution of the bending deformation (curvature) $\kappa(x)$.

Since we know the deformations, we can compute displacements and rotations here and there, such as the rotation θ_b at joint b and the horizontal displacement u_c at c. For the rotation θ_b, we apply a couple of unit magnitude at b (you decide on the sense) and find appropriate forces/couples for the reactions at a and d and their corresponding moment distribution m^{θ_b}. This structure is restrained from moving at a and d, so we can apply all possible forces there (see black arrows in Fig. 6.17b).

Two instances are shown in Fig. 6.17b1, b2. The white arrow is the applied load of unit magnitude and the black arrows equilibrate the unit load. We can use any one of the virtual moment distributions in Fig. 6.17b1, b2. When plugged into the unit-load formula along with the real deformations $\kappa(x)$ in Fig. 6.17(1), they all should yield $\theta_b = 0.0355PL^2/EI$. The result is in radians and PL^2/EI is indeed without units.

Similarly, for the horizontal displacement at c we apply a unit horizontal force at c, with the ensuing equilibrium moment distribution m^{u_c}. The equilibrium distribution in Fig. 6.17c1 is an obvious choice. Since we happen to know the real moment

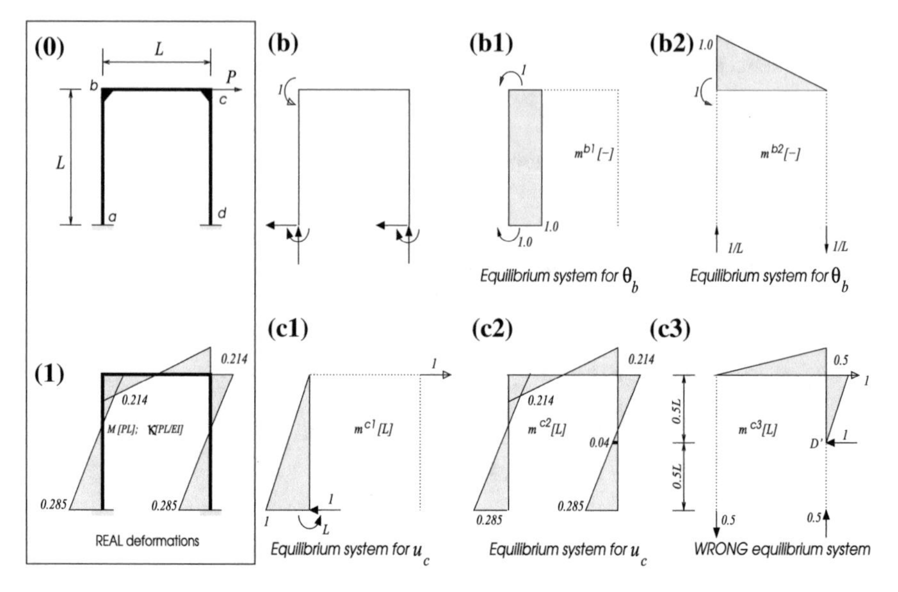

Fig. 6.17 Redundant frame

distribution $m(x)$ due to a force P at b (see Fig. 6.17(1)), this distribution being obviously in equilibrium with P, it follows that $m^{u_{c2}}(x) = m(x)/P$ in Fig. 6.17c2 is also an equilibrium solution for a unit force at c. The integration will be longer

$$u_c = \int_a^b \kappa\, m^{u_{c1}}\, \mathrm{d}x = \int_a^b \kappa\, m^{u_{c2}}\, \mathrm{d}x + \int_b^c \kappa\, m^{u_{c2}}\, \mathrm{d}x + \int_c^d \kappa\, m^{u_{c2}}\, \mathrm{d}x \quad (6.22)$$

so, give it some thought before embarking with complicated equilibrium distributions.

Finally, the system in Fig. 6.17c3 is a valid equilibrium system but of little use to compute u_c. Although the black arrows equilibrate the unit force, the black arrow at d' (mid-way between c and d) will add noise to the virtual work equation $u_c - u_{d'} = \int_x \kappa m^{c3} \mathrm{d}x$, assuming that $u_{d'}$ is positive to the right.

6.6 Summing Up

The unit-load method is a technique for computing a displacement or a rotation along the structure once the distribution of the deformations, that is, curvature $\kappa(x)$ for bending elements and axial strain $\epsilon(x)$ for trusses, is known. It is based on the principle of virtual work, where the deformations of the compatible system are the actual deformations of the structure, and the virtual equilibrium system comprises special tailored loads in equilibrium with a unit load in the direction of the sought displacement or rotation.

Part II
The Structures

In Part II we consider entire structures, that is, interconnected structural elements supported in a proper manner and subjected to external loads. We describe the different sorts of structural elements, connectors and supports in Chap. 7, followed in Chap. 8 by an example of the analysis of a simple structure by the two main methods of analysis.

In Chap. 9, we provide guidelines for predicting the graphic form of the expected $nsm(x)$ distributions and reactions without actually performing the analysis.

Chapter 7
Types of Structures

Structures are composed of prismatic shaped elements, such as bars, beams, and columns, rigidly or partially connected to one another or to the ground. We will classify these structures by the type of internal loads found inside the structure when external loads are applied to them or when they are subjected to temperature fluctuations. The most general type of skeletal structures are frames, a linear subset of which are beams. Trusses, which have only normal internal forces, constitute a class apart.

This chapter also discusses the type of connections that can exist between elements and between the structure and ground.

7.1 The Frame, the Beam and the Truss

Skeletal structures come in several types and flavors, and there are several ways to classify them. We will group them according to the type of internal loads they produce. This leads to three well-known archetypes:

- frames (including arches), internal loads nsm
- beams (including grids), internal loads sm
- trusses (ideal and real), internal forces n

as shown in Fig. 7.1. The three types of structures look different but the adherence of any particular structure to one of the groups depends not only on its appearance but also on the external loading. What differentiates the structures is the type of internal loads found in them.

Frames (nsm) are the most versatile structures. General structures under general loadings are frames. They constitute the lion's share of the skeletal structures. At every section along the structure one can expect the three internal forces: normal forces n, shear forces s and bending moments m.

Arches ($n - sm$) likewise carry all three types of internal forces, but their curved form is purposely designed to reduce the magnitude of the bending moments at the

M.B. Fuchs, *Structures and Their Analysis*,
DOI 10.1007/978-3-319-31081-7_7

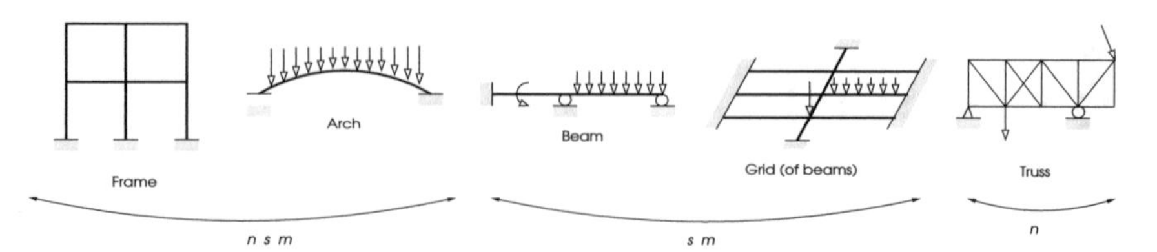

Fig. 7.1 Classification of skeletal structures

expense of the compressive normal forces when subjected to gravitational loads, as when used as the main supporting structure in bridges. An arch can in principle support all types of loading but they are usually designed to carry vertical forces: masonry in ancient times, or speeding trains and crawling traffic in contemporary applications. The compressive normal forces in its arch-like form allows it to resist the loads with relatively small bending moments.

The ancients found that for their building material (stone), reducing the bending moment was imperative. Flexion has compression along the outer fibers on one side of the beam but tension on the opposite side; hence the arch solution with its high compression forces along the span of the structure. (Reducing the bending moments is arguably still the main purpose of structural design.)

Beams (sm) are more restricted; they carry primarily bending moments m and shear forces s. Beams are straight linear structures on a set of supports and can be subjected to point couples, lateral point forces and lateral distributed forces. The forces should not have an axial component, that is, a component parallel to the beam axis, lest it induces normal internal forces. This is important. Restricting the beam to couples and to forces perpendicular to the beam axis allows us to focus on the main task of the beam, which is to resist bending moments. If a beam is subjected to external loads which also cause normal internal forces, it can always be treated as a frame element.

Grid of beams (sm) are typically used to cover large spaces. These are plane structures usually made of beams arranged along an orthogonal grid. One way of studying such structures is to consider each separate beam as being submitted to the directly applied loads and to the interaction forces with the orthogonal elements attached to it. We will assume that the beams have open sections, such as I-sections, which have low torsional stiffness. When subjected to lateral forces the internal forces in every beam will be primarily shear and bending moments.

Trusses (n), the third type of structure, are rather apart from the others (see Fig. 7.2). To call a structure a truss several conditions must concur. The elements must be slender and straight, they usually follow a triangulated pattern, and the loading is restricted to forces applied at the connections between elements (nodes). (If couples or distributed forces are present or if the forces are applied directly on the elements the structure is not a truss.) With these limitations the bending moments are known to be small, and the only internal forces of any consequence will be the normal forces, which complies with the requirements for belonging to the truss family.

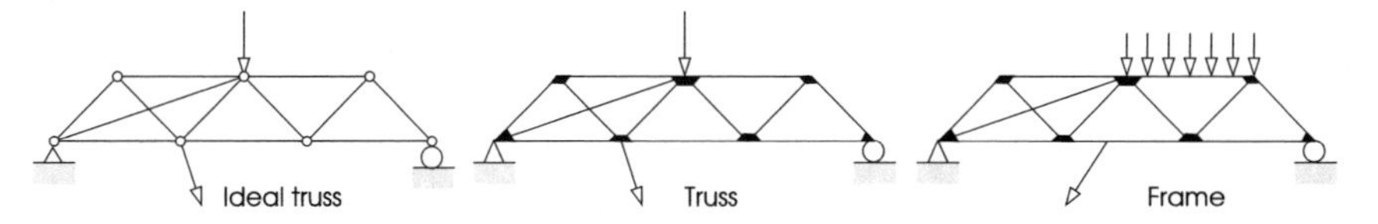

Fig. 7.2 Ideal and not ideal trusses

We sometimes consider *ideal trusses* where the elements are frictionless pin-connected to one another and to the reactions (Fig. 7.2a). In such ideal structures the shear forces and bending moments are identically zero, and the only non-zero internal forces are strictly normal forces. These are for the most part theoretical constructs with very few practical examples. In practice trusses are not ideal because the elements are rigidly connected instead of pin-connected. But if the elements are long and slender enough the bending moments are relatively small and are often neglected. Consequently, the structure can be treated as a truss (Fig. 7.2b).

Note, the same structure, when submitted to lateral forces along the elements (Fig. 7.2c) is not a truss because the elements will experience bending. It will be a frame, as we will call any skeletal structure which is neither a beam (or grid of beams) nor a truss.

All this assumes, of course, that the structures are stable, an issue which we will address shortly.

7.2 Full Connectivity

Consider the equilibrium configuration of the two-storey frame in Fig. 7.3 which is subjected to a horizontal thrust at node *e* with the initial structure in the background. When you look at point g on the upper beam of the deformed structure you will notice that the line there is continuous and so is the slope or *first* derivative. In mathematical terms we say that the curve is C1-continuous at g which means also that the tangent to the right of g and the tangent to the left at that point are identical (see column 'elastic' at point g in Fig. 7.3). This excludes the occurrence of a kink, an abrupt change of slope at a point in the structure, as shown in column 'plastic' at point g. If this happens the structure has locally exceeded the elastic limit and has failed. It will not return to its original shape after removal of the loads. We tacitly assume that the structures are strong enough to resist the applied loads in an elastic manner, and that plastic deformation will never occur. In other words, the structures will remain C0-continuous (they will not break) but also C1-continuous in the deformed configuration (continuous slope[1]).

[1] However, the deformed structures will usually not be C2-continuous. Indeed the second derivative, or curvature, changes abruptly when a couple is applied. A sudden change of curvature is difficult to observe with the naked eye but it often is present in the equilibrium configuration.

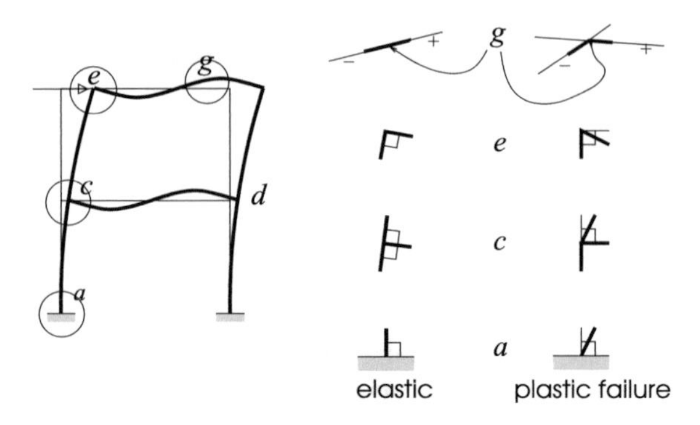

Fig. 7.3 Continuity

Now what about junction e? The slope of line ceg is already discontinuous at e in the initial configuration. We will therefore broaden our definition of C1-continuity and require that the tangents on both sides at a point on any line of the structure have the same relative angle between them, both in the deformed state and in the initial one. This satisfies the continuity at g. Initially the angle between left and right was 180° and it has to remain so in the deformed shape. But now it also satisfies junction e on cef. Initially the relative angle was 90° and it will remain so in the equilibrium configuration. This junction moves and rotates under loading but locally it will do so as a rigid body (see column 'elastic' in Fig. 7.3).

The same observation is valid for junction c. All the curves passing through c must retain their relative angles at c. For instance, the angles acd and ecd at c are right angles in the original and in the equilibrium configuration.

At a you may imagine a piece of structure inside the wall. Consequently, at a the relative angle stays at 180°. This means that for a built-in member, the tangent must remain perpendicular to the wall in the deformed configuration.

On the other hand, all the cases in column 'plastic' of Fig. 7.3 are examples of failures due to plastic flow. Although the elements may not have broken, the structures will never reassume their original configuration. This should not occur in well-designed structures unless the loading went beyond specifications. In no case will the structures in this book exceed the elastic limit.

7.3 Partial Connections Between Elements (Releases)

The structure in Fig. 7.3 is rather specific. It has all its members fully connected to its neighborhood, be it other members or the ground. Not all structures have such firm connections at all the junctions. Some members can be partially connected to their neighbors or to the ground.

Consider the elements rigidly attached to a node in Fig. 7.4. We assign numbers 1 and 2 respectively to the horizontal and vertical translations and 3 to the rotation

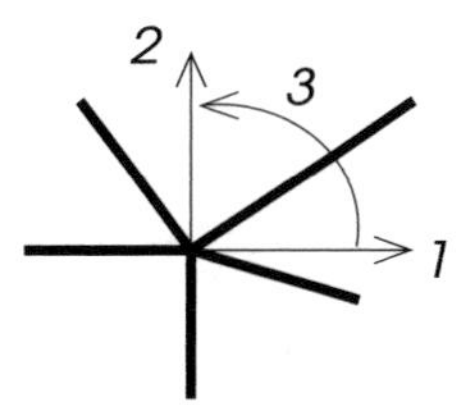

Fig. 7.4 Rigid connections

Table 7.1 Local releases

No.	Name	Releases	Description
1	Telescopic connection	Normal release	Independent axial translations
2	Lateral guide	Shear release	Independent lateral translations
3	Hinge	Moment release	Independent rotations

of the node (junction). The extremities of members which are fully connected to the node translate and rotate together with the node. However, elements can also be partially connected to nodes.

There exist three elementary partial connectors (Table 7.1).

The accepted symbols for these partial connectors is shown in Fig. 7.5a. Columns (b) of the figure shows the open degrees of freedom. Column (c) gives examples of instances where one side of the connector is fixed. This is, in fact, usually the case. Finally, column Fig. 7.5d is for particular cases or alternative symbols.

Two members joined by means of a *telescopic connection* (row 1 in Fig. 7.5) will have independent axial displacements (1) on both sides of the connector (as shown in column (b)). However, the lateral translation (2) and rotation (3) will be the same.

Similarly, when two members are attached by means of a *lateral guide* (row 2 in Fig. 7.5) the lateral translations (2) on both sides of the connector can be different. However, the axial translation (1) and the rotation (3) at the connection will be the same for both members.

Finally, a *hinge* (row 3 in Fig. 7.5) will allow for independent rotations (3) at the connection but the extremities will translate (1, 2) in the usual fashion.

In columns (c) and (d) we have a classical hinge where the translations are fixed but the hinge allows rotation of the element with respect to ground.

There exist connections with two releases. These double releases do not occur very frequently, except for the combination 2 and 3 (last line in Fig. 7.5). This is called a *roller*. We use such rollers very often in bridges (see ‘ideal truss’ in Fig. 7.8) to allow for free thermal expansion.

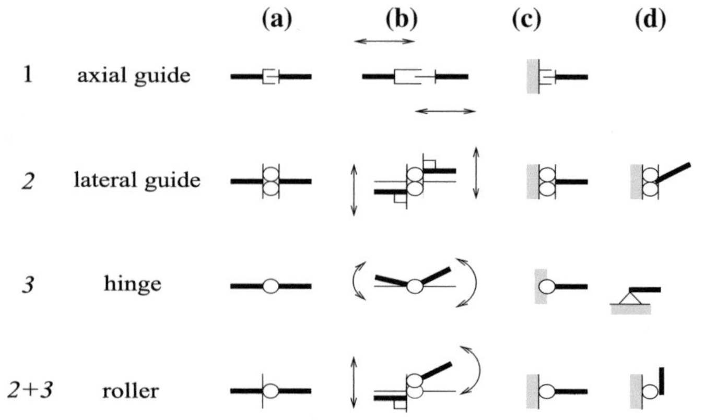

(a) (b) (c) (d)

1 axial guide

2 lateral guide

3 hinge

2+3 roller

Fig. 7.5 Connectors

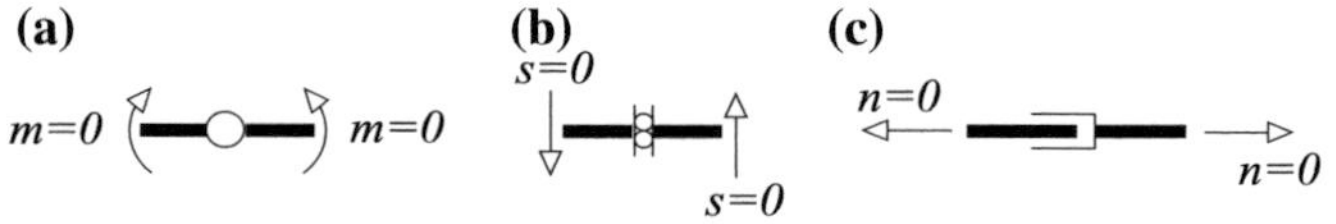

(a) $m=0$ $m=0$ (b) $s=0$ $s=0$ (c) $n=0$ $n=0$

Fig. 7.6 Zero moment, shear, normal force

Releases: Partial connections are also called *releases* (see column 'Releases' in Table 7.1 and Fig. 7.6). It is evident that at a cross-section in the vicinity of an axial guide, Fig. 7.6c, the normal force must be zero ($n = 0$). Indeed, if the axial force to the right of the release, for instance, were different from zero, the part would simply move in the direction of the force, which contradicts the concept of equilibrium. Hence the axial guide necessarily *releases* or annuls the adjacent normal force. Similarly, in the close neighborhood of a lateral guide the shear force is necessarily zero ($s = 0$), which makes it a *shear (force) release* (Fig. 7.6b).

For same reasons a hinge ($m = 0$) is a *(bending) moment release* (Fig. 7.6a). Indeed, if the moment to the right in Fig. 7.6, for instance, were different from zero, the right winglet connected to a frictionless hinge would rotate indefinitely around the hinge.[2]

We have mentioned that these partial connections can also be placed between a member and the ground. Here the physical aspects of the releases is more evident. One side of the connector does not move (it is grounded). So the extremities of the members attached to it will be fixed, except for the particular movement which was released.

[2]It does rotate, but once the equilibrium configuration is reached, the structure should be at rest and hence the bending moment should be zero.

7.4 Partial Connections Between Elements and the Ground (Supports)

Partial connections between elements exist also between elements and ground. There they define the type of supports.

In Fig. 7.7 we show some classical supports for structures. The numbers in the first column indicate the degrees of freedom (dofs) which are free (released, open, etc.). In those directions external forces can be applied (see white arrows in the second column). In the direction of the movements which are *fixed* the ground will apply reactive *forces* (black arrows in second column).

Finally, in Fig. 7.8 we depict a few typical structures with partial connections between members and between members and the ground. The first example is perhaps the most typical: a *simply supported beam*. The beam is hinged at a and has a roller at b.

Members in structures need not be straight. Arches are typical structures with curved members and the one in the figure is a very special one. The two members are hinged to the ground and to one another. Hence its name, *three-hinged arch*. This arch was very popular at one time because, being statically determinate (see further on) it deforms freely under temperature variations, without creating internal forces. It was used, and still is to some extent, for covering large areas such as railway platforms, exhibition halls, airport terminals, etc.

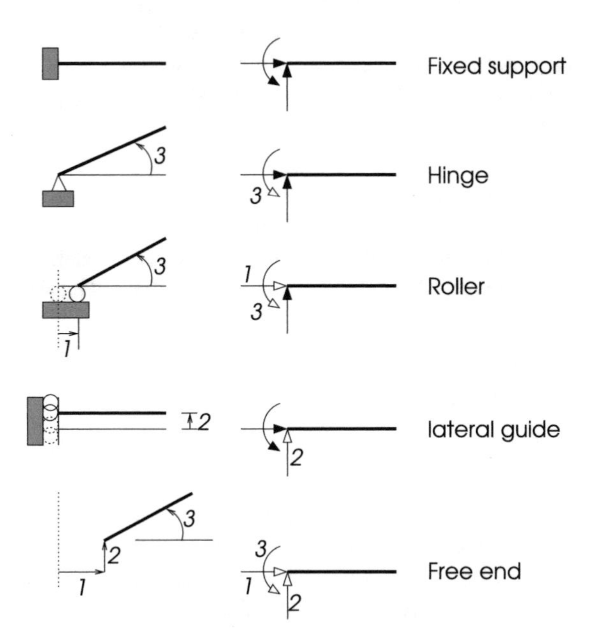

Fig. 7.7 Type of supports

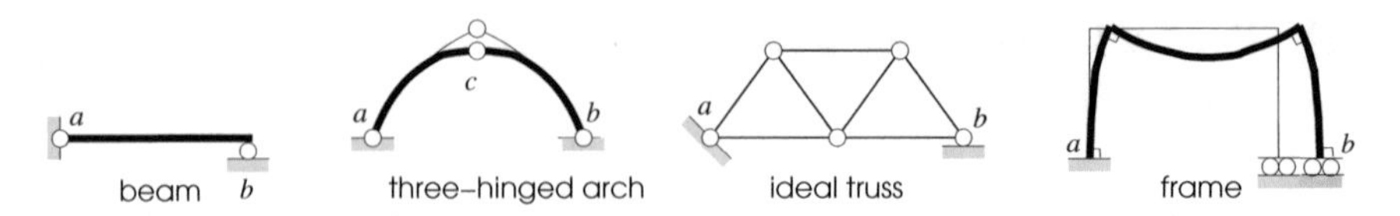

Fig. 7.8 Typical structures

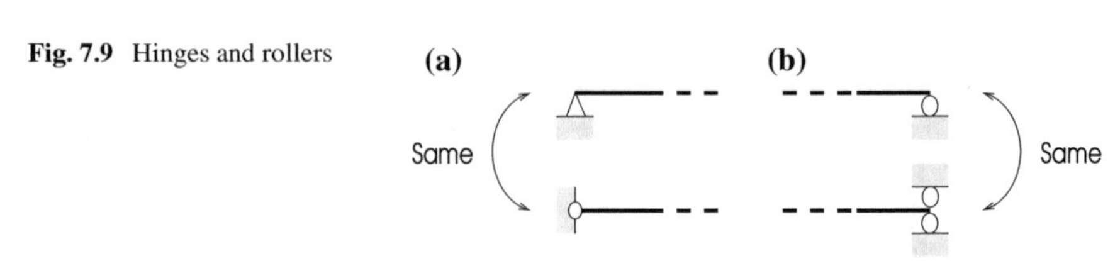

Fig. 7.9 Hinges and rollers

Next we have an *ideal truss* as in Fig. 7.2. Remember that in an ideal truss all the members are straight and hinged, and the loads are forces applied at the connections. The truss is hinged at a and lies on a horizontal roller at b.

The last structure in Fig. 7.8 is a frame, built-in or *clamped* at a and with a lateral guide at b. At b the structure can only move horizontally without rotation. A possible distorted shape under side forces is also depicted. All members must keep a 90° alignment with their neighbors at the connections, as indicated. You will note that the lateral guide behaves almost as a clamped support. However, it allows for a horizontal translation, keeping the alignment at 90°.

Although double releases are not very frequent (except for the roller) triple releases, that is, connections which have a normal release, a shear release and a moment release (1, 2, 3) occur all the time. These are simply free ends (last example in Fig. 7.7).

Note: In Fig. 7.9a we have two representations of a hinged support. They are identical. In both cases the support prevents horizontal and vertical translation but allows rotation. We will be using both.

Finally, the simple roller in Fig. 7.9b is the same as the double one below. Both prevent vertical movement in both directions. The simple roller will be used herein.

7.5 Summing Up

We have introduced the frame, beam and truss, three typical structures which we will learn to analyze. The versatility of these structures depends on the type of connections between their elements and between the structures and the ground.

We will, in the sequel, have ample opportunity to study the structural response of such structures and their internal forces.

Chapter 8
Structural Analysis

Analyzing a structure is synonymous to computing its internal forces. There are several approaches to the analysis of a structure but they can all be reduced to two: the force method and the displacement method. Whilst the former was historically paramount and is still important for simplified models of structures, the displacement method is nowadays the main analysis tool.

In this chapter we will analyze a simple structure by both methods to highlight their main features.

8.1 The Two Main Methods of Analysis

The aim of structural analysis is to compute the internal forces in structures including the reactions, and when required also the displacements or deformed shape. There exist two main methods of analysis: the Force or Flexibility method and the Displacement or Stiffness method.

The *Force method*, which determines only the internal forces is the traditional method of analysis and has been in use since the inception of structural analysis. It is suitable for small and mid-sized problems, and it requires a fair deal of structural insight to reduce the problems to tractable ones. If you happen to do some mental structural analysis while sipping a coffee at breakfast, the chances are that you will be thinking along Force method lines.

The *Displacement* method, on the other hand, computes the displacements first, even if you are not interested in them, and then the internal forces. With the advent of the computer in the second half of the 20th century the Stiffness (Displacement) method became the technique of choice. Its underlying principles are not that recent, but since the Displacement method requires a great deal of computation, it is not well suited for hand calculations; and its acceptance had to wait for algorithmic implementations. Computer programs are invariably based on the Displacement method and they can handle large and, indeed, very large structures in the blink of an eye. We will introduce the two methods for the simple case of a rigid beam on five supports,

M.B. Fuchs, *Structures and Their Analysis*,
DOI 10.1007/978-3-319-31081-7_8

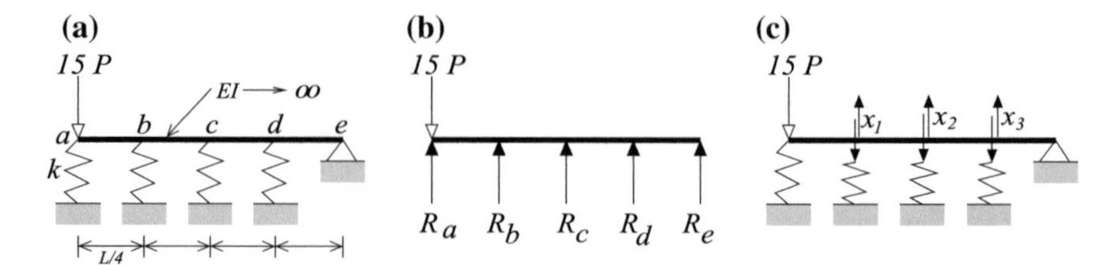

Fig. 8.1 Rigid beam on elastic supports

four elastic and one hinged. This may not sound like much of a structure but in truth, elastic structures are not much more than a spatial arrangement of various types of elastic springs.

In Fig. 8.1a we show a rigid beam hinged at the right extremity and supported by four equidistant elastic supports. A force $15P$ is applied to the beam at extremity a. The beam is said to be rigid because it is so over-designed that it will not deform to any significant degree under the applied force. The only elasticity of the structure resides in the elastic supports. They have each a stiffness k and consequently, if one compresses a spring with a force F it will contract by an amount $\delta = F/k$. (The spring is said to be linear because δ is a linear function of F.)

Analyzing a structure consists in first calculating the internal forces. In the present case, we will limit ourselves to obtaining the reactions by the two analysis techniques starting with the Force approach.

8.2 Analysis by the Force Method

8.2.1 Determining the Statical Redundancy

As a preamble to the Force method we need to determine the degree of static redundancy of the structure. In other words, can we calculate the internal forces using equilibrium equations only (statically determinate structures), or are we missing equations to do the job (redundant structures)? The number of missing equations is the degree of redundancy of the structure.

One way to do this is to disconnect the beam from all its supports and consider the equilibrium of the beam under the action of the external load and the forces that the five supports apply to it. The beam is now a free body subject to force $15P$ and the reactions R_a, R_b, R_c, R_d and R_e (Fig. 8.1b).

Five unknowns means we need five linear equations to compute them. Two are simple to come by: the equilibrium equations. For instance, the sum of the vertical forces must be zero $-15P + R_a + R_b + R_c + R_d + R_e = 0$ and the sum of the moments of all the forces with respect to some point, say e, must also be zero (we customarily take the counterclockwise direction as the positive sense in a moment equation) $(15P - R_a)L - R_b(0.75L) - R_c(0.5L) - R_d(0.25L) = 0$. The third equilibrium equation, the sum of the horizontal forces is zero, is automatically satisfied in our case (with beams we always make sure that there are no axial components.)

We have thus two equations for five unknowns shown in matrix form below

$$\begin{bmatrix} 1 & 1 & 1 & 1 & 1 \\ -L & -0.75L & -0.50L & -0.25L & 0 \end{bmatrix} \begin{Bmatrix} R_a \\ R_b \\ R_c \\ R_d \\ R_e \end{Bmatrix} = \begin{Bmatrix} 15P \\ -15PL \end{Bmatrix} \tag{8.1}$$

which is manifestly not enough to solve the equations. We are missing $(5 - 2 =)$ three equations. The structure is said to be statically redundant with redundancy $R = 3$; which means that when we try to analyze the structure using equilibrium equations only, we will be missing three equations.

This is where the Force method comes in handy. It will provide R additional equations: the compatibility equations.[1]

8.2.2 *Releasing the Structure*

We select the intermediate reactions R_b, R_c and R_d as the redundant reactions which will be the unknowns x_1, x_2 and x_3. To do this, we *release* the structure by disconnecting the corresponding springs from the beam, and replacing them by the forces applied by the springs to the beam (the reactions x_i, $i = 1, 2, 3$) and by the beam to the springs (the same x_i but in the opposite sense). The released structure is now subjected to $15P$ and to the, as yet unknown, equal and opposite forces x_i, seven forces in total, as shown in Fig. 8.1c. It should be clear that the released structure is statically determinate and can be analyzed using equilibrium equations only.

It is obvious that for just any set of values for the reactions, the springs will not connect to the beam. Let us, for instance, assume very large values for the x_i, as in the case of the figure. We notice that there are gaps between the extremities of the springs and the connecting points on the beam. In fact, for all sets of forces x_i, except for the values of the reactions, there will be discontinuities. (Why shouldn't there be?) Only for the exact values of the reactions will the gaps close.

We have here the three conditions for computing the x_i; the x_i should be such as to close the three gaps. These three conditions are called the compatibility equations. (We can always write as many compatibility equations as unknown redundants.)

[1] If the structure is statically determinate, that is, if the redundancy is zero, such as for a beam on two supports, there is no need for the Force method to compute the redundants: there simply are no redundants. The Force method is applicable only to compute the redundant forces.

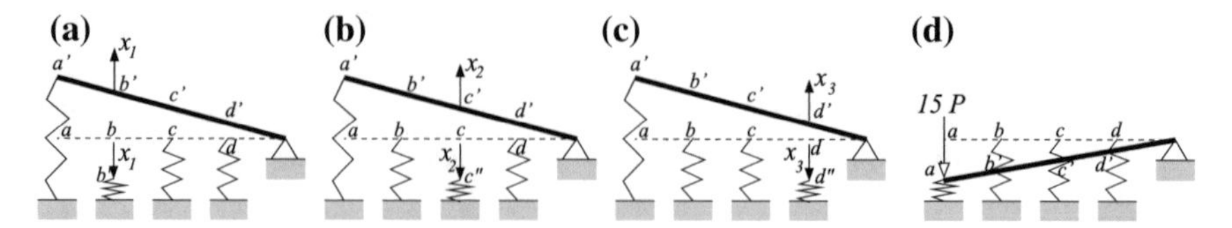

Fig. 8.2 Released beam

8.2.3 The Compatibility Equations

We now compute the gaps or discontinuities as a function of the reactions and the external load, and then we will write the condition for the gaps to be zero. A typical gap is the superposition of the discontinuities due to the three pairs of forces x_i and to P.

Let us start with the discontinuities due to the pair of forces x_1 (Fig. 8.2a). The released structure is a simple beam on two supports (support a is elastic and support e is hinged) with three unconnected springs. One x_1 pulls the beam up and the other x_1 compresses spring b. The gaps are then $bb' + bb''$, cc' and dd'. The beam being rigid the displacements of the beam at the intermediate springs can be expressed in terms of aa' (triangle ebb', for instance, is similar to triangle eaa'): $bb' = (3/4)aa'$, $cc' = (2/4)aa'$ and $dd' = (1/4)aa'$. It is easy to show that the reaction at a is $3x_1/4$ putting the spring in tension. Consequently, spring a elongates by $aa' = 3x_1/4k$. Noting that spring b is compressed by $aa'' = x_1/k$, the discontinuity at b due to x_1 is $\delta_1^1 = (9/16 + 1)x_1/k = (25/16)x_1/k$, and the discontinuities at c and d are respectively $\delta_2^1 = (6/16)x_1/k$ and $\delta_3^1 = (3/16)x_1/k$. (Note, δ_i^j is the discontinuity at i due to the pair of forces x_j.)

We can repeat this for computing the discontinuities due to the pair of forces x_2 (Fig. 8.2b). The beam is pulled up by x_2 at c' and the spring at c'' is compressed by x_2. Since the load is at the middle if the beam, it is obvious that the reaction at a is $2x_2/4$ putting the spring in tension and causing an elongation $aa' = 2x_2/4k$. The displacements of the beam at the intermediate springs are still $bb' = (3/4)aa'$, $cc' = (2/4)aa'$ and $dd' = (1/4)aa'$. Keeping in mind that x_2 also compresses the spring at c by $cc'' = x_2/k$ the gaps are now $\delta_1^2 = (6/16)x_2/k$, $\delta_2^2 = (4/16 + 1)x_2/k = (20/16)x_2/k$ and $\delta_3^2 = (2/16)x_3/k$.

We leave it as an exercise to obtain the third set (column) of discontinuities, $\delta_1^3 = (3/16)x_3/k$, $\delta_2^3 = (2/16)x_3/k$ and $\delta_3^3 = (17/16)x_3/k$ (see Fig. 8.2c).

So far we have determined the discontinuities due to the x_i. We need to complete the picture with the discontinuities due to the applied force $15P$ (Fig. 8.2d). The external force is right on top of the spring at a, and the latter takes the entire thrust of the force.[2] Thus the spring contracts by $aa' = -15P/k$. The gaps at the intermediate springs are therefore $\delta_1^0 = -(3/4)15P/k$, $\delta_2^0 = -(2/4)15P/k$ and

[2] This in only true for a determinate beam. A force on top of a flexible support in a redundant beam will affect the entire beam.

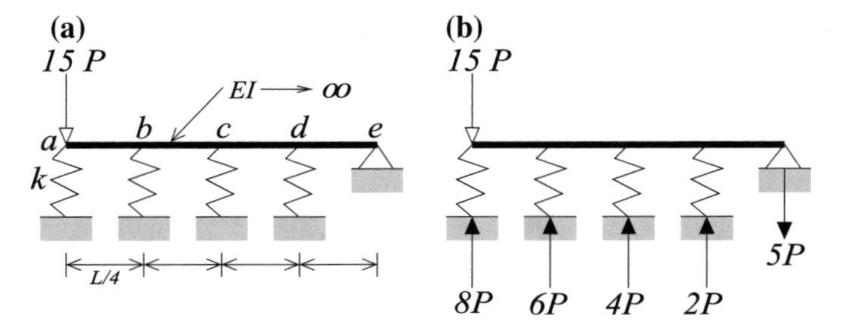

Fig. 8.3 Results

$\delta_3^0 = -(1/4)15P/k$. In this treatise, we use subscript (or superscript) '0' to indicate 'due to external causes'. The sign of the δ_i^0 is negative because the gap is an overlap instead of a distancing.

We are almost done. The total gap at b, that is, the sum of the four individual ones must be zero $\delta_1 = \delta_1^1 + \delta_1^2 + \delta_1^3 + \delta_1^0 = 0$ (compatibility requirement). The same is valid for the total gap c, the second intermediate spring, $\delta_2 = \delta_2^1 + \delta_2^2 + \delta_2^3 + \delta_2^0 = 0$ and for the gap at the third spring d. Collecting all the terms, the three equations called *compatibility equations* become in matrix form

$$\frac{1}{k}\frac{1}{16}\begin{bmatrix} 25 & 6 & 3 \\ 6 & 20 & 2 \\ 3 & 2 & 17 \end{bmatrix}\begin{Bmatrix} x_1 \\ x_2 \\ x_3 \end{Bmatrix} = \frac{15P}{k}\frac{1}{4}\begin{Bmatrix} 12 \\ 8 \\ 4 \end{Bmatrix} \tag{8.2}$$

the solution of which gives the reactions of the redundant elastic supports $x_1 = 6P$, $x_2 = 4P$ and $x_3 = 2P$. With these values we can compute the remaining reactions, at a and e, using equilibrium (8.1). The final result is drawn in Fig. 8.3b.

This concludes the analysis by the Force method. We now turn our attention to the other great method of analysis to see how the Displacement method would tackle the same problem.

8.3 Analysis by the Displacement Method

8.3.1 Determining the Kinematical Degrees of Freedom

Because the beam is rigid and hinged at the right extremity, it is clear that one way or another the equilibrium position of the beam, that is, the position at which the structure will be at rest, is the configuration shown in Fig. 8.4a, where the only unknown parameter is the slope of the beam or the angle of rotation θ at e, measured in radians. The structure is said to have one kinematic degree of freedom (d.o.f. or simply dof in structural jargon). It means that given θ we know the deformed shape

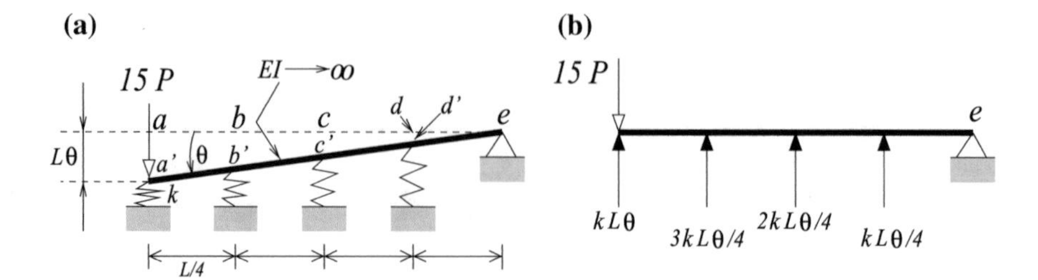

Fig. 8.4 Degrees of freedom

of the entire structure and hence almost everything we need. For instance, $L\theta$ is the vertical displacement of the left extremity.[3] The compressive force in spring a is therefore $kL\theta$, and this is also the value of R_a. By simple geometry (Fig. 8.4a, we find that the following reaction is $R_b = (3/4)kL\theta$, and so on (see Fig. 8.4b).

$$R_a = kL\theta \qquad R_b = (3/4)\ kL\theta \qquad R_c = (2/4)\ kL\theta \qquad R_d = (1/4)\ kL\theta \tag{8.3}$$

All we need for the solution is the value of θ.

8.3.2 The Equilibrium Equations

The displacements at the dofs are determined by the equilibrium equations. This is obtained by writing the condition that the rotational dof at e is in equilibrium, in other words, that the sum of the moments of all the forces with respect to e is zero (see Fig. 8.4b).

Assuming the anticlockwise sense as positive, we have

$$[15P - kL\theta]L - [(3/4)kL\theta](3/4)L - [(2/4)kL\theta](2/4)L - [(1/4)kL\theta](1/4)L = 0$$

which yields the *equilibrium equation*

$$\frac{30}{16}\ kL\ \theta = 15P \tag{8.4}$$

and the equilibrium value of the slope $\theta = 8(P/kL)$. Any other value of θ will not equilibrate the applied force.

[3]In truth the vertical displacement is $L \sin\theta$, but we consider very small displacements for which, in this case, $\sin\theta \to \theta$.

The rest is immediate. Introducing θ in (8.3) gives the reactions $R_a = 8P$, $R_b = 6P$, $R_c = 4P$ and $R_d = 2P$, and the reaction at e is $5P$ down, which is what we obtained with the Force method (Fig. 8.3).

Small displacements. You will have noticed that the equilibrium equations were written for the beam in the original position as in Fig. 8.4b. In truth, we should have written the equilibrium equation in the displaced position (Fig. 8.4a) because this is where we have equilibrium. However, this would complicate the equation and, since θ is small, the result is for all practical purposes the same. Writing the equilibrium equations in the original configuration is standard procedure in structural analysis, again, thanks to the assumption of small displacements.

8.4 Summing Up

Evidently, the displacement method seems to be much shorter and much simpler than the force method. Nevertheless, for most structures there are many open degrees of freedom, leading to large sets of equations. Even in the present example, if the beam were flexible we would have had nine degrees of freedom with nine equilibrium equations to solve instead of one; not a problem for a computerized procedure but rather tedious for hand calculations. The redundancy, on the other hand, is the same if the beam is flexible or not, so that with the force method we remain with three compatibility equations. It will, however, take longer to write the equations (to determine its coefficients).

Chapter 9
Qualitative Analysis and Design

Before embarking on a full analysis of a structure, it is often useful to have an idea of what the final result should look like. Qualitative analysis can give you the general form of the nsm-diagrams without actually performing an analysis, that is, without going through the calculations that provide the values of the internal forces or reactions.

For complex structures under complicated loading this may be easier said than done. But our aim is also didactic. It will assist us in performing and especially understanding the results of a real analysis. We will be mainly concerned with redundant beams and with some elementary frame examples. The method is to a large extent visual, and it uses an holistic approach in the sense that we employ everything we know about structural analysis, especially equilibrium, in order to improve the approximate result we are working on.

The key to performing a qualitative analysis is Hooke's law for a beam element in bending $m = (EI)\,\kappa$, where EI is considered constant.

9.1 An Introductory Example

Consider the propped cantilever beam in Fig. 9.1a fixed at a, with a roller at b, subjected to uniform distributed forces along its span. Here and elsewhere we assume that the cross-section of the beam is also uniform.

(a) We draw an assumed deflected shape of the beam under the loading taking into account the boundary conditions, that is, how the beam is attached to the supports. This means that the beam, when sagging under the load, keeps the deflections zero at the extremities a and b, and the slope of the beam at a, horizontal. Beyond that, we draw the simplest possible curve we can think of, such as the thick line in Fig. 9.1a.

M.B. Fuchs, *Structures and Their Analysis*,
DOI 10.1007/978-3-319-31081-7_9

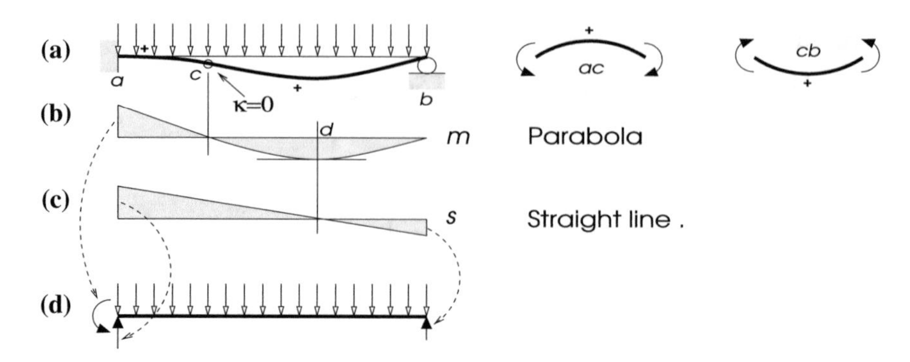

Fig. 9.1 Qualitative analysis of a propped cantilever

(b) Next we try to determine the sign of the curvature along the span. We notice two regions: close to the root the upper fibres are obviously elongated, so we indicate this by setting a + sign along he upper fibres, and for the rest of the beam the lower fibres are elongated. The + sign is along the bottom fibres. Somewhere in-between the curvature changes sign (which we guess to be at section c).
The two sketches to the right show a typical curvature for segment ac and a typical curvature for segment cb. Now, curvatures engender bending moments by means of Hooke's law for flexion of beams, $m \leftarrow EI\,\kappa$ (black couples in the sketches). Consequently, along ac we have a negative bending moment and negative curvature $\kappa < 0$, and for segment cb the moments and curvatures are both positive ($\kappa > 0$). At c the gradually changing curvature passes through zero. This is indicated by a virtual hinge (the beam does not have a hinge there but it has a zero bending moment.)

(c) We can now proceed to draw an outline of the bending moment. The bending along ab is a parabola (constant distributed forces), it is zero at c ($\kappa = 0 \rightarrow m = o$) and at b (the free end), and along ac the tensile fibers are at the top so the moment is drawn above the beam. The parabola shown is the only possible solution.

(d) The shear is a straight line (constant distributed forces). Recalling that the shear is the slope of the moment function, the line passes through zero at d where the slope is horizontal. Since the bending moment diagram is sloping down along ad the shear diagram must be there above the line as drawn in Fig. 9.1c.

(e) The signs of the force reactions are deduced from the shear diagram and the fixing moment is provided by the moment diagram.

We note that R_a, the reaction at a, is larger than R_b. Structures have a tendency to channel the flow of forces to the stiffer parts of the structure. The fixed reaction at a is stiffer than the roller support at b, and more of the uniform distributed forces is taken by the left support than by the right one.

9.2 The Basic Tools

Beams and Frames

- The first ingredient is a decent rendering of the equilibrium configuration. We do our best to draw the simplest deflected shape of the structure under the applied forces, taking into account the boundary conditions, that is, the way the structure is attached to the reactions.
- We then proceed by determining segments with curvature of the same sign and positioning points with expected change of sign curvature.
- We usually assume that the elements do not elongate.
- We then apply all the known rules regarding the nsm-diagrams and the relation between them.
  - Since the loading is limited to point loads and distributed forces of constant magnitude the bending moment (between point forces) is either linear or parabolic, and the shear force is correspondingly constant or linear depending on whether the segment is free of distributed forces or not.
  - The normal force in frames is piecewise constant (can be zero).
  - The sign of the reactions can be inferred from the nsm-diagrams.
- It often happens that following the steps leads to impossible results. It may be necessary to retrace steps, redraw the equilibrium shape or refine the position of some zero curvature points (inflexion points) to improve the approximation.

Truss-Beams. For truss-beams, qualitative analysis tries to determine the sign (tension or compression) of the bars of the structure. This is primordial information for the design of the truss, since compression members are prone to buckling and will require a substantial moment of inertia. In broad outline the procedure is as follows:

- We start by modeling the truss as a beam and deducing the equilibrium configuration or deflected shape.
- We next proceed with determining an approximation of the bending moment and shear force distribution and the sign of the reactions.
  - The sign of the bars composing the upper and lower chords (the flanges) depends on the bending moments.
  - The sign of the diagonals are determined from the shear distribution.
  - The sign of the posts is usually deduced from equilibrium considerations of the joints.
- The method may not always enable us to find the sign of all the bars and may sometimes require additional considerations.

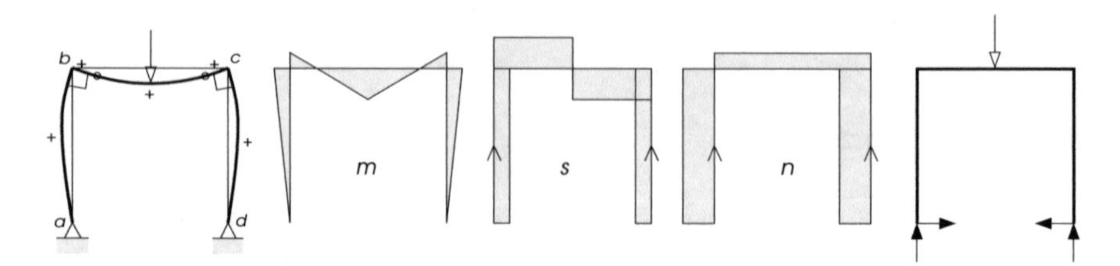

Fig. 9.2 Qualitative analysis of a portal frame

9.3 Qualitative Analysis of Portal Frame

The symmetric portal frame on hinged supports in Fig. 9.2 is subjected to a vertical force at mid-span of the beam. The beam is rigidly connected to the columns, consequently the 90° angle at b, and c must be maintained in the equilibrium configuration (heavy line on the left of the sketch). We next position three + signs in obvious locations of tension, that is, along the extrados of the columns and the intrados of the beam.

Turning our attention to the m-distribution, we note that the moment diagram along column ab is a straight line from zero at a (hinged support) to some m_b with tension along the outer fiber. The bending moment m_b on the beam therefore also has tension on the outer fiber. Consequently, there is a change of sign of the curvature along the beam; hence the two (virtual) hinges along the beam and the + signs near the corners b and c. The shear diagram follows suit.

At this point it is advantageous to determine the direction of the reactions (rightmost part of Fig. 9.2). The vertical reactions are each half the applied force. The horizontal reactions are easily determined from the s-diagram. All this shows that the columns and the beam are all in compression (see n-distribution).

9.4 Qualitative Design of Truss-Beams

The typical displaced shape (equilibrium configuration) for the beam on two supports in Fig. 9.3a has tension along the upper fibers for the entire span. The bending moment is continuous along ac, linear for segments ab and bc, and zero at the extremities. Since the bending moment has tension along the upper side and with zero bending moments at the extremities, we can expect an m-distribution as shown in Fig. 9.3b, from which the related s-distribution follows (Fig. 9.3c).

The sketch of the displaced shape varies only slightly with the actual cross-sections or type of beam. Let us assume that we are dealing with the 5-cell truss-beam shown in Fig. 9.3d. (Note, the reactions were deduced from the shear diagram.) We would like to show that it is possible to draw some general conclusions regarding the future design of the truss-beam.

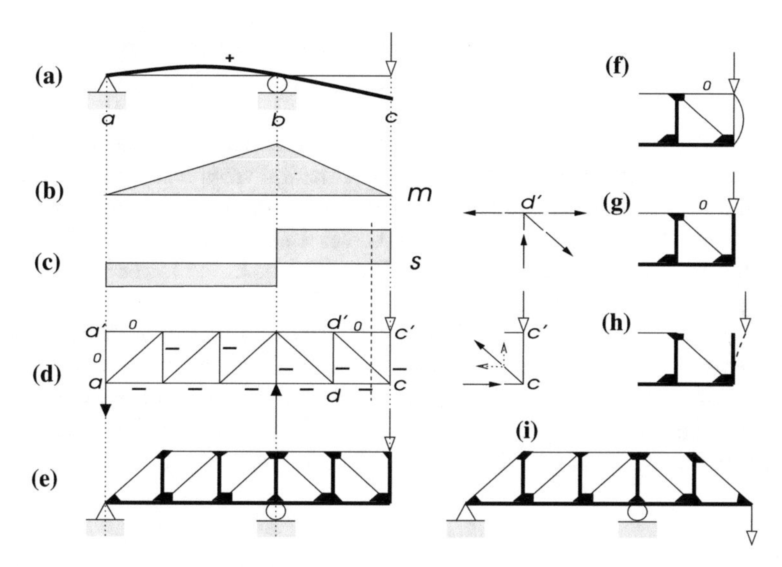

Fig. 9.3 Qualitative analysis and design of a truss-beam

The main task is to distinguish between tension bars and compression bars. Indeed, tension bars can be slender and will fail only if the material reaches its tension bearing capacity. Compression bars will be more massive because they are prone to Euler buckling and should therefore be provided with cross-sections with substantial moments of inertia (I), which will prevent early lateral bending.

The upper and lower chords depend on the bending moments.

In Fig. 9.3d, we have signified the compression bars with a − sign. Following the curvature pattern the entire lower flange is in compression and the upper flange in tension. Next we notice that all the diagonals are in tension. This follows from the s-diagram. Consider for instance the fictitious section of the truss between d and c. The diagram in (d) shows the forces acting on the '+'-facet of the section. The shear diagram indicates that the shear points up. This is only possible with tension in the diagonal $d'c$. Indeed, such a tension force has a vertical component pointing up. The sign in the other diagonals is determined in a similar manner from the shear diagram.

The forces in the diagonals are governed by the shear diagram.

Finally, the sign of the vertical elements (posts) is determined from the equilibrium of carefully selected nodes. Bar cc' is obviously in compression. All the other posts are also in compression. Indeed, nodes with a diagonal in tension have the post in compression. The diagram in the row of Fig. 9.3c, for instance, shows the equilibrium of joint d'. Since the diagonal is in tension (black arrow) the post is necessarily in compression (white arrow). We also find that three elements have zero internal forces.

Having determined the compression bars, we need to prevent there Euler buckling as in Fig. 9.3f. A bending-resistant element (substantial cross-section) will do the job (see Fig. 9.3g).

All this leads to the design shown Fig. 9.3e where we have removed bars with zero internal forces which are not required from a structural viewpoint. Interestingly, the zero-force bar $d'c'$ should be kept in place to prevent the instability shown in Fig. 9.3h. Even a slender bar will help because straight elements are very stiff axially.

Finally, the external force could have been applied to the lower-chord node c' resulting in design Fig. 9.3i.

Note: In the instability depicted in Fig. 9.3h, the beam buckles to the right, therefore a cable would be enough to prevent the failure. The post can in principle also buckle to the left thus requiring a rod with compressive axial stiffness. However, due to the shape of the equilibrium configuration (see Fig. 9.3a) the displacement depicted in Fig. 9.3h will prevail, and $d'c'$ can be a cable.

9.5 Illustrated Examples

9.5.1 Continuous Beam

The continuous or multi-span beam in Fig. 9.4a is fixed at a and d, has two intermediate supports and is loaded by a constant distributed force along span ab. We know that along ab the bending moment m is parabolic and s is linear (uniform distributed forces), and along spans bc and cd, m is linear and s is constant(no distributed forces).

The key for obtaining a representative sketch of the sm distributions and the reactions is to draw an approximation of the equilibrium shape of the beam (heavy line in figure). We start by noting that the slopes must be zero at the extremities and that the beam must hug the supports at b and c.

A tentative curve is usually similar to what is drawn in the figure. Next, we position the + signs in segments where the tensile side of the curved beam is obvious. Starting from a, the beam sags under the load with tension along the top fibers, but then has to curve up so as to pass over the support at b. But then it has to go down to pass over c. Then again join d with a horizontal slope. This defines four points where $\kappa = 0$.

One of the dilemmas is to decide whether a zero curvature point is to the right or the left of an intermediate support. It helps to remember that a $\kappa = 0$ point corresponds to a zero bending moment, and if you have more than one zero curvature point on a segment where the moment is linear (such as bc or cd), the moment will be zero on the entire span. This does not tally with a curved segment of the beam.

With all this the moment diagram is rather obvious, taking into account that the effect of a local loading fades with the distance from the loading. The shear diagram (Fig. 9.4c) follows easily from the moment diagram. Remember that: the shear is above the line when the moment slopes down, and vice versa.

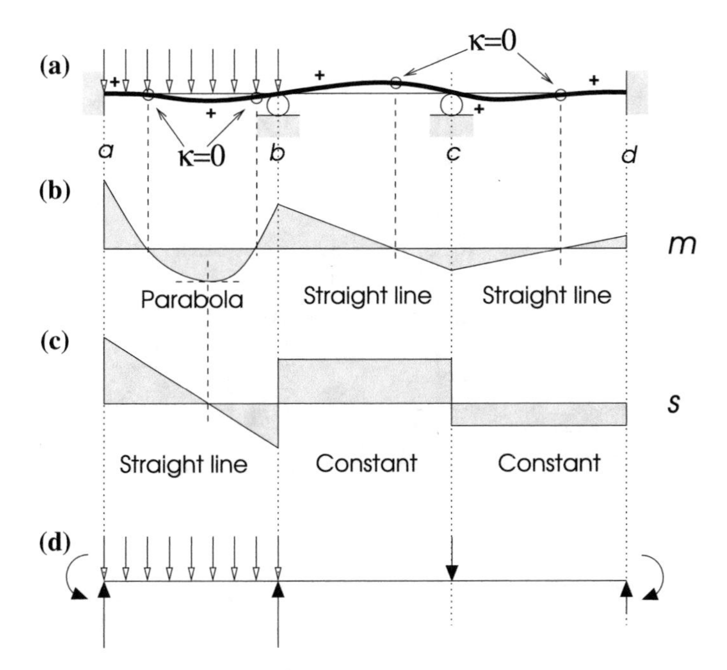

Fig. 9.4 Qualitative analysis of a continuous beam

Finally, a discontinuity in the m-diagram is an indication of a point couple (in this case reactions applied by the walls), and a discontinuity in the shear diagram results from a point force (support reactions in the present example).

9.5.2 Continuous Beam on Five Supports

The continuous beam on five supports in Fig. 9.5a is loaded by a point force on bc and a constant distributed force along cd. The bending moment m is linear and s is constant on unloaded segments, that is, ab, de, and segments to the left and to the right of the point force on span bc. On cd (uniform distributed forces), m is parabolic and s is linear.

We start by tentatively drawing a displaced shape. The curve should pass over the supports and sag under the loads on bc and cd. The loading is not symmetric and the slope at c is not necessarily horizontal.

A simple curve obeying these restrictions is shown in Fig. 9.5a. Note that the intensity of the loading is not given, so the curve is only a typical approximation.

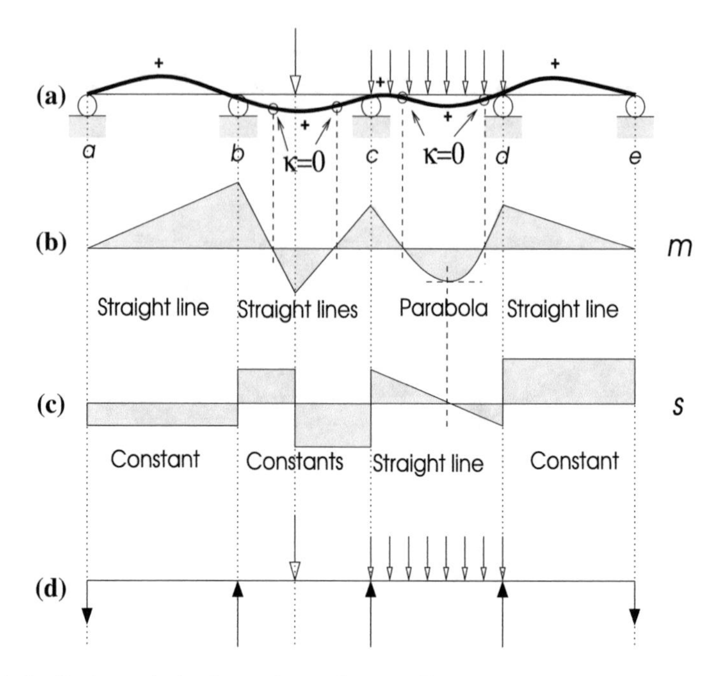

Fig. 9.5 Qualitative analysis of a continuous beam on 5 supports

Next we position the + signs along the tensile side of the beam in locations where the tensile side of the curved beam is obvious. In regions where the curvature changes sign, we position a point of zero curvature to the best of our understanding (an inflexion point). This may require some thought. For instance, somewhere in the vicinity of b there should be such a $\kappa = 0$ point. It cannot be to the left of the support because that would mean a zero bending moment on span ab. This does not comply with the curvature of ab. Consequently, the inflexion point is to the right of b.

Having positioned the four inflexion points, the bending moment distribution follows suit, making allowance for a zero bending moment wherever the curvature is null. The moment diagram is continuous (its slopes are not) because no point-couples are applied by the reactions. Similarly, reactive forces exist wherever there is a discontinuity in the shear diagram (Fig. 9.5d).

9.5.3 *Fixed Portal Frame with Distributed Forces*

The symmetric portal frame with fixed supports in Fig. 9.6 is subjected to uniform distributed forces along the beam. This example is by-and-large similar to the portal frame treated earlier, except for the distributed force on the beam replacing the point force and the fixed connection to the ground instead of the hinges. The beam is rigidly

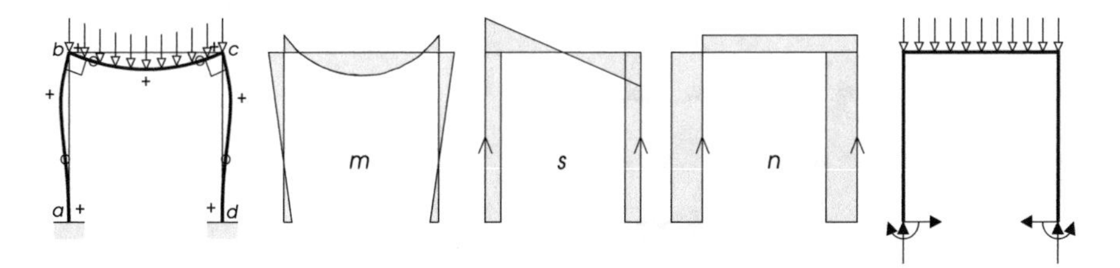

Fig. 9.6 Qualitative analysis of a portal frame under distributed forces

connected to the columns, consequently the 90° angle at b and c must be maintained in the equilibrium configuration (heavy line on the left of Fig. 9.6).

We next position three + signs in obvious locations of tension, that is, along the extrados of the columns and the intrados of the beam. Here we will need two more + signs at the root of the columns on the intrados because the deflected shape must change the sign of the curvature, in order to connect to the ground with a vertical slope. This entails a virtual hinge for each column.

Turning our attention to the m-distribution, we note that the moment diagram along columns ab is a straight line passing through the virtual hinge (κ is zero there and so is the bending moment) to some m_b with tension along the outer fiber. The bending moment m_b on the beam has therefore also tension on the outer fiber. Consequently, there is a change of sign of the curvature along the beam. Hence the two (virtual) hinges along the beam and the + signs near the corners b and c. The bending moment on the beam is a parabola and the shear is linear with a zero value at mid-span, where the bending distribution is extremum. The n-diagram is similar to the case with a point force. Finally, the sense of the reactions at a and d can be read off the nsm-diagrams at the extremities.

9.5.4 Four-Cell-Truss-Beam on Three Supports

The typical displaced shape (equilibrium configuration) for beam ae on three supports and the related qualitative moment and shear diagrams are given in Fig. 9.7a–c.

Assume that the beam is the 4-cell truss on three supports shown in Fig. 9.7d, use bending components for the compression elements of the truss (to impede local buckling), and remove the elements without a structural function.

The moment diagram is piecewise linear on segments ab, bc, ... (segments without distributed loads), and the corresponding shear is piecewise constant. The bending moments are zero at the extremities, the distribution is continuous (no point couples) and the m-diagram is thus obvious. The shear diagram follows suit.

Assume that the structure is the truss-beam on three supports shown in Fig. 9.7d.

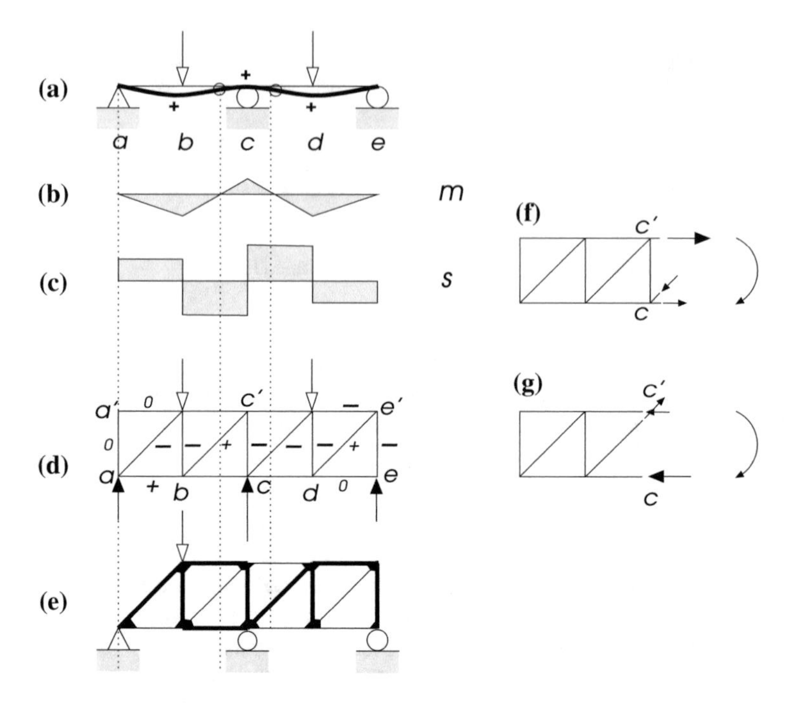

Fig. 9.7 Qualitative analysis and design of a 4-cell truss-beam

The sense of the reactions and the sign of the diagonal elements follow immediately from the shear diagram. It is also easy to find the elements with zero internal forces. The sign of the vertical posts can be determined from equilibrium of the end-nodes. Finally, the bars of the two outer cells are also easy to come by.

The two inner cells are less obvious, because the moment diagrams change signs somewhere within each cell (see dotted lines). The horizontal equilibrium of d in Fig. 9.7d requires cd to be in tension. Furthermore, a total section of the truss a little to the right of bar cc' (sketch Fig. 9.7f) shows that bar $c'd'$ is also in tension as mandated by the negative bending moment there.

Finally, the horizontal equilibrium of node b' sets bar $b'c'$ in compression and a total section immediately to the left of post cc' (see Fig. 9.7g) sets bc in compression also on account of the bending moment.

The final design is shown schematically in Fig. 9.7e. Note that bars aa' and $a'b'$ can be removed, but bar de is necessary because of the stability of ee' (a bar in compression), although de has zero internal forces.

Note: It is instructive to realize that we have gathered significant information on the behavior of this structure without having performed any formal analysis.

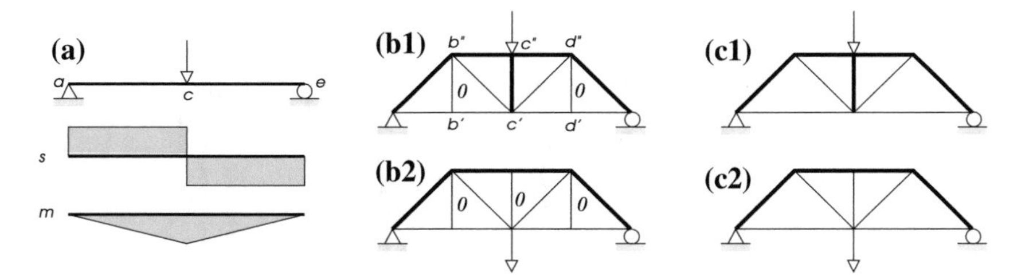

Fig. 9.8 Truss-beam with zero-force members

9.5.5 *Simple Truss-Beam*

A simple beam, centrally loaded by a point force, and its corresponding qualitative shear and moment diagrams are shown in Fig. 9.8a. Assuming that the beam has the form of the simple truss depicted in Fig. 9.8b, it is easy to determine the tension and compression bars shown in the figure. This didactic example can help in determining the function of some zero-force bars in preventing Euler buckling, in particular, the importance or not of the vertical posts $b'b''$, $c'c''$ and $d'd''$.

In Fig. 9.8b1 the point force is applied along the upper chord, and in Fig. 9.8b2 the force is along the lower chord. The vertical equilibrium of nodes b', c'' and d' shows clearly that the vertical posts have all zero forces except post $c'c''$ in the upper truss which is in compression, hence the heavy line for that post.

In Fig. 9.8c we have removed the zero-force bars without any apparent structural function. In all cases the outer posts have been removed. Interestingly, in Fig. 9.8c2, post $c'c''$ has been retained. Although it carries no force it halves the buckling length of compression bar $b''d''$.

9.5.6 *Simply Supported Portal Frame*

A simply supported portal frame, centrally loaded by a point force, and its corresponding qualitative bending moment diagram are shown in Fig. 9.9a, b. This is a classical didactic example which does not exactly belong to this chapter on qualitative analysis but it can help in understanding the behaviour of a frame.

The roller at d precludes a horizontal reaction there and consequently also at a. Therefore the columns ab and dc will be moment free, that is, bending free. They will experience only compression.

As a result the columns will remain straight and only bc will bend. In order to maintain orthogonal corners at b and c the only possible equilibrium configuration is the one shown qualitatively in Fig. 9.9c.

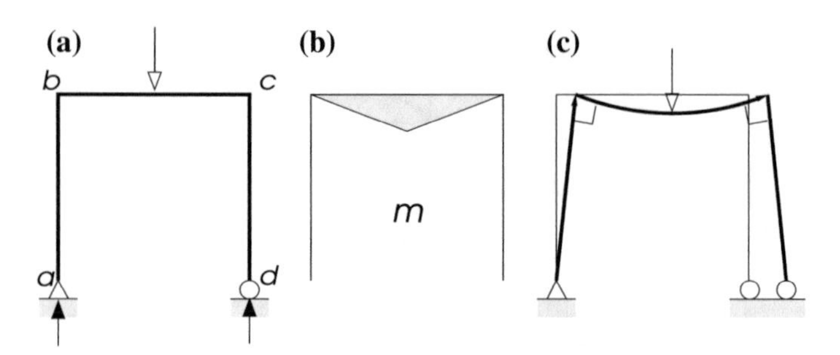

Fig. 9.9 Simply supported portal frame

Question: When the structure reaches the equilibrium configuration the vertical reactions should cause bending in the columns in contradiction to what is depicted in Fig. 9.9b. The apparent paradox is explained by the small displacements assumption. The side motion of the frame is so small that the bending moments are negligible (second order effect.)

9.6 Summing Up

In standard cases we can often predict qualitatively the distribution of the internal forces $nsm(x)$ on the basis of the estimated deformed shape of bending elements. It is shown that locating zones of curvatures with the same sign allows us to approximate the bending moment diagrams and ensuing shear force diagrams and determine the sense of the reactions.

In the case of truss-beams a technique is demonstrated to determine the elements in compression before analysis. The designer has thus a general impression of the truss were compressive elements are heavier in order to prevent early buckling.

Part III
Flexibility

Part III deals with the first of the two main methods of analysis: the flexibility or force method. We start with the reciprocity theorems and the flexibility matrix in Chap. 10.

Since the force method applies only to statically redundant structures we continue with a Chap. 11 which discusses methods of determining the degree of redundancy of a structure. The chapter also gives properties of redundant and statically determinate structures. This is followed by Chap. 12 which constitutes the core of the method.

Chapter 13 explains the effects of heating and related prestressing on structures.

Chapter 10
Flexibility Coefficients

In this chapter we will apply virtual work using equilibrium systems and compatible systems taken from a *same* real structure. For every deformed shape of the structure we can derive an equilibrium system and a compatible system. Taking the equilibrium system from one deformed shape and the compatible system from another deformed shape, and vice versa, will eventually lead to Betti's reciprocal theorem and to Maxwell's flexibility coefficients.

10.1 Virtual Work with Related ϵ and n

Consider the free-floating beam of bending stiffness $EI(x)$ reproduced twice in Fig. 10.1. The figure relates to a beam but it represents general elastic structures. We impose an arbitrary displacement $v^i(x)$ in one case and displacement $v^j(x)$ in the other, as shown in the figure. Now, curve v^i has curvatures $\kappa^i(x) = \mathrm{d}^2v^i/\mathrm{d}x^2$. These curvatures cause internal bending moments $m^i(x) = EI(x)\,\kappa^i(x)$ (Hooke's law). The bending moments are in equilibrium with applied loads $q^i(x) = \mathrm{d}^2m/\mathrm{d}x^2$. What are these forces? They are the auto-equilibrated forces which produced the displacements $v^i(x)$.[1] A beam does not deform just like that. In order to deform the beam according to $v^i(x)$, we need to apply loading $q^i(x)$.

We have just built an equilibrium system i composed of $q^i(x)$ and $m^i(x)$ and a compatible system i made up of $v^i(x)$ and $\kappa^i(x)$, with the peculiarity that the moments are related to the curvatures by Hooke's law. This relation between κ and m is essential. We can repeat the procedure for case j which leads to the equilibrium and compatible systems in Table 10.1.

We now write two virtual work equations using equilibrium from i and compatibility from j and conversely:

[1]Depending on the nature of $v(x)$, the loads can comprise point forces and couples as well as distributed loads. Also, you can add a rigid-body motion to the imposed displacements without affecting the curvatures and the forces.

M.B. Fuchs, *Structures and Their Analysis*,
DOI 10.1007/978-3-319-31081-7_10

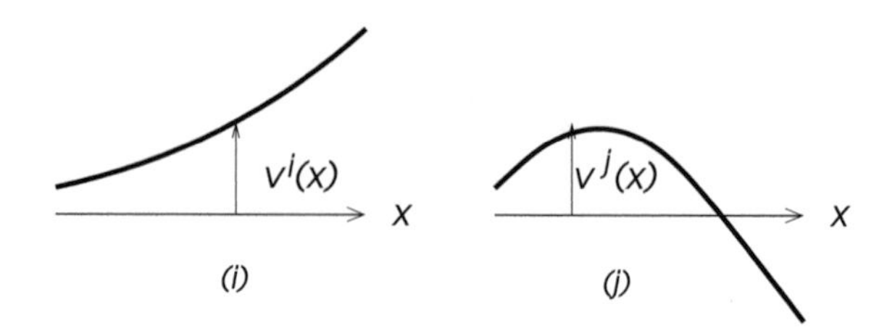

Fig. 10.1 Two arbitrary displacements of a beam

Table 10.1 Two equilibrium and compatible systems

Case	Equilibrium system		Compatible system		Note
i	$q^i(x)$	$m^i(x)$	$v^i(x)$	$\kappa^i(x)$	$m^i(x) = EI(x)\,\kappa^i(x)$
j	$q^j(x)$	$m^j(x)$	$v^j(x)$	$\kappa^j(x)$	$m^i(x) = EI(x)\,\kappa^i(x)$

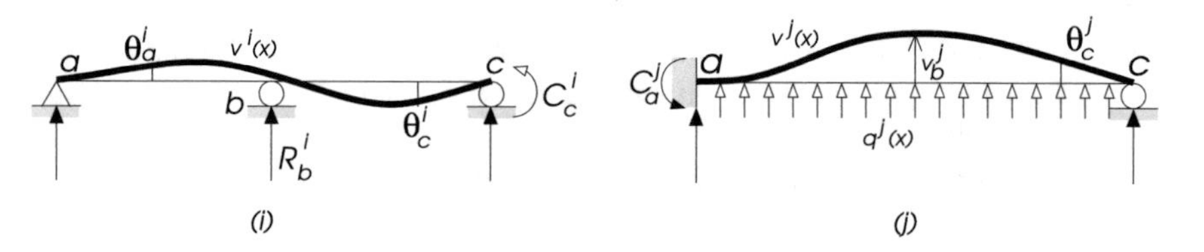

Fig. 10.2 Two arbitrary displacements and loadings

$$
\int_x q^i\, v^j\, \mathrm{d}x = \int_x m^i \kappa^j\, \mathrm{d}x \;=\; \int_x (EI\,\kappa^i)\,\kappa^j\, \mathrm{d}x
$$

$$
\int_x q^j\, v^i\, \mathrm{d}x = \int_x m^j \kappa^i\, \mathrm{d}x \;=\; \int_x (EI\,\kappa^j)\,\kappa^i\, \mathrm{d}x
$$

The last term in the equations is the important one. Because we are using the same structure in both sets of systems, we can replace m by $EI\kappa$. Clearly, since $\int_x EI\kappa^i\kappa^j\,\mathrm{d}x = \int_x EI\kappa^j\kappa^i\,\mathrm{d}x$ we find $\int_x q^i\, v^j\,\mathrm{d}x = \int_x q^j\, v^i\,\mathrm{d}x$, in other words,

Betti: $(\text{loads})^i \times (\text{displacements})^j = (\text{loads})^j \times (\text{displacements})^i$

This is Betti's reciprocal theorem.

The property is valid for any linear elastic structure, be it a beam, frame or truss. It holds for any two sets of forces and related displacements as long as the structure is the same in both cases. The supports need not be the same as in (Fig. 10.2).

In case i the *same* beam is supported at a, b and c and subjected to a couple C^i_c at extremity c. The reactions are R^i_a, R^i_b and R^i_c and the equilibrium shape is $v^i(x)$. In case j the beam is in a propped cantilever configuration. Under the action of the distributed forces $q^j(x)$ it assumes an equilibrium shape in the form of $v^j(x)$. The reactions here are the couple at the root, C_a^j, and the reactions R_a^j and R_c^j. If, for the sake of the argument, we disregard the supports, we will find the same beam displaced in two different ways by two different sets of forces. This is exactly the situation for which the reciprocal theorem was written.

The point of interest is that in the equilibrium sets one needs to include *all* the loads, applied loads as well as reactions, because some reactions of case i may move through the displacements j and vice versa. Assuming an up-direction for forces and displacements and anti-clockwise sense for couples and rotations as positive, the reciprocal law becomes $R_b^i \times v_b^j + C_c^i \times (-\theta_c^j) = C_a^j \times \theta_a^i + \int_{\text{beam}} q^j(x)v^i(x)\,\mathrm{d}x$, where we note that $\theta_c^j \equiv (\mathrm{d}v^j/\mathrm{d}x)_c$ and $\theta_a^i \equiv (\mathrm{d}v^i/\mathrm{d}x)_a$.

10.2 Betti and Maxwell Theorems

10.2.1 Betti's Theorem

Strictly speaking, the Italian civil engineer Enrico Betti (1823–1892) has enunciated his reciprocal theorem for two identical structures, with identical supports, under different loadings as for the portal frame in Fig. 10.3.

Consider the linear elastic frame in Fig. 10.3 subjected to an (*external* loading)^i. These loads cause (displacements)^i, deformations $\boldsymbol{\epsilon}^i(x) \equiv \{\epsilon^i(x)\,\kappa^i(x)\}^T$, and internal forces $\boldsymbol{n}^i(x) \equiv \{n^i(x)\,m^i(x)\}^T$ and (reactions)^i at the supports. (They also engender shear forces but these will not come into consideration in Euler–Bernoulli elements). Likewise, the same structure under a different (*external* loading)^j has (displacements)^j, deformations $\boldsymbol{\epsilon}^j(x)$, internal forces $\boldsymbol{n}^j(x)$ and (reactions)^j.

Now, the (external loads)^i, the (reactions)^i and $\boldsymbol{n}^i$ are an equilibrium system i, and the (displacements)^i in conjunction with the deformations $\boldsymbol{\epsilon}^i$ are a compatible system i. We have in parallel similar systems j.

We will now write a virtual work equation with the equilibrium system i and the compatible system j:

$$\text{(external forces + reactions)}^i \times \text{(displacements)}^j = \int_x (n^i\,\epsilon^j + m^i\,\kappa^j)\,\mathrm{d}x$$

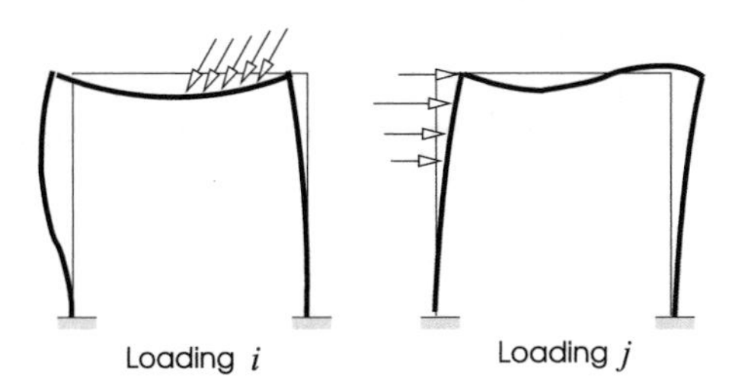

Fig. 10.3 Betti's theorem

We should note that since the supports are the same in both cases, there are no displacements^j corresponding to the reactions^i. So this relation reduces to

$$(\text{external forces})^i \times (\text{displacements})^j = \int_x (n^i \epsilon^j + m^i \kappa^j)\,\mathrm{d}x = \int_x (EA\,\epsilon^i \epsilon^j + EI\,\kappa^i \kappa^j)\,\mathrm{d}x$$

and also

$$(\text{external forces})^j \times (\text{displacements})^i = \int_x (n^j \epsilon^i + m^j \kappa^i)\,\mathrm{d}x = \int_x (EA\,\epsilon^j \epsilon^i + EI\,\kappa^j \kappa^i)\,\mathrm{d}x$$

where we have also introduced Hooke's law for bars $n = EA\kappa$ and Hooke's law for beams $m = EI\kappa$ to produce the last terms.

Now since the last term in both relations are equal we obtain the original form of Betti's law of reciprocity. (Note the term *external.*)

$$(\textit{external}\,\text{loads})^i \times (\text{displacements})^j = (\textit{external}\,\text{loads})^j \times (\text{displacements})^i \tag{10.1}$$

We recall that this is valid for two loadings on the same structure including the same supports.

10.2.2 Degrees of Freedom

Maxwell's theorem is a downsized version of the theorem of Betti. It applies reciprocity to the case where the loads and displacements i and j consist each of one unit load, either a force or a couple, and corresponding displacement and rotation. But first we need to put some order in the definition of a degree of freedom. We have been using the concept throughout the text, but it deserves additional consideration.

Simple Degrees of Freedom

In its simplest form a degree of freedom, sometimes referred to in short as *dof* or *dofs* in plural, is a vector attached to a point on a structure. As such it is defined by its position and direction. Typical examples are the translational degrees of freedom i, j, k in the structures in Fig. 10.4 and the rotational degree of freedom l in the frame and beam. When loaded, every structure, being a flexible entity, will experience some displacement at an arbitrary degree of freedom j in the direction j. (The displacement along j is the component of the displacement at j projected in the direction j.)

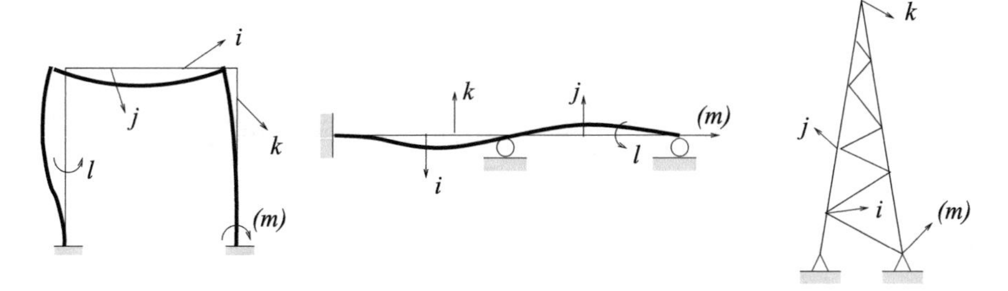

Fig. 10.4 Examples of degrees of freedom

Since beams, by definition, have only lateral displacements, the only possible degrees of freedom of beams are lateral vectors and rotations. Similarly, trusses will have only translational degrees of freedom at the nodes.

Open and Closed Degrees of Freedom

By extension, we often define dofs at zero displacement locations such as (m). We then distinguish between *open* or *free* dofs i, j, k, l and *closed* or *fixed* dofs (m). Free degrees of freedom, or even worse, fixed degrees of freedom do not sound right, but still, the terms are used. The frame is built-in at the supports, the rotation (m) is thus prevented. Likewise, the horizontal displacement (m) of the right extremity of the beam is zero, because we assume inextensible beams. Finally, the truss being reacted on by hinged supports, the degree of freedom (m) is fixed.

We can define as many degrees of freedom in a structure as we possibly need. We can also attach several degrees of freedom at a same point.

10.2.3 Maxwell's Theorem

Consider the two degrees of freedom defined for the propped cantilever in Fig. 10.5a. When a unit load, a couple of magnitude 1 in this case, is exerted at j (Fig. 10.5b), the structure deforms and in particular there is a displacement at i. The component along i of this displacement at i due to a unit load at j is the flexibility coefficient f_{ij} or f_i^j. Likewise, we have the flexibility coefficient f_{ji} or f_j^i which is the displacement at j (along j) due to a unit load at i.

Maxwell theorem states

$$\boxed{\text{Maxwell: } f_{ij} = f_{ji} \text{ or also } f_i^j = f_j^i}$$

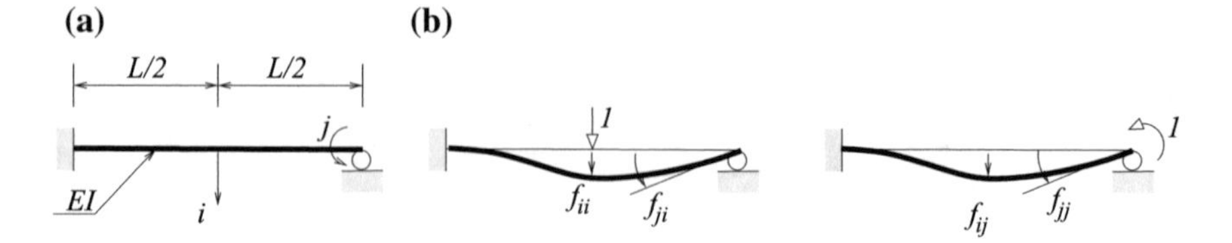

Fig. 10.5 Influence coefficients. **a** Degrees of freedom, **b** influence coefficients

Indeed, when we apply Betti's theorem for the specific case of the unit loads at i and j (let us call them 1_i and 1_j) we have $1_i \times f_{ij} = 1_j \times f_{ji}$ and since the unit loads have magnitude 1, this proves Maxwell's relation.

Although this symmetry, $f_{ij} = f_{ji}$ is embedded in structural theory it nevertheless is a startling result. In the case of the propped cantilever in Fig. 10.5 f_{ij} is a displacement due to a unit couple and it is equal to f_{ji}, a rotation/slope due to unit force. You will also note that the units concord. If the couple at j had real dimensions the displacement at i would have been a *length*. But since the couple is dimensionless the units of f_{ij} are $length/(force \times length) = 1/force$. On the other hand, if the force at i was given in *force* units the rotation at j would have been dimensionless (radians). Now that the force is dimensionless the units of f_{ji} are also $1/force$.

As shown in Fig. 10.5, there exist coefficients of the type f_{ii} which is the displacement at i due to a unit load at i.

The name *influence* coefficient for the f_{ij} is appropriate. It expresses how the displacement at i is influenced by a load at j. They are also called *flexibility* coefficients. The larger f_{ij}, the larger the displacement at i due to a unit load at j, in other words, the more flexible the structure.

10.3 Computing Influence or Flexibility Coefficients

How does one compute an influence coefficient? Since these coefficients are displacements we will use the basic technique for computing displacements, that is, the unit-load method.

In a preparatory step, we analyze the structure under the unit load 1_i at i, and we determine the distribution of internal forces $\boldsymbol{n}^i(x)$ and hence the deformations $\boldsymbol{\epsilon}^i(x)$. We next analyze the same structure for a unit load 1_j at j, and we likewise determine $\boldsymbol{n}^j(x)$ and $\boldsymbol{\epsilon}^j(x)$.

For the displacement f_{ij} we use the deformations due to 1_j, that is, $\boldsymbol{\epsilon}^j$ and a system of internal forces in equilibrium with 1_i (we are computing the displacement along i). One such equilibrium system is $\boldsymbol{n}^i(x)$. Consequently, the unit-load method gives

$$f_{ij} = \int_x \boldsymbol{n}^i(x)^T \boldsymbol{\epsilon}^j(x)\,\mathrm{d}x \tag{10.2}$$

and since the structure is linear elastic, $\boldsymbol{n}^j(x) = [\boldsymbol{EA}(x)]\,\boldsymbol{\epsilon}^j(x)$, or in inverse form $\boldsymbol{\epsilon}^j(x) = [\boldsymbol{EA}(x)]^{-1}\boldsymbol{n}^j(x)$ which when introduced in (10.2) yields

$$f_{ij} = \int_x \boldsymbol{n}^i(x)^T\, [\boldsymbol{EA}(x)]^{-1}\boldsymbol{n}^j(x)\,\mathrm{d}x \tag{10.3}$$

This is a very simple equation. Indeed, $[\boldsymbol{EA}]$ is a diagonal matrix and its inverse is also diagonal with, in the general case, entries $1/EA$; $1/GA$ $1/EI$. Expanding (10.3) gives

$$f_{ij} = \int_x \left(\frac{n_i\,n_j}{EA} + \frac{s_i\,s_j}{GA} + \frac{m_i\,m_j}{EI}\right)\mathrm{d}x \tag{10.4}$$

Incidently, this equation shows clearly the symmetry of f_{ij}.

The total displacement is thus the sum of the displacement due to axial, shear and bending deformations. For the actual computing we make allowances for the approximations that were discussed in the unit-load method: For flexural structures (beams, frames) the shear deformation is assumed zero, and the displacement due to bending deformations is much larger than the one due to axial deformation. We neglect the latter and f_{ij} depends only on the bending moments (10.5).

For trusses, there is no bending so the only term is the one involving the normal forces (10.6).

$$\text{(Frames and beams)}\quad f_{ij} = \int_x \frac{m_i\,m_j}{EI}\,\mathrm{d}x \tag{10.5}$$

$$\text{(Trusses)}\quad f_{ij} = \int_x \frac{n^i\,n^j}{EA}\,\mathrm{d}x = \sum_r \frac{n_r^i\,n_r^j}{E_rA_r}L_r \tag{10.6}$$

where the summation (index r) runs over all the bars of the truss. Note the discretization is valid only for trusses with constant E_rA_r per element.

In fact, the integration for frames is performed element by element and then summed, so even for frames we have $f_{ij} = \sum_r \int_x (m_r^i\,m_r^j/E_rI_r)\mathrm{d}x$.

10.4 Flexibility Matrices

We can position many degrees of freedom on a structure and for any pair of degrees we can determine their influence coefficients. On the beam of Fig. 10.6a for which we have positioned four degrees of freedom which define a set of 16 influence coefficients $f_{ij}, i, j = 1..4$. Quantities with a double index can conveniently be arranged in matrix form

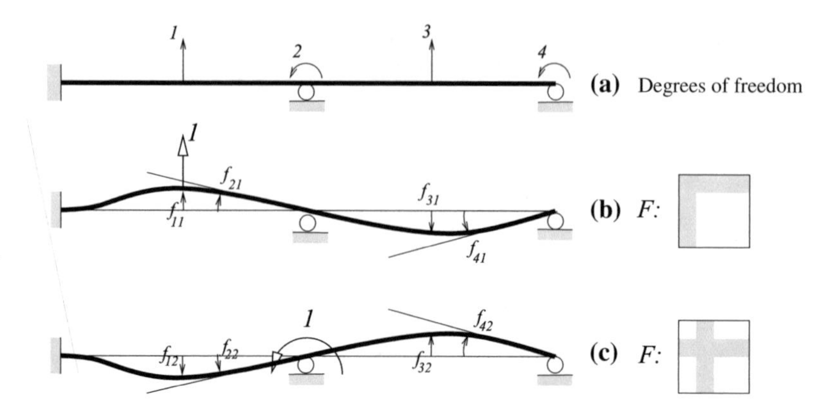

Fig. 10.6 Beam flexibility coefficients

$$\boldsymbol{F} = \begin{bmatrix} f_{11} & f_{12} & f_{13} & f_{14} \\ f_{21} & f_{22} & f_{23} & f_{24} \\ f_{31} & f_{32} & f_{33} & f_{34} \\ f_{41} & f_{42} & f_{43} & f_{44} \end{bmatrix} \tag{10.7}$$

which we call a flexibility matrix[2] usually referred to as $\boldsymbol{F}$. This is *a* flexibility matrix and not *the* flexibility matrix of the structure. For the same structure we could have defined other degrees of freedom leading to a different flexibility matrix $\boldsymbol{F}$.

In Fig. 10.6b we have indicated the physical interpretation of the flexibility coefficients f_{i1}, $i = 1..4$. These are the displacements at the degrees of freedom due to a unit force applied at 1. These coefficients are the entries of the first column of $\boldsymbol{F}$. Likewise in Fig. 10.6c we find f_{i2}, $i = 1..4$, the displacements at the degrees of freedom due to a unit couple at 2. They populate the second column of $\boldsymbol{F}$, and so forth for the remaining columns.

Matrix $\boldsymbol{F}$ being symmetric the transposed of the columns are the entries of the corresponding rows as is visualized in the figure.

Another simple example which leads to a 2×2 flexibility matrix is the truss in Fig. 10.7a where two degrees of freedom are defined at the free node b. Note, the degrees of freedom are in the prolongation of the bars. This leads to unexpected displacements. Applying a unit force at dof 1 we get $n^1_{bc} = 1$ and a zero force in ab, ($n^1_{ab} = 0$) (Fig. 10.7b1).

If $n^1_{ab} = 0$ then $\epsilon^1_{ab} = 0$. In other words, bar ab does not elongate. Consequently node b can move only along the dashed line (perpendicular to ab). So although we have a down force, and although there are no internal forces in ab this bar has a structural purpose. It pulls b to the left until it reaches its equilibrium position b'. The flexibility coefficient f_{11} is, by definition, the projection of bb' on the line bc. Coefficient $f_{21} = 0$ for the same reason. Assuming that the bars are of uniform stiffness EA, and since $L_{bc} = L$ we get $f_{11} = L/EA$.

[2] We also use F_{ij} for the components of the flexibiliy matrix.

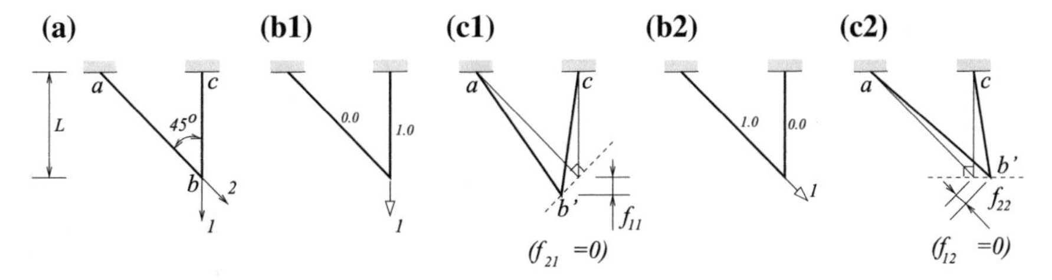

Fig. 10.7 Truss flexibility coefficients

Applying a similar reasoning for the influence coefficients under a unit force along dof 2 (Fig. 10.7b2, c2) we obtain the influence matrix of the truss for these two degrees of freedom

$$\boldsymbol{F} = \frac{L}{EA}\begin{bmatrix} 1.0 & 0.0 \\ 0.0 & 1.41 \end{bmatrix} \tag{10.8}$$

This flexibility matrix happens to be diagonal, and half of its coefficients are zero, but do not jump to conclusions. In a short while we will emphasize that $\boldsymbol{F}$ matrices are usually very populated.

Note: If you are puzzled by the apparent contradiction between ab not elongating when a unit force is applied to dof 1 and the fact that ab' is manifestly longer than ab (Fig. 10.7b2), remember that we are dealing with very, very small displacements. We can make b' approach b until we are satisfied that $ab' = ab$ is not a blatant exaggeration.

10.5 Properties of Flexibility Matrices

This section is partly based on a paper by Argyris.[3]

1. *Flexibility matrices are symmetric*:
   This is Maxwell all over again. Indeed, an alternative way of writing $f_{ij} = f_{ji}$ is

$$\boldsymbol{F} = \boldsymbol{F}^T \tag{10.9}$$

   Consider all the coefficients in *row* i of $\boldsymbol{F}$, that is f_{ij}, $j = 1..N$ and next all the coefficients in *column* i of $\boldsymbol{F}$ (f_{ji}, $j = 1..N$). Clearly the coefficients in the same position j in both arrays are equal. Consequently, corresponding row and column vectors of $\boldsymbol{F}$ are the transpose of one another (see Fig. 10.6b, c).
2. *The diagonal entries are positive*:
   This follows from (10.4), be setting $j = i$

[3] JH Argyris, Energy Theorems and Structural Analysis, Part I. General Theory, *Structures* 347–394, Oct 1954.

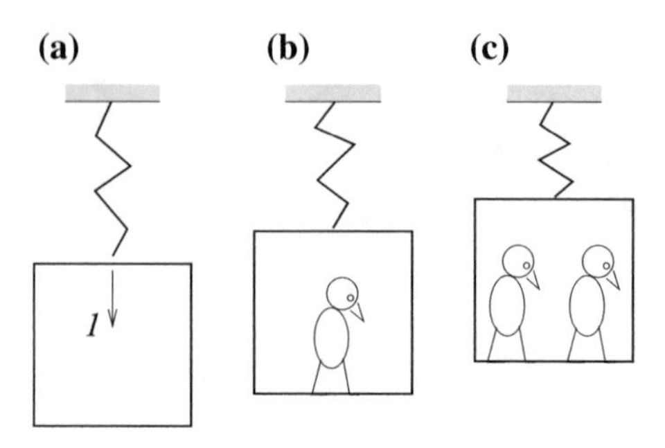

Fig. 10.8 Negative main diagonal flexibility coefficient

$$f_{ii} = \int_x \left(\frac{n_i\, n_i}{EA} + \frac{s_i\, s_i}{GA} + \frac{m_i\, m_i}{EI} \right) \mathrm{d}x = \int_x \left(\frac{n_i^2}{EA} + \frac{s_i^2}{GA} + \frac{m_i^2}{EI} \right) \mathrm{d}x \qquad (10.10)$$

Indeed every component in the integrand is individually positive and a sum of positive numbers is positive. If i happens to be a 'fixed' degree of freedom, all the f_{ji} including f_{ii} are of course zero. But this is a marginal case. For our purposes we will remember

$$f_{ii} > 0 \qquad (10.11)$$

It goes without saying that a structure with an $f_{ii} < 0$ would hardly be possible. If the degree of freedom of the elevator in Fig. 10.8a had $f_{11} < 0$, then stepping into the cabin would shorten the cable (spring in the figure) and pull us up (b). The heavier the load the higher we would go (c). Fortunately, our structural models preclude negative f_{ii}.

Keep in mind that f_{ij}, $j \neq i$ can be negative. In Fig. 10.6, for instance, we had $f_{21} < 0$. Under a unit force at 1 the beam at 2 rotated clockwise, that is, in the opposite sense of degree of freedom 2.

3. *Matrix* $\boldsymbol{F}$ *is positive definite*:
   A symmetric matrix $\boldsymbol{A}$ is said to be positive definite if when pre- and post-multiplied by *any* non-zero vector $\boldsymbol{b}$, the triple product $\boldsymbol{b}^T\boldsymbol{A}\boldsymbol{b}$ is *always* positive ($\boldsymbol{b}^T\boldsymbol{A}\boldsymbol{b} > 0$). One will note that the triple product, also called quadratic form, $\boldsymbol{b}^T\boldsymbol{A}\boldsymbol{b}$ is, of course, a scalar, $(1 \times N) \times (N \times N) \times (N \times 1) = 1 \times 1$. Also, we assume that matrix $\boldsymbol{A}$ and vector $\boldsymbol{b}$ are of the same length N.
   The flexibility matrix is such a positive definite matrix. What we usually pre- and post-multiplies the matrix with is a vector $\boldsymbol{f}$ of forces applied along the degrees of freedom.

$$\boldsymbol{F} > 0; \qquad \boldsymbol{f}^T\boldsymbol{F}\boldsymbol{f} > 0 \qquad (10.12)$$

   The short form on the left is symbolic writing for the positiveness of $\boldsymbol{F}$ and the triple product, or quadratic form, to the right is its definition.
   To show the positiveness of flexibility matrices, let us consider a structure on which we have defined dofs and let $\boldsymbol{f}$ and $\boldsymbol{\delta}$ be forces and corresponding displacements at the dofs. Inside the structure we have internal forces $\boldsymbol{n}$ and deformations

$\boldsymbol{\epsilon}$. Using the equilibrium system $\boldsymbol{f}$ and $\boldsymbol{n}$ and the compatible system $\boldsymbol{\delta}$ and $\boldsymbol{\epsilon}$, we can write the virtual work equation

$$\boldsymbol{f}^T \boldsymbol{\delta} = \int_x \boldsymbol{n}(x)^T \boldsymbol{\epsilon}(x)\, \mathrm{d}x \tag{10.13}$$

(Note, the reactions are also part of the equilibrium system but they will not perform any work for displacements $\boldsymbol{\delta}$). Remember that in this case the internal forces and deformations are related by Hooke's law $\boldsymbol{\epsilon}(x) = [\boldsymbol{EA}]^{-1}\boldsymbol{n}(x)$, which when introduced in (10.13) yields

$$\boldsymbol{f}^T \boldsymbol{F} \boldsymbol{f} = \int_x \boldsymbol{n}(x)^T \, [[\boldsymbol{EA}(x)]^{-1}]\, \boldsymbol{n}(x)\, \mathrm{d}x \tag{10.14}$$

Now the right-hand side of (10.14) is positive, because the integrand, $n^2/EA + s^2/GA + m^2/EI$, is positive for any x, hence $F > 0$.

4. ***F** is a full matrix*:
When we apply a force, unit or otherwise, to a structure it moves into an equilibrium configuration, and every degree of freedom we may have defined on the structure will experience a displacement. Granted, some coefficients will be more significant than others, some may even turn out to be zero, but as a rule the displacements at the dofs, that is, the f_{ij} coefficients will be non-zero. In other words, $\boldsymbol{F}$ is a full matrix.
We emphasize this because a close parent of the flexibility matrix, the stiffness matrix $\boldsymbol{K}$ is often mostly empty, as we shall see in the sequel.
5. *Coefficient f_{ij} depends only on i and j*:
This may seem obvious but we are mentioning it to differenciate these coefficients from the stiffness matrices we will meet later. If we have defined degrees of freedom 1 and 2 on a structure, we obtain the 2×2 flexibility matrix $\boldsymbol{F}$ (10.15).

$$\boldsymbol{F} = \begin{bmatrix} f_{11} \, f_{12} \\ f_{21} \, f_{22} \end{bmatrix}; \qquad \boldsymbol{F}' = \begin{bmatrix} \boldsymbol{f}_{11} \, \boldsymbol{f}_{12} \, f_{13} \, f_{14} \\ \boldsymbol{f}_{21} \, \boldsymbol{f}_{22} \, f_{23} \, f_{24} \\ f_{31} \;\; f_{32} \, f_{33} \, f_{34} \\ f_{41} \;\; f_{42} \, f_{43} \, f_{44} \end{bmatrix} \tag{10.15}$$

When we later add degrees of freedom 3 and 4, in the ensuing 4×4 flexibility matrix $\boldsymbol{F}'$, the left upper 2×2 matrix (bold in (10.15)) is still $\boldsymbol{F}$. As we will see, in stiffness matrices, adding a degree of freedom 3 will affect all the stiffness coefficients.

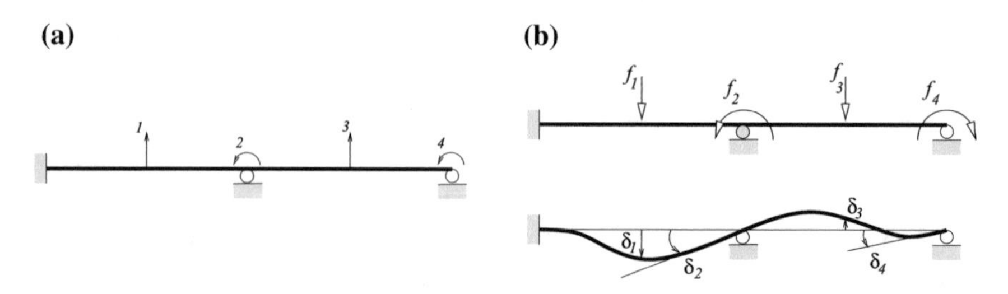

Fig. 10.9 Loads and displacements at degrees of freedom. **a** Degrees of freedom, **b** real loads and displacements

10.6 Forces and Displacements at the Degrees of Freedom

Consider the beam in Fig. 10.9a with four degrees of freedom defined along its span, and let $\boldsymbol{F}$ (10.7) be the corresponding flexibility matrix. We now apply four real forces $\boldsymbol{f}$ along the degrees of freedom. The structure moves into its equilibrium configuration and we find four real displacements $\boldsymbol{\delta}$ along the same degrees of freedom. It is clear that the displacements $\boldsymbol{\delta}$ are linearly related to the forces $\boldsymbol{f}$ by means of the flexibility matrix (coefficients F_{ij})

$$\boldsymbol{\delta} = \boldsymbol{F}\boldsymbol{f} \qquad \text{or} \qquad \begin{Bmatrix} \delta_1 \\ \delta_2 \\ \delta_3 \\ \delta_4 \end{Bmatrix} = \begin{bmatrix} F_{11} & F_{12} & F_{13} & F_{14} \\ F_{21} & F_{22} & F_{23} & F_{24} \\ F_{31} & F_{32} & F_{33} & F_{34} \\ F_{41} & F_{42} & F_{43} & F_{44} \end{bmatrix} \begin{Bmatrix} f_1 \\ f_2 \\ f_3 \\ f_4 \end{Bmatrix} \tag{10.16}$$

This results from superposition. Indeed, the first column of $\boldsymbol{F}$, which we shall call $\boldsymbol{F}_1$, has the displacements at the degrees of freedom due to a unit force applied to 1. When we apply at 1 a force of magnitude f_1 we obtain the displacements $f_1\boldsymbol{F}_1$. When we apply at 2 a force of magnitude f_2, we obtain the displacements $\boldsymbol{\delta} = f_2\boldsymbol{F}_2$, and so on. If we apply simultaneously forces at all the degrees of freedom, we have $\boldsymbol{\delta} = \sum_i^4 f_i\boldsymbol{F}_i$ or

$$\begin{Bmatrix} \delta_1 \\ \delta_2 \\ \delta_3 \\ \delta_4 \end{Bmatrix} = f_1 \begin{Bmatrix} F_{11} \\ F_{21} \\ F_{31} \\ F_{41} \end{Bmatrix} + f_2 \begin{Bmatrix} F_{12} \\ F_{22} \\ F_{32} \\ F_{42} \end{Bmatrix} + f_3 \begin{Bmatrix} F_{13} \\ F_{23} \\ F_{33} \\ F_{43} \end{Bmatrix} + f_4 \begin{Bmatrix} F_{14} \\ F_{24} \\ F_{34} \\ F_{44} \end{Bmatrix} \tag{10.17}$$

which is what appears in (10.16).

10.7 Generalized Degrees of Freedom

We can broaden the scope of (10.16) by using generalized quantities for the displacements and forces in $\boldsymbol{\delta} = \boldsymbol{F}\boldsymbol{f}$. In the original version of (10.16), the entries of $\boldsymbol{f}$ were independent forces along the degrees of freedom, and in $\boldsymbol{\delta}$ we had corresponding displacements.

What if the N forces in $\boldsymbol{f}$ were dependent on N' $(N' < N)$ scalars $\boldsymbol{f}'$ by a linear relation of the type

$$\boldsymbol{f} = \boldsymbol{H}\boldsymbol{f}' \qquad (N \times 1) = (N \times N')\,(N' \times 1) \tag{10.18}$$

where $\boldsymbol{H}$ is a $(N \times N')$ rectangular matrix of constants.

The components of $\boldsymbol{f}'$ are not forces but they are very much related to forces. Indeed for every vector $\boldsymbol{f}'$ we get through (10.18) a vector of forces $\boldsymbol{f}$. This is why the term generalized forces is appropriate. What we are looking for is an expression of the type $\boldsymbol{\delta}' = \boldsymbol{F}'\boldsymbol{f}'$ where $\boldsymbol{\delta}'$ will be some generalized displacements, and $\boldsymbol{F}'$ will have the properties of a generalized flexibility matrix.

In the first step we introduce (10.18) into (10.16) which yields $\boldsymbol{\delta} = \boldsymbol{F}\boldsymbol{H}\boldsymbol{f}'$. Next, we pre-multiply both sides of this equation by the transpose matrix $\boldsymbol{H}$. Defining the generalized displacements $\boldsymbol{\delta}'$ by

$$\boldsymbol{\delta}' = \boldsymbol{H}^T\,\boldsymbol{\delta} \qquad (N' \times 1) = (N' \times N)\,(N \times 1) \tag{10.19}$$

we get

$$\boldsymbol{\delta}' = \boldsymbol{F}'\boldsymbol{f}' \tag{10.20}$$

where the ‘generalized’ flexibility matrix $\boldsymbol{F}'$ is given by the congruent transformation

$$\boldsymbol{F}' = \boldsymbol{H}^T\,\boldsymbol{F}\,\boldsymbol{H} \qquad (N' \times N') = (N' \times N)(N \times N)(N \times N') \tag{10.21}$$

You will have noted that the components of $\boldsymbol{\delta}'$ are not displacements. They are linear combinations of displacements. Hence we call them generalized displacements.

It can be shown that the basic properties of a flexibility matrix are maintained in a generalized flexibility matrix $\boldsymbol{F}'$: the matrix is symmetric $\boldsymbol{F}'^T = \boldsymbol{F}'$, its diagonal entries are positive $F'_{ii} > 0$ and the matrix is positive definite $\boldsymbol{F}' > 0$.

10.8 Illustrated Examples

10.8.1 Propped Cantilever with Hinge

More often than not flexibility coefficients will be computed for statically determinate structures. You may be under the impression that we are studying flexibility

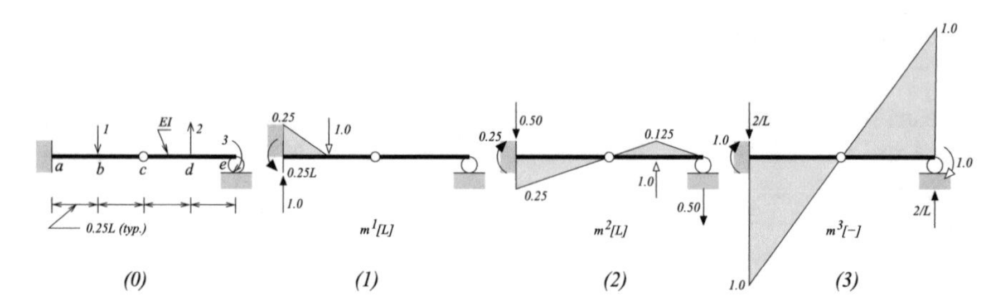

Fig. 10.10 Propped cantilever with hinge

coefficients for pure academic reasons. Although there is some truth in that, these coefficients will come in handy, very soon, when we will 'at last' be ready for the Force or Flexibility method of structural analysis.

On the statically determinate[4] beam of uniform bending stiffness EI in Fig. 10.10(0) we have defined three dofs for which we can compute the corresponding 3×3 flexibility matrix. The structure is determinate because without the hinge we have a propped cantilever of redundancy 1, the roller being superfluous for stability. The internal hinge reduces the redundancy by one, to yield $R = 0$.

The internal forces can hence be computed based on equilibrium and for the computation of the flexibility coefficients we need, in the case of beams, only the moment distributions. These are given in Fig. 10.10(1–3) for a unit load applied at dofs 1–3.

The absence of distributed forces reduces the moment functions to one or more segments of straight lines. The key for determining the moments is by noting that in the close vicinity of the hinge the bending moment is zero, $m_b = 0$.

For a unit (dimensionless) force at dof 1 the reaction R_e at the roller must be zero, otherwise we have at c a moment of magnitude $R_e \times 0.5L$ which is impossible. Without that reaction there is no bending moment along be. Hence we obtain for m^1 the typical bending moment of a cantilever of span $L/4$.

With a unit force along dof 2 the beam can be divided into two unloaded segments, ad and de. The moment distribution is thus a broken line composed of two linear segments, which is zero at the hinge c and at the roller e. The only possible configuration is the line shown in Fig. 10.102 and its companion flipped about the beam axis by 180°. The diagram can be finalized by the value of the moment at d. Considering the free-body ce with the unit force at d, we find that we need a reaction $R_e = -0.5$ to get a zero moment at extremity c. Hence the moment at the unit force $m_d^2 = -0.5 \times 0.25L = -0.125L$.

[4]As we shall see in a statically determinate structure we can compute $nsm(x)$ by means of equilibrium only.

The last diagram, m^3, is the easiest to determine. It is a straight line (the entire beam is without loads) passing through zero at the hinge and equal to 1.0 (the value of the applied couple) at e (tension above).

Having the moments, all that is left for determining $\boldsymbol{F}$ is to calculate the integrals $\int_x m^i m^j/EI\,\mathrm{d}x$ for $i, j = 1, 3$, keeping in mind that the matrix is symmetric.

$$\boldsymbol{F} = \frac{L^3}{EI}\begin{bmatrix} 0.0052 & -0.0065 & -0.026/L \\ & 0.013 & 0.0573/L \\ \text{Symm} & & 0.3333/L^2 \end{bmatrix} \tag{10.22}$$

To compute $F_{23} = (1/EI)\int_a^e m^2 m^3\,\mathrm{d}x$, for instance, using the formula, we divide the beam into segments ad and de such as to have integrals of products of linear functions. For the integrals, we use the values $g_a = 0.25L$, $g_d = -0.125L$, $g_e = 0$ and $h_a = 1$, $h_d = -0.5$, $h_e = -1$, which with $L_{ad} = 0.75L$, $L_{de} = 0.25L$ yields $F_{23} = F_{32} = 0.0573L^2/EI$.

Note the different units in the matrix. It follows from mixing translational and rotational dofs in a same matrix.

10.8.2 Portal Frame

The portal frame of piecewise uniform bending stiffness in Fig. 10.11(0) is a statically determinate structure for which we have determined 2 degrees of freedom. We could easily have increased the number of degrees of freedom. In fact the set of degrees of freedom is unlimited.

In Fig. 10.11(1, 2) we have indicated the moment distributions $m_i(x)$, $i = 1, 2$, resulting from a unit load applied in turn at each of the degrees of freedom. Remember that with statically determinate structures there is only one equilibrium solution. This means that, by inspection, when you find reactions in equilibrium with the external

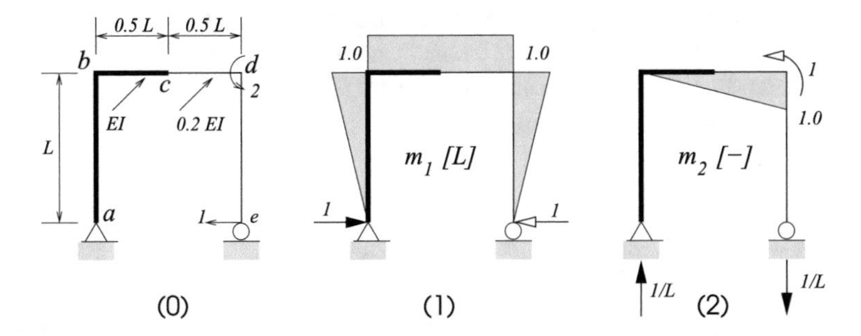

Fig. 10.11 Simply supported portal frame

loads (the loading in Fig. 10.11(1) is an obvious case) there is no need to look further. This is our lucky day. We happened to have stumbled on the only equilibrium solution.

You will have noticed in Fig. 10.11(0) that the cross-sections are uniform along *ac* with a stiffness *EI*. Along *cg* the stiffness is also uniform but that part is more flexible (0.2 *EI*). The variation in stiffness does not affect the forces (internal and reactions) because the structure is statically determine which by definition means that the forces are calculated on the basis of the equilibrium equations. In these equations the stiffnesses do not appear. But the stiffnesses have bearing on the displacements, and therefore also on the flexibility coefficients.

Using the formula for F_{12}, for instance, we have only segments *bc* and *cd* to care about. The only new thing is the stiffness which is different for the two segments.

$$\boldsymbol{F} = \frac{L^3}{EI}\begin{bmatrix} 5 & -2/L \\ -2/L & 1.5/L^2 \end{bmatrix} \tag{10.23}$$

10.8.3 Simple Truss

Consider the simple truss in Fig. 10.12a with the two indicated degrees of freedom. The bars are all of length *L* with normal stiffness *EA*. Applying forces of unit magnitude at the degrees of freedom yields the internal forces shown in Fig. 10.12b, c.

For the influence coefficient f_{21}, for instance, we need to sum up only the lower chord bars (10.6)

$$f_{21} = \frac{0.289 \times 1.0}{EA}L + \frac{0.289 \times 1.0}{EA}L = 0.578\,\frac{L}{EA}$$

measured in *length*/*force*.

Note: You will realize that, under a lateral force at mid-span, node *c* moves to the right! We may have thought otherwise (node *b* sags dragging *c* along to the left). What happens in reality is that this truss behaves like the simple beam on two supports shown in Fig. 10.12d, where *abc* is along the lower fiber and *aghc* is along the upper fiber. Both in the truss and in the beam the lower fibers are in tension and the upper ones in compression. The displacement f_{21} is to some extent representative of the extension of the lower fibers.

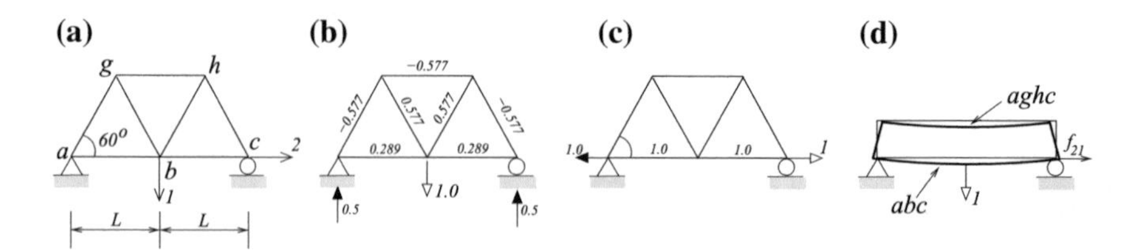

Fig. 10.12 Simple truss

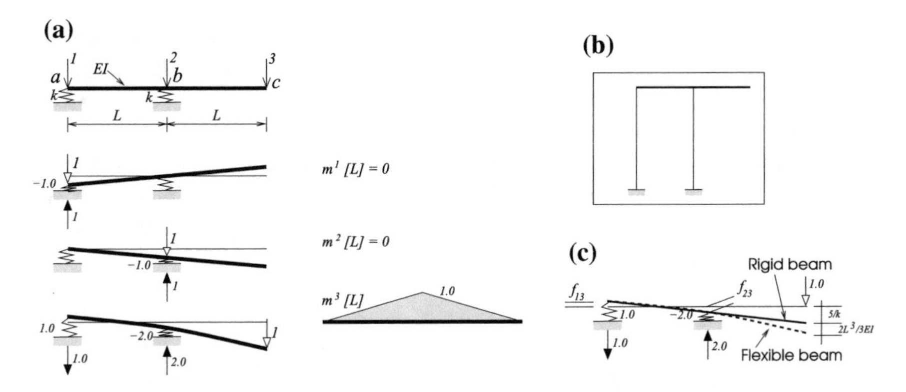

Fig. 10.13 Beam on two springs. **a** Degrees of freedom, **b** alternative structure

We will also note that following Maxwell, $f_{12} = f_{21}$, that is, under a force along dof 2 there is a lateral displacement along dof 1, which is also a startling result. Finally, the flexibility matrix is

$$\boldsymbol{F} = \frac{L}{EA} \begin{bmatrix} 0.0183 & 0.058 \\ 0.058 & 0.02 \end{bmatrix} \tag{10.24}$$

10.8.4 Beam on Elastic Supports

For the (statically determinate) beam on elastic supports in Fig. 10.13a we have defined three degrees of freedom along the lateral directions at the supports and at the free end. The flexural stiffness of the beams is EI and the stiffness of both springs is k. The structure can also be construed as a beam supported by two rods or truss elements in Fig. 10.13b. (If $E'A'$ and L' are the characteristics of the rods then $k = E'A'/L'$). But this is where the analogy ends. Springs are much more flexible than rods and their deformation is to be taken into consideration even in the presence of beam bending.

The formula for the influence coefficients (10.5), (10.6) is here

$$f_{ij} = \int_a^c \frac{m^i\, m^j}{EI}\, \mathrm{d}x + \frac{n_a^i\, n_a^j}{k_a} + \frac{n_b^i\, n_b^j}{k_b} \tag{10.25}$$

The integral represents the contribution of the deformation of the beam to the flexibility (displacement), and the following terms are the contribution of the deformation of the two springs. When we apply a unit force at degree of freedom 1, all that happens is the spring at a has a compressive force of unit magnitude and contracts. The other reaction is zero and the beam moves as a rigid body. Consequently, the analysis results are $m^1(x) = 0$, $n_a^1 = -1.0$, $n_b^1 = 0$.

Similarly, a unit force at dof 2 gives $m^2(x) = 0$, $n_a^2 = 0$, $n_b^2 = -1.0$. For the third unit force the beam deforms. We have $m^3(x)$ as shown in the figure and also $n_a^3 = 1.0$, $n_b^3 = -2.0$. The integration is easy and after arranging the terms we obtain the flexibility matrix

$$\boldsymbol{F} = \frac{1}{k}\begin{bmatrix} 1 & 0 & -1 \\ 0 & 1 & 2 \\ -1 & 2 & \left(5 + \frac{2}{3}\frac{kL^3}{EI}\right) \end{bmatrix} \quad \boldsymbol{F}_{EI\to\infty} = \frac{1}{k}\begin{bmatrix} 1 & 0 & -1 \\ 0 & 1 & 2 \\ -1 & 2 & 5 \end{bmatrix} \quad \boldsymbol{F}_{k\to\infty} = \begin{bmatrix} 0 & 0 & 0 \\ 0 & 0 & 0 \\ 0 & 0 & \frac{2}{3}\frac{L^3}{EI} \end{bmatrix} \tag{10.26}$$

Having factored out $1/k$ (measured in *length/force*) all the entries of $\boldsymbol{F}$ are nondimensional. Only in f_{33} do we meet bending deformations. In fact, the ratio expresses the play between the spring stiffness k and the the beam stiffness EL/L^3. The total displacement at 3 is due to the flexibility of the springs $5/k$ and to the flexibility of the beam $2L^3/3EI$.

In (10.26) we also indicate the flexibility matrix in two limiting cases: a rigid (infinitely stiff) beam (and finite k), $EI \to \infty$ (see Fig. 10.13c) and a flexible beam on rigid supports $k \to \infty$. Note, the deformed shape of the beam is the same as in the case of flexible supports. The two support points are different. Also, no matter whether the supports are flexible or rigid, the reactions are the same. This follows from the structure being statically determinate.

10.8.5 Generalized Flexibility Matrix

Let us consider again the beam on flexible supports, and for the sake of clarity we will be using an infinitely rigid beam. The flexibility matrix is $\boldsymbol{F}_{EI\to\infty}$ in (10.26). We now assume that for some reason the forces are not independent but they are a function of two independent scalars f_1' and f_2' in the following manner:

$$\begin{Bmatrix} f_1 \\ f_2 \\ f_3 \end{Bmatrix} = \begin{bmatrix} 1 & 1 \\ 3 & 0 \\ 2 & -1 \end{bmatrix} \begin{Bmatrix} f_1' \\ f_2' \end{Bmatrix} \tag{10.27}$$

This equation is some arbitrary instance of (10.18). The generalized displacements are (10.19)

$$\begin{Bmatrix} \delta_1' \\ \delta_2' \end{Bmatrix} = \begin{bmatrix} 1 & 3 & 2 \\ 1 & 0 & -1 \end{bmatrix} \begin{Bmatrix} \delta_1 \\ \delta_2 \\ \delta_3 \end{Bmatrix} \tag{10.28}$$

Now performing the triple product $\boldsymbol{H}^T\boldsymbol{F}_{EI\to\infty}\boldsymbol{H}$, we get the generalized flexibility relation $\boldsymbol{\delta}' = \boldsymbol{F}'_{EI\to\infty}\boldsymbol{f}'$

$$\begin{Bmatrix} \delta_1 + 3\delta_2 + 2\delta_3 \\ \delta_1 - \delta_3 \end{Bmatrix} = \begin{bmatrix} 50 & -16 \\ -16 & 8 \end{bmatrix} \begin{Bmatrix} f_1' \\ f_2' \end{Bmatrix} \tag{10.29}$$

What does all this mean? In this arbitrary case, not very much. But it will help, in the near future to have defined the concept of generalized flexibility matrix.

10.8.6 Truss-Beam on Elastic Supports

In Fig. 10.14a we have depicted a simple truss (diagonals at 45°) on elastic supports k at a and b. This structure is similar to the previous example, the beam being replaced here by this simple truss.

The influence coefficients for the three degrees of freedom are here (10.6)

$$f_{ij} = \sum_{m=1}^{7} \frac{n_m^i n_m^j}{E_m A_m / L_m} + \frac{n_a^i n_a^j}{k_a} + \frac{n_b^i n_b^j}{k_b} \tag{10.30}$$

Note the similarity with (10.25). The integral over the beam length is replaced by a summation over the seven bars of the truss. The bending of the beam for a unit force at 3 is here the forces in the bars as shown in Fig. 10.14a3. Clearly, the forces in the springs (the reactions) are the same as in the case of the beam.

10.8.7 Comparing the Beam and the Truss

It is instructive to look at the beam and truss under the same loadings. For a unit force at a and b both structures move as rigid bodies through the deformations of the springs.

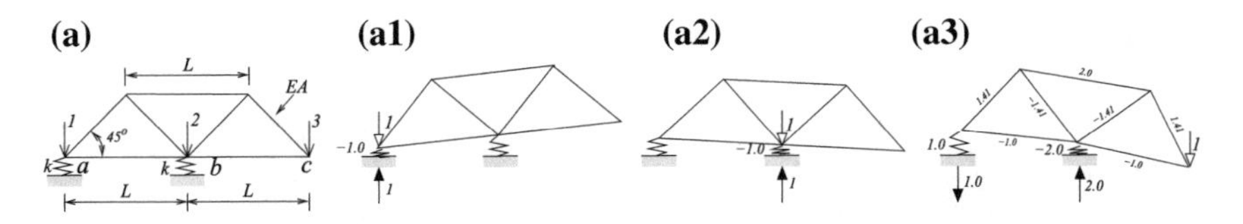

Fig. 10.14 Truss on springs. **a** Degrees of freedom

For the third loading (Fig. 10.15a1, a2) the truss and the beam have symmetric deformation patterns, although tilted, due to the differential spring deformations. Note that, similar to the beam, the lower bars of the truss (lower fibers in beams) are in compression (−1.0, −1.0) and the upper bars (fibers) are in tension (1.41, 2.0, 1.41).

The flexibility matrices are essentially the same except for f_{33}. Assuming identical EA's for all the bars we get

$$(f_{33})_{\text{truss}} = \frac{1}{k}\left(5 + 11.66\,\frac{k}{EA/L}\right); \qquad (f_{33})_{\text{beam}} = \frac{1}{k}\left(5 + 0.666\,\frac{k}{EI/L^3}\right) \tag{10.31}$$

For convenience we have added the f_{33} coefficient of the beam.

When we compare the tip displacement of the structures under a tip force we can learn a few interesting comparative properties. Let us assume that instead of elastic supports we have rigid supports ($k \to \infty$), and that we are designing a beam and a truss to have the same tip deflection $(f_{33})_{\text{beam}} = (f_{33})_{\text{truss}}$. From (10.31) we get $0.666\,L^3/EI = 11.66\,L/EA$ or

$$A_{\text{truss}} = 17.5\,\tfrac{I}{L^2} \tag{10.32}$$

This is the relation between the cross-sectional area A of the members of the truss, the moment of inertia I of the beam and the typical length L, for the beam and the truss to have the same stiffness at the extremity c.

To gain insight into relation (10.32), we assume that the beam has a rectangular cross-section $b \times h$ with a height to base ratio $h/b = 4$ and a height to half-span ratio $h/L = 0.2$ (see Fig. 10.15b). The cross-sectional area of the beam is thus $A_{\text{beam}} = bh = 0.01\,L^2$. The moment of inertia of the beam is $I = bh^3/12 = 3.33\,10^{-5}L^4$, which when introduced in (10.32) yields $A_{\text{truss}} = 5.833\,10^{-4}L^2$. For trusses the shape of the cross-section does not influence the axial stiffness. Let us assume a square $a \times a$ cross-section. With $a^2 = 5.833\,10^{-4}L^2$ we get $a = 0.0242\,L$. This should be compared with the small side of the cross-section of the beam $b = 0.05\,L$.

To emphasize this result compare the cross-section of the truss element with the one of the beam in Fig. 10.15b. Granted the truss elements are longer than the beam.

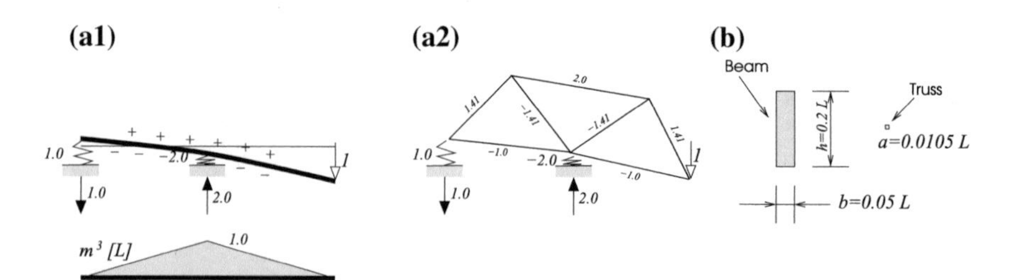

Fig. 10.15 Comparing beam and truss

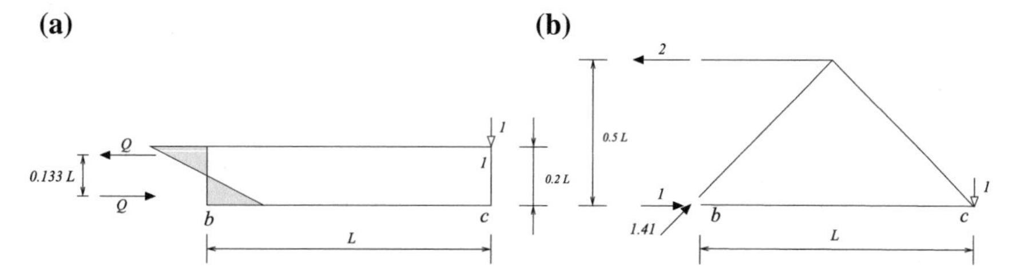

Fig. 10.16 Comparing the stresses

Taking the lengths into account we find that the ratio of truss weight over the beam weight, for a same material of course, is

$$\frac{W_{\text{truss}}}{W_{\text{beam}}} = \frac{(3 + 4 \times 0.707)\, L\, A_{\text{truss}}}{2\, L\, A_{\text{beam}}} = 0.17 \tag{10.33}$$

The beam is about 6 times heavier than the truss.

This is simple accounting. Let us have a look at the bending moment at the support b (a little to the right of b). Evidently, span bc is a cantilever and when a unit force is applied at c the bending moment at b is L (tension in the upper fibers). How do the beam and truss cope with this moment?

In Fig. 10.16 we have depicted the overhang bc for the beam (a) and the truss (b). The drawings are somehow to scale. For the truss it is simple. The moment L (remember, dimensionless forces) is taken by a couple of forces of magnitude 2 distant by $0.5\,L$ (the lower chord has a compressive force of $1 + 1.41 \cos 45° = 2$). For the beam bending, we will get the familiar symmetric (assuming that the cross-section is symmetric) triangular distribution of longitudinal stresses. The tensile and compressive parts can be replaced by equivalent forces Q positioned at $1/3$ of the triangles height. Consequently, we understand that the moment L is taken by a couple $Q \times 0.133\,L$. An estimate of the forces at play in the beam is therefore $Q = 1/0.133\,L = 7.5\,L$ (as compared to $2\,L$ for the truss.) The beam develops higher stresses and higher strains, hence its increased flexibility.

Notwithstanding all this, the beam is still the most widely used structural element. One of the reasons is that structural design is more often than not cost driven rather than weight driven. A beam may be using up more material than a truss but it is much easier (and cheaper) to manufacture, transport and erect (think of what is involved in connecting all the elements of a truss). Also, beams take up less space than trusses which may often play in favor of the former design.

Moreover, in designing the truss we did not take into consideration the possible *buckling* of the compressive bars. To prevent this Euler buckling we would have to increase the cross-sections of the bars in compression well beyond the A_{truss} that is given in (10.32).

Table 10.2 Calculating F_{12}

Coefficient	Interval	g_1	g_2	h_1	h_2	Result
F_{12}	ab	−0.188	0.156	−0.5	0.25	0.0118
F_{12}	bc	0.156	0	0.25	1.00	0.0195

10.8.8 Propped Cantilever

Consider the propped cantilever of uniform cross-section for which we have defined two degrees of freedom as indicated in Fig. 10.17a. The structure is redundant, and we have not yet covered the subject of analysis of redundant structures. So we assume that we are given the moment distributions $m^i(x)$ for a unit force at i and $m^j(x)$ for a unit couple at j.

In this and in most other examples which we will solve, the structure can be broken down into segments for which both m^i and m^j are linear functions of x, in the present case, segments ab and bc. The partition is numerically convenient because we have a formula for the integral of two linear functions $g(x)$ and $h(x)$ defined by their end values on an interval $[x_1, x_2]$: $g_1,\ g_2,\ h_1,\ h_2$, respectively,

$$(1/L_{12}) \int_{x_1}^{x_2} g(x)h(x)\mathrm{d}x = \frac{1}{3}\,(g_1h_1 + g_2h_2) + \frac{1}{6}\,(g_1h_2 + g_2h_1) \tag{10.34}$$

where $L_{12} = x_2 - x_1$ is the element length.

The four influence coefficients are thus $F_{ij} = \int_a^b (m^i m^j/EI)\mathrm{d}x + \int_b^c (m^i m^j/EI)\mathrm{d}x$. The details for calculating F_{12}, for instance, by means of (10.34) are given in Table 10.2 with $F_{12} = F_{12}(ab) + F_{12}(bc)$.

$$\boldsymbol{F} = \frac{L}{EI}\begin{bmatrix} 0.0092\,L^2 & 0.0313\,L \\ 0.0313\,L & 0.25 \end{bmatrix} \tag{10.35}$$

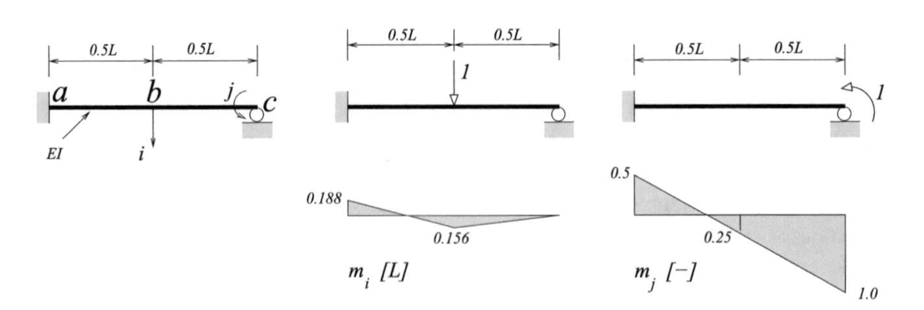

Fig. 10.17 Propped cantilever

10.9 Summing Up

This chapter introduced basic tenets of structural theory starting with Betti's theorem which lies at the heart of structural analysis. We have defined the pervasive degrees of freedom and the notion of influence or flexibility coefficients which can become the symmetric components of flexibility matrices. After summarizing the properties of flexibility matrices, the last part of the chapter presents flexibility matrices in a wide range of typical structures.

Chapter 11
Redundancy

From an analysis aspect, determining the degree of static redundancy of a structure is a prerequisite of the force method. As such, this chapter can be considered as a first step in the force method. For the displacement method we do not need to know the degree of redundancy.

Redundancy also tells us how far the structure is removed from being statically determinate, a condition which is notoriously unsafe. As such, the redundancy is also important from a design viewpoint irrespective of the method of analysis. It tells us something regarding the behavior of the structure. Recognizing the redundancy, that is, safety, stiffness, prestressing issues and the like, or designing for determinacy in order to prevent locked stresses, will be discussed at the end of the chapter.

11.1 Typical Redundant Structures

In Fig. 11.1 we have typical examples of a determinate and redundant structures. The simple beam to the left is statically determinate. The reactions can easily be computed (two unknowns which are computed by two equilibrium equations: the vertical and rotational equations). From there the $s(x)$ and $m(x)$ follow suit. The degree of redundancy is $R = 0$.

The following two examples depict redundant structures. In these cases we cannot obtain the internal forces $s(x)$ and $m(x)$ using equilibrium only. For the beam on three supports we cannot compute the reactions. Indeed, we have three unknown reactions and still only two equilibrium equations. We have one unknown too many, hence the degree of redundancy is $R = 1$. The structure is said to be externally redundant.

The last example is more interesting. The frame on simple supports is also redundant although we can compute the reactions using equilibrium. However, we cannot proceed to compute the internal forces with only equilibrium equations at our disposal. We shall see that the degree of redundancy of that structure is $R = 3$. The structure is said to be internally redundant.

M.B. Fuchs, *Structures and Their Analysis*,
DOI 10.1007/978-3-319-31081-7_11

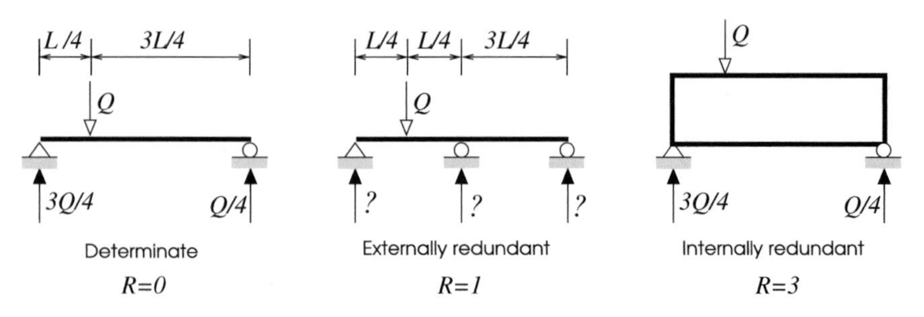

Fig. 11.1 Typical determinate and redundant structures

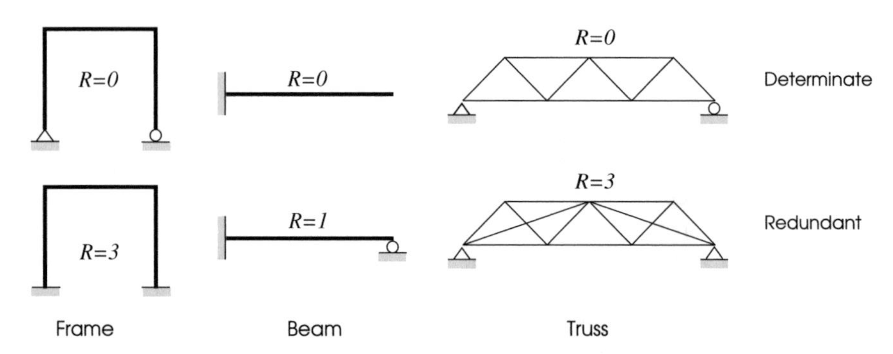

Fig. 11.2 Degree of redundancy

11.2 Definition of Statical Redundancy

A structure is said to be statically *determinate* if we *can* compute the internal forces $\boldsymbol{n}(x)$, that is, $n(x)$, $s(x)$ and $m(x)$, at any section x along the structure by means of *equilibrium equations only*. Conversely, if we *cannot* compute $\boldsymbol{n}(x)$ using only equilibrium equations, the structure is statically *redundant*.

The degree of redundancy R is the number of equations we are missing to compute $\boldsymbol{n}(x)$. In the first row of Fig. 11.2, we show three typical statically determinate structures $R = 0$ (we have all the equations we need), and in the second row we can see corresponding redundant examples. For the frame and the truss we are missing three equations, and in the propped cantilever example ($R = 1$) we are one equation short. You will have noticed that the redundant designs seem stronger than the determinate ones: the frame has now both columns built in; to the beam we have added a simple support at the right extremity, and the truss now has two additional elements, and the roller was replaced by a hinge. There is thus a numerical aspect to redundancy—the number of missing equilibrium equations—and a physical aspect to redundancy—safety and stiffness. In this chapter we deal with numerical considerations first.

11.3 Counting the Redundancy

We will first describe a systematic method for establishing the degree of redundancy R of a structure by *counting* the number of independent unknown internal forces U and the number of equilibrium equations E, and base the redundancy R on the difference between the two

$$R = U - E \tag{11.1}$$

In many cases we can arrive to the same result by *inspection*, less systematic but closer to the mechanical behavior of the structure. We start with the counting.

11.3.1 Elements and Nodes

Consider the portal frame in Fig. 11.3a and let us decompose it into three elements ab, bc and cd, and four nodes a, b, c and d, making sure, for the sake of argument, that the sections are a little removed from the reactions and the corners.

The structure is the sum of its elements. Nodes are dimensionless points at the extremities of the elements. Note, support a is clamped whereas reaction d is hinged (the column can rotate at d).

- *The elements govern the number of unknowns U*
  The central idea is to acknowledge that, if we are given the internal forces at one extremity of a line element we can compute the internal forces anywhere along the element using equilibrium. Thus, for element bc if $\boldsymbol{n}_b$, that is nsm_b at extremity b, is given, we can compute $\boldsymbol{n}(x)$ at any section x along bc. In other words, an element carries three unknowns. If these three unknowns at one extremity become known, we will be able to work out the internal forces anywhere along the element.
  Consequently, the portal frame in the example carries 3 (elements) × 3 (unknowns) per element, that is, $U = 9$ unknowns in total for the entire structure.

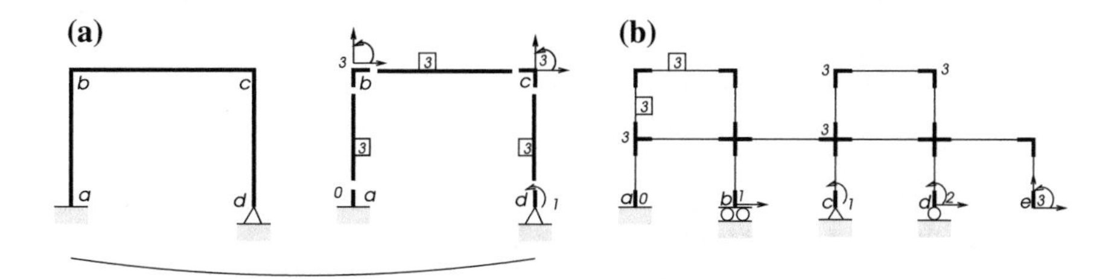

Fig. 11.3 Counting the redundancy in frames

With this we have written equilibrium equations for the entire structure, except for the nodes which must also be in equilibrium.

- *The 'open' degrees of freedom at the nodes provide the number of equations E.* Every node has nominally three degrees of freedom. Degrees of freedom are either open ('free', a bit confusing) or closed ('fixed'). When the structure has reached its equilibrium configuration (it does not move), the node is necessarily in equilibrium. In other words, the forces applied to the node in the direction of the open degree of freedom are in equilibrium. This is an equation we can write.
  Node b, for instance, has 3 free degrees of freedom. It is subjected to the internal forces of extremity b of bc and also to the internal forces of extremity b of ab. It can also have some external forces applied to it. All this must be in equilibrium. Node b provides therefore 3 equations. This is true for nodes b and c. At each of them one can write 3 equations.
  Node a and d are different. At node a all the degrees of freedom are closed. It is automatically in equilibrium.
  An alternative way of considering fixed degrees of freedom is as follows. At node a we could write three equilibrium equations, however, we also have three (unknown) reactions there. So if we decide to write the three equilibrium equations, we introduce three new unknowns, the reactions, in addition to the intrinsic unknowns of every element. This is not of much help. Therefore, node a does not provide *useful* equilibrium equations. We say that node a provides 0 equations.
  Finally, node d, is a hinge. Only the rotational degree of freedom is free. The translational degrees of freedom are fixed. When we write the three equilibrium equations, $\sum F_x = 0$ and $\sum F_y = 0$ do not give much because the unknown vertical and horizontal reactions at d will come into play (we provide two equations but we also introduce two new unknowns). So these equations are not effective. However, $\sum M_d = 0$ is an effective equilibrium equation. This condition states that node d is hinged and the ground cannot apply a couple there. Consequently, we will find that the (internal) bending moment is zero in a section of the beam very close to d. (If there were a bending moment, there would be nothing to balance it.)
- *The redundancy R is the number of unknowns U minus the number of equations E.* The number of Equations is the sum of the 'free' degrees of freedom at the nodes. The number of Unknowns is the number of elements times the number of unknowns per element. The redundancy is in fact the number of equations we are missing to compute the unknowns.
  For the portal frame the unknowns are $U = 3 + 3 + 3 = 9$ (three elements with three intrinsic unknowns each). The number of equations counting clockwise from a is $E = 0 + 3 + 3 + 1 = 7$. The redundancy is thus $R = U - E = 2$ (we are missing two equilibrium equations). The Force method will eventually provide two additional equations known as compatibility equations.

Another simple example is the two-storey frame in Fig. 11.3b. In the figure the nodes are blackened and the elements remain lightly shaded. The 13 elements carry $U = 13 \times 3 = 39$ unknowns.

The nine nodes connecting the floors to the columns have each three free degrees of freedom. At ground level we find five different types of nodes. Node a is fixed (zero degrees of freedom). Node b is a horizontal guide and allows only horizontal translation (1). Node c is a hinge and has only rotational freedom (1). Node d is a roller (2). Only the vertical degree of freedom is fixed. Node e is didactically important. It reminds us that we must position a node at both extremities of every element even if it is a free end. Since all the degrees of freedom are open, node e provides three equations. The total number of equations is $E = (9\times 3 =)27+0+1+1+2+3 = 34$.

This frame is thus redundant of degree $R = 39 - 34 = 5$. If we try to compute the internal forces with equilibrium conditions only we will be missing five equations. The Force method will provide those missing in the form of five compatibility equations.

11.3.2 Other Decompositions

There are many ways to divide a structure into nodes and elements, the one chosen in Fig. 11.3a is only one option. In Fig. 11.4 we give additional examples for decomposing the same portal frame. All lead, fortunately, to $R = 2$.

You will note that elements need not be straight. In the first example of Fig. 11.4 the frame was decomposed into two segments, including the broken line *abc*, and in the second example we have one broken line *abcd*. Here also, if the internal forces applied to the node at one extremity are known, we will be able to compute the internal forces at any section x.

The method is simple. We should however be careful always to have nodes at both extremities of the elements, *including free ends*; forgetting a free-end node is a common error.

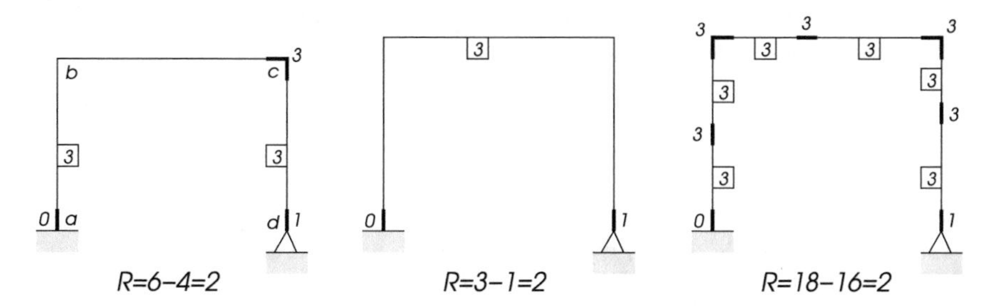

Fig. 11.4 Decompositions into nodes and elements

11.3.3 Internal Releases

We are now familiar with partially fixed nodes at the supports. Such supports can be construed as partially released. Some structures may also have internal releases embedded in one or more elements.

The elementary releases are the *hinge* or moment release, the *lateral guide* or shear release and the *longitudinal guide* (*telescopic connection*) or normal force release (see Fig. 11.5).

It is imperative to position a node at such a release in order not to have elements with embedded releases. Leaving a release inside an element only complicates matters.

Now, *every such release introduces a new equilibrium equation at the node* and thus increases E by 1 (and reduces the redundancy by 1). As shown in Fig. 11.5a–c, for the three basic releases we need separate equilibrium equations along degrees of freedom 3 and 4. If you consider, for instance, the guided release in Fig. 11.5b, it is clear that the $\sum F_y = 0$ condition must be written separately for dof 3 attached to the left part, and for dof 4 attached to the right part. (For a regular solid node there is only one equation in the lateral direction, the sum of the lateral forces along 3 must be zero. Adding an equilibrium equation reduces the redundancy by 1 ($U-(E+1) \rightarrow R-1$).)

The same prevails for the hinge. Here the couples along 3 for the left part and the couples along 4 for the right part must separately be zero, otherwise the element extremities rotate unchecked about the hinge. Finally, for the telescopic connection, the normal forces along 3 for the left and along 4 for the right must separately be zero. All such nodes are therefore worth four equations.

Consider, for instance, the arches in Fig. 11.6. The number of unknowns is indicated by an arrow to the corresponding element, and the number of equations appears next to its node. Starting from the left we have an arch built in at both extremities. With one element and zero dofs the redundancy is $R = 3 - 0 = 3$.

Introducing a hinge at the left reaction adds a dof and enables writing a moment equation about the reaction point. The redundancy is now $R = 3 - 1 = 2$. An additional hinge to the right gives $R = 3 - 2 = 1$.

When we now remove the ability to apply a horizontal reaction in the left support, we are left with a roller support which can apply a vertical reaction only (two dofs). Consequently, we have now a statically determinate structure, $R = 3 - 3 = 0$. (You will have recognized a simply supported arch.)

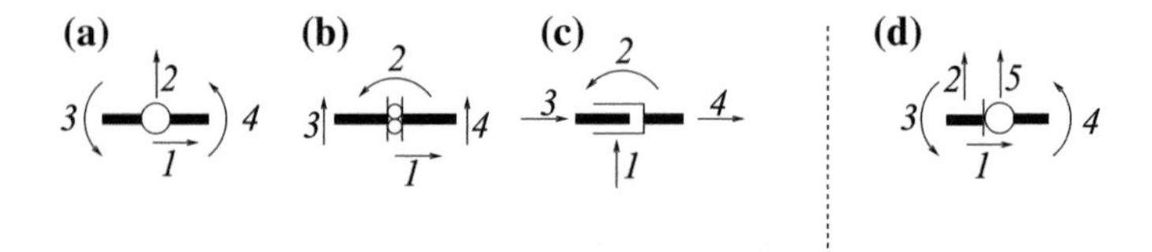

Fig. 11.5 Internal releases. **a** Hinge. **b** Guide. **c** Telescopic. **d** Roller

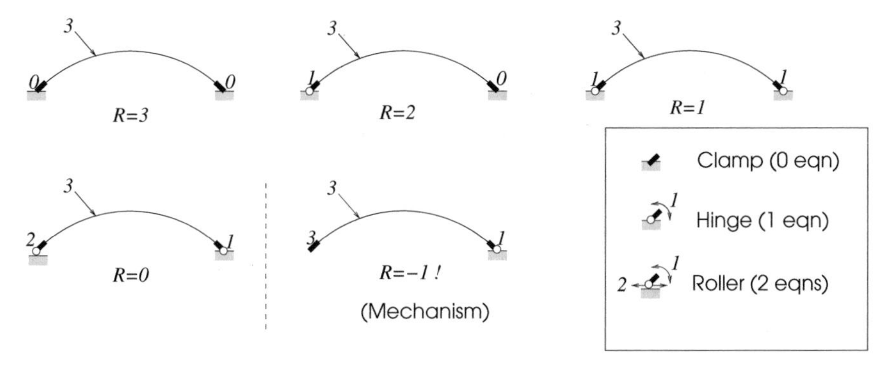

Fig. 11.6 Redundancy in arches

Don't add another release. For instance, if we remove all the reactions from the left we get a free end, which still requires a node worth three equations for the three dofs. Adding the dof to the right yields $E = 4$. This produces a negative redundancy ($R = 3 - 4 = -1$), that is, a mechanism (indeed the thing dangles freely about the right hinge).

11.3.4 Frames with Truss Elements

Framed structures often have truss elements (rods, bars, cables) embedded in the assembly. Such elements being long and slender have relatively low or no bending stiffness but are very stiff axially (for cables we assume tension). The purpose of such elements is often to add an axial stiffness in the direction of the element axis. Bending is a relatively flexible mode of deformation when compared to axial extension. For this reason frame structures are sometimes reinforced with axial elements, or truss elements. Due to their low bending stiffness when compared to the surrounding frame such elements draw little bending but substantial axial forces. Consequently, we often assume that these elements have zero bending and shear, and the only internal force is the normal force n.

So *this element carries only one unknown* instead of the usual three. This has to be remembered when determining the redundancy of the structure.

Consider the case of the emblem above the doorstep to tailor Rich's shop (Fig. 11.7a). The initial idea was to support the sign by means of a cantilever beam, however the arrangement proved too flexible. Instead of using a heavier beam the structure was stiffened by means of a rod or a cable, also called a truss element.

The redundancy of the new structure is nominally $R = 3$. Indeed, two elements each worth three unknowns give $U = 2 \times 3 = 6$, which with three dofs at the extremity results in $R = 6 - 3 = 3$. However, due to the thinness of the truss element its bending rigidity is very small but its axial stiffness is very effective. Consequently,

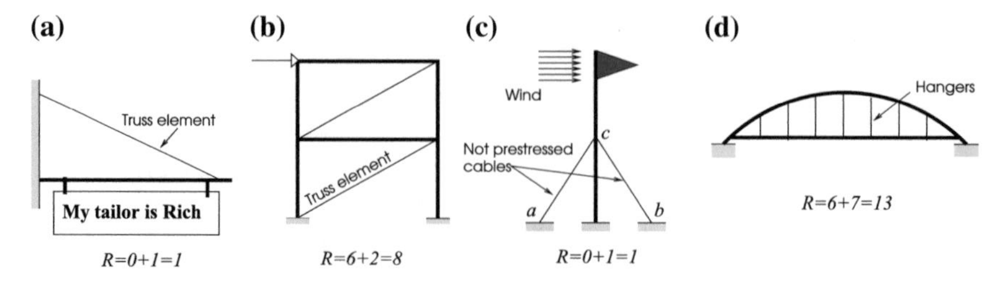

Fig. 11.7 Combining bending and axial elements

we ignore the bending effect ($s = m = 0$) of the truss element and assume that it will carry only axial forces. The diagonal element has, therefore, only one static unknown and the redundancy is only $R = (3 + 1) - 3 = 1$.

The 2-story, 2-column frame in Fig. 11.7b proved also too flexible under a side load. It was braced by two truss elements to reduce the side-sway. Dividing the structure into six frame elements, two truss elements and four nodes gives $R = ((6 \times 3) + (2 \times 1)) - (4 \times 3) = 8$. Incidentally, if the side load were to change direction the truss elements would be in compression, and there is always the danger of Euler buckling of either element. As with all other failure modes, we assume here that the loading and design are such that buckling and other failure modes do not occur.

The mast in Fig. 11.7c is similar, however here they have used cables instead of bars. A cable is a non-linear element. In tension it has axial stiffness and it behaves much like a bar, if we disregard the issue of slack in the cable. In compression the cable is nonexistent. Its stiffness depends therefore on its internal force, and this is why two cables are used. When the wind blows in the direction of the figure, cable ac is in tension and cable cb is of no use. But since we have little control of the wind direction, we need both cables. Without pre-tension we will always have only *one* active cable. With two bending elements, one axial element and two free nodes (one at c and one at the top of the mast) the redundancy is $R = (2 \times 3 + 1) - 2 \times 3 = 1$. It is in fact the barber's cantilever rotated by 90°.

The bow-string bridge in Fig. 11.7d is a good example for introducing the method for determining redundancy by inspection, because the counting approach is rather tedious. We have 14 free nodes at the connections of the hangers to the arch plus two nodes connecting the beam (roadway) to the arch. Consequently $E = 16 \times 3 = 48$. All these nodes connect 18 frame elements plus seven hangers, that is, $U = 18 \times 3 + 7 \times 1 = 61$. (The hangers connecting the roadway to the arch are typical extensional elements.) The redundancy is therefore $R = 61 - 48 = 13$.

On the other hand, with the method for finding the redundancy by inspection we could argue as follows. It is easy to determine that the frame without hangers is six times redundant. With the seven hangers we are adding seven unknown axial forces, hence $R = 6 + 7 = 13$.

11.4 Redundancy by Inspection

Engineers do not always determine the degree of redundancy of a structure by counting nodes and elements. With time you will look at a structure and say, "Oh! this one is easy. The redundancy is x or y." We may exaggerate here in order to make the point, but in practice the degree of redundancy is often found by inspection. The technique is based on the number of 'releases' we have to introduce into the structure in order to bring it down to a state of known redundancy, often of redundancy zero (determinate). Remembering the redundancy of a few typical structures, especially the popular tree, helps.

11.4.1 The Simple Tree

The simple tree in Fig. 11.8a is statically determinate. The structure has a fixed trunk with branches extending from it. To show that $R = 0$, start with the trunk in Fig. 11.8b. Here $R = 0$, obviously. It is a cantilever. Remember to put a node at both ends of every element. Next, we show in Fig. 11.8c that inserting a node in an element also adds an element, and the operation does not change the redundancy (we add three unknowns and three equations). Finally, branching off as in Fig. 11.8d also adds both an element and a node, consequently R remains unchanged. It is clear that we can construct any simple tree in this manner, and since $R = 0$ for the tree in Fig. 11.8b, it will remain so for the tree in Fig. 11.8a. (Note a branch need not be straight.)

However, if a branch loops, that is, if it grows back into the tree, the structure becomes redundant because when we connect an element between two existing nodes we add three unknowns (for the element), but we get no extra equations. Hence the redundancy increases by 3. For instance, when we move from Fig. 11.8d, e, we add

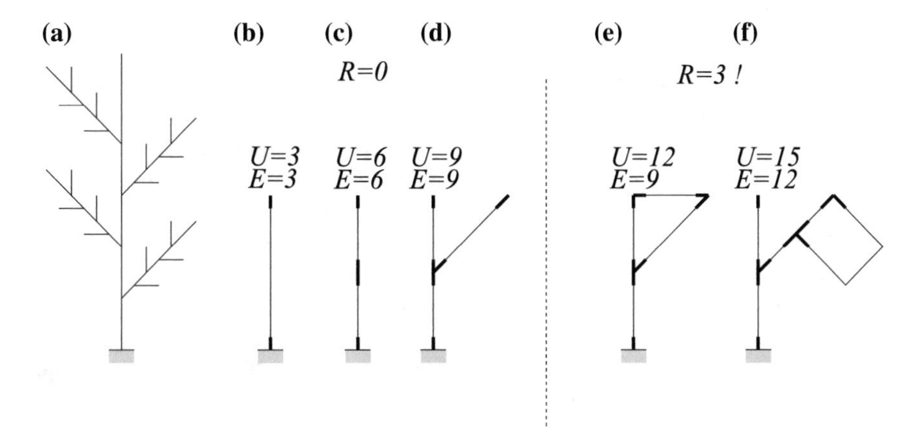

Fig. 11.8 Trees are statically determinate

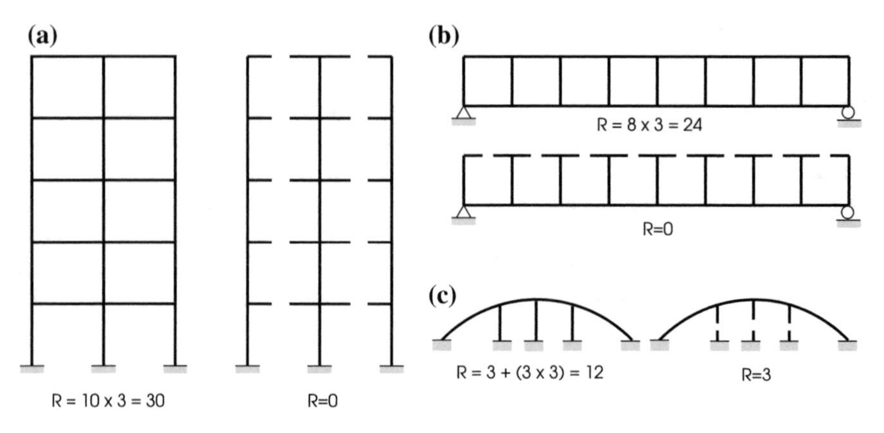

Fig. 11.9 Redundancy by releasing

an element thus increasing U by 3 but the dofs (E) remain unchanged. The same argument is valid for Fig. 11.8f.

The simple tree need not be rooted in the ground. All it needs is three proper reactions. It can, for instance, sprout from a statically determinate beam. Other typical structures are the clamped arch or *portal frame* ($R = 3$) and the *simple beam* ($R = 0$). With experience, we become acquainted with redundancy from our examination of other structures, which will help in finding the redundancy of a new structure we have to deal with.

Before concluding, we leave it to the reader to realize that adding a tree to a structure does not affect the redundancy of the structures.

11.4.2 Recognizing the Redundancy by Releasing the Structure

We introduce 'releases' in the structure until a structure with a known redundancy is generated. The number of releases plus the redundancy of the released structure is the redundancy of the original one. For the five-storey building in Fig. 11.9a, ten total cuts (releases) of the beams will reduce it to three separate trees. These tree structures are statically determinate and their internal forces can be found by equilibrium considerations only; however, we are missing the values of the 30 internal forces (three per release or *nsm* at every cut). Consequently, $R = 10 \times 3 = 30$.

A similar approach can be used to determine the redundancy of the Vierendeel beam in Fig. 11.9b.[1] Cutting open the cells along the upper (or lower) flanges produces a tree (more like bushes) on a simply supported beam. The result is statically

[1] A Vierendeel beam is often used in place of a similar truss-beam. The shear force which is taken in trusses by the diagonal braces is here transferred through shear in the flanges.

determinate. However, at the eight total releases we have 8×3 static unknowns, hence $R = 24$.

Finally, the structure in Fig. 11.9c is of redundancy $R = 12$. Indeed, three total cuts of the columns reduce the system to three trees (columns) and a doubly clamped arch from which three trees sprout. The arch has a redundancy of 3 (its trees do not modify it), the three cuts are worth three unknowns each, consequently the structure is redundant of degree $R = 3 + 3 \times 3 = 12$.

11.5 Beams

A beam is a structure which can be modeled as a straight line, and is subjected to lateral forces and couples without any axial components. This is usually enough to guarantee that the reactions have only lateral components and that $n = 0$ in the structure. If these conditions are not met the structure will be treated as a frame with all possible types of internal forces n, s, m.

By removing anything axial (including displacements) we are streamlining the analysis. In a beam analysis, every element carries two static unknowns, the shear and the moment, and every free node has two degrees of freedom, a lateral translation and a rotation. Having learned our trade with frames one should have no problem in determining the redundancy in beams. You assign two equations per node and two unknowns per element (instead of three and three as with frames).

11.5.1 Counting Unknowns and Equations

Consider the beam in Fig. 11.10a0. We have two elements. Every element carries two static unknowns, say s and m at the left end ($n = 0$). The number of unknowns is thus $U = 2 \times 2 = 4$. How many equations can we write at the nodes? The fixed end is safe; no equations. The roller supports allow horizontal translation and rotation until an

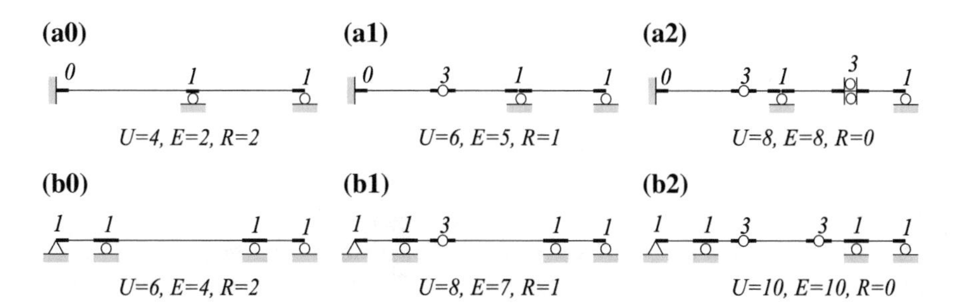

Fig. 11.10 Counting equations and unknowns

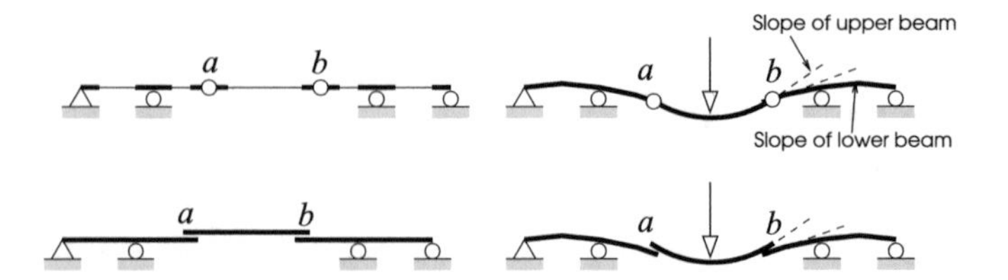

Fig. 11.11 Physical interpretation of hinge in a continuous beam

equilibrium configuration is reached (the vertical displacement is barred). But since we have a beam we are not concerned with the axial (horizontal) displacements. Consequently, a roller support provides one (the rotational equilibrium) equation. The nodes have thus $E = 2 \times 1$ equations and the redundancy of the beam is $R = 4 - 2 = 2$.

With a hinge in the first span (Fig. 11.10a1) we have three elements ($U = 3 \times 2 = 6$). The hinge is worth three equations, one less than for frames (the two rotations on both sides of the hinge and the common lateral translation). This yields $E = 3 + 1 + 1 = 5$ equations. We are missing the $R = 1$ equation. Note, introducing a hinge has reduced the redundancy by 1.

Finally, an additional lateral guide in the second span (Fig. 11.10a2) generates a statically determinate structure (four elements, eight equations).

The continuous beam in Fig. 11.10b0 has $R = 2$ redundancies. Introducing one hinge, then two hinges in the middle span (Fig. 11.10b1, b2) reduces the redundancy to $R = 1$ and $R = 0$ respectively.

Note: The determinate arrangement in Fig. 11.10b2, repeated in the first line of Fig. 11.11, could be a model for the structure in the second line of the same figure.

It is composed of two symmetrically located simply supported beams, both with an overhang, plus a third beam *ab* resting on the two beams. The connection between the central and the two side beams at *a* and *b* is, for all practical purposes, a hinged connection. When loaded by a lateral force, the beams have a common vertical displacement at *a* and *b* but different slopes, which is exactly what a hinge does.

So-called Gerber beams are designed along similar lines. Nominally redundant they can be considered statically determinate by positioning hinges at locations of zero bending moment.

11.5.2 Redundancy by Inspection

We propose memorizing three simple cases (Fig. 11.12): these are (a) the simply supported beam; (b) the cantilever beam (both statically determinate); and (c) the fixed-ends beam, which has redundancy $R = 2$. (See the second row in the figure to count the redundancy.) Bear in mind that with beams we assume that there are no axial phenomena.

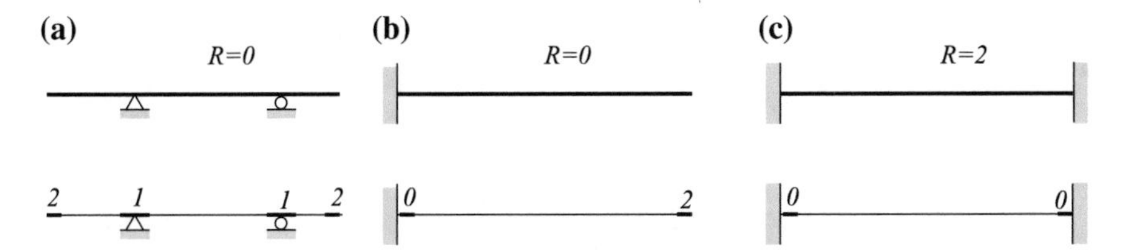

Fig. 11.12 Beam redundancy by inspection

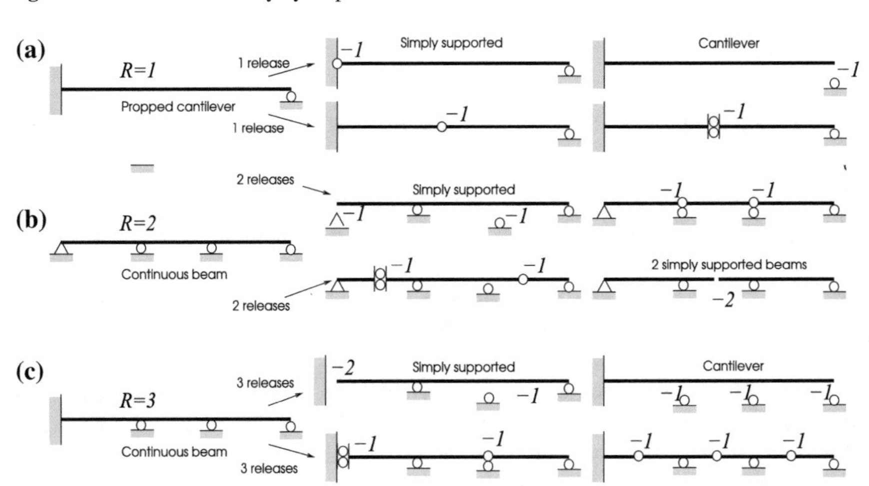

Fig. 11.13 Beams by inspection; more examples

Given a beam, one can now try to release it until an arrangement of known redundancy is obtained.

The first example is the propped cantilever in Fig. 11.13a. Cantilevers are rather flexible arrangements, and it is often necessary to support the free end, hence the propped cantilever design. Its redundancy is $R = 1$ since one release will yield a statically determinate arrangement.

Several alternative releases are shown in the figure. Inserting a hinge in the wall creates a simply supported beam (a hinge to the left and a roller to the right). The wall prevents rotation and vertical displacement, and with a hinge we have now allowed the rotation at the wall.

Next, one can disconnect the roller support from the beam to yield a cantilever, but we don't know the value of the reaction, hence the structure is statically redundant of degree 1. Note, a hinge or a guide can be introduced anywhere along the span also yielding $R = 0$.

The second example is the continuous beam on four supports in Fig. 11.13b. Its redundancy is $R = 2$ since two releases produce statically determinate structures, either by removing supports or by introducing hinges and guides, each one reducing the redundancy by one. In one alternative, two releases, a hinge and a guide, were introduced at the same point along the central span (this is in fact a total cut), which resulted in two simply supported beams, each statically determinate, of course.

Finally, the beam in Fig. 11.13c has $R = 3$, since three releases are necessary to produce a statically determinate system.

Continuous beams are usually released by hinges at intermediate supports.

11.6 Trusses

11.6.1 Counting Unknowns and Equations

The structure depicted in Fig. 11.14a is typical for several one-span railway bridges. Although its members are rigidly connected, we will consider it in a first approximation as a truss. As indicated earlier, in a truss a member is worth one static unknown, the normal force, and every free node has two degrees of freedom (the rotational degree of freedom is always satisfied in trusses since all the forces pass through the node).

In the example, the structure has 50 bars and 20 free nodes (we assume that the X braces are not connected in the center) plus the roller support which has one degree of freedom when in a truss, the horizontal displacement. (Note, in a frame the roller has two degrees of freedom, a displacement parallel to the ground and a rotation). With $E = 40 + 1$ the static redundancy is $R = U - E = 9$.

When one diagonal is removed from every cell (Fig. 11.14b) we have $R = 41 - 41 = 0$. The truss being statically determinate one can readily calculate the normal forces using equilibrium equations only. Had we considered a frame analysis the redundancy would have been $R = (50 \times 3) - (20 \times 3 + 1) = 89$!

Finally, the arch in Fig. 11.14c is a triangulated pattern on two hinged supports (four force reactions, no couples). Counting bars and open degrees of freedom yields $R = 1$.

11.6.2 Redundancy by Inspection

The one recognizable statically determinate truss is the *simple truss*, an example of which is given in Fig. 11.15a5. One possible description of a simple truss follows its

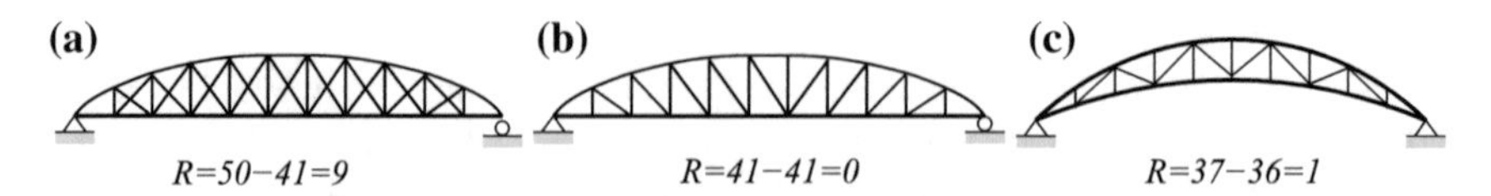

Fig. 11.14 Truss bridges

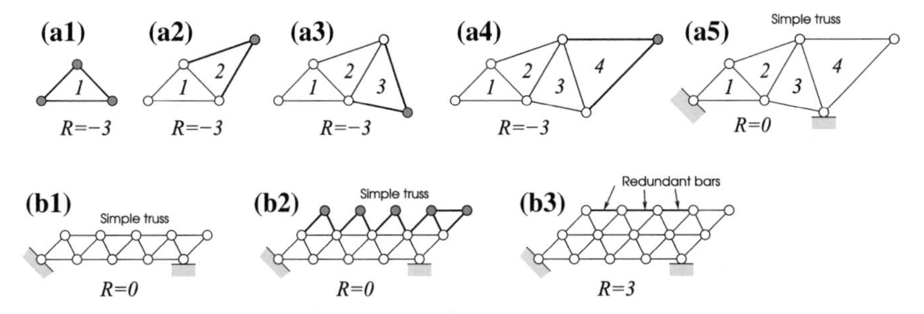

Fig. 11.15 Redundancy of triangulated patterns

method of assembly. We start with a triangle (Fig. 11.15a1), three nodes and three elements, that is $U = 3,\ E = 6$ and consequently $R = -3$. A negative redundancy is not very encouraging, but we are not done yet. We now expand the structure by adding *one new node and two new members* at the same time, as in Fig. 11.15a2, a4. This operation does not change the redundancy. Indeed, adding a node adds two equilibrium equations and adding two members adds two unknowns. Since both U and E are increased by 2 in each step, redundancy is unchanged and we remain with $R = -3$. Finally, we close three degrees of freedom by providing three reactions, as in Fig. 11.15a5. We now have three less equilibrium equations to write, and the redundancy increases by 3, consequently we obtain $R = 0$. If we can construct our truss by this method our structure is a simple truss and it is statically determinate.

We should be cautious with this. Not every triangulated structure can be assembled this way. A common pitfall is the truss in Fig. 11.15b3. It has the looks of a simple truss but it is not. The first two layers of nodes in Fig. 11.15b1 compose a simple truss. We can now add the third layer of nodes according to the rules, starting from left, and still have a simple truss (Fig. 11.15b2). But to complete the structure, we now must connect three bars to *existing nodes*. This breaks the basic rule of adding a node and two bars. We cannot add a bar to two existing nodes and maintain the redundancy level because the additional bar is redundant. Indeed, structure Fig. 11.15b2 is perfectly stable without the additional bars.

Another way to detect redundancy is to realize that when you add a node and two bars you are not very dependent on the fabrication tolerances of the bars. If the bars have slightly different lengths the truss can still be assembled although the nodes will have small differences in coordinates. But when you add a bar between two existing nodes the bar must have the same length as the distance between he nodes. There is no room for tolerance.

Determining the redundancy by inspection would be to release the truss until a simple truss is obtained. For example, removing one diagonal in each of the nine braced cells of the truss in Fig. 11.14a yields a simple truss $R = 0$, but we have nine static unknowns thus $R = 9$. Similarly, the truss in Fig. 11.14b is a simple truss hence $R = 0$. Finally, replacing one hinged support by a roller in truss Fig. 11.14c yields a simple truss, but we now have a static unknown, the corresponding horizontal reaction.

Table 11.1 Redundancy

Redundant	Determinate	Mechanism
$R > 0$	$R = 0$	$R < 0$

11.7 Mechanisms and Conditionally Stable Structures

What happens when the redundancy is negative, $R < 0$? Such systems are not really structures but rather mechanisms. These are structures which incorporate at least one mechanism somewhere along the line. *You can displace part of the structure without deforming any element.* This is the hallmark of unstable structures (Table 11.1).

One man's happiness is another man's nightmare. I know many engineers who cannot imagine a world without such moving things. Structural engineers could very well do without them. If we are not careful enough, structures can wind up as mechanisms. So how can we detect a possible mechanism in a structure?

One way is to count the degrees of freedom and the static unknowns, and if $R < 0$ the elements unfortunately comprise a mechanism.

The dual approach uses the property that

> we can displace a mechanism without deforming the elements.

This is never possible in a sound structure, but does well in a mechanism. We will use this for detecting mechanisms by inspection.

11.7.1 Mechanism Versus Structure

The typical mechanism in Fig. 11.16a0 was designed to transform a rotational displacement into a translation and vice versa. Counting equations and unknowns (Fig. 11.16a1) yields a negative redundancy ($R = -1$). In the figure a truss analysis was considered. As a frame, we get the same redundancy (starting from left $E = 1+4+2 = 7$, $U = 2 \times 3 = 6$). In other words, there are not enough elements, or there are too many degrees of freedom. When we replace the roller by a hinge as in Fig. 11.16b0 the counting (Fig. 11.16b1) gives $R = 0$. We now have a structure which can sustain loads.

By inspection, we notice that for the mechanism we can displace the system without deforming the elements. Indeed, in Fig. 11.16a2 we can get a new shape even if the elements are rigid bodies. This is never possible with a stable structure. As shown in Fig. 11.16b2, to move the node one element elongates while the other contracts.

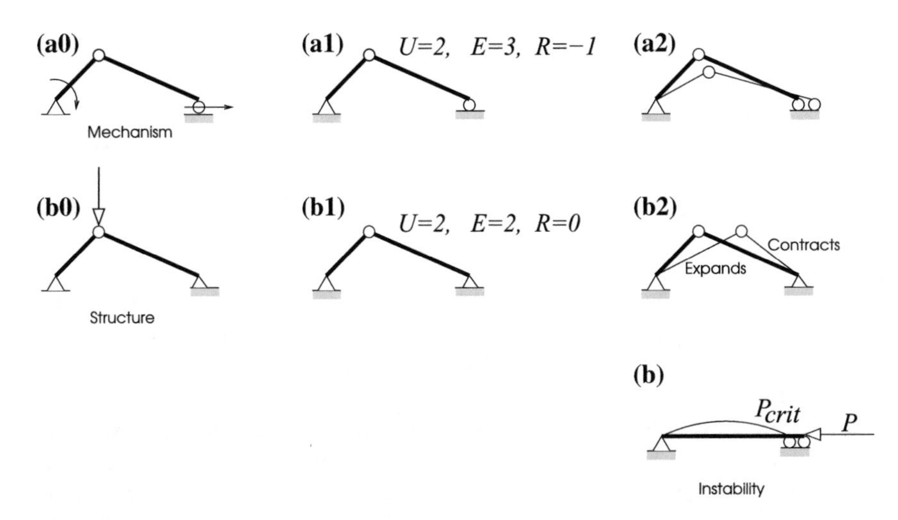

Fig. 11.16 Mechanism and structure

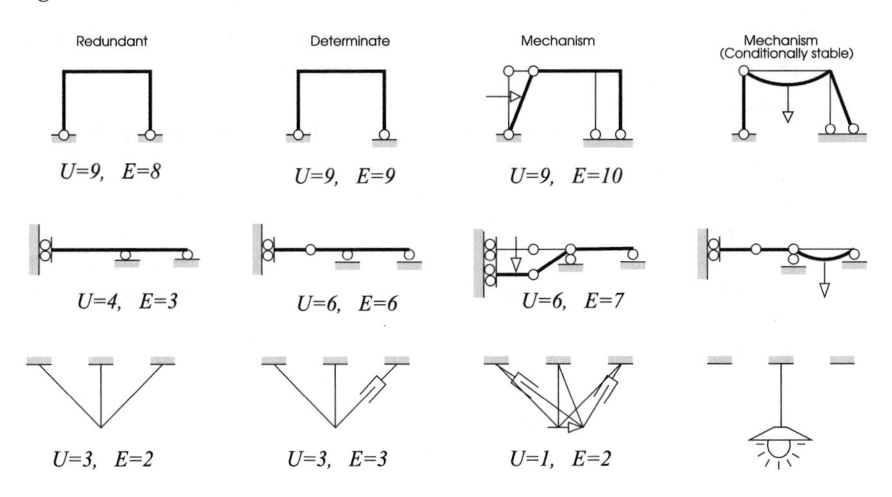

Fig. 11.17 Mechanisms

Note: The instability depicted in Fig. 11.16b is of a different kind. Above a critical compressive force this perfectly sound bar in compression buckles. This important phenomenon can be studied by non-linear analysis and is not considered herein.

Additional examples appear in Fig. 11.17. The frame, beam and truss of the first column are all redundant, $R = 1$. Introducing one release makes them determinate. (In the truss we have introduced an axial release which effectively removes the element.)

Now, introducing a second release reduces the redundancy to $R = -1$, and indeed we have mechanisms. Note that in all cases one can displace part of the structure without deforming the elements. The rigidity of an assemblage stems from the fact

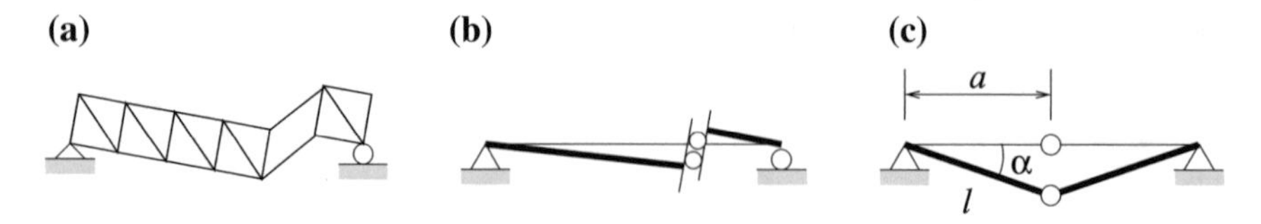

Fig. 11.18 Mechanisms by inspection

that in order to create a displacement we need to deform (stretch or bend) at least one element. This requires substantial forces and merits the name structure. If you can move the construction without deforming anything, that is, with zero forces, it is not a structure. It is merely a mechanism.

In Fig. 11.18, we show a few more examples of $R = -1$ mechanisms, which are easily detected by checking whether or not we can displace all or part of the structure without deforming any of its members. One of the cells of the truss in Fig. 11.18a is a non-triangulated cell, which calls for caution. The arrangement is indeed faulty since there is nothing to take the shear force (the major purpose of the diagonal elements is to carry the shear). This truss is conceptually a beam on two simple supports, with a shear release along its span as in Fig. 11.18b. The beam itself, which is represented by a line in the model, happens to be a truss beam.

A similar case is shown in Fig. 11.18c where a mid-span hinge makes the beam a mechanism. Although when counting unknowns (4) and equations (4) the structure seems to be statically determinate, it is a mechanism. Indeed, the beam has no resistance to small displacements of the type shown in the figure (more on this later).

Note, if the members of the structure in Fig. 11.18a were rigidly connected at the nodes, the deformation pattern shown is impossible, unless the truss has failed at the joints of the problematic cell, and this is exactly what we should be wary of. As a truss the structure cannot support any loads (imagine frictionless hinges at the nodes.) Consequently, as a frame, we cannot count on the normal forces to do the job, and the assembly will work in bending mainly. It is therefore a frame and should be designed accordingly with bending (and shear) resisting elements and connections.

11.7.2 Conditionally Stable Structures

In the last column of Fig. 11.17, we have depicted 'structures' which exhibit a mechanism but are still capable of carrying some class of loads. In other words, the system is unstable for some loads but it is perfectly stable under other loads. The frame for instance is a mechanism for side loads, but for a carefully located lateral force it can behave as a structure, that is, it has to deform in order react the applied load.

The beam is a similar example. The first span is undoubtedly unusable. The second span, however, is a simply supported beam and can carry forces. Finally, the one-bar truss in the last example is something most of us have an example of somewhere at home: a light bulb hanging from the ceiling. Under the weight of the electric fixture

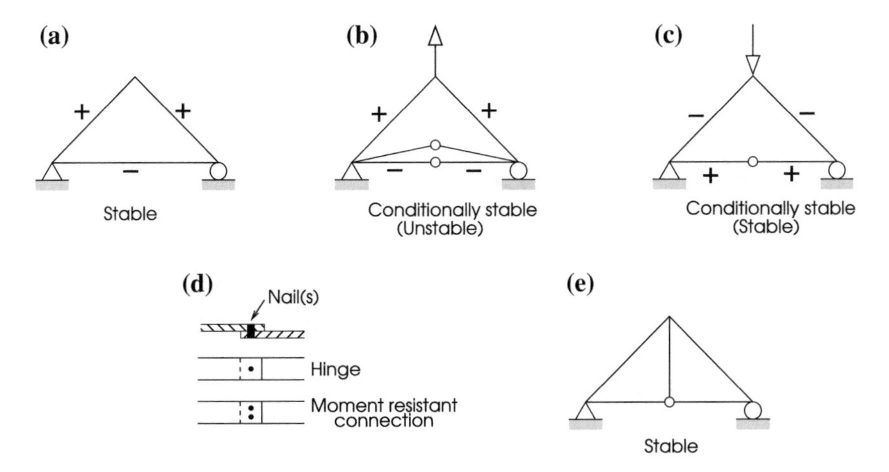

Fig. 11.19 Conditionally stable trusses

it behaves as a decent structure. But when there is a draft in the room (side forces) it swings back and forth without offering any resistance (a mechanism).

A different didactic example is shown in Fig. 11.19. The three-bar truss in Fig. 11.19a is a simple truss and is, of course, stable. Now let us assume that the horizontal element was assembled from two joists united by one nail as in Fig. 11.19d. This makes it a hinged connection, most probably not frictionless, but a hinge nevertheless. This mechanism ($R = -1$) is conditionally stable. Indeed, for an upload (Fig. 11.19b) the horizontal bars are in compression and will eventually buckle (the hinge will get out of alignment), thus loosing their ability to provide the necessary axial rigidity. Under a download (Fig. 11.19c) the bars will be in tension and nothing unusual is to be expected.

Granted, a radical solution is to add a post as in Fig. 11.19e. One can also use two nails instead of one (Fig. 11.19d). This creates a moment resistant connection, thus preventing early buckling.

11.7.3 $R \geq 0$ is not always sufficient for stability

In many cases we may very well determine the redundancy of a structure by counting degrees of freedom and static unknowns, and in truth we do not even have a structure. There is the widely used text-book example of the circular 'beam' on roller supports, depicted in Fig. 11.20a, formally of redundancy $R = 1$. We should consider it a frame, $R = (4 \times 3) - (4 \times 2 + 3)$. But, in that configuration all the reactions pass through the center of the circle. As a result, the reactions cannot apply a moment with respect to the center. Therefore, if the applied loads are also concentric, all is well (Fig. 11.20a1). But if they are not, and if they have a moment with respect to the

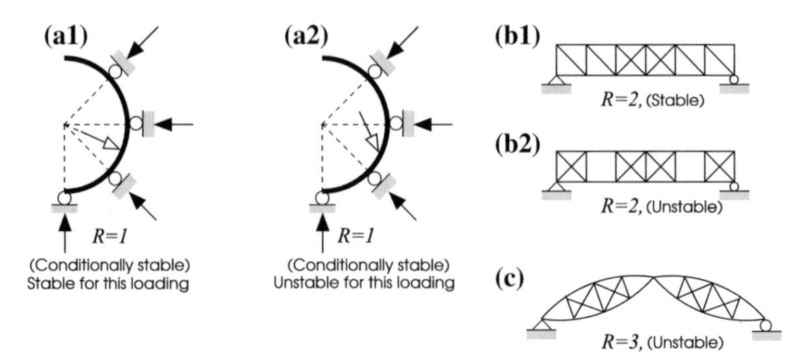

Fig. 11.20 $R > 0$ mechanisms

center of the circle (Fig. 11.20a2), the beam will never be in rotational equilibrium, no matter what the reactions. This structure, although formally redundant, is only conditionally stable.

The truss in Fig. 11.20b1 has two redundant bars and is stable. Move the bars in cells 2 and 5 to 1 and 6, and you get a mechanism with the same number of bars and degrees of freedom (Fig. 11.20b2). This example is the quintessence of the problem. The method counts bars and nodes but has no information on the connectivity, that is, which bar is connected to which nodes. Hence,

the condition $R > 0$ for stability is only a necessary condition.

We also need to make sure that the connectivity of the bars makes sense.

A last example is shown in Fig. 11.20c. A bars and nodes count will detect nothing erroneous with the structure. However, although $R = 3$ the 'structure' is a mechanism. Apply a vertical force at the apex and it will flatten out. $R > 0$ is not enough for stability. It tells you that there are enough members but is says nothing regarding the connectivity, that is, where the members are located.

11.7.4 Large Internal Forces in Quasi Non-stable Structures

Structures are not always either stable or unstable. It may occur that a badly designed structure is almost unstable, that is, the structure functions but it is in the danger zone. Such quasi-unstable structures may develop unacceptably large internal forces.

The frame in Fig. 11.21a, for instance, is stable. Notice that it is in equilibrium under a clockwise couple PL and an anti-clockwise couple $(P/\alpha) \times \alpha L = PL$. However, if $\alpha \rightarrow 0$, equilibrium is maintained but the forces at a and d become excessively large. This is an indication of instability or pending instability.

Indeed, for $\alpha = 0$ the structure (Fig. 11.21b) is unstable because it can move infinitesimally as a rigid body (in this case an infinitesimal rotation about a). The

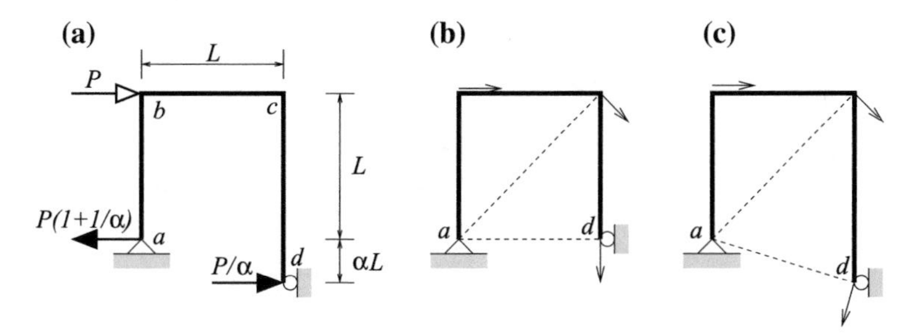

Fig. 11.21 Quasi unstable frame. **b** unstable, **c** stable

arrows indicate the infinitesimal displacements of nodes b, c and d, and we note that the displacement at d is compatible with the boundary condition there.

On the other hand, for $\alpha \neq 0$ (Fig. 11.21c) a rigid body rotation about a is not possible because the displacement at d would have a horizontal component, thus violating the boundary condition at d, and indeed the structure is stable.

11.8 Properties of Statically *Redundant* structures

11.8.1 Redundant Structures Are Safer

Determinate structures are notoriously unsafe for the simple reason that there is no redundancy in the design ($R = 0$.) Redundant structures are safer, and their relative safety increases with R, that is, the more redundant the structure the more comfortable we are to use it.

When hanging over an abyss we all prefer to be attached by two or more ropes rather then by one, and we are hanging over an abyss every time we board an elevator. Even if one cable is enough (no redundancy, $R = 0$) to carry the weight of the elevator shown in Fig. 11.22a1, we would rather use the elevator to the right.

The reason is obvious. If something would happen to the cable in the one-cable configuration the passenger would find himself in a rather awkward situation. The three-cable design is much better (two cables are redundant, hence $R = 2$). Remove one or even two cables and we still have a structure. Safety rules indeed make provision for many cables in elevators. The models for the structures in Fig. 11.22a1 are given in Fig. 11.22a2, where the stiffnesses of the springs represent the longitudinal stiffnesses of the cables.

The beam on three supports in Fig. 11.22b0 has redundancy $R = 1$. A redundancy $R > 0$ indicates that from a structural viewpoint there is something superfluous. Indeed, the sagging of a support, for instance (Fig. 11.22b1) leaves us with a more flexible structure, but it still is a structure. Similarly the partial damage stylized by the inclusion of a hinge in Fig. 11.22b2 leaves the structure still capable of sustaining loads. (A hinge may represent a weak cross-section which can carry shear but cannot

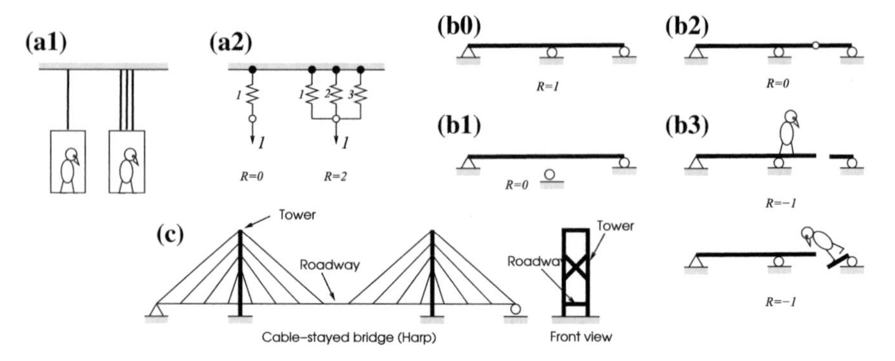

Fig. 11.22 Safety of redundant structures

transfer a bending moment.) Even the total failure of the beam, as in Fig. 11.22b3, is conditionally stable. Depending on where we are standing we can feel a structure under our feet.

Finally, the cable-stayed bridge of 'harp' type in Fig. 11.22c is redundant mainly due to the multiplicity of its cables. And indeed we can loose a cable and still have a well functioning bridge. In fact, one cited advantage of this type of bridge over suspension bridges is the ease of maintenance of its cables. We can remove a faulty cable and replace it by a new one while the bridge remains operational.

In short, statically determinate structures have exactly what it takes to be a structure, nothing less but also nothing more. As determinate structures are on the verge of collapse, tinker with one of its components and you may find yourself standing on a mechanism. Redundant structures, on the other hand, are more reliable.

11.8.2 Redundant Structures Are Stiffer

This property does not need much convincing. The cantilever beam in Fig. 11.23 is determinate, and the corresponding propped cantilever is redundant of degree $R = 1$. Indeed, the simple support is redundant from a statical view point. However, it makes the structure stiffer (and safer, of course). Under a vertical force along its span, the redundant design will sag much less than the cantilever.

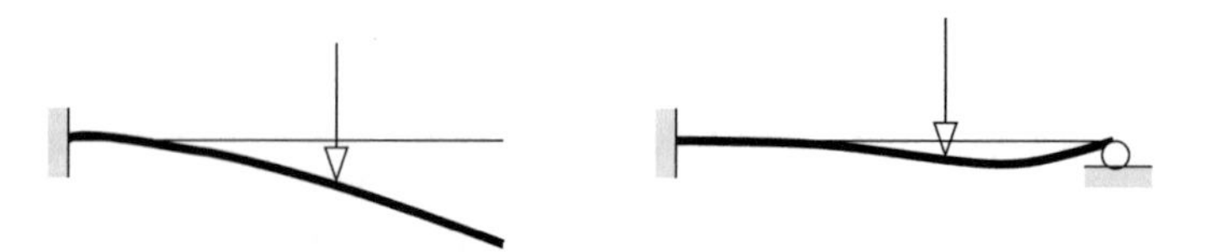

Fig. 11.23 Stiffness of redundant structures

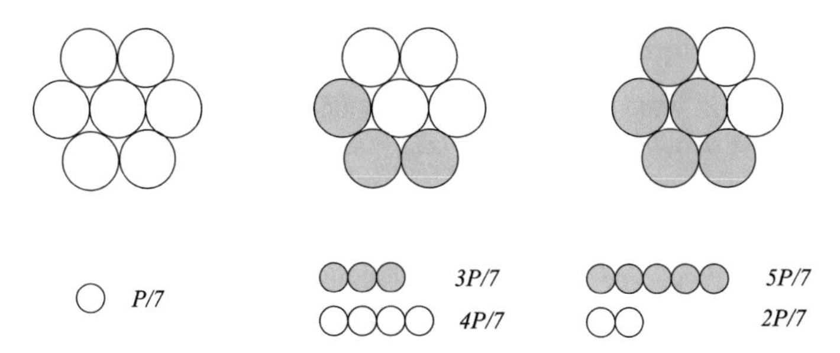

Fig. 11.24 A cable with several strands as a redundant structure

11.8.3 Flow of Forces

The property of *flow of forces* in statically redundant structures permeates the entire field of structures.

Forces tend to flow through the stiffer paths in redundant structures.

The example of the cable is almost intuitive. When a force P is applied to a twisted cable with, say, seven identical strands, every strand carries $P/7$ as shown in Fig. 11.24 (left).

We may well imagine that the cable is divided into two sub-cables, one with three strands and the other with four. Obviously they will carry $3P/7$ and $4P/7$ respectively. And which is the stiffest of the two? The one with four strands. Similarly, if in a thought exercise we group the cables in a $5-2$ pattern, the stiffer sub-cable will now carry $5P/7$ and the more flexible one $2P/7$. Consequently, the load P does not flow equally into the two sub-cables; more force flows into the stiffer one. By distributing the stiffnesses of a statically redundant structure we can in fact tune (design) the internal forces in the structure.

This does not happen in statically determinate structures. There the internal forces are controlled by equilibrium, irrespective of the stiffnesses distribution.

11.9 Properties of Statically *Determinate* Structures

Statically determinate structures, also called isostatic structures are the exception. Look about you and, except for structurally unimportant fixtures such as the chandelier dangling overhead, almost every structure in view will be hyperstatic or statically redundant. The main reason for this state of affairs is that statically determinate structures are notoriously unsafe assemblies. Failure of one element may cause the loss of the entire structure.

However, statically determinate structures are important for two main reasons:

1. They develop internal forces (stresses) only under applied loads.
2. They are the vehicle by means of which structures are analyzed in the Force method.

11.9.1 No External Forces Means No Internal Forces

This is probably the most significant difference between determinate and redundant designs beyond the safety and stiffness issues. In Fig. 11.25a. we have a system of four springs attached to two degrees of freedom where forces can be applied. The lateral distances between the springs are assumed very small. In fact everything works in one longitudinal direction and we have a uniaxial structure.

This arrangement has redundancy $R = 2$. We have four members with one unknown each ($U = 4$), and we have two open degrees of freedom at the nodes ($E = 2$). By inspection we also note that springs 3 and 4 are superfluous, and the remaining $R = 0$ structure (see Fig. 11.25d) can readily support forces applied at the degrees of freedom.

Without external forces the equilibrium equations of the redundant case (Fig. 11.25b, c) and the determinate arrangement are respectively

$$\begin{bmatrix} 1 & -1 & -1 & 0 \\ 0 & 1 & 1 & 1 \end{bmatrix} \begin{Bmatrix} n_1 \\ n_2 \\ n_3 \\ n_4 \end{Bmatrix} = \begin{Bmatrix} 0 \\ 0 \end{Bmatrix} ; \quad \begin{bmatrix} 1 & -1 \\ 0 & 1 \end{bmatrix} \begin{Bmatrix} n_1 \\ n_2 \end{Bmatrix} = \begin{Bmatrix} 0 \\ 0 \end{Bmatrix} \tag{11.2}$$

where n_i is the force in spring i, positive in tension. The coefficient matrices of the redundant and determinate assemblies are respectively rectangular (more unknowns than equations) and square which is the essence of determinacy: the number of equilibrium equations equals the number of unknown internal forces with solutions

$$\begin{Bmatrix} n_1 \\ n_2 \end{Bmatrix} = \begin{Bmatrix} -n_4 \\ -(n_3 + n_4) \end{Bmatrix} ; \quad \begin{Bmatrix} n_1 \\ n_2 \end{Bmatrix} = \begin{Bmatrix} 0 \\ 0 \end{Bmatrix} \tag{11.3}$$

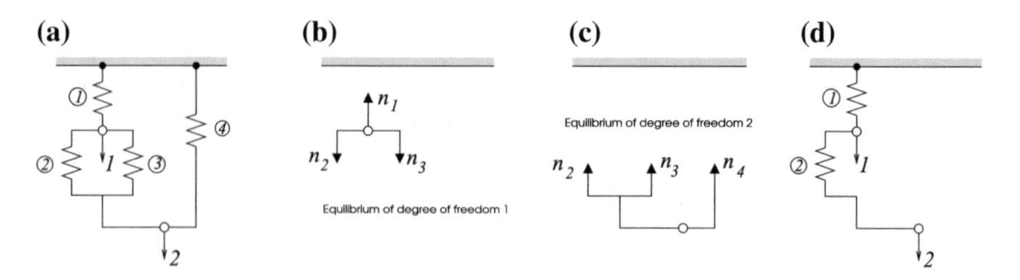

Fig. 11.25 Equilibrium of redundant system

The redundant case has a plethora of possible equilibrium solutions depending on the values assigned to n_3, n_4. For instance, with $n_3 = 1$ and $n_4 = -1$, we get $n_1 = 1$ and $n_2 = 0$ which is a perfectly sound equilibrium solution (Fig. 11.25). Nothing of this sort happens in the determinate case. The system $\boldsymbol{An} = \boldsymbol{0}$ with a non-singular square matrix $\boldsymbol{A}$ has only $\boldsymbol{n} = \boldsymbol{0}$ as a solution, no matter what.

Consequently, zero external loading yields zero internal forces in a statically determinate structure. This seems rather obvious, but we will see that this property is often counter-intuitive, in particular when a structure is heated or when we plan to prestress it. This is only possible if the structure is hyperstatic.

11.9.2 Heating Will Not Produce Internal Forces

This is an immediate corollary of the previous property.

Without going into too much detail at this point let us make clear that the deformation of a structural component or of an entire structure can have two origins:

(a) mechanical deformations due to external loads (forces and couples); but also
(b) thermal deformations due to temperature variations inside the structure.

A structure in the desert in mid-summer will heat up during the day and cool down at night, with a temperature range of the order of 40 °C. Similarly, a satellite orbiting the Earth will heat up in the sun and then cool down as it is shielded by the Earth's shade. In space the temperature may change over a range of 100°.

When an element heats up it expands. Consider the uniform rod in Fig. 11.26a1. Under the action of external axial applied forces P, the rod will expand with uniform strain ϵ until it reaches an equilibrium position where at every section there is an internal force $n = P$, and the strain is then $\epsilon = P/EA$. The total mechanical elongation of the rod is $e_M = \epsilon L = PL/EA$.

Now assume that the rod is heated such that the temperature increases uniformly by T (Fig. 11.26a2). Here the internal structure of the atoms changes and the rod assumes a slightly longer shape. What was once a rod of length L is now a rod of length $(1 + \alpha)L$ where the coefficient of thermal expansion α is a property of the material. Nothing else has changed. There are no external forces applied and

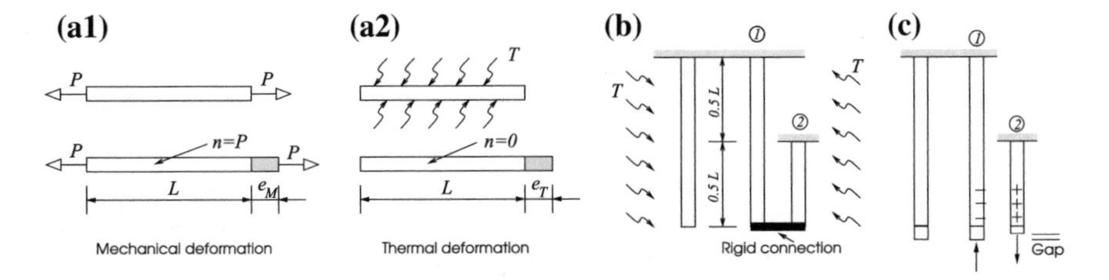

Fig. 11.26 Heated rod

consequently the internal force is $n = 0$ everywhere. The total elongation, that is, the increase in length of the element is $e_T = \alpha L$.

Consider now the two structures in Fig. 11.26b, the one-element determinate structure on the left and the two-element redundant one ($U = 2, E = 1$) rigidly connected on the right. Note the difference in lengths of the constituents of the redundant arrangement, and also that the rigid connection compels the extremities to move in unison. Now let us heat both structures uniformly and increase the temperature by T (no external forces applied). The determinate bar will elongate freely and there will of course be no internal forces.

The redundant arrangement will however encounter some difficulties when its elements try to expand. Let us temporarily remove the rigid connection between the rods extremities (Fig. 11.26c). Clearly, the shorter rod will elongate by half as much as the longer one, and a gap will appear at the tip. With the rigid connection this is not possible. What will happen in a real structure is that bar 2 will compress bar 1 and bar 1 will put bar 2 in tension until the gap is bridged (see black arrows). We have internal forces in the structure without applying external ones. In the final configuration the deformations will be the superposition of the mechanical deformation (due to the internal forces n in both bars) and the thermal deformations $e = e_M + e_T$.

For our purposes this is enough to establish that under temperature changes internal forces may appear in redundant structures, but never in determinate ones.

11.9.3 No Prestressing Possible

Prestressing is the technique by which internal loads appear in a structure prior to any external loading. Usually prestressing is a voluntary procedure by means of which the designer can modify the distribution of internal loads (usually forces) in a structure in a beneficial manner. We were concerned with such procedures earlier, but at this juncture we mention it to emphasize that this is a method reserved for redundant arrangements only. In determinate structures it is impossible to lock internal forces. Only external forces engender internal forces (excluding prestressed concrete.)

Consider the cantilever beam and the doubly built-in element in Fig. 11.27a. Since the cantilever is determinate, no surprises are possible. What you see is what you get. We do not see external forces, *ergo*, there are no internal forces.

In a redundant structure we can never be sure. The $R = 3$ beam in Fig. 11.27a could be stressed. When it gets very cold the beam may shrink and the walls would then pull on it to keep it in place, creating $n > 0$ internal axial forces. Or the walls may not remain exactly perpendicular to the element, thus creating internal moments. All these assumed displacements need not be discernable by the naked eye, but they can produce appreciable internal loads.

The prestressing shown in Fig. 11.27b is due to a misalignment of the supports. The determinate beam (with the hinge) has no problem in accommodating the discrepancy without deforming its elements. In the redundant case, the misalignment causes a deformation of the beam and thus internal forces. In the configuration the intermediate

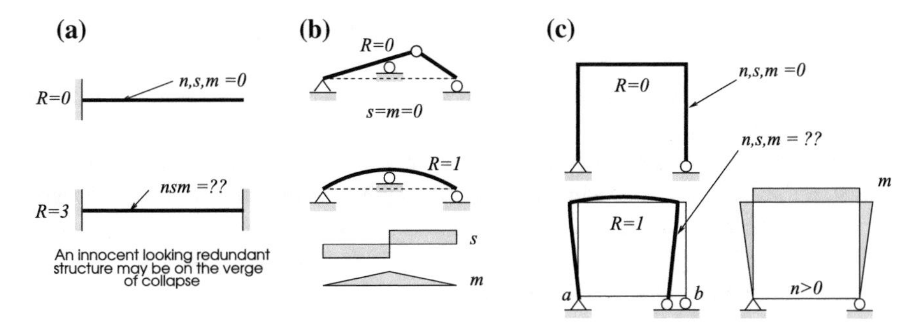

Fig. 11.27 Prestressing is for redundant structures only

support will produce an upward reaction (and corresponding downward reactions at the extremities), yielding the type of s- and m-distributions shown in the figure.

Finally the determinate portal frame in Fig. 11.27c has no external forces applied to it and therefore guaranteed zero internal forces. On the other hand, the same frame with a cable joining ends a and b has a redundancy $R = 1$ and can be prestressed. In the example the frame was prestressed by joining a and b with a cable whose length was shortened in a controlled manner. The final shape (in an exaggerated displacement scale) and the ensuing internal bending moment distribution are shown in the figure.

11.9.4 Non-sensitivity to the Exact Dimensions of the Elements

This property is related more to the compatibility issue of redundant structures. Let us assume that we have ordered from a manufacturer three bars to form the ideal truss in Fig. 11.28a0. Two bars have arrived as specified, but bar ad has a length larger than $\sqrt{2}L$.

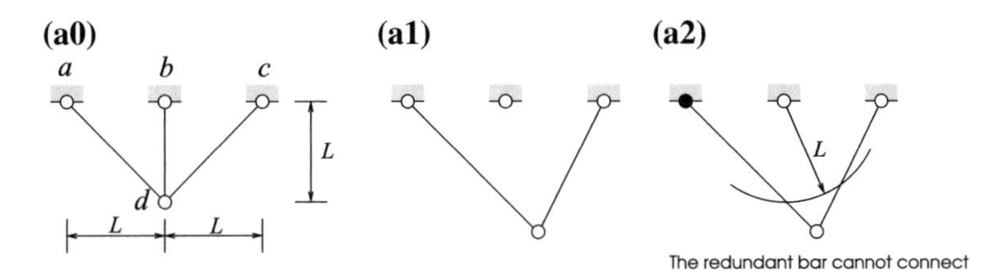

Fig. 11.28 Connecting a redundant bar

The problem was not discovered at delivery, and when assembling bars *ad* and *cd*, to form a simple truss, nothing unusual happened. A sharp-eyed observer may have noticed the lopsided shape of the structure, but the problem is entirely aesthetic (Fig. 11.28a1).

When trying to connect bar *bd* however, it became clear that something went wrong. Indeed, *bd* is now a redundant bar (connecting two existing nodes), and since the distance between the nodes is not equal to L, it cannot be attached.

This will not happen with determinate trusses. We can always assemble the bars even if the lengths are not exactly as planned.

11.10 Summing Up

We have described how to determine the degree of statical redundancy R of plane structures, in particular frames, beams and trusses. This can be done by counting the number of missing equilibrium equations and/or also by checking how far the structure is removed from being statically determinate. It was emphasized that $R > 0$ is a necessary but not a sufficient condition for stability.

In the last paragraphs of the chapter we have compared the main structural aspects of redundant and determinate structures.

Chapter 12
The Force Method

The Force method is the first of the two main analysis methods presented in this book. It is intended for relatively simple structures of modest redundancy. Although the method is quite general it leads to heavy computations for complex structures. Historically it was practically the only method available until the advent of computational techniques. This paved the way to the systematic and relatively simple Displacement method which is intended for computerized solutions.

The Force method has retained its position in structural analysis because it is intuitive and it lends itself to a better understanding of the behaviour of a structure and its response to applied loads and applied deformations such as heating.

12.1 Note

At the outset we need to clarify two aspects of the method:

The Force method is relevant to statically *redundant* structures only.

This is not a severe restriction because except for some stylized examples such as a beam on 2 supports, or a simple truss, *all* structures (bridges, towers, buildings, aircrafts ...) are redundant.

By definition, we cannot compute the internal forces in a redundant structure by means of equilibrium equations only. To analyze the structure we need to consider all three sets of equations: equilibrium, deformations-displacements and elasticity.

This is were the Force method steps in.

It replaces the analysis of a structure of redundancy R with $R + 1$ analyses of a statically determinate 'released' structure under $R + 1$ different loadings. We already know how to analyze determinate structures.

The Force method does *not* compute displacements.

M.B. Fuchs, *Structures and Their Analysis*,
DOI 10.1007/978-3-319-31081-7_12

The Force or Flexibility method produces the internal forces distributions $\boldsymbol{n}(x)$ but not the displacements. Having obtained the internal forces, and if we are interested in some displacement or slope we can always use the unit-load method, for instance. At any rate computing displacements is not part and parcel of the force method.

Before launching into the subject, we start with one last preliminary point regarding the method: *releasing the redundant structures*, a concept central to the Force method.

12.2 Released Structures Subjected to the Released Loads

Releasing a structure means mentally performing cuts in the structure, such as partially or entirely disconnecting supports from the structure, introducing hinges and lateral guides in its elements and the like. All these operations reduce the redundancy of the structure.

The Force method relies on the premise that when you apply the released loads, which were once internal, as equal and opposite external loads at the releases, the released structure and the original one have identical internal forces, that is, same $\boldsymbol{n}(x)$ distributions and same reactions. The previous sentence seams more complicated than it actually is. Examples of releasing $R = 1$ and $R = 2$ beams will clarify the concept.

12.2.1 $R = 1$ *Propped Cantilever*

The propped cantilever in Fig. 12.1a1 is subjected to a uniform distributed force of magnitude $q = Q/L$, where L is the span of the beam. To explain we have been given the results of the analysis: the reactions and the corresponding shear and bending moment diagrams in the box of Fig. 12.1a. Note the linear shear and parabolic moments with the extremum moment where the shear is zero. One elementary release will reduce the redundancy by 1 and yield a statically determinate structure. There are many ways to achieve this result, six of which are shown in columns (b) and (c) of Fig. 12.1.

(b1) The structure is released by disconnecting the roller at b from the beam and instead the reaction $3Q/8$, where $Q = qL$, is applied. Note, internal forces come in pairs. So, $3Q/8$ is applied to the beam (this is what the roller does to the beam) and an opposite $3Q/8$ is applied to the roller, a little removed for clarity's sake. This is the force applied by the beam to the roller.

(b2) The moment at the root is released (set to zero) by means of a hinge. We then apply the value of the bending moment, $QL/8$, on both sides of the hinge, on one side to the beam and on the other to the wall.

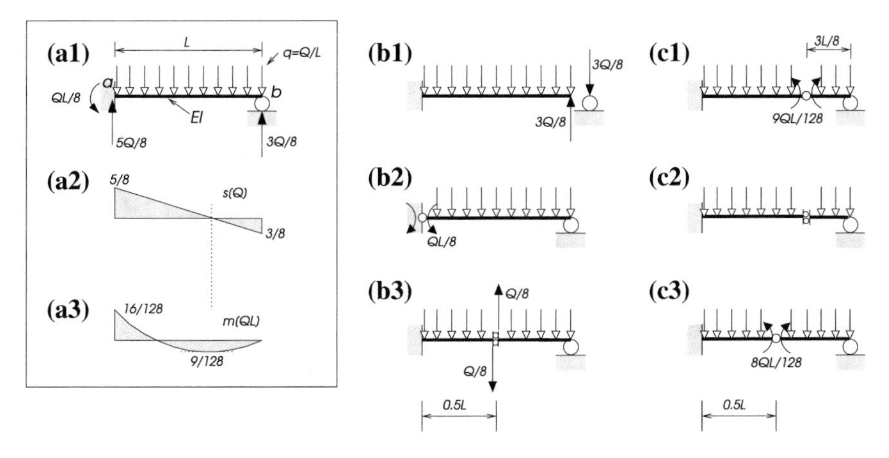

Fig. 12.1 Released propped cantilever

(b3) The shear at mid-span is released with the help of a lateral guide to which the shear force is applied in the form of the pair of external forces $Q/8$.

(c) The cases shown in column (c) are self-explanatory however, case (c2) merits a comment. Releasing the shear at $5/8$ of the span indeed sets the shear to zero, but since the actual shear is anyway null at that station (see (a2)) there is nothing to apply there.

The main import of this exercise is that releasing a structure and applying the corresponding internal fores at the releases (in pairs) produces a determinate structure with the same internal forces as the original redundant one.

12.2.2 $R = 2$ Continuous Beam

This example is similar. Consider the twice redundant continuous beam in Fig. 12.2a. Any two supports would be enough for stability, so the remaining two are superfluous, hence $R = 2$.

Here also we are given the reactions and shear and moment distributions which are indicated in the figure. In particular, at supports c and d the bending moments are respectively 0.164 PL and 0.053 PL. By introducing hinges at these supports, as in Fig. 12.2b, and applying equal and opposite corresponding couples near the hinges we obtain a statically determinate beam with essentially the same internal forces, including the reactions, as the original structure. Three alternative released structures are also given in the figure.

In Fig. 12.2c supports c and d were disconnected from the beam and replaced by their respective reactions.

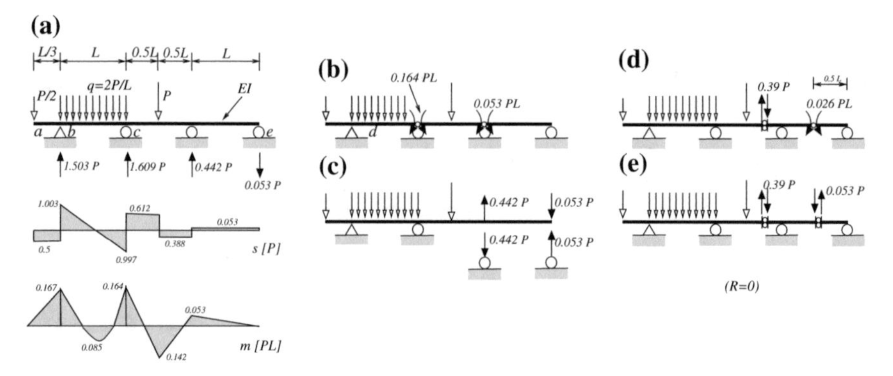

Fig. 12.2 Released continuous beam

In Fig. 12.2d the structure was released by a lateral guide on span cd and a hinge on de and replaced respectively by the corresponding shear force and bending moment. In Fig. 12.2e the shear force in spans cd and de was released by 2 lateral guides. Here also, we apply the corresponding actual shear forces on both sides of the release.

The above released structures to which the internal forces were applied as external ones are now statically determinate structure with the same reactions, internal forces $s(x)$ and $m(x)$, and indeed equilibrium shape as the redundant beam. In other words, when adding the released loads at the releases the analysis of the released and the original structures are the same.

12.3 The Essence of the Force Method

When we release a redundant structure to make it determinate and apply the redundant loads at the releases, in conjunction with the applied loads of course, we get exactly the same internal forces as the original structure. The released structure, being determinate, can be analyzed by means of equilibrium equations which is something we know how to do.

Unfortunately the redundant or released loads are not known. In other words, for a redundancy R, we have R unknown loads denoted $x_1\ x_2 \ldots\ x_R$ which we often aggregate in vector $\boldsymbol{x}$. One way or another we need R equations to compute these unknown loads.

The Force method does precisely that.

The Force method provides R *compatibility equations* to compute the R redundants.

These linear equations are based on the deformation equations and on Hooke's law. Hooke's law implies that the results depend on the stiffnesses (essentially the axial stiffness EA and bending stiffness EI) of the elements composing the structure, a property which sets redundant structures apart from determinate ones.

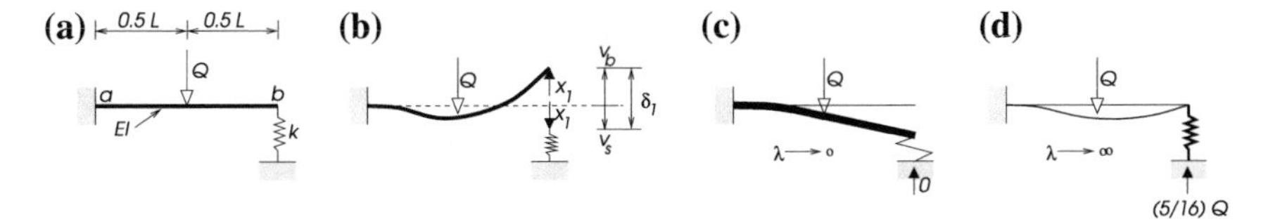

Fig. 12.3 Cantilever with spring

12.3.1 *R* = 1 *Cantilever Beam with Elastic Spring*

Consider beam ab of length L and of uniform bending stiffness EI, built in at a, supported at b by a linear spring of stiffness k and subjected to a point-force Q at mid-span, as depicted in Fig. 12.3a. The redundancy of the structure is $R = 1$. Indeed, without the spring the structure is statically determinate (a cantilever) and the spring carries one static unknown, the force in the spring, hence $R = 1$.

There are several ways to release a structure. We choose to make the structure determinate by disconnecting the spring from the beam. The result is a cantilever beam and a spring, both determinate arrangements. But for the released structure to be a simulation of the original redundant one, we need to apply the interaction forces to the fictitious section. Since they are, as yet, unknown let us apply equal and opposite forces of unknown magnitude x_1 to the section.[1]

Let us assume that under the action of the external force and some arbitrary value of x_1, the deformed released structure will assume a shape similar to that in Fig. 12.3b.

We may not have drawn an exact replica of the deformed beam; however, one thing is almost certain, the tip of the beam will not connect to the spring at b. There will be a discrepancy or gap δ_1 at b. In fact, only for the exact value of the redundant force will the discrepancy disappear. The Force method uses this condition $\delta_1 = 0$, the compatibility equation, to compute x_1.

From Fig. 12.3b we have $\delta_1 = v_b + v_s$, where v_b and v_s indicate *b*eam and *s*pring displacements respectively. (Note that the positive sense of the vertical displacements of the beam and of the spring are in opposite directions.) These displacements can either be computed, for instance, by the unit-load method, which is what we will usually be doing for didactic reasons, or taken from tables and handbooks as in the present case. The beam displacement due to x_1 and Q is $v_b = (L^3/EI)(x_1/3 - 5Q/48)$ and the spring contraction is of course $v_s = x_1/k$. The minus sign in the expression of v_b is there because v_b is positive upwards and Q points down.

The compatibility equation which is the condition for the gap between the beam and the spring at b to close is $\delta_1 = v_b + v_s = 0$ or

$$\delta_1 = \frac{L^3}{EI}\left(\frac{1}{3}x_1 - \frac{5}{48}Q\right) + \frac{1}{k}x_1 = 0 \tag{12.1}$$

[1] The redundants are assigned a sequential number. Here we have only one redundant, hence subscript 1.

from which we get the value of the redundant

$$x_1 = \frac{5\,\lambda}{16\,\lambda + 48}\,Q \quad \text{with} \quad \lambda = \frac{k}{EI/L^3} \tag{12.2}$$

This is the reaction at b and also the compressive force in the spring. Having the reaction at b, we can now analyze a statically determinate structure (the released structure, that is, the cantilever) under known loads: Q and x_1 at the tip.

This concludes this first example of the Force method, but before proceeding, a note on the non-dimensional parameter λ and its influence on the solution. Parameter λ measures the relative stiffness of the spring k with respect to the stiffness of the beam EI/L^3. This is the hallmark of a redundant structures.

In redundant structures the internal forces depend on the stiffnesses.

If the beam is very stiff, $EI/L^3 \to \infty$, and/or the spring is very flexible, $k \to 0$, we have in the limit a cantilever beam *without* a spring at b as in Fig. 12.3c. It is of no consequence to a massive beam if there is a tiny spring at its extremity or not, and with $\lambda \to 0$ in (12.2) we indeed get $x_1 = 0$. If there is no support there is no reaction. In fact, only the first half of the beam bends in a cantilever mode. The second part is stress free and moves along as a rigid body.

On the other hand, if the beam is very flexible, $EI/L^3 \to 0$, and/or the spring is very stiff, $k \to \infty$, we have in the limit a beam with a rigid support at b (Fig. 12.3d). Dividing the numerator and the denominator of the expression of x_1 in (12.2) by λ and having $\lambda \to \infty$ we get the reaction of a propped cantilever with a central point force, $x_1 = (5/16)\,Q$.

So, the value of the reaction at b can be anywhere in the interval $[0, (5/16)\,Q]$ as a function of the relative stiffnesses. Depending on how we distribute the stiffnesses we get different results. It is not until we reach redundancy that structures start to behave like that. Statically determinate structures are more like dead wood. Whatever the play of stiffnesses, the results will depend only on the equilibrium equations.

For instance, let us take an intermediate value for the relative stiffness, say, a ratio λ which will produce a reaction $x_1 = 0.2Q$ at b, that is with (12.2), $0.2 = 5\lambda/(16\lambda + 48)$, or $\lambda = 5.33$. It is easy to establish that the beam under forces Q and $0.2Q$ will deflect downwards at b by an amount $v_b = (0.45/12)\,L^3/EI$ and the spring will contract by $v_s = 0.2Q/k$. With $\lambda \equiv kL^3/EI = 5.33$ these two amounts are equal.

12.3.2 $R = 2$ Continuous Beam

A second example for the Force method is the $R = 2$ beam in Fig. 12.2, which we reproduce for convenience in Fig. 12.4a1.

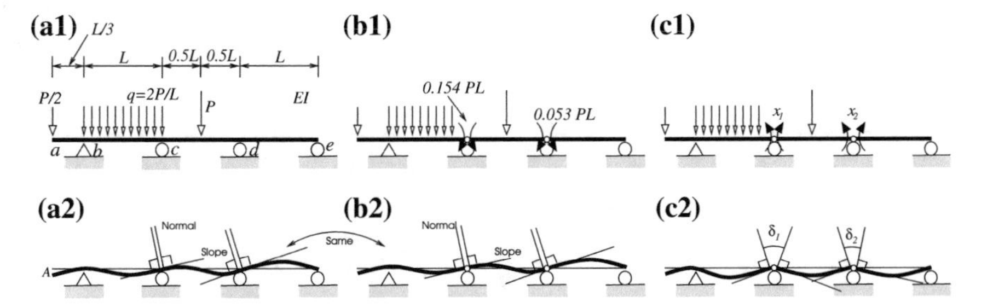

Fig. 12.4 Continuous beam compatibility equations

We introduce $R = 2$ releases in the structure, thus establishing a statically determinate structure. There are many ways to do this (four alternatives were given in Fig. 12.2). Here we opt for releasing the bending moments at supports c and d.

When we apply the bending moments as external loads at the releases, as in Fig. 12.4b1, the structures in (a) and (b) have identical internal forces. In particular they have the same equilibrium configuration. (Compare the thick lines in Figs. 12.4a2, b2.)

The problem is that these $R = 2$ bending moments are unknown. If we had these couples there would be no problem, since we would be confronted by a statically determinate structure where all the forces are known. As said, there exist $R = 2$ compatibility equations to compute $\boldsymbol{x}$. The idea behind the compatibility equations is rather simple.

Let x_1 and x_2 be the unknown bending moments at the releases (Fig. 12.4c1). Let us guess some values for the bending moments. The displaced shapes of the beam in Fig. 12.4c1 and the original one in Fig. 12.4a1 will not be the same. This is particularly evident at the releases.

Note that, whereas in the original beam the slopes at c and d are continuous, this is no longer the case in the released beam. Unless we are very lucky, for any pair of values of the unknowns, we will have a discontinuity at the releases.

One way to visualize the slope discontinuity is to draw normals to the slopes on both sides of the releases (as in Fig. 12.4a2 or Fig. 12.4b2). Discontinuous slopes create angles between the normals, δ_1 and δ_2. Continuous slopes have parallel normals (Fig. 12.4a2, b2), that is, a zero intercept angle, $\delta_1 = \delta_2 = 0$.

When we apply arbitrary moments at the releases, we get discontinuities at the releases, $\delta_1 \neq 0$, $\delta_2 \neq 0$. When we apply the exact moments at the releases, we have zero discontinuities, $\delta_1 = \delta_2 = 0$). We can use this as a means of calculating the value of the released (redundant) loads, by imposing that x_1 and x_2 should be such that

$$\delta_1(x_1,\ x_2) = 0\,, \quad \delta_2(x_1,\ x_2) = 0 \qquad \text{or} \qquad \boldsymbol{\delta}(\boldsymbol{x}) = \boldsymbol{0} \tag{12.3}$$

where $\boldsymbol{\delta}$ is the vector of discontinuities.

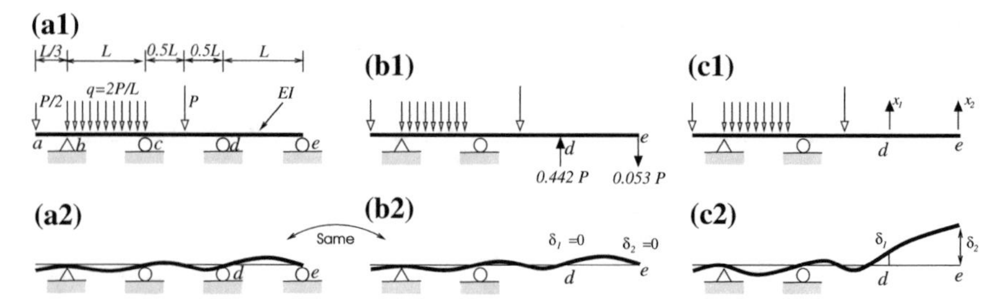

Fig. 12.5 Different release

These are the $R = 2$ compatibility equations which will allow us to compute the $R = 2$ redundants x_1 and x_2.[2]

Same Beam with Different Releases

We can of course select a different way for releasing the structure. Disconnecting supports d and e from the beam is another alternative (Fig. 12.2c). This releases at each support two equal and opposite forces: the vertical force that the ground applies to the beam (the actual reaction) and the opposite vertical force that the beam exerts on the ground. As indicated in Fig. 12.5b1, applying the external loads plus the exact reactions at d and e of the released structure is equivalent to applying the external loads to the original one. In particular we get the same displacements (see Fig. 12.5a2, b2). It would be nice to know these redundants.

When arbitrary loads x_1 and x_2 are exerted at the releases, the released structure behaves differently from the redundant one. We get, in particular, discontinuities δ_1 and δ_2 at the releases. Here the discontinuities measure the distances between the beam at d and e and the supports.[3] The discrepancies $\boldsymbol{\delta} = \{\delta_1\ \delta_2\}$ are displacements of one 'end' of the release relative to the second end.

As in the previous example, the compatibility equations (12.3) should allow us to compute the redundants $\boldsymbol{x}$. We will perform the actual computations in the sequel.

12.4 Non-valid Releases

When we introduce R *valid* releases in a structure of redundancy R you obtain a determinate stable structure. But we can inadvertently position the releases wrongly and produce anything but a stable determinate structure.

[2]It is a common misconception that the slopes at the supports are null, in fact it is the slope discontinuities at the supports which are null. There is no apparent reason for horizontal slopes over the supports.

[3]Since the supports are fixed, the discontinuities are also absolute displacements of the beam. But this is the exception.

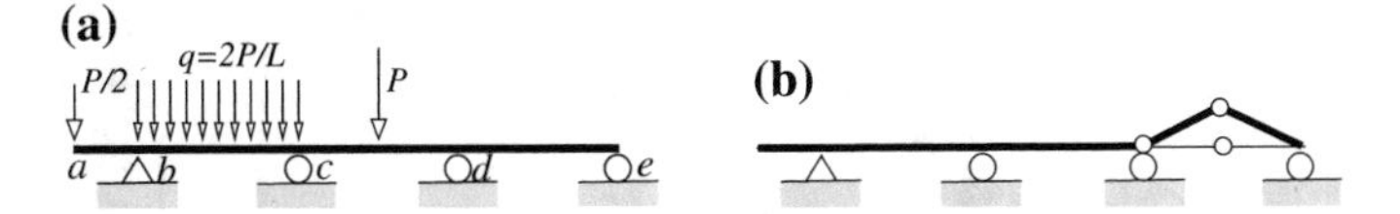

Fig. 12.6 Non-valid release

In the case of the continuous beam a non-valid release is, for instance, two hinges in the last span (Fig. 12.6). Note that from a until the first hinge we have a stable beam on three supports, and from there on we have a mechanism. Indeed we can displace it without deforming the elements.

The hinges have left us with a shorter beam of redundancy $R = 1$, supports b, c and d, and an addendum in the form of a mechanism, which is a far cry from what we need. So we always need to apply engineering judgment to release the structure in a sound manner.

12.5 Formalizing the Compatibility Equations

Consider the case were the beam was made determinate by releasing the bending moments on top of supports c and d. We seek the expressions $\delta_1(x_1, x_2, f^0)$ and $\delta_2(x_1, x_2, f^0)$ which when equated to zero (the compatibility equations) should provide $\boldsymbol{x}$. (Note, f^0 represents external causes, the point and distributed forces in our case.)

For the time being let us define at the releases four degrees of freedom as shown in Fig. 12.7. We can apply there couples $\boldsymbol{c} = \{c_1\, c_2\, c_3\, c_4\}$, and also compute the slopes $\boldsymbol{\theta} = \{\theta_1\, \theta_2\, \theta_3\, \theta_4\}$. It is clear from Fig. 12.7b that $\delta_1 = \theta_1 + \theta_2$ and $\delta_2 = \theta_3 + \theta_4$.

Every slope is the result of the four couples and of the external loads. We can write by superposition four equations of the type $\theta_i = \theta_i^1 + \theta_i^2 + \theta_i^3 + \theta_i^4 + \theta_i^0$ where θ_i^j is slope i due to c_j, and θ_i^0 is slope i due to the external causes.

Now $\theta_i^j = f'_{ij}\, c_j$, where f'_{ij} is the slope at i due to a unit force at j, in other words, the influence coefficient between degrees of freedom i and j. (We are using the f'_{ij} nomenclature in anticipation of an f_{ij} influence coefficient.) Consequently, we have $\theta_i = f'_{i1}c_1 + f'_{i2}c_2 + f'_{i3}c_3 + f'_{i4}c_4 + \theta_i^0$. In matrix form we can write the expressions for all the θ_i incorporated in vector $\boldsymbol{\theta}$:

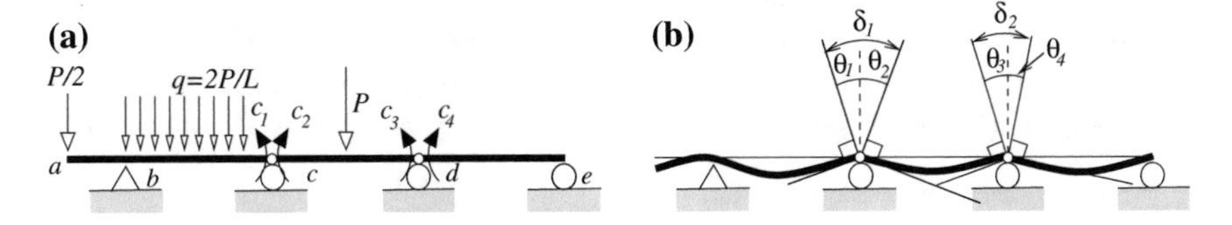

Fig. 12.7 Slope discontinuities at the releases

$$\boldsymbol{\theta} = \boldsymbol{F}'\,\boldsymbol{c} + \boldsymbol{\theta}^0 \qquad \text{or} \qquad \begin{Bmatrix} \theta_1 \\ \theta_2 \\ \theta_3 \\ \theta_4 \end{Bmatrix} = \begin{bmatrix} f'_{11} & f'_{12} & f'_{13} & f'_{14} \\ f'_{21} & f'_{22} & f'_{23} & f'_{24} \\ f'_{31} & f'_{32} & f'_{33} & f'_{34} \\ f'_{41} & f'_{42} & f'_{43} & f'_{44} \end{bmatrix} \begin{Bmatrix} c_1 \\ c_2 \\ c_3 \\ c_4 \end{Bmatrix} + \begin{Bmatrix} \theta_1^0 \\ \theta_2^0 \\ \theta_3^0 \\ \theta_4^0 \end{Bmatrix} \tag{12.4}$$

In our case the couples are not independent. They always come in equal and opposite pairs: $c_1 = c_2 = x_1$ and $c_3 = c_4 = x_2$. This is in matrix form

$$\boldsymbol{c} = \boldsymbol{H}\,\boldsymbol{x} \qquad \text{or} \qquad \begin{Bmatrix} c_1 \\ c_2 \\ c_3 \\ c_4 \end{Bmatrix} = \begin{bmatrix} 1 & 0 \\ 1 & 0 \\ 0 & 1 \\ 0 & 1 \end{bmatrix} \begin{Bmatrix} x_1 \\ x_2 \end{Bmatrix} \tag{12.5}$$

The first line, for instance, is $c_1 = [1\,0]\,\{\boldsymbol{x}\}$ or $c_1 = x_1$, the second line is $c_2 = [1\,0]\,\{\boldsymbol{x}\}$ or $c_2 = x_1$, and so on.

On the other hand, it is easy to see that the discrepancies $\boldsymbol{\delta}$ are related to the slopes by

$$\boldsymbol{\delta} = \boldsymbol{H}^T\,\boldsymbol{\theta} \qquad \text{or} \qquad \begin{Bmatrix} \delta_1 \\ \delta_2 \end{Bmatrix} = \begin{bmatrix} 1 & 1 & 0 & 0 \\ 0 & 0 & 1 & 1 \end{bmatrix} \begin{Bmatrix} \theta_1 \\ \theta_2 \\ \theta_3 \\ \theta_4 \end{Bmatrix} \tag{12.6}$$

The first line, for instance, is $\delta_1 = [1\ \ 1\ \ 0\ \ 0]\ \{\boldsymbol{\theta}\}$ or $\delta_1 = \theta_1 + \theta_2$ Note, that the coefficients matrix in (12.6) is the transpose of the matrix in (12.5).[4]

Pre-multiplying both sides of (12.4) by $\boldsymbol{H}^T$ yields with (12.5) and (12.6)

$$\boldsymbol{\delta} = \boldsymbol{F}\,\boldsymbol{x} + \boldsymbol{\delta}^0 \qquad \text{or} \qquad \begin{Bmatrix} \delta_1 \\ \delta_2 \end{Bmatrix} = \begin{bmatrix} f_{11} & f_{12} \\ f_{21} & f_{22} \end{bmatrix} \begin{Bmatrix} x_1 \\ x_2 \end{Bmatrix} + \begin{Bmatrix} \delta_1^0 \\ \delta_2^0 \end{Bmatrix} \tag{12.7}$$

where $\boldsymbol{F}$ is a generalized flexibility matrix $\boldsymbol{F} = \boldsymbol{H}^T\,\boldsymbol{F}'\,\boldsymbol{H}$ for the generalized degrees of freedom at x_1 and x_2. Finally, the compatibility equations are the linear relations $\boldsymbol{\delta} = \boldsymbol{0}$ or

$$\boldsymbol{F}\,\boldsymbol{x} + \boldsymbol{\delta}^0 = \boldsymbol{0} \tag{12.8}$$

In the case of our example, these are two equations in the two unknowns, x_1 and x_2. In the general case we obtain R equations in the R redundants $\boldsymbol{x}$. We then solve these equations. Having $\boldsymbol{x}$ we wind up with a statically determinate structure under the external and the R pair of loads $\boldsymbol{x}$, which are known. Voila!

From here on we navigate in charted waters. We can either solve the whole thing from scratch, which should be easy (maybe lengthy, but still easy), the structure being determinate, or make use of intermediate results, using superposition, as we will see in the sequel. (We shall see that in practice we compute directly the generalized

---

[4]This congruency is not fortuitous. It has to do with Betti's symmetry and it permeates the entire field of structural theory.

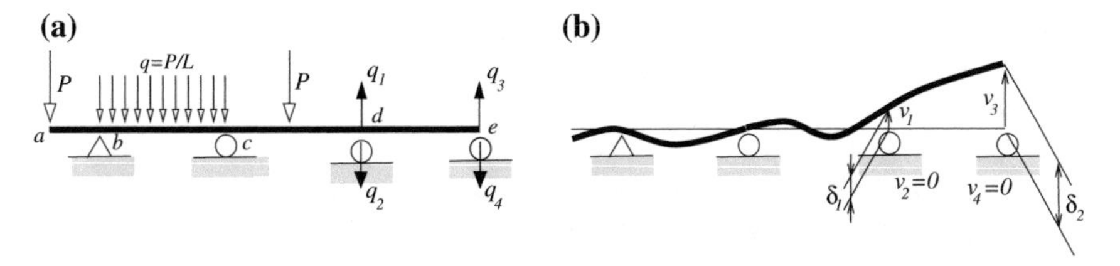

Fig. 12.8 Displacement discontinuities at the releases

influence matrix $\boldsymbol{F}$. The roundabout way through $\boldsymbol{H}$ and $\boldsymbol{F}'$ is to show that $\boldsymbol{F}$ is an influence matrix.)

If we had chosen the reactions in Fig. 12.8 as redundants we would proceed in a similar manner. As visualized in Fig. 12.8a, we define four degrees of freedom at the releases, where we apply the forces $\boldsymbol{q} = \{q_1\, q_2\, q_3\, q_4\}$. Degree of freedom 1 for instance is attached to the beam at d and degree of freedom 2 is attached to the (fixed) support at d. In Fig. 12.8b, we find corresponding displacements $\boldsymbol{v} = \{v_1\, v_2\, v_3\, v_4\}$ and, yes, the supports being rigid, $v_2 = v_4 = 0$, but we leave them in the formulation in order to show that the approach is quite general. For instance, $\delta_1 = v_1 + v_2$ and $\delta_2 = v_3 + v_4$ which is in fact $\delta_1 = v_1$ and $\delta_2 = v_3$.

Now the displacements are due to the four forces $\boldsymbol{q}$ and the external loads. System (12.4) becomes $\boldsymbol{v} = \boldsymbol{F}'\, \boldsymbol{q} + \boldsymbol{v}^0$ (a different $\boldsymbol{F}'$ of course). You will have noticed that only f'_{11}, f'_{13}, f'_{31}, f'_{33} are different from zero, which shows that we are somewhat overdoing it.

Next (12.5) and (12.6) become $\boldsymbol{q} = \boldsymbol{H}\, \boldsymbol{x}$ and $\boldsymbol{\delta} = \boldsymbol{H}^T\, \boldsymbol{v}$ (the same $\boldsymbol{H}$) which leads to the same form of the compatibility equations (12.8).

12.6 Implementation of the Compatibility Equations

The main burden of analyzing a structure by the Force method is to compute the coefficients of matrix $\boldsymbol{F}$ and the components of vector $\boldsymbol{\delta}^0$ in the compatibility equations $\boldsymbol{F}\, \boldsymbol{x} + \boldsymbol{\delta}^0 = \boldsymbol{0}$.

It should be clear that $\boldsymbol{F}$ is a flexibility matrix in generalized coordinates (pairs of equal and opposite coordinates.) Indeed $f_{ij}\, x_j$ is the discrepancy at i due to x_j, that is, the sum of two displacements at i due to two equal and opposite loads at j. Consequently, f_{ij} is the discrepancy at i due to two equal and opposite *unit* loads at j. Likewise δ_i^0 is the discontinuity (sum of two opposite displacements) at i due to external causes.[5]

We can thus apply, almost blindly, what we already know regarding the flexibility coefficients to the compatibility equations.

[5]Note, $\boldsymbol{F}$ is symmetric.

1. In a first instance, we analyze the released (statically determinate) structure R times under R equal and opposite unit loads at the releases. This yields R distributions of internal loads $\boldsymbol{n}^i(x)$, $i = 1, \ldots, R$. Recall that $\boldsymbol{n}^i(x) \equiv \{n^i(x)\, s^i(x)\, m^i(x)\}^T$.
2. Next we analyze the same released structure once more, this time under the applied loads. The resulting deformations will be named $\boldsymbol{\epsilon}^0(x)$. Here, $\boldsymbol{\epsilon}^0(x) \equiv \{\epsilon^0(x)\, \gamma^0(x)\, \kappa^0(x)\}^T$.
3. We then compute the flexibility coefficients $f_{ij} = \int_x \boldsymbol{n}_i^T [\boldsymbol{EA}]^{-1} \boldsymbol{n}_j \, \mathrm{d}x$ noting that we need to compute only $R(R+1)/2$ coefficients since, by symmetry, $f_{ji} = f_{ij}$. (The superscripts were given subscript status.)
4. Finally, the discrepancy at the releases due to external causes is $\delta_i^0 = \int_x \boldsymbol{n}_i^T \, \boldsymbol{\epsilon}_j^0 \, \mathrm{d}x$ (from the unit-load method).

With regard to the numerical computations, we recall that $\gamma = 0$ (Euler–Bernoulli beams) and, since the axial stiffness of flexural elements is very large, we have $\int_x (n_i n_j / EA) \, \mathrm{d}x \ll \int_x (m_i m_j / EI) \, \mathrm{d}x$, and similarly in the δ terms, and therefore we use only the moments terms in the computations

Consequently the coefficients for beams and frames are

$$F_{ij} = \int_x \frac{m^i \, m^j}{EI} \, \mathrm{d}x \,, \qquad \delta_i^0 = \int_x m^i \, \kappa^0 \, \mathrm{d}x \tag{12.9}$$

were we have dropped the (x) for clarity reasons.

12.7 Solved Examples

We will be looking at some examples of typical redundant beams, frames, trusses and other structures. There is nothing new under the sun. Things keep repeating themselves. All we really need to do is to determine the degree of redundancy of the structure, to define a suitable released structure and to draw correct moment diagrams (or compute the bar forces in the case of trusses) for the unit pair of redundants and the applied external loads.

From there on it is all downhill. We calculate the coefficients of the compatibility equations, solve the equations to get the real values of the redundants and then analyze the released structure under the external and released loads.

12.7.1 *R* = 1 *Propped Cantilever*

The $R = 1$ propped cantilever of stiffness EI in Fig. 12.9a is subjected to a point-force Q at 2/3 of the span L. It is released by a hinge at $L/3$ from the wall (Fig. 12.9b).

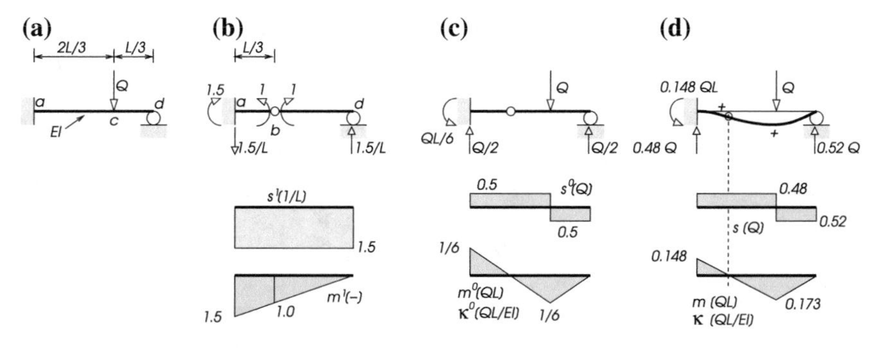

Fig. 12.9 Propped cantilever

The bending moments due to unit couples applied on both sides of the hinge, m^1, and the curvature due to the applied force, κ^0, are depicted respectively in Fig. 12.9b, c.

For m^1, we note that the beam, being free of distributed forces, the shear s^1 is constant and the bending moment m^1 is linear. We need two points to position a straight line: at d the bending moment is zero (the free end) and at b its value is 1, the applied unit couples. Hence, the moment at the root is 1.5, non-dimensional. This linear moment diagram may take some convincing. If necessary, try to equilibrate bd with a unit couple at the left extremity b and segment ab with a unit couple also at b, the right end.

For κ^0 we first compute the bending moment distribution m^0. It is composed of a straight line along ac passing through zero at the hinge and of a straight line along cd connecting to zero at the extremity. All we need to finalize the diagram is the value of the moment at c, m_c^0, and noting that the reaction at d is necessarily $Q/2$ (why? because the bending moment in the vicinity of the hinge must be zero) we get $m_c^0 = QL/6$. Finally, from Hooke's law we have $\kappa^0 = m^0/EI$.

Evaluating the integrals (12.9) we get $F_{11} = 0.75\,L/EI$ and $\delta_1^0 = -0.0092\,QL^2/EI$ and the compatibility equation $(0.75\,L/EI)\,x_1 - 0.0092\,QL^2/EI = 0$ yielding the moment at 1/3 of the span $x_1 \equiv m_b = 0.0123\,QL$.

For the final diagrams, we can either start from scratch with a determinate beam (the released beam) with all applied forces known, or get the results from superpositioning of the results of the released beam under x_1 at the hinges (which is in fact x_1 times the results due to unit couples at the hinges) and the released beam under the applied loads. The bending moment distribution, for instance, is $m(x) = x_1\,m^1(x) + m^0(x)$ and the shear is $s(x) = x_1\,s^1(x) + s^0(x)$. The same is valid for the reactions, such as, the vertical reaction at the roller at d is $R_d = x_1\,R_d^1 + R_d^0$ and for the displacements, for that matter.

In the present case, we need the values of the linear moments at the extremities of the segments. For example the bending moment at the root $m_a = x_1\,m_a^1 + m_a^0$ that is $m_a = (0.0123 \times 1.5 - 0.167)\,QL = 0.148\,QL$. The diagrams are shown in Fig. 12.9d. Note, the curvature and moment distributions are identical, because EI is uniform, but with different units, QL for the moments and QL/EI for the curvatures.

To conclude the analysis, we have drawn the equilibrium shape in the form of the simplest smooth curve starting at a with a zero slope, joining the roller at d, with curvature signs taken from $\kappa(x)$. The region with zero curvature is indicated by a ring (a virtual hinge). Starting from a we find tension ('+' signs) along the upper fibers where the bending moment is negative and tension along the lower fibers where the moment is positive.

Note, after all that work, and after ensuring that the computations are correct, what we need to look at *first* are the bending moments, and in particular the regions of large bending moments. In our example, if trouble is to be expected it will most probably occur under the force with the root in a close second place. The beam hardly moves there, zero displacement and zero slope, but this is not what counts.

A beam suffers most from bending moments.

Near the virtual hinge, on the other hand, the beam has moved substantially, but the bending moments are very small, and this is the safest region of the beam.

12.7.2 $R = 2$ Beam on Mixed Supports

The beam in Fig. 12.10a, on rigid supports at a and c and elastic supports at b and d is submitted to a point-couple PL at c. Note the difference in elastic stiffness at b and d. We assume $k = EI/L^3$ (both sides have dimensions *force/length*).

The $R = 2$ structure is released by hinges at b and c (Fig. 12.10b). This type of release is not intuitive but it is the simplest. The released structure is thus loaded by the point-couple PL and by the released bending moments x_1 and x_2. In order

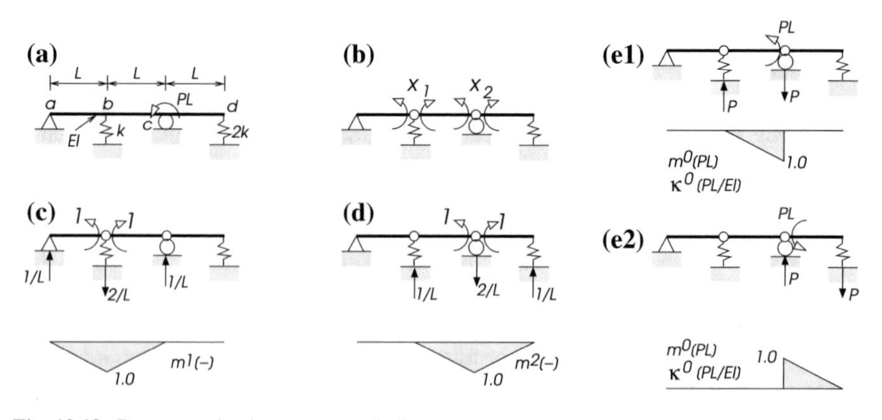

Fig. 12.10 Beam on mixed supports: analysis

to determine the unknown $\boldsymbol{x}$ we write the compatibility equations $\boldsymbol{F}\ \boldsymbol{x} + \boldsymbol{\delta}^0 = 0$. For those interested the equations require that there should be a continuous slope at b and c. As mentioned earlier, one does not need to understand the meaning of the compatibility equations in order to use them.

To compute the coefficients of $\boldsymbol{F}$ and $\boldsymbol{\delta}^0$ one determines the bending moments m^1 and the forces in the springs n_b^1 and n_d^1 due to unit loads at d.o.f. 1, the bending moments m^2 and the forces in the springs n_b^2 and n_d^2 due to unit loads at d.o.f. 2, and the curvatures κ^0 and the extensions e_b^0 and e_d^0 due the external couple PL. The unit bending moment has the classical hat-shape.

For κ^0 we need to be careful in this case. The external cause is a point couple and we must make a choice whether to apply PL to the left (Fig. 12.10e1) or to the right (Fig. 12.10e2) of the hinge. In both instances the couple is anti-clockwise and although the curvatures are different they yield the same final result. In the sequel we use the curvatures and extensions of Fig. 12.10e2. The redundant couple x_2 will therefore be the bending moment to the left of support c.

The expression for the coefficients is $F_{ij} = \int_a^d m^i m^j / EI \mathrm{d}x + n_b^i n_b^j / k + n_d^i n_d^j / 2k$ and $\delta_i^0 = \int_a^d m^i \kappa^0 \mathrm{d}x + n_b^i e_b^0 + n_d^i e_d^0$. Note, the elongations of the elastic supports are the forces in the springs (the reactions, positive in tension) divided by the respective stiffness of the springs.

From Fig. 12.10c, d, e2 we find that the non-zero forces (reactions) and elongations in the springs are $n_b^1 = 2/kL$, $n_b^2 = -1/kL$, $n_d^2 = -1/2kL$, $e_d^0 = P/2k$.

After integrations, and noting that $k = EI/L^3$ we obtain the compatibility equations

$$\frac{L}{6EI}\begin{bmatrix} 28 & -11 \\ -11 & 13 \end{bmatrix}\begin{Bmatrix} x_1 \\ x_2 \end{Bmatrix} + \frac{L}{6EI}\begin{Bmatrix} 0 \\ -5PL \end{Bmatrix} = \begin{Bmatrix} 0 \\ 0 \end{Bmatrix} \tag{12.10}$$

which yields the bending moments over supports b and d: $x_1 = 0.226PL$ and $x_2 = 0.576PL$.

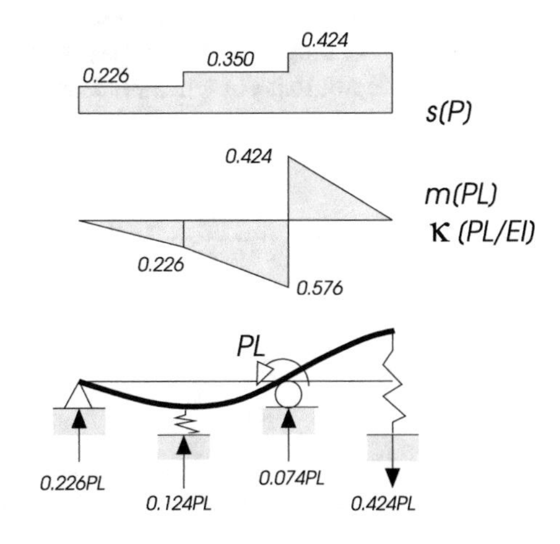

Fig. 12.11 Beam on mixed supports: results

At this point we have a statically determinate structure, the released structure, with all the applied forces known. This concludes the main steps of the force method. What is left is to draw the final diagrams, the displaced shape and to compute the reactions (Fig. 12.11).

One will note that the curvature changes sign at support c without passing through a point of zero curvature. That is, the second derivative of the deflected shape is discontinuous but the first derivative, the slope, is continuous.

12.7.3 $R = 2$ Beam: Releasing the Moments over Two Supports

Using the release of the $R = 2$ continuous beam in Fig. 12.7a, by introducing hinges at supports c and d as in Fig. 12.7b we show in Fig. 12.12 the m^1, m^2 moment distributions for unit couples at c and d and the deformations κ^0 due to the external forces. The deformations are obtained by computing the bending moment m^0 due to the applied forces and using Hooke's law $\kappa^0 = m^0/EI$. Note, we have a separate distribution for the concentrated (κ_1^0) and for the distributed forces (κ_2^0). The loading integrals are thus made up of two terms $\delta_i^0 = \int_x m^i(x)\,\kappa_1^0(x)\,\mathrm{d}x + \int_x m^i(x)\,\kappa_2^0(x)\,\mathrm{d}x$.

$$\frac{L}{EI}\begin{bmatrix} 0.666 & 0.167 \\ 0.167 & 0.666 \end{bmatrix}\begin{Bmatrix} x_1 \\ x_2 \end{Bmatrix} + \frac{PL^2}{EI}\begin{Bmatrix} 0.118 \\ 0.0625 \end{Bmatrix} = \begin{Bmatrix} 0 \\ 0 \end{Bmatrix} \tag{12.11}$$

the solution of which is the bending moments at the supports $x_1 = -0.164\,PL$ and $x_2 = -0.053\,PL$.

The final shear and moment distributions are given at the start of this chapter in Fig. 12.2a. The bending stiffness being uniform the curvature distribution is identical to the moment distribution in different units, $\kappa = m/EI$.

A notable part of the procedure is the moment distributions under unit couples on both sides of the hinge. At first it may seem difficult to obtain the (flat) pyramidal shape, but it is not that complicated. For m^1, for instance, on span ab the moment

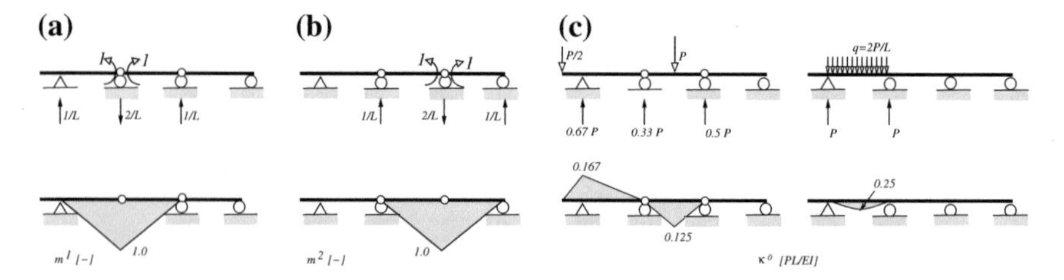

Fig. 12.12 Releasing the moments. **a** Moments due to unit couples at 1. **b** Moments due to unit couples at 2. **c** Curvature distribution due to external loads (P *left*, q *right*). Final curvature is the sum of the two

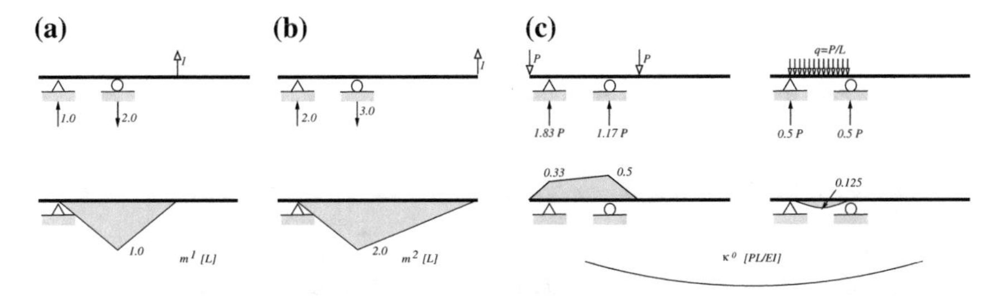

Fig. 12.13 Disconnecting the supports. **a** Moments due to unit forces at 1. **b** Moments due to unit forces at 2. **c** Curvatures due to external loads Forces P (*left*) + distributed q (*right*) Final curvatures are the sum of the two

is zero at a (the free end) and it equals 1 at b (there is a unit couple applied there), and in between the variation is linear (no distributed forces). On bc the moment goes similarly from 1 to zero. We don't even need the reactions to compute the moments

12.7.4 Same R = 2 Beam: Disconnecting Two Supports

Selecting the reactions at c and d as unknown redundants x_1 and x_2 as in Fig. 12.8, we get the moment distributions shown in Fig. 12.13 which lead, after calculating the integrals (12.9), to the compatibility equations

$$\frac{L}{EI}\begin{bmatrix}0.667 & 1.5\\ 1.5 & 4.0\end{bmatrix}\begin{Bmatrix}x_1\\ x_2\end{Bmatrix}+\frac{PL}{EI}\begin{Bmatrix}-0.215\\ -0.451\end{Bmatrix}=\begin{Bmatrix}0\\ 0\end{Bmatrix} \tag{12.12}$$

yielding the reactions $x_1 = 0.442\,P$ and $x_2 = -0.053\,P$.

It is now evident that the idea behind the compatibility equations has already been forgotten. We do not need to know that the ith row of the compatibility equations says that the discontinuity at release i due to the redundants and due to the external loads is zero. Instead, we follow a simple sequential procedure and get the results. Such techniques are what successful engineering is all about.

We will see in a short while that the Displacement method is even more routine work. It is almost idiot-proof and this is why the Displacement method is so suited for computerized implementations.

12.7.5 Beam on Many Supports—Three Moments Equations

This is classical material. Consider the beam on $N+2$ supports shown in Fig. 12.14a. The redundancy of this structure is of course N; two supports suffice to carry any load.

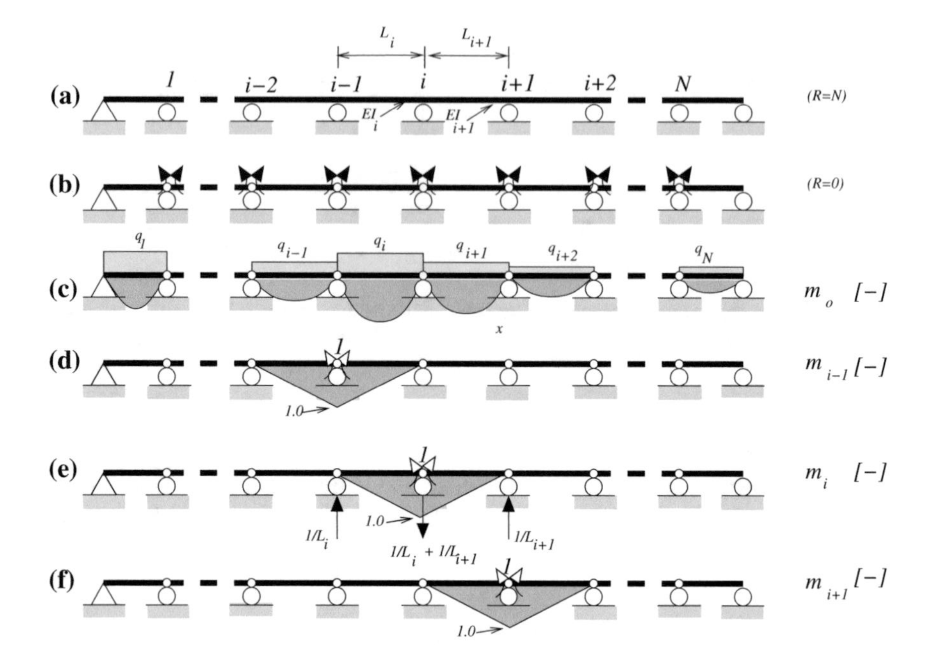

Fig. 12.14 Three moments—releasing the moments over the supports

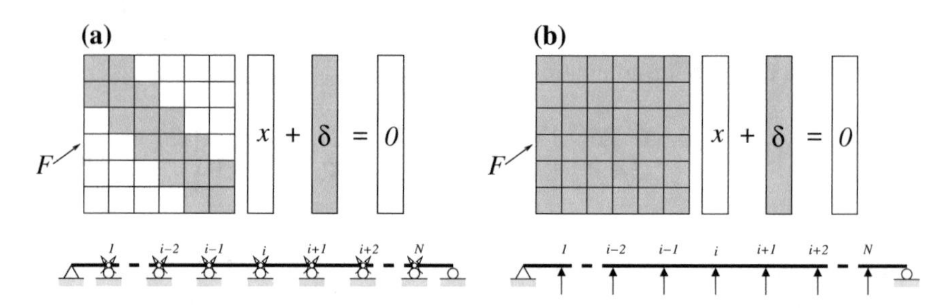

Fig. 12.15 Three moments—flexibility matrix

When releasing this kind of structure, intuition is sometimes a poor counselor. What comes first to mind is to release the structure by removing the N intermediate supports and selecting the reactions as the R redundants $\boldsymbol{x}$ (see Fig. 12.15b, right). There is nothing very wrong with this but it will lead to a full matrix $\boldsymbol{F}$, and we will have to compute a host of flexibility coefficients.

However, setting hinges at the field supports, as in Fig. 12.14b, left, and taking the bending moments at the supports as the redundants $\boldsymbol{x}$ is better. This leads to a banded matrix $\boldsymbol{F}$ (three coefficients per line/column) and also $\boldsymbol{\delta}_o$ is easier to compute.

Introducing Hinges Over the Supports

The repetitive nature of the structure lends itself to developing general expressions for the coefficients of $\boldsymbol{F}$ and $\boldsymbol{\delta}^0$, the three-moments equations. In Fig. 12.14c we have limited the external loads to a uniformly distributed force q_i, constant per span. Since the released structure can be construed as the juxtaposition of simply supported beams, the bending moment in span i is a parabola with maximum moment $(m_{\max})_i^0 = 0.125(q_i L_i^2)_i$, where q_i and L_i are respectively the magnitude of the distributed forces and the length of span i. We note that span i is the span in front of redundant support i.

In Fig. 12.14d, e, f we have depicted the bending moment distributions due to unit moments at three consecutive supports, $i-1,\ i,\ i+1$ respectively. In Fig. 12.14e, we have also indicated the reactions. Clearly, the bending moments are triangular with a peak value of 1, and cover only the adjoining span (this is the hinged releases' secret).

The general flexibility and load coefficients are

$$f_{ij} = \int_0^L \frac{m_i(x)\,m_j(x)}{EI(x)}\,\mathrm{d}x \quad \delta_i^0 = \int_0^L m_i(x)\,\kappa^0(x)\,\mathrm{d}x \quad \text{with} \quad \kappa^0(x) = m^0(x)/EI(x)$$

where L and $EI(x)$ are respectively the length and flexural rigidity of the beam.

Clearly, $f_{ij} = 0$, but for the diagonal entry f_{ii} and for the adjacent coefficients $f_{i,i-1}$ and $f_{i,i+1}$. Beyond this the product is zero since each function is zero where the other is not. Integration easily gives

$$fi,i-1 = \frac{1}{6}\left(\frac{L}{EI}\right)_i \quad fi,i = \frac{1}{3}\left[\left(\frac{L}{EI}\right)_i + \left(\frac{L}{EI}\right)_{i+1}\right] \quad fi,i+1 = \frac{1}{6}\left(\frac{L}{EI}\right)_{i+1} \tag{12.13}$$

This yields a diagonal matrix of the type in Fig. 12.15a.

A typical line has the form $f_{i,i-1}x_{i-1} + f_{i,i}x_i + f_{i,i+1}x_{i+1} + \delta_i^0 = 0$, and since the x_i are moments, hence the name 'three moments equations'. The first row has only two coefficients since $f_{13} = 0$ and similarly for the last row $(f_{N,N-2} = 0)$.

As for the loading vector $\boldsymbol{\delta}^0$, it all depends on the specific case. The integration for a typical component ranges only over the adjacent spans

$$\delta_i^0 = \int_{L_i} m_i\,\kappa^0\,\mathrm{d}x + \int_{L_{i+1}} m_i\,\kappa^0\,\mathrm{d}x \tag{12.14}$$

because $m_i = 0$ elsewhere.

Removing the Supports

Had we continued with the redundants as in Fig. 12.16b, we would have obtained the basic bending moments shown in the diagrams Fig. 12.16d–f.

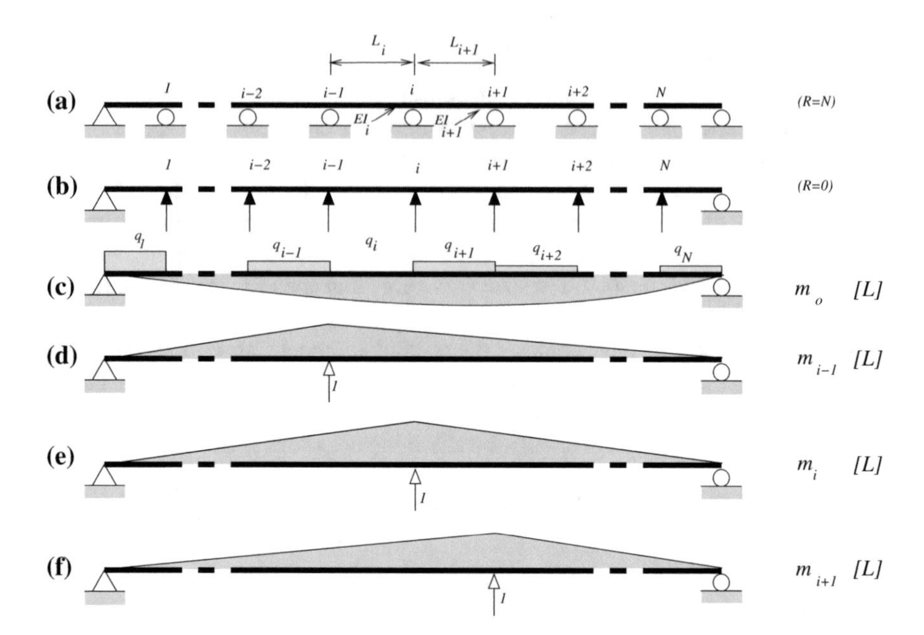

Fig. 12.16 Three moments—releasing the reactions

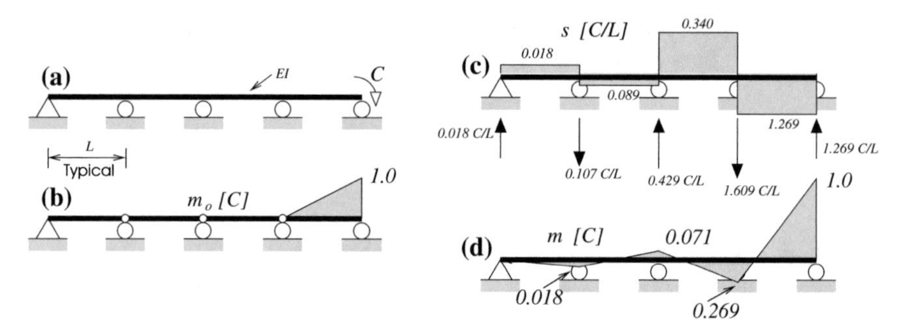

Fig. 12.17 Continuous beam with end couple

It is clear that none of the flexibility coefficients will be zero, thus leading to a full matrix as in Fig. 12.15b. Moreover, m^0 is often not simple to establish. From every viewpoint, the equations in Fig. 12.15a or Fig. 12.14 are preferable.

Example

As a example, consider the $R = 3$ continuous and uniform beam (EI) with equal spans L subjected to an end-couple C shown in Fig. 12.17a. The particular m^0, also $\kappa^0 = m^0/EI$, is given in Fig. 12.17b. With the help of the formulas (12.13) and after computing the loading vector (12.14), we get the compatibility equations

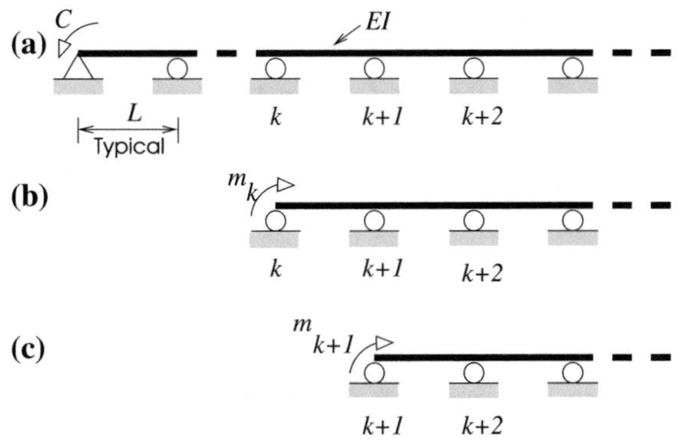

Fig. 12.18 Semi-infinite beam with end couple

$$\frac{L}{EI}\begin{bmatrix} 2/3 & 1/6 & 0 \\ 1/6 & 2/3 & 1/6 \\ 0 & 1/6 & 2/3 \end{bmatrix}\begin{Bmatrix} x_1 \\ x_2 \\ x_3 \end{Bmatrix} + \frac{LC}{EI}\begin{Bmatrix} 0 \\ 0 \\ -1/6 \end{Bmatrix} = \begin{Bmatrix} 0 \\ 0 \\ 0 \end{Bmatrix} \tag{12.15}$$

the solution of which is $x_1 = 0.018C$, $x_2 = -0.071C$ and $x_3 = 0.268C$. The results of the analysis are shown in Fig. 12.17c for the reactions and shear diagram and in Fig. 12.17d for the moments. Note the typical (and rather fast) decaying of the alternating moments. You may also verify that the reactions are statically equivalent to a couple of magnitude C with a zero resultant.

Semi-infinite Beam

Incidently, there exists an interesting application to semi-infinite repetitive beams, an example of which is shown in Fig. 12.18a. The beam has identical spans running from the left end all the way to infinity. Can one get the moment distribution to, say, a couple C at the left end?

We will reason in the following manner.[6] Consider three consecutive supports k, $k+1$, $k+2$, and let us cut the structure to the left of support k (Fig. 12.18b) and to the left of support $k+1$ (Fig. 12.18c).

The point is that both structures are identical semi-infinite beams with a couple at the left end. Consequently the ratios of two consecutive moments in both beams should be the same

$$\frac{m_{k+1}}{m_k} = \frac{m_{k+2}}{m_{k+1}} = \lambda \tag{12.16}$$

or $m_{k+1} = \lambda m_k$ and $m_{k+2} = \lambda^2 m_k$ where λ is the constant decay factor. Introducing these values in a typical three moment equation for equal spans and equal stiffnesses $m_k + 4m_{k+1} + m_{k+2} = 0$ yields a quadratic equation in λ, the only physically sound solution of which is

$$\lambda = -(2 - \sqrt{3}) = -0.268 \tag{12.17}$$

[6] M Ryvkin, Oral communication.

Starting with the applied couple C at support $k = 1$, we get the decaying sequence $m_k/C = (-0.268)^{k-1}$, that is, 1, −0.268, 0.0718, −0.0192… This is a very fast decay with the magnitude of the bending moment at the fourth support already less than 2 % of the applied couple.

12.7.6 Beam Stiffened by Underlying Truss

A simply supported uniform beam of flexural stiffness EI and length L is subjected to a force Q at mid-span. The displacement under the load is deemed excessive, and it is decided to stiffen the structure by adding a truss under it, as shown in Fig. 12.20a making the structure once redundant.

Redundancy An often easy way to establish the redundancy is to follow its assembly. Starting with beam ac, there is no difficulty in mounting bars ad and cd of the truss (see Fig. 12.19a). If the lengths of these bars were not exactly what they should be, node d will have a little offset. The geometry may not be exactly what was designed but there are no obstacles during assembly. This is typical for determinate designs, in other words, up to this point $R = 0$.

But now nodes b and d are fixed and you want to connect bar $b'd'$ (Fig. 12.19b). If the length of bar $b'd'$ is not exactly equal to the distance bd there will be a problem. Even if we manage to squeeze the bar in, chances are that we will have pre-stressed the structure. Since bar $b'd'$ cannot be freely assembled, it proves that the bar is redundant and we have thus $R = 1$.

You can always determine redundancy by enumeration (Fig. 12.19c). Since the structure is 2-dimensional we look at it as a frame. The structure has four nodes (a b c and d), two frame elements and three truss elements. The number of unknowns (in small circles) is $U = 2 \times 3 + 3 \times 1 = 9$ and the number of equations at the nodes (in small squares) is $E = 1 + 3 + 2 + 2 = 8$. (We recall that node d is connected only to truss elements, which have no moments about d, hence the moment equation is automatically satisfied and we have only two non-trivial equilibrium equations there.) Consequently, $R = U - E = 1$.

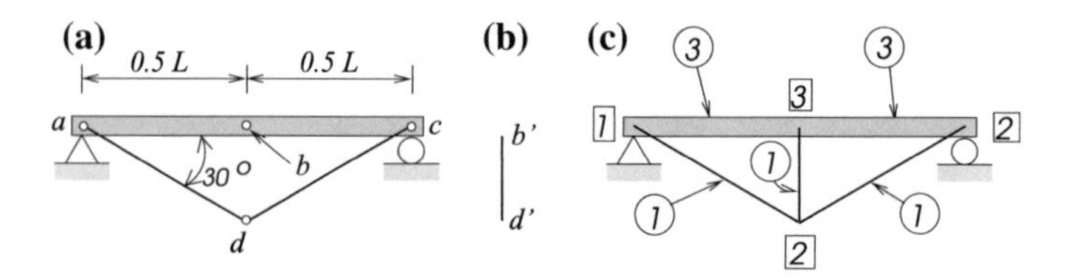

Fig. 12.19 Beam stiffened by underlying truss

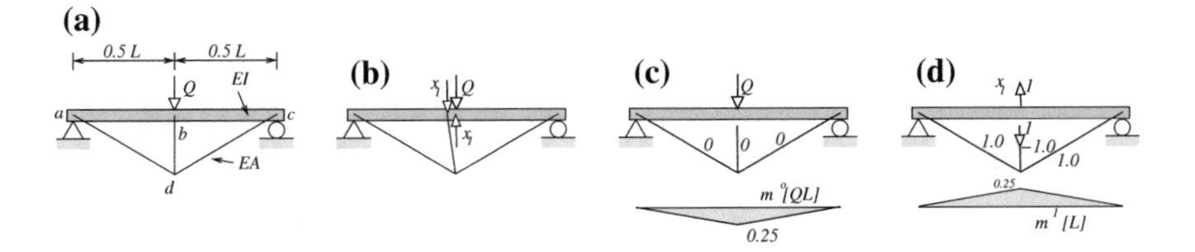

Fig. 12.20 Stiffened beam: analysis

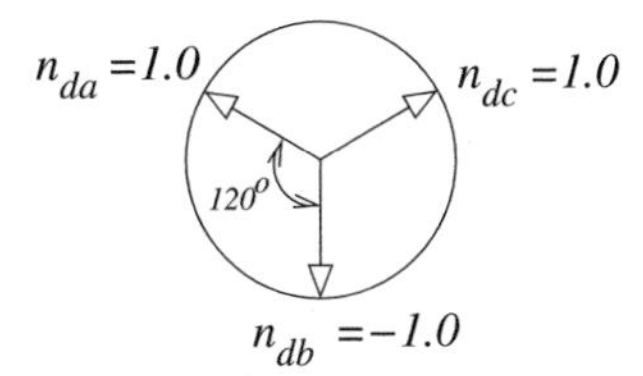

Fig. 12.21 Equilibrium of node d

Release To release this $R = 1$ structure element, bd is disconnected from the beam at b. This creates a determinate assembly with as loads Q and the pair of equal and opposite axial forces x_1 (truss element) applied at the release, positive in compression.)
Compatibility From here on we follow the standard procedure. The analysis of the structure under Q and $x_1 = 1$ is given in Fig. 12.20c, d.

The forces in the truss under Q are zero ($n_j^0 = 0$), and interestingly, the magnitude of the forces in the truss due to $x_1 = 1$ is $|n_j^1| = 1$, as can be seen in Fig. 12.21 (because of the 30° inclination of the bars). Noting that $L_{da} = L_{dc} = L/\sqrt{3}$ and $L_{db} = L_{da}/2$ we obtain for the coefficients of the compatibility equation

$$F_{11} = \frac{1}{48}\frac{L^3}{EI} + \frac{5\sqrt{3}}{6}\frac{L}{EA} \qquad \delta_i^0 = -\frac{1}{48}\frac{QL^3}{EI}$$

which yields the now familiar form

$$x_1 = \frac{\beta}{\beta + 69.3}\,Q \quad \text{with} \quad \beta = \frac{AL^2}{I} \tag{12.18}$$

The saturation load (for $\beta \to \infty$) is Q which means that the force applied by the truss to the beam annuls the external load for large values of β. This makes sense because when $\beta \to \infty$ the truss acts as a rigid support at b.

We also note that if $\beta = 69.3$, we have $x_1 = Q/2$. This is a relatively high value for β to reach half the saturation load. It would mean that this arrangement for stiffening the beam with a truss is not very effective. For that value of β the displacement under Q is half of what it was initially.

12.7.7 $R = 2$ Portal Frame

The frame in Fig. 12.22a is supported by a horizontal guide at a and built in at d. The bending stiffness EI is uniform, as usual,

elements designed to take bending are very stiff axially

consequently we assume $EA \to \infty$.

Note, the horizontal guide prevents the rotation and the vertical displacement at a, but the horizontal displacement is an open degree of freedom. The structure is twice redundant because without the support at a we have a tree which is a statically determinate structure, and the two reactions at a are therefore redundant. The reactions correspond to the prevented displacements: the vertical reaction and the bending moment at the root. Hence $R = 2$.

We have released the structure by introducing hinges at a and d as shown in Fig. 12.22b. The rotations are now also degrees of freedom. Support d became a hinge and since only the vertical displacement is prevented at a that support is a roller. The released frame is a statically determinate structure subjected to the unknown couples x_1 and x_2 on both sides of the hinges at a and d, respectively, and to the applied force P. Note, at a, one couple is applied to the column while the other goes straight to the support.

Following the standard procedure we draw the moment diagrams m^0, m^1 and m^2, due to respectively the applied force P and unit couples at releases 1 and 2, also shown in Fig. 12.22. This leads to two compatibility equations in the two unknown redundants.

$$\begin{Bmatrix} \delta_1 \\ \delta_2 \end{Bmatrix} \equiv \begin{bmatrix} f_{11} & f_{12} \\ f_{21} & f_{22} \end{bmatrix} \begin{Bmatrix} x_1 \\ x_2 \end{Bmatrix} + \begin{Bmatrix} \delta_1^0 \\ \delta_2^0 \end{Bmatrix} = \begin{Bmatrix} 0 \\ 0 \end{Bmatrix} \tag{12.19}$$

We remind the reader that δ_1, for instance, is the slope with respect to the vertical of column ab at a, and the condition is that the slope should be zero in the released structure as is the case in the original one. But you need not bother too much about the rationale behind the equations. The compatibility conditions come about automatically if we follow the procedure correctly.

Having computed the coefficients in (12.19) using (12.9) and keeping in mind that the columns and beam have different flexural stiffness ($EI_{\text{Beam}} = 5\ EI_{\text{Column}}$),

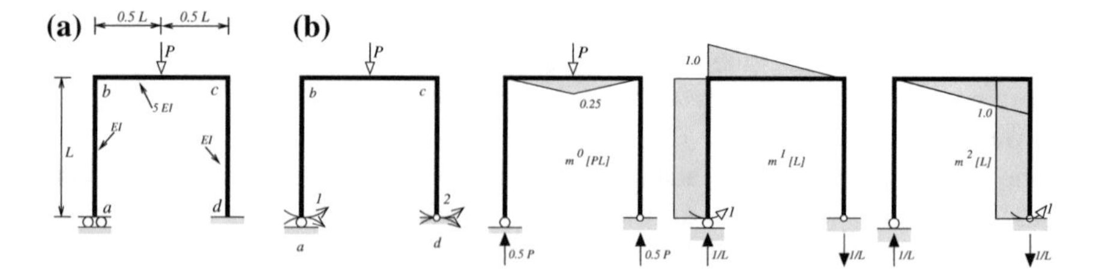

Fig. 12.22 Portal frame: analysis. **a** R = 2 structure. **b** Released structure

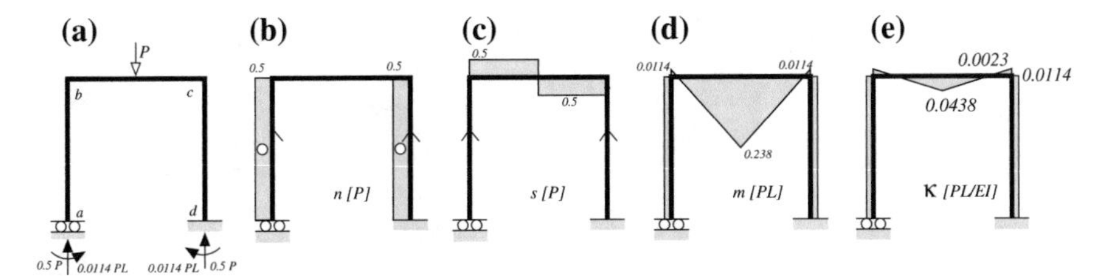

Fig. 12.23 Portal frame: results. **a** External loads and reactions. **b** Normal forces. **c** Shear forces. **d** Bending moments. **e** Curvature

we obtain

$$\frac{L}{30EI}\begin{bmatrix} 32 & -1 \\ -1 & 32 \end{bmatrix}\begin{Bmatrix} x_1 \\ x_2 \end{Bmatrix} + \frac{PL^2}{80EI}\begin{Bmatrix} -1 \\ 1 \end{Bmatrix} = \begin{Bmatrix} 0 \\ 0 \end{Bmatrix} \tag{12.20}$$

with the symmetric results $x_1 = -x_2 = 0.01136\,PL$. This leads to the final diagrams n, s, m and κ shown in Fig. 12.23. Here the curvature and bending moments are different drawings because the bending stiffness is not uniform.

A few notes:

- Although the curvature is strictly symmetric the displacements are not because supports a and d are different (the boundary conditions).
- As always for symmetric problems the shear diagram is anti-symmetric. The actual shear is of course symmetric as can be seen in Fig. 12.24a, b.
- Finally, note the bending moments in Fig. 12.23d. The moment distribution on element bc is almost identical to the one of a simply supported beam on two supports. There we would have zero moments at the ends and 0.25 PL at mid-span. You will notice that the diagram is translated in the up direction by 0.012 PL thus creating moments at the joints and reducing the field moments accordingly. Such translation of the bending moment is a frequent phenomena in redundant structures and it warrants a second look.

Modeling a Frame Consider the beam on two supports loaded at mid-span by a concentrated force P with rotational springs at the supports shown in Fig. 12.25b. The beam has a uniform flexural stiffness EI and the springs have a rotational stiffness k_θ. A rotational spring, as depicted in Fig. 12.25d, is a structural component which when subjected to a bending moment m assumes a relative rotation θ of its extremities, and obeys the elastic constitutive law $m = k_\theta \theta$ (compare with $n = k\delta$ for an extensional spring). In some sense a rotational spring is a constrained hinge in allowing relative rotations but also by creating internal moments when it does rotate.

The beam with two rotational springs is twice redundant. But is it is also a symmetric structure under symmetric loading. Consequently, as in Fig. 12.26a we can consider only half the structure to which half the load is applied. At the axis of symmetry (now the wall) everything anti-symmetric must be zero. A vertical displacement retains symmetry and must be allowed, but a rotation is necessarily anti-symmetric and prohibited. Hence the guided support.

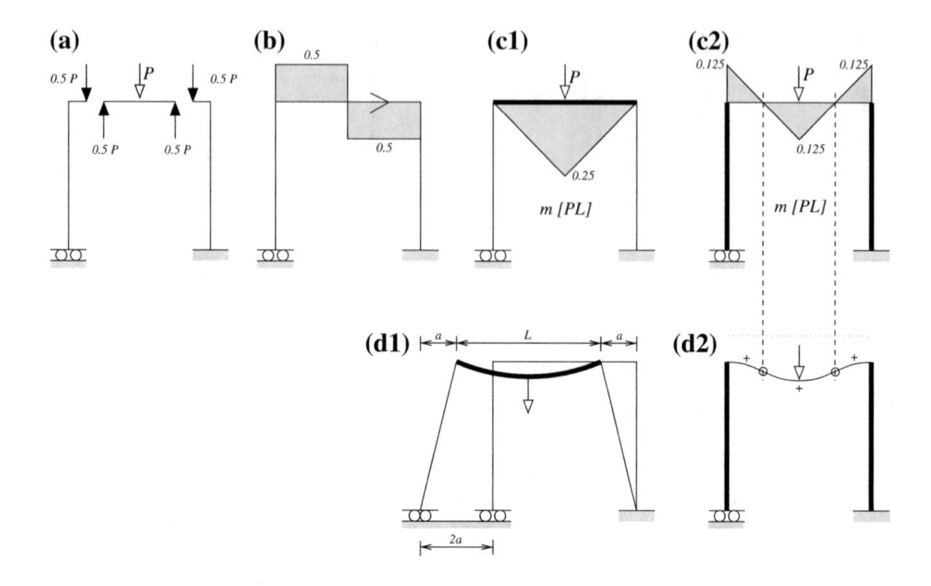

Fig. 12.24 Portal frame: limit cases. **a** Shear forces. **b** Shear diagram. **c1** Bend. moments. **c2** Bend. moments. **d1** Displaced shapes. **d2** Equilibrium configurations

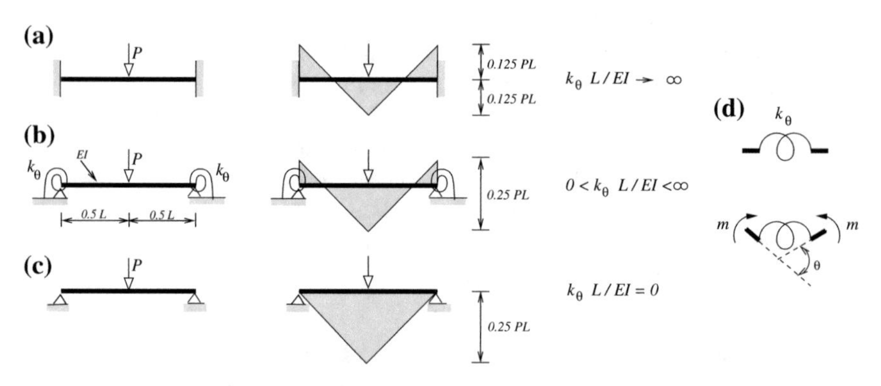

Fig. 12.25 Simply supported beam with rotational springs

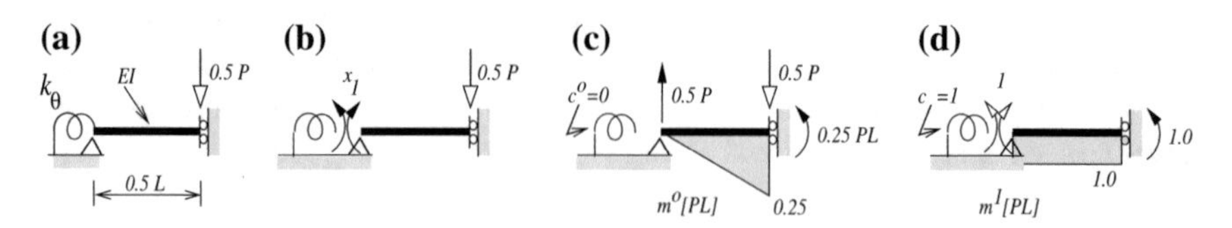

Fig. 12.26 Beam with rotational springs: analysis

This structure is now once redundant. When we disconnect the spring from the beam we obtain a determinate structure with one unknown internal moment x_1 (Fig. 12.26b). In order to find the value of x_1, we analyze the structure under the external loads (Fig. 12.26c) and under unit and opposite couples at the section (Fig. 12.26d). You will have noticed that for the external loads analysis there is no couple applied to the spring $c^0 = 0$, and under unit couples we have $c^1=1$ (dimensionless). The (scalar) compatibility equation is $F_{11}\,x_1 + \delta_{10} = 0$ with

$$F_{11} = \int_0^{0.5L} \frac{(m_1(x))^2}{EI}\,ds + \frac{(c_1)^2}{k_\theta} = \frac{0.5L}{EI} + \frac{1}{k_\theta} \tag{12.21}$$

$$\delta_{10} = \int_0^{0.5L} \frac{m^1(x)m^0(x)}{EI}\,ds = \frac{(0.25L)^2 P}{EI}$$

which yields

$$x_1 = \frac{\beta}{2+\beta}\,\frac{-PL}{8} \quad \text{with} \quad \beta = \frac{k_\theta L}{EI} \tag{12.22}$$

We note that x_1 is always negative (tension in the upper fibres at the supports) but that the magnitude, as expected in redundant structures, depends on the relative importance of the stiffnesses, that is, the ratio of the spring stiffness k_θ and the flexural rigidity of the beam EI/L.

The limit cases are $EI/L \to \infty$ (the beam is very rigid in bending and/or very short) and/or $k_\theta = 0$ (the springs are very weak) so that the structure is effectively a beam on simple supports, yielding $x_1 = 0$ (Fig. 12.25c), and $EI/L \to 0$ or $k_\theta \to \infty$ (the springs are very rigid and the beam feels built in at both extremities) for which $x_1 = -PL/8$. The bending moments and the equilibrium configurations of the limit cases can be seen in Fig. 12.24.

In the general case the bending moment is thus between these limits

$$\text{(Simply supported)} \quad 0 \geq \; x_1 \; \geq -\frac{PL}{8} \quad \text{(Fixed supports)}$$

When we plot (see Fig. 12.27) the moment as a function of the relative rigidities parameter EI/Lk_θ (note that it is non-dimensional), we get the typical variation of the force in a redundant element as a function of its stiffness. As the stiffness increases from zero, the element starts 'pulling' loads into it. But the phenomenon tapers off into a convergence to the asymptote, a rigid spring, which when combined with a hinged support becomes a rigid wall.

The horizontal element of the frame in Fig. 12.23a can therefore be modeled as a beam on rigid supports attached to two rotational springs. In the vertical direction the beam 'feels' as if on rigid supports (elements designed for bending are very rigid axially), but it also seems to have rotational springs at the extremities. Indeed, when the beam bends it has also to bend the columns (Fig. 12.28), hence the elastic rotational resistance.

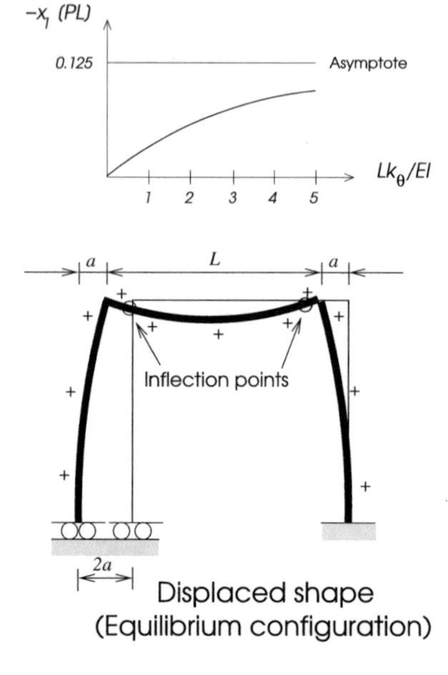

Fig. 12.27 Redundant bending moment versus relative stiffnesses

Fig. 12.28 Portal frame: displacement

In this particular example, the bending moment diagram of beam bc (Fig. 12.23d) has end-moments 0.0114 PL, and looking for the corresponding value in Fig. 12.27, we must come to the conclusion that the resistance that the beam feels at its extremities is rather small, and for all practical purposes the columns behave like simple supports almost without rotational springs. If we knew that from the start we could have effortlessly predicted a bending moment of 0.25 PL at mid-span instead of the computed 0.24 PL moment.

12.7.8 $R = 2$ Frame

The frame in Fig. 12.29a is built-in at a, has a horizontal guide at d and an applied force at c. The redundancy of the structure is $R = 2$. Indeed, the reactions at d are not necessary for stability. We will in this case release the structure by removing the support at d and replace it by the unknowns x_1 and x_2 (Fig. 12.29b.) (The horizontal guide prevents vertical displacement and rotation at d.) In order to set up the 2 compatibility equations $\boldsymbol{F}\,\boldsymbol{x} = \boldsymbol{p}$ we need the curvature distribution κ^0 of the released structure under the applied force P and the bending moment distributions m^1 and m^2 under unit forces at the redundants respectively, $x_1 = 1$ and $x_2 = 1$ (see Fig. 12.29c, d, e). You will have noted that under P, leg db has no internal forces and the frame behaves in fact as a cantilever. Also, the curvature distribution is the moment distribution divided by the bending stiffness $\kappa^0 = m^0/EI$.

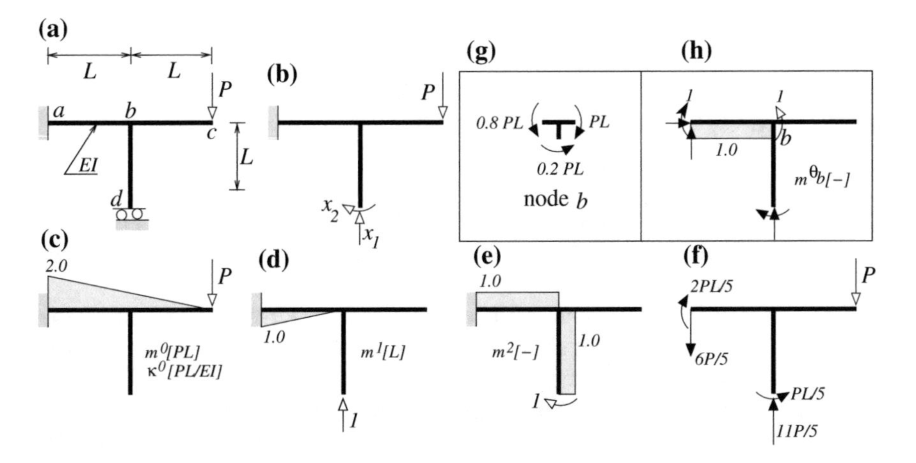

Fig. 12.29 Redundant frame: analysis

Calculating the coefficients of the compatibility equations $F_{ij} = \int_{\text{frame}} m^i m^j / EI \mathrm{d}x$, $\delta_i^0 = \int_{\text{frame}} m^i \kappa^0 \,\mathrm{d}x$ yields

$$\frac{1}{6}\frac{L}{EI}\begin{bmatrix} 2L^2 & -3L \\ -3L & 12 \end{bmatrix}\begin{Bmatrix} x_1 \\ x_2 \end{Bmatrix} + \frac{PL^2}{6EI}\begin{Bmatrix} -5L \\ 9 \end{Bmatrix} = \begin{Bmatrix} 0 \\ 0 \end{Bmatrix} \tag{12.23}$$

In practice one ignores the symbols of the units to obtain a regular system of 2 algebraic equations

$$\frac{1}{6}\begin{bmatrix} 2 & -3 \\ -3 & 12 \end{bmatrix}\begin{Bmatrix} x_1 \\ x_2 \end{Bmatrix} + \frac{1}{6}\begin{Bmatrix} -5 \\ 9 \end{Bmatrix} = \begin{Bmatrix} 0 \\ 0 \end{Bmatrix} \tag{12.24}$$

the solution of which is $x_1 = 11/5$ and $x_2 = -1/5$ where forces (x_1) are given in P and couples (x_2) are in PL. We can compute the reactions at a (see Fig. 12.29f) and having all the loads applied to the frame we can now draw the $nsm(x)$ diagrams. The moment diagram can also be obtained by superposition: $m(x) = (11P/5)\, m^1(x) - (PL/5)\, m^2(x)$.

The curvature diagram in Fig. 12.30a, which is identical to the bending moment diagram but in other units, is the basis for drawing the displaced shape or equilibrium configuration (Fig. 12.30d). We draw he simplest possible configuration which

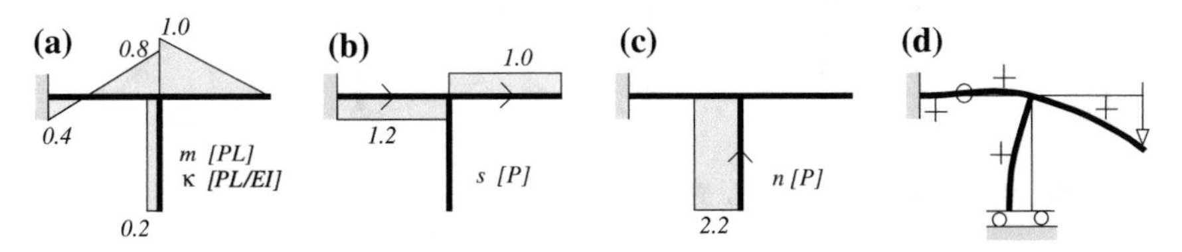

Fig. 12.30 Redundant frame: results

complies with the boundary conditions and the sign of the curvatures (note the '+' sign where we have extension and the little circle where the curvature is zero). This is the one and only configuration where the structure can produce the nsm diagrams which effectively set everything in equilibrium.

Equilibrium and Rotation of Node b

To complete this exercise we show the rotational equilibrium of node b in Fig. 12.29g (see figures within squares.) The couples were deduced from the bending moment diagram in Fig. 12.30a.

Finally we use the unit load method to compute the rotation of node b (see Fig. 12.29h.) We apply a unit couple at b and select from the five possible reactions (black arrows in the figure) any simple combination that will equilibrate the couple. As shown, a simple choice is the couple at a. The resulting bending moment m^{θ_b}, so called because it was built to compute θ at b, is given in the figure. We find $\theta_b = \int_{\text{frame}} \kappa\, m^{\theta_b} \mathrm{d}x = -PL^2/5EI$.

12.7.9 $R = 1$ Cable-Braced Frame

Our next example is a frame $abcd$ hinged at a with a roller at d as shown in Fig. 12.31a. This frame of uniform stiffness EI was statically determinate, but under a horizontal force P at d the structure proved to be too flexible (excessive horizontal displacement at d). The lateral motion was therefore restrained by adding a cable of stiffness EA along the diagonal bd.

Redundancy: A cable is a structural element which is very stiff in tension but has no flexural rigidity. The cable here is also redundant (the structure without the cable is a perfectly sound frame). Now, a central property of redundant elements is that they draw loads in proportion to their relative stiffness. Having no bending stiffness the cable will therefore not carry bending moments but it will have normal forces because of its longitudinal stiffness. Its only static unknown is thus the axial force n_{bd}. Adding one redundant to an otherwise statically determinate assembly produces a redundancy $R = 1$.

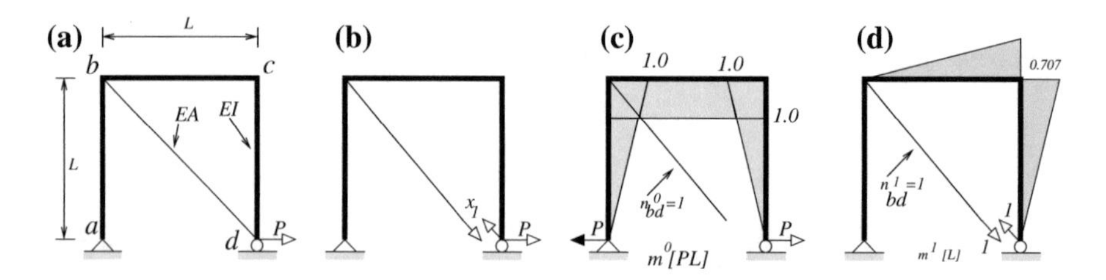

Fig. 12.31 Braced portal frame: analysis. **a** External loads. **b** Released structure

Release: In Fig. 12.31b we have released the structure by disconnecting the cable from d and defining the redundant axial force x_1 as positive in tension. The determinate structure is now subject to P and x_1. It is important to note that the released structure is composed of the actual frame which is an assemblage of flexural elements and of the cable bd which is an extensional element. Consequently, we are in need of the curvature κ^0 in the frame and the extension ϵ^0 in the cable due to the external forces (superscript '0'), and to the bending moments m^1 in the frame and normal forces n^1 in the cable due to the unit forces at degree of freedom 1 (the '1' superscript refers to the degree of freedom and not to the unit loads).

Compatibility: The m^0 distributions in the frame and n^0 in the cable due to the external load and m^1 and n^1 due to unit opposite forces at the cut are given respectively in Fig. 12.31c, d. The κ^0 and ϵ^0 distributions are the same as the corresponding m^0 and n^0 with a different scale ($\kappa^0 = m^0/EI$ and $\epsilon^0 = n^0/EA$).

Note, for the frame we make allowance only for the flexural deformations κ, and neglect the deformation due to the normal extensions of the frame elements. Components designed for bending are relatively bulky and therefore very stiff axially. For the cable, however, the axial deformations, ϵ_{bd}, should not be neglected. Clearly $\epsilon^0_{bd} = 0$ in a free dangling cable and $n^1_{bd} = 1$ (a cable with a unit force applied to it has unit internal forces).

The coefficients of the compatibility equation $\delta_1 = F_{11}\,x_1 + \delta^0_1 = 0$ are

$$\begin{aligned} F_{11} &= \int_a^d \frac{(m_1(x))^2}{EI}\,\mathrm{d}x + \frac{L_{bd}}{EA} = 0.333\,\frac{L^3}{EI} + 1.41\,\frac{L}{EA} \\ \delta^0_1 &= \int_a^d \frac{m^1(x)m^0(x)}{EI}\,\mathrm{d}x = -0.589\frac{PL^3}{EI} \end{aligned} \tag{12.25}$$

which leads to[7]

$$x_1 = \frac{\beta}{\beta + 4.24}\,1.77\,P \qquad \text{with} \qquad \beta = \frac{AL^2}{I} \tag{12.26}$$

(In fact $\beta = EAL^2/EI$.)

The Cable: Here again we get the dependency of the load in the redundant cable as a function of the relative stiffness of the cable. Keeping the frame constant, for $\beta = 0$ (there is no cable) x_1 is of course zero. After dividing the numerator and denominator in (12.26) by β, we note that as β increases the force in the cable increases until a saturation value $x_1 = 1.77P$ for $\beta \to \infty$, when the cable becomes a rigid link between b and c.

To make life simple, let us assume a value for β such that the tensile force in the cable is $x_1 = 1.41\,P$.

What kind of a cable would this be? Setting $x_1 = 1.41\,P$ in (12.26, left) gives $\beta = 16.91$. If we assume, for instance, that the frame elements have a square cross-section

[7] x is the running coordinate along the structure; x_i is a static unknown.

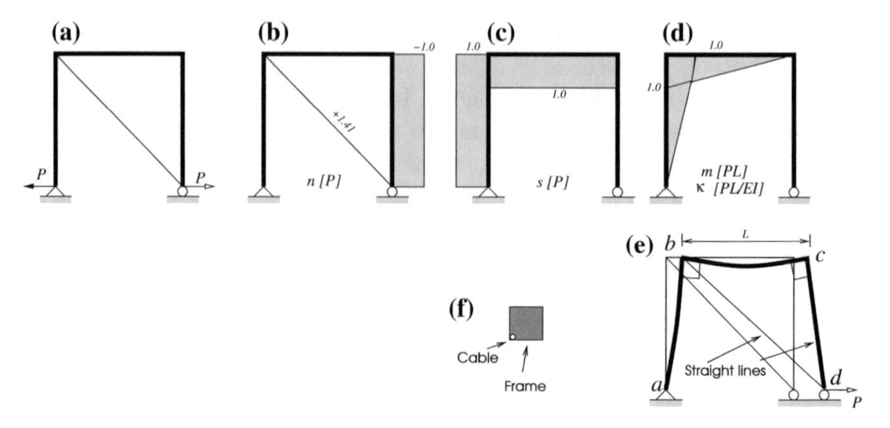

Fig. 12.32 Braced portal frame: results. **a** External loads and reactions. **b** Normal forces. **c** Shear forces. **d** Bending moments and curvature. **e** Cross-sections. **f** Equilibrium configuration

$a \times a$ ($I = a^4/12$) and the cable has a diameter d ($A = 0.25\pi d^2$) and $L = 10\,a$ we get, after introducing in (12.26, right) $d/a = 0.133$ (see Fig. 12.32f), which shows that you do not need much of a cable to almost produce the saturation load.

With this value of β the resulting internal forces, shown in Fig. 12.32, have simple numerical values. We call your attention to the deformed shape in Fig. 12.32e. Since the curvature is null along cd, that element does not deform (although bd is in compression, the column does not contract in a significant way on account of its large axial stiffness). The cable on the other hand, elongates (relatively significant axial flexibility).

Next, although the elements deform, the corners remain with their original angles, 90° in this case. As discussed earlier, this has to do with the deformations remaining elastic. A change of angle implies plastic flow, which signals local failure. The frame will not reassume its original configuration upon removal of the loads.

Finally, in the deformed configuration, the vertical reactions at d should engender bending in cd. However, the displacements were exaggerated in the drawing. In practice the displacements are small enough to neglect these second-order bending moments.

Note: A cable is a structural element that should be used with some caution in an analysis. In the first instance it is a non-linear element, in the sense that its constitutive law depends on the stresses. In a regular bar, Hooke's law $n = EA\epsilon$ is valid both in tension and in compression (the latter until the bar buckles, but in this entire treatise we ignore this buckling effect). For a cable in tension, we have $n = EA\epsilon$ in tension but in compression the stiffness is zero $n = 0\,\epsilon$. Consequently when we analyze a structure with cables we should make sure that nowhere is there a cable in compression.

Also, a cable has slack. When you pull on the cable of length L with a force P, the elongation will be larger than the expected $\epsilon\,L = PL/EA$ because the cable will start by straightening out. In other words, the apparent stiffness starts from a value much smaller than EA and increases with the force in the cable. In fact, more often than not, cables are pre-strained (straightened) when used in structures. This also allows compression in the cables, but we won't deal with this here.

12.7.10 Fixed Circular Arch with Applied Couple

Our next example is the circular arch shown in Fig. 12.33a. The arch of radius R is fixed at both ends a and f, and has a flexural stiffness $5EI$, except for segment ab, which is of stiffness EI. Segment ab has an intercept angle $\theta = 0.25\pi$ or 45°.

The redundancy is of course $R = 3$, one support would suffice, and the structure is released by introducing hinges at the supports and at the crown d, turning it into a three-hinged arch. The structure is submitted to a counterclockwise concentrated couple C at b. The basic moment distributions $m^0(x)$, $m^1(x)$, $m^2(x)$ and $m^3(x)$ are obtained by subjecting the released structure respectively to the external load C and to unit and opposite couples at both sides of the releases a, d and f.

In Fig. 12.33b–e, we have indicated the reactions in the respective cases. Clearly, the reactions must create an opposite couple to C, and since the reactions are hinged, only equal and opposite forces can be considered for the couple. In Fig. 12.33b1, c1, segment db is without loads and is hinged at both ends. Consequently the reaction must pass through its extremities, which fixes the direction of the reaction. The same is valid for segment ad in Fig. 12.33e1.

In the case of Fig. 12.33d, d1 because of symmetry only horizontal reactions can exist. (If there were vertical ones the thing would take off vertically.) Once the direction of the reactions is given, the magnitude can be determined by simple statics (Fig. 12.33d1).

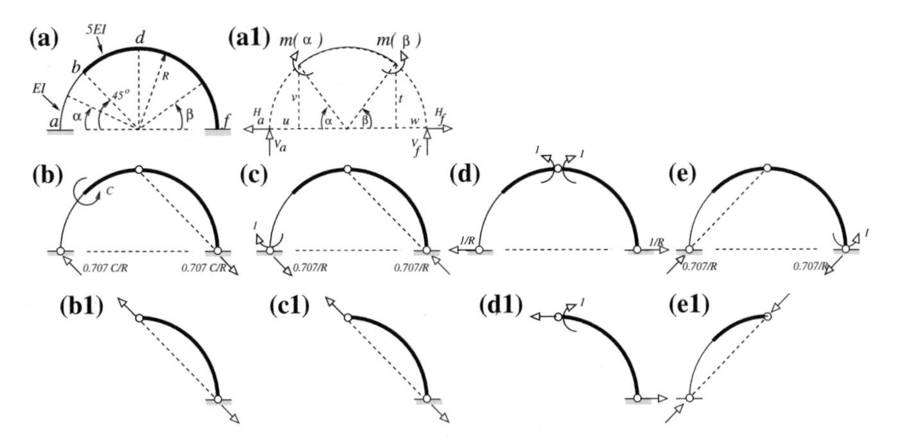

Fig. 12.33 Circular arch: analysis

Note, the reactions are all subsets of the general case shown in Fig. 12.33a1. For numerical convenience, we will be giving the expression for the internal moments starting from left and from right as a function of α and β respectively. Remember that at $\alpha = 0.25\pi$ there is a sudden change in the flexural stiffness and also an applied couple C. We have in terms of the reactions $m(\alpha) = V_a u + H_a v$ and $m(\beta) = V_f w + H_f t$ or

$$\begin{aligned} m(\alpha) &= V_a R(1 - \cos\alpha) + H_a R \sin\alpha + C_a \; ; \quad 0 \le \alpha \le \pi/4 \\ m(\alpha) &= V_a R(1 - \cos\alpha) + H_a R \sin\alpha + C_o \; ; \quad \pi/4 \le \alpha \le \pi/2 \\ m(\beta) &= V_f R(1 - \cos\beta) + H_f R \sin\beta + C_f \; ; \quad 0 \le \beta \le \pi/2 \end{aligned}$$

The specific values of the reaction components and constants are

m^i	V_a	H_a	V_f	H_f	C_a	C_o	C_f
m^0 $[C/R]$	0.5	0.5	−0.5	0.5	0	−1	0
m^1 $[1/R]$	−0.5	−0.5	0.5	−0.5	1	1	0
m^2 $[1/R]$	0.0	1.0	0.0	1.0	0	0	0
m^3 $[1/R]$	0.5	−0.5	−0.5	−0.5	0	0	1

The coefficients g_{ij} $(i = 1, 2, 3; \;\; j = 0, 1, 2, 3)$ of the compatibility equations are

$$g_{ij} = \int_0^{0.25\pi} \frac{m^i(\alpha) m^j(\alpha)}{EI} R\mathrm{d}\alpha + \int_{0.25\pi}^{0.5\pi} \frac{m^i(\alpha) m^j(\alpha)}{5EI} R\mathrm{d}\alpha + \int_0^{0.5\pi} \frac{m^i(\beta) m^j(\beta)}{5EI} R\mathrm{d}\beta \tag{12.27}$$

where it will be recognized that the loading vector is $\delta_i^0 = g_{io}$, and the coefficients of the flexibility matrix are $f_{ij} = g_{ij}; i, j = 1, 2, 3$.

$$\begin{aligned} g_{ij} = & \frac{R^3}{EI} \int_0^{0.25\pi} \{[V_a^i V_a^j (1 - \cos\alpha)^2 + (V_a^i H_a^j + V_a^j H_a^i)[(1 - \cos\alpha) + \sin\alpha] H_a^i H_a^j \sin^2\alpha \\ & + (V_a^i C_a^j + V_a^j C_a^i)(1 - \cos\alpha) + (H_a^i C_a^j + H_a^j C_a^i) \sin\alpha] + C_a^i C_a^j\} \mathrm{d}\alpha \\ & + \frac{R^3}{5EI} \int_{0.25\pi}^{0.5\pi} \{[V_a^i V_a^j (1 - \cos\alpha)^2 + (V_a^i H_a^j + V_a^j H_a^i)[(1 - \cos\alpha) + \sin\alpha] H_a^i H_a^j \sin^2\alpha \\ & + (V_a^i C_o^j + V_a^j C_o^i)(1 - \cos\alpha) + (H_a^i C_o^j + H_a^j C_o^i) \sin\alpha] + C_o^i C_o^j\} \mathrm{d}\alpha \\ & + \frac{R^3}{5EI} \int_0^{0.5\pi} [V_f^i V_f^j (1 - \cos\beta)^2 + (H_f^i V_f^j + H_f^j V_f^i) \sin\beta (1 - \cos\beta) \\ & + H_f^i H_f^j \sin^2\beta] \, \mathrm{d}\beta \end{aligned} \tag{12.28}$$

After integration over α and β we obtain the compatibility equations

$$\frac{L}{EI} \begin{bmatrix} 0.4939 & 0.2031 & -0.1030 \\ 0.2031 & 0.4283 & 0.0030 \\ -0.1030 & 0.0030 & 0.1283 \end{bmatrix} \begin{Bmatrix} x_1 \\ x_2 \\ x_3 \end{Bmatrix} + \frac{LC}{EI} \begin{Bmatrix} 0.1058 \\ 0.0899 \\ -0.0043 \end{Bmatrix} = \begin{Bmatrix} 0 \\ 0 \\ 0 \end{Bmatrix} \tag{12.29}$$

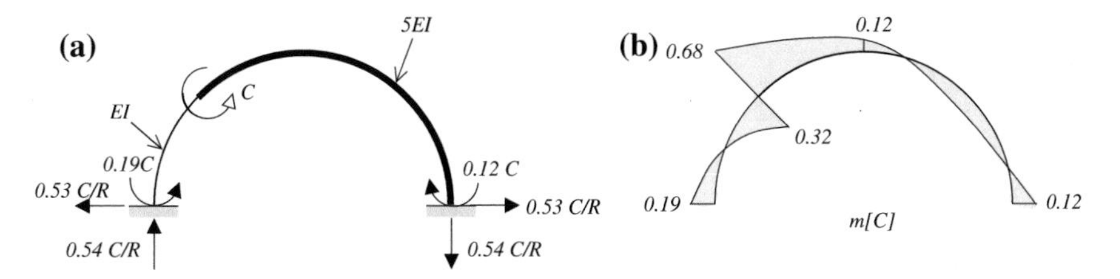

Fig. 12.34 Circular arch: reactions and bending moments

yielding the solution $\boldsymbol{x} = C\{-0.1892 \quad -0.1195 \quad -0.1159\}^T$. The reactions and bending moments distribution are shown in Fig. 12.34a, b respectively.

12.7.11 Grid of Beams Under Lateral Force

The grid of beams in Fig. 12.35a has a longitudinal beam 0 and three orthogonal transverse beams at equal intervals. All beams are of length $2a$, they are simply supported at the extremities, and their bending stiffness is EI. Beam number 3 is subjected to a lateral force Q as indicated.

Note that the grid is a plane structure. It is called a *grid* because the loading is out of plane. If the structure were loaded by in-plane forces it would be called a *frame*.

We will assume that the connection between the longitudinal and lateral beams is such that they apply vertical loads to one another. In other words, the flexion of one beam does not cause torsion in the other. This assumption is plausible either when the beams are connected only in the lateral translational direction or when the beams have a low torsional rigidity (open sections such as I-beams). The assumption is very important in the present context because, as we shall see, it reduces the degree of redundancy of this structure to $R = 3$.

When we disconnect the lateral beams from the longitudinal one we obtain four simply supported beams, that is, a statically determinate arrangement. The redundancy is equal to the number of pairs of interaction forces that are released at the junctions.

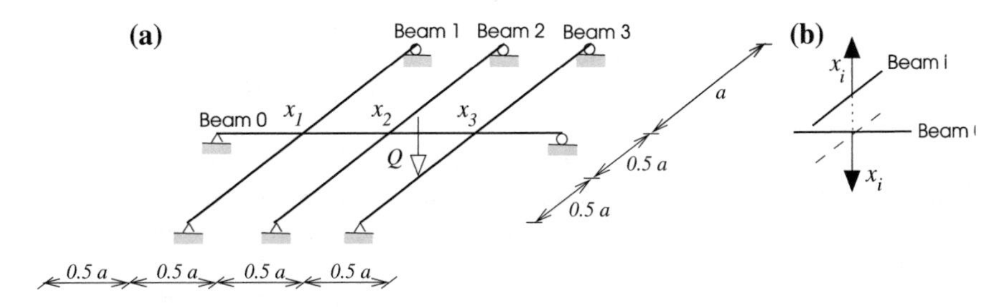

Fig. 12.35 Plane grid of beams

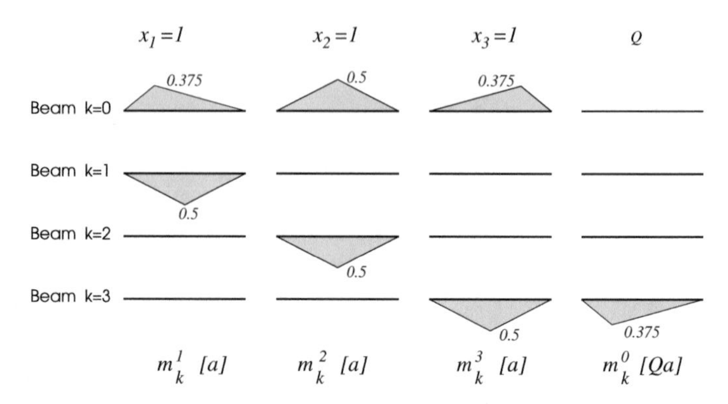

Fig. 12.36 Grid of beams: analysis

As indicated, we will assume that at every junction $i = 1, 2, 3$ we expose only a pair of vertical interaction forces x_i. These $R = 3 \times 1 = 3$ redundant interaction forces are defined positive as shown in Fig. 12.35b. (In the general case there would be three pairs of interaction forces at every connection, thus making the redundancy $R = 3 \times 3 = 9$.)

We proceed by applying to the structure the unit and opposite forces x_i, $i = 1, 2, 3$ at the releases and the external forces Q. We obtain in each case four bending moment distributions m_k^i $(i = 0, 1, 2, 3)$ for the four beams $k = 1, 2, 3, 4$, where $i = 0$ refers to the external load. In Fig. 12.36, we have drawn the ending moments m_k^i for unit values at the redundants (columns 1 to 3) and for the external force (column 4). You will note that many $m_k^i = 0$. For instance, when you apply x_2 at the junction between beams 0 and 2, only these beams are affected. Also, the applied force is exerted on beam 3 only. The coefficients of the compatibility equations are

$$F_{ij} = \sum_{k=0}^{3} \int_0^{2a} m_k^i \, m_k^j \, \mathrm{d}x \quad ; \quad \delta_{i0} = \sum_{k=0}^{3} \int_0^{2a} m_k^i \, \kappa_k^0 \, \mathrm{d}x \tag{12.30}$$

with $\kappa_k^j = m_k^j/(EI)_k$, $j = 0, 1, 2, 3$, which leads to the following compatibility equations

$$\frac{1}{96}\frac{a^3}{EI}\begin{bmatrix} 25 & 11 & 7 \\ 11 & 32 & 11 \\ 7 & 11 & 25 \end{bmatrix}\begin{Bmatrix} x_1 \\ x_2 \\ x_3 \end{Bmatrix} + \frac{1}{96}\frac{Qa^3}{EI}\begin{Bmatrix} 0 \\ 0 \\ 11 \end{Bmatrix} = \begin{Bmatrix} 0 \\ 0 \\ 0 \end{Bmatrix} \tag{12.31}$$

the solution of which is $\boldsymbol{x} = Q\ \{0.0805\ 0.1547\ \ -0.5306\}^T$. One can now get the reactions and draw the moment and shear diagrams on all four beams. (Note, in line with our convention for beams, we have assumed that there are no axial internal forces.)

To get an idea of how this grid supports the applied force, we have schematically depicted the reactions in Fig. 12.37a. Clearly, most of the load is taken by the nearest

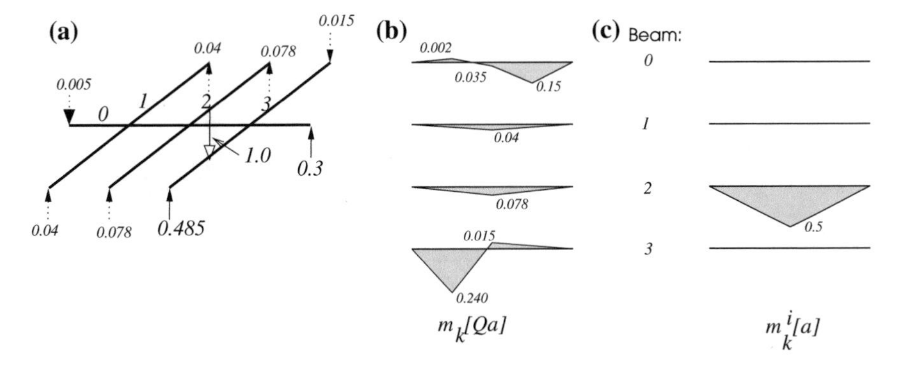

Fig. 12.37 Grid of beams: results

supports (0.3 Q and 0.485 Q) which with what can be learned from the bending moment diagrams in Fig. 12.37b would indicate that, for this particular loading beams 1 and 2 are rather superfluous. But then, a structure is seldom designed for a single loading condition.

We would like to complete this example by computing the vertical displacement v_2 at node 2 (the junction of beam 2 with beam 0). We can use the unit-load method since we know the curvature distribution of the structure $\kappa_k(x) = m_k(x)/EI$ of all four beams of the grid $k = 0, 1, 2, 3$.

We apply a unit force at 2 and look for an equilibrium solution. We can support the unit force with beam 0, or beam 2, or a combination of the two. Let us put the unit force on beam 2. The resulting moment distribution m_k^2 is shown in Fig. 12.37c. For the displacement we integrate only over beam 2. This yields $v_2 = 0.026\ Qa^3/EI$.

12.7.12 R = 2 Truss

Consider the truss in Fig. 12.38a.

Redundancy: The structure is redundant of degree $R = 2$. Indeed, when we remove one diagonal in each cell we obtain a simple truss ($R = 0$). And since a truss member carries only an axial force the redundancy is $R = 2$.

Release: We are not really removing elements but rather disconnecting one end of the redundant bars by fictitious cuts as in Fig. 12.38b. The result is a simple truss plus two dangling bars. Strictly speaking this structure is conditionally stable. No forces can be applied to the disconnected extremity of the bar unless the force is axial.

If no forces are applied at the free extremity, the force in the bar is zero and if an axial force is applied then the internal force is equal to that force. In either case the force is known which maintains the statical determinacy.

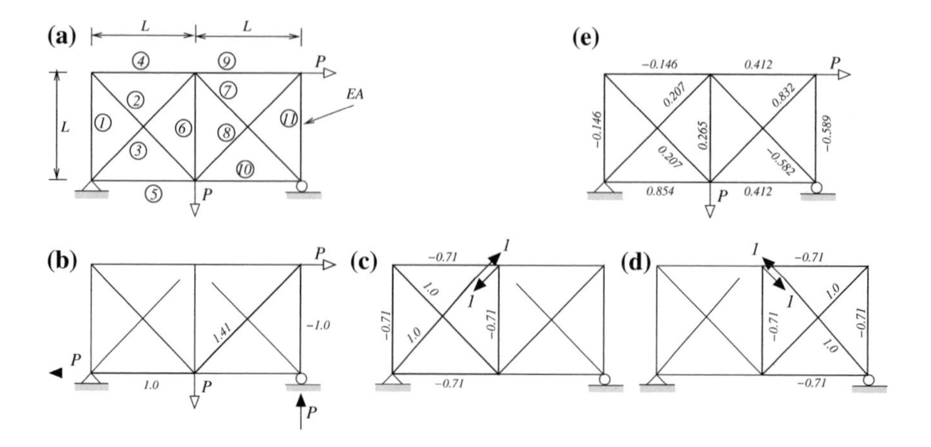

Fig. 12.38 Redundant truss: analysis. **a** External loads. **b** $n^0[p]$. **c** $n^1[p]$. **d** $n^2[p]$. **e** Results n[p]

Compatibility: It is always the same story. We apply the external loads to the released structure and then unit forces at the cuts 1 and 2, and then perform three analyses to get the internal forces in the bars, which in this case are normal forces $\boldsymbol{n}^0$, $\boldsymbol{n}^1$ and $\boldsymbol{n}^2$, where the latter, for instance, has components n_k^2, $k = 1 \ldots 11$. The distribution of internal forces are shown in Fig. 12.38b, c, d respectively.

The coefficients of the compatibility equations are $f_{ij} = \sum_{k=1}^{11} n_k^i n_k^j / (EA/L_k)_k$ and $\delta_i^0 = \sum_{k=1}^{11} n_k^i \epsilon_k^0 L_k$, where the axial strain is given by Hooke's law $\epsilon_k^0 = n_k^0 L_k/(EA)_k$, which produces the compatibility equations

$$\frac{L}{EA}\begin{bmatrix} 4.83 & 0.50 \\ 0.50 & 4.83 \end{bmatrix}\begin{Bmatrix} x_1 \\ x_2 \end{Bmatrix} + \frac{PL}{EA}\begin{Bmatrix} -0.707 \\ 2.707 \end{Bmatrix} = \begin{Bmatrix} 0 \\ 0 \end{Bmatrix} \tag{12.32}$$

yielding $x_1 = 0.207\,P$ and $x_2 = -0.582\,P$.

The final result, which can be obtained by superposition ($\boldsymbol{n} = \boldsymbol{n}^0 + x_1\boldsymbol{n}^1 + x_2\boldsymbol{n}^2$), or by analyzing the statically determinate released structure under the original external forces and $0.207\,P$ and $-0.582\,P$, is given in Fig. 12.38e.

Note: Except for some simple cases, trusses should be analyzed by computerized methods, in particular, the Displacement method which we will study further on. The present example, as well as some others, serve mainly the purpose of emphasizing the generality of the Force method.

12.8 Summing Up

We have concluded the first of the two basic methods of analysis of structures: the Force method. It assumes that we know how to analyze determinate structures, and it provides a systematic approach to deal with redundant structures. In essence, we transform the redundant structure into a determinate one by assuming (unknown) internal forces at the R redundancies. We then write R compatibility equations to enforce continuity of the structure. These linear equations allow us to compute the value of the redundants. We now have a determinate structure with all forces known, which is an exact replica of the redundant one.

The following chapter will apply the same technique to redundant structures submitted to applied deformations such as temperature variations and prestressing.

Chapter 13
Applied Strains and Initial Stresses

In the present chapter we show how a redundant structure under *temperature variation* and similar events is analyzed. The force method is used in its standard form, but instead of applying loads we apply deformations. As we will show, this leads to some simplifications but also to intriguing final deformation distributions.

13.1 Initial Internal Forces

Initial internal forces (initial stresses) $\boldsymbol{n}^{\text{init}}$ are internal forces which are present in the structure without any external forces applied to it. In other words, we could be contemplating a structure without any loads applied to it, and still it may be suffering from substantial internal forces. Such internal forces are called initial forces or initial stresses.

When a statically determinate structure is without external loads, we can be assured that it will be free of any internal forces. This is almost by definition. In a determinate structure the internal forces can be obtained from the equilibrium equations, and these give zero internal forces for zero external forces. Therefore:

Initial stresses can exist in statically *redundant* structures only.

As we shall see, the way initial stresses can be locked in a redundant structure is by subjecting its elements to *applied strains* or applied deformations, $\boldsymbol{\epsilon}^a$, that is, ϵ^a for axial elements and κ^a for bending elements.

13.2 Mechanical Strain Versus Applied Strain

Consider the fiber of length L in Fig. 13.1. The material is characterized by its modulus of elasticity E and by its coefficient of thermal expansion α. Both are determined experimentally in a laboratory.

M.B. Fuchs, *Structures and Their Analysis*,
DOI 10.1007/978-3-319-31081-7_13

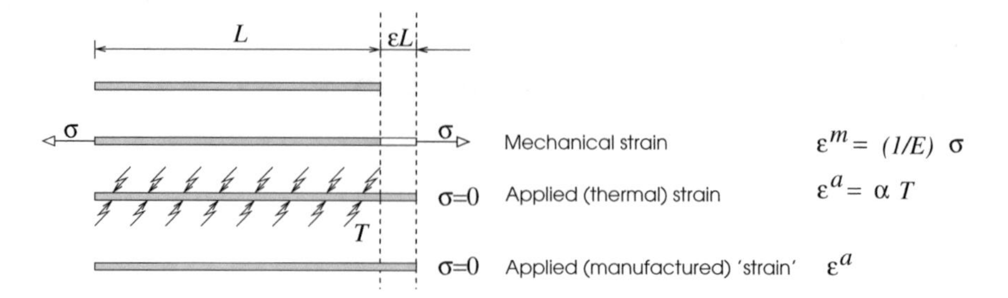

Fig. 13.1 Strained fibers

Mechanical Strain

Hooke's law states that a fiber of length L when subjected to axial stresses σ elongates by an amount ϵL (see Fig. 13.1), where the strain ϵ is related to the stress by Hooke's famous law $\epsilon = (1/E)\,\sigma$. Such strain, which results from a stress, is called a *mechanical strain*, sometimes denoted by ϵ^m when other types of strains are involved.

Mechanical strains result from stresses by Hooke's law.

Applied Strain

A fiber can elongate for other reasons, the most common being an increase in temperature (heating). If the coefficient of thermal expansion of the material of the fiber is α, the fiber when heated by an amount T will increase in length by an amount $\alpha\,T$. We say that the fiber was subjected to applied strain $\epsilon^a = \alpha\,T$, and its length is now $L(1+\epsilon^a)$. But the fiber is not stressed. It is simply longer.

A heated fiber is simply longer. It is free of any stresses.

A generalization of the concept of thermal expansion is a fiber which we thought was of length L, but in truth it was supplied of length $(1+\varepsilon^a)L$. In all respects its length has 'expanded' as if it had been heated. We call this strain 'manufactured' strain.

In both cases of change in length, thermal or manufactured, ε^a is called an applied strain. In all instances of applied strain the fiber does not feel stressed at all. It is simply longer than its nominal length.

13.3 Applied Strains in Axial and Bending Elements

Now consider the structural element of length L and height h in Fig. 13.2a. Here and elsewhere the datum length L is measured at a datum temperature.

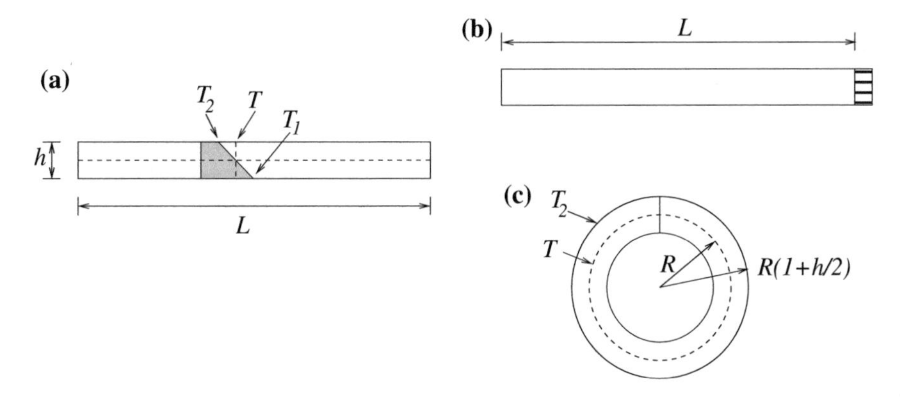

Fig. 13.2 Constant curvature

We now assume that along its length the element has experienced a uniform differential temperature increase. (Uniform means that the same temperature increase has occurred along the entire span, and it is differential in the sense that across the section the temperature varies linearly between the upper and lower fibers.) The temperature of the bottom fiber has increased by T_1 and the temperature of the top fiber by T_2, with $T_1 > T_2$ as in the figure. In between, the temperature variation is linear.[1]

As indicated in Fig. 13.2a, the temperature increase can conveniently be decomposed into an average temperature increase $T = (T_1 + T_2)/2$ and a differential distribution. Denoted by $\Delta T = (T_1 - T_2)$, this distribution varies linearly from $-\Delta T/2$ for the upper fiber to $+\Delta T/2$ at the bottom fiber, with zero temperature change at mid-level (dashed axis).

- Under T the element will remain straight and expand (or contract if $T < 0$) freely as in Fig. 13.2b. The axial thermal deformation ('strain') will of course be $\epsilon_a = \alpha T$ (13.1), where we recall that α is the coefficient of thermal expansion of the material.
- When subjected to the differential temperature variation, the lower fibers of the element expand, the upper ones contract, and the element curves upwards. The point to bear in mind is that the curvature of the curve is forcibly constant, the temperature distribution being the same along the entire element. The segment becomes therefore circular.

To compute this curvature let us assume that ΔT and L are such that the curved beam closes upon itself to produce the ring shown in Fig. 13.2c. The radius of the circle formed by the middle fiber we call R. As the middle fiber does not expand, its length is the same as the length of the element, $L = 2\pi R$. Equating the length of the lower fiber of the beam to the length of the outer fiber of the ring, $L(1+\alpha\Delta T/2) = 2\pi(R+h/2)$, gives $h = \alpha R\Delta T$. And since the curvature is, by definition the inverse of the radius of curvature, $\kappa = 1/R$, we obtain the local curvature due to a temperature differential

[1] We are not dealing with heat transfer. The temperature distribution has reached a steady state.

$$\epsilon^a = \alpha T, \qquad \kappa^a = \frac{\alpha \, \Delta T}{h} \tag{13.1}$$

where the superscript 'a' indicates that these strains are applied. They are not related to internal forces but originate from other sources, in general thermal effects. We can now generalize Hooke's law for axial and bending elements respectively

$$\epsilon = \frac{n}{EA} + \epsilon^a \qquad \text{or} \qquad n = EA\,(\epsilon - \epsilon^a) \tag{13.2}$$

$$\kappa = \frac{m}{EI} + \kappa^a \qquad \qquad m = EI\,(\kappa - \kappa^a) \tag{13.3}$$

where ϵ and κ are *total* deformations, n/EA and m/EI are *mechanical* deformations, and ϵ^a and κ^a are *applied* deformations. One of the most important aspects of analysis with applied strain is that having, for instance, obtained the bending moment m at a given location the curvature at that point will not be m/EI but rather $m/EI + \kappa^a$.

Note: Equation (13.1) describe the applied strain due to temperature. Temperature is one source of applied strain. There are others. In order to facilitate things we will not talk about temperature effects and then use (13.1) to get to the applied strains. Instead, we will analyze structures whose elements are subjected to applied strain, whatever the origin.

13.3.1 The Case of Statically Determinate Structures

When we heat a statically determinate structure it will deform, but it will not develop internal forces nor reactions. Consider a beam in a desert on an early summer morning (Fig. 13.3a1). The top fibers are heated by the sun but the bottom fibers may be at a lower temperature on account of the cool sand. In between we assume a linear variation. As a result the beam expands and curves with a constant curvature $-\kappa^a$, and will rest on two supports a and b.

When we write the equilibrium equations of the beam, we have $R_a + R_b = 0$ and $L R_b = 0$, that is, $R_a = R_b = 0$ (L is the beam length). We find that since there are no external forces, there are no reactions nor internal forces. The structure is indeed warmer but since no external forces are applied, no internal forces will appear irrespective of whether the structure is painted green or cooled.

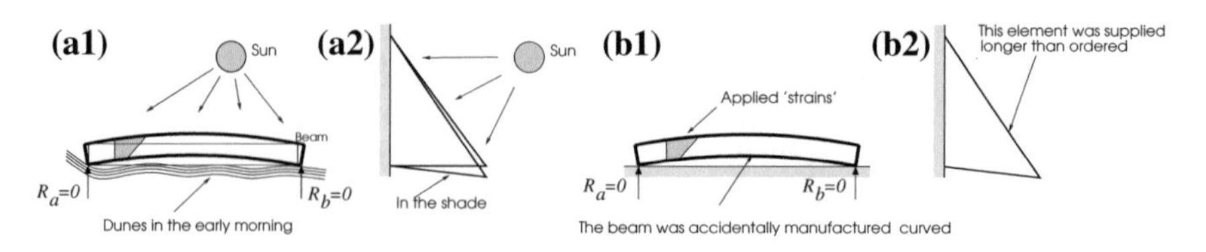

Fig. 13.3 Heated beam and heated bar

In a statically determinate structure there cannot be initial stresses.

The bending moments are thus also zero. But there is (constant) curvature so the deformed shape is a circular arc of radius $R = 1/\kappa^a$.

In Fig. 13.3a2 the diagonal bar of a statically determinate truss is uniformly heated. The second bar is in the shade and escapes the sun's rays. The heated bar elongates, the other one remains unchanged, and as a result the free node moves down (the horizontal bar rotates as a rigid body) in order to accommodate the longer bar. But there are no internal forces in the bars, nor are there reactions.

In short, statically determinate structures, when heated (partially or fully), deform but remain free of stresses.

13.3.2 Displacements of Determinate Structures Under Applied Strain

Recall that the unit-load method for computing a displacement component is $d_i = \int_x \epsilon(x)^T \boldsymbol{n}^i(x)\,\mathrm{d}x$, where $\epsilon(x)$ are the final or total strains in the structure. As we shall see, in the general case of redundant structures with applied strains, the structure has locked stresses which by virtue of Hooke's law become 'mechanical' stresses. Total strains are the sum of the applied and the mechanical strains (even in the absence of external loads). Not so with determinate structures. In the absence of external loads there are no internal forces and, therefore, no mechanical strains. The applied strains are then the final ones. This makes life a bit easier.

How would we, for instance, compute the slope θ_a at the left extremity of the beam in Fig. 13.3b1, when the applied strain due to the heating is, say, $\kappa^a = \bar{\kappa}$? In Fig. 13.4a we have drawn the final curvatures $\kappa(x) = \bar{\kappa}$, which are here the applied ones, and the virtual bending moments $m^{\theta_a}(x) = 1 - x/L$, which are in equilibrium with a unit couple at a. Assuming that the beam is of length L, we get $\theta_a = \int_o^L \bar{\kappa}\,(1 - x/L)\,\mathrm{d}x = 0.5\,\bar{\kappa}\,L$ (expressed in radians). The rotation at a is thus in the sense of the unit couple.

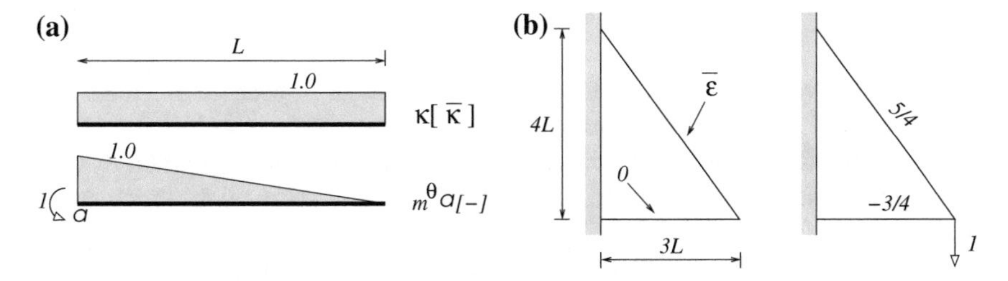

Fig. 13.4 Displacements of statically determinate structures under applied strain

In a similar vein, the vertical displacement of the free node in the truss in Fig. 13.3b is given by $v = \sum_j n_j^v \, \epsilon^a \, L_j$, where j runs over the two bars of the truss, and n_j^v are the forces in the bar due to a vertical unit force at the node. The final strains are simply the applied ones ϵ^a, that is, zero in the horizontal bar and $\bar{\epsilon} = \alpha T$ in the oblique bar. Noting that the truss (and the wall) in Fig. 13.4b form a Pythagorean triangle (3 4 5), we have $v = (-3/4)(0)(3L) + (5/4)\bar{\epsilon}(5L) = 6.25\,\bar{\epsilon}\,L$. It is easy to verify that the horizontal component of the displacement is zero (the bar rotates as a rigid body about its left extremity.)

13.3.3 Extending the Scope of Applied Strain

We have already pointed out that a uniformly heated element is just a longer element than the one we had planned. We have mounted an element of length L in our structure, but it changed length due to an environmental factor (the sun). The datum length of our element is suddenly $(1 + \epsilon^a)\, L$ with all the ensuing consequences. (In redundant structures the penalty may be annoying and in extreme cases devastating.)

But we do not have to resort to the sun's rays to create applied strain. It can simply be the result of a manufacturing error or some other cause. We may have ordered a straight beam of length L, and the manufacturer may have delivered a curved one (Fig. 13.3b1); or the diagonal rod in the truss in Fig. 13.3b2 may have been supplied longer than ordered. Since we can model the discrepancy as some sort of distortion κ^a or ϵ^a, it makes no difference if the strains are due to a change in temperature or due to some error in manufacture. It does not make any difference if the element was curved on the ground (before mounting) or if all this happened after it was mounted because of differential heating.

Here also the final strain is the applied 'strain' since there is no mechanical strain and what was done in the previous section is also valid here.

13.4 Applied Strain in Redundant Structures

13.4.1 Internal Forces Due to Applied Strain

Applied strain gets really interesting in redundant structures, because in a redundant structure it will cause internal forces. The following example is a clear illustration that in a redundant structure we can have internal forces without applied external loads, whereas in a statically determinate arrangement, no external forces means no internal forces.

Indeed, consider the statically determinate truss in Fig. 13.5a and the symmetric statically redundant truss in Fig. 13.5b. Both have two degrees of freedom at the free node.

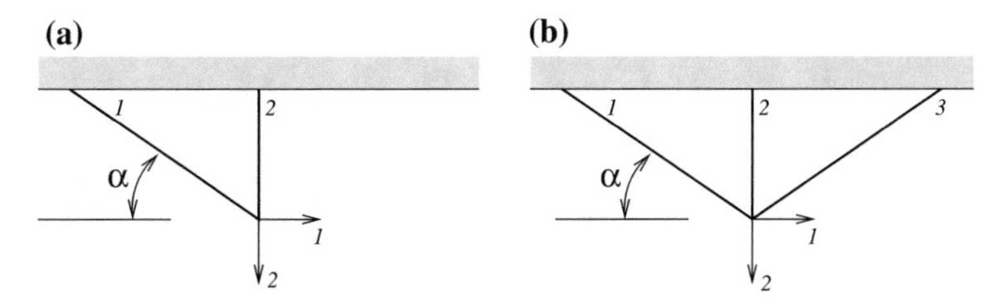

Fig. 13.5 Determinate-Redundant truss

Assuming positive internal forces in tension, the two equilibrium equations $\mathbf{Q}\,\mathbf{n} = \mathbf{o}$ at the free node are respectively

$$\begin{bmatrix} \cos\alpha & 0 \\ \sin\alpha & 1 \end{bmatrix} \begin{Bmatrix} n_1 \\ n_2 \end{Bmatrix} = \begin{Bmatrix} 0 \\ 0 \end{Bmatrix}, \quad \begin{bmatrix} \cos\alpha & 0 & -\cos\alpha \\ \sin\alpha & 1 & \sin\alpha \end{bmatrix} \begin{Bmatrix} n_1 \\ n_2 \\ n_3 \end{Bmatrix} = \begin{Bmatrix} 0 \\ 0 \end{Bmatrix} \tag{13.4}$$

In the determinate case the coefficient matrix $\boldsymbol{Q}$ is, by definition, square, and since the determinant of the coefficient matrix is different from zero, $|\mathbf{Q}| \neq 0$, in other words, the structure is stable, the only solution is $\mathbf{n} = 0$, that is, there are no internal forces. In the redundant case however there are, again by definition, more unknowns than equations. The coefficient matrix is rectangular, and we can get a non-trivial solution $\mathbf{n} \neq 0$. Indeed,

$$\begin{bmatrix} \cos\alpha & 0 \\ \sin\alpha & 1 \end{bmatrix} \begin{Bmatrix} n_1 \\ n_2 \end{Bmatrix} = -n_3 \begin{Bmatrix} -\cos\alpha \\ \sin\alpha \end{Bmatrix}$$

For every value of n_3, we can write $n_1 = n_3$ and $n_2 = -2\sin\alpha\, n_3$. For instance, $\boldsymbol{n} = \{2,\ -4\sin\alpha,\ 2\}^T$ is a possible auto-equilibrated set of forces. So with redundant structures, internal forces can exist in the absence of external forces, and we shall see that these occur when there is applied strain even without external forces.

13.4.2 A Simple Example

In Fig. 13.6a, we show two identical bars of nominal length L and axial stiffness EA, rigidly joined at both ends. The arrangement is fixed at the left extremity and free to move horizontally at the right extremity. We assume that everything happens axially. (The two bars are factually collinear. They were drawn a little apart for clarity's sake.) This is thus a structure with two bars (U = 2), one dof ($E = 1$) the horizontal displacement of the free node, and the redundancy is therefore $R = U - E = 1$.

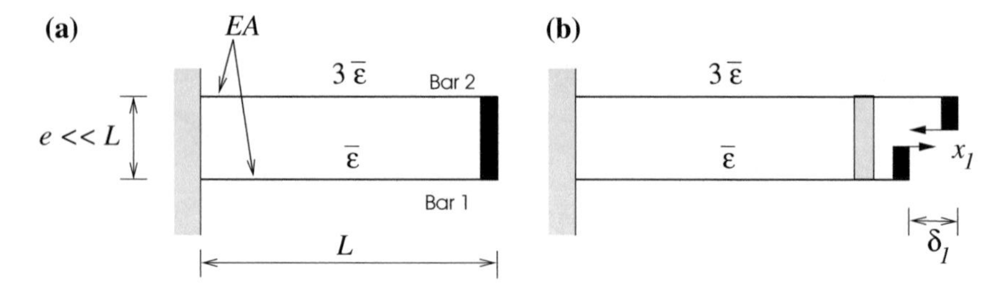

Fig. 13.6 Uniaxial example

The bars are subjected to applied strain $\varepsilon_1^a = \bar{\epsilon}$ and $\varepsilon_2^a = 3\bar{\epsilon}$. This is equivalent to stating that if the bars were not connected they would be of different lengths: $L_1 = L(1 + \bar{\epsilon})$ and $L_2 = L(1 + 3\bar{\epsilon})$, where L_i is the length of bar i. The applied strains could be of thermal origin where bar 1 was heated by T causing a dilatation of $\bar{\epsilon} = \alpha T$ per unit length and bar 2 was heated three times more with a dilatation $3\bar{\epsilon}$ (α is the coefficient of thermal expansion of the material). The difference in length could also be due to a manufacturing discrepancy. Bars 1 and 2 were simply supplied with different lengths.

It is easy to visualize that we will have to pull on bar 1 and push on bar 2 to connect the two bars. Having succeeded in connecting the bars what would the 'initial' internal forces be?

To compute the initial forces n_1 and n_2, we proceed in a classical fashion: determining the redundancy and releasing the structure, writing and solving the compatibility equation to compute the internal forces.

In Fig. 13.6b we have released the structure by disconnecting the two bars at the right extremity, thus exposing the redundant (axial) interaction forces x_1 which were chosen positive as indicated in the figure. The forces in the bars under the unit loads ($x_1 = 1$) are $n_1^1 = 1, n_2^1 = -1$ (no units), and the strains in the bars due to the applied strains are—$\epsilon_1^o = \bar{\epsilon}$, $\epsilon_2^o = 3\bar{\epsilon}$.

The compatibility equation is $\delta_1 = F_{11}\, x_1 + \delta_1^o = 0$ with $F_{11} = Ln_1^1(n_1^1/EA) + Ln_2^1(n_2^1/EA) = 2L/EA$ and $\delta_1^o = Ln_1^1\epsilon_1^o + Ln_2^1\epsilon_2^o = -2L\bar{\epsilon}$, yielding $x_1 = EA\bar{\epsilon}$. The axial forces in the bars, in fact the initial stresses, are a direct consequence of x_1. The final strains, however, are the sum of the mechanical strains n_i/EA and the applied strains ε_i^a:

$$n_1^{\text{init}} = EA\,\bar{\varepsilon} \qquad \varepsilon_1^{\text{init}} = \frac{EA\,\bar{\varepsilon}}{EA} + \bar{\varepsilon} = 2\,\bar{\varepsilon}$$

$$n_2^{\text{init}} = -EA\,\bar{\varepsilon} \qquad \varepsilon_2^{\text{init}} = \frac{-EA\,\bar{\varepsilon}}{EA} + 3\bar{\varepsilon} = 2\,\bar{\varepsilon}\,.$$

These strains are called by extension, initial strains. Note that in this case the final strains are (luckily) equal, because the two bars must, of course, have the same lengths at all times.

13.4.3 The General Case

In the general case an R-redundant structure can be subjected to applied strain $\varepsilon^a(x)$ along some or all of its members. The applied strains vector has two components; axial strain, causing extension, and curvature strain causing flexion, $\varepsilon^a(x) = \{\varepsilon^a(x) \ \ \kappa^a(x)\}^T$.

We release the structure in any appropriate manner. The determinate arrangement is now subjected to R redundant forces $\boldsymbol{x}$ and to the applied strain. The compatibility equations are $\boldsymbol{F}\,\boldsymbol{x}+\boldsymbol{\delta}^o = \boldsymbol{0}$, were the coefficients of the flexibility matrix are standard $F_{ij} = \int_x (n^i\, n^j/EA + m^i\, m^j/EI)\,\mathrm{d}x$ and the loading vector is $\delta_i^o = \int_x (\,n^i\,\varepsilon^o + m^i\,\kappa^o\,)\,\mathrm{d}x$, where ε^o and κ^o are the deformations of the released structure due to external causes, in the present case, the applied strain (remember, the released structure is determinate). When you apply strain to a statically determinate structure the resulting stress is zero, $\boldsymbol{n}^o = 0$ and the resulting strain is the applied strain, $\varepsilon^o(x) = \varepsilon^a(x)$ and thus

$$\delta_i^o = \int_x (\,n^i\,\varepsilon^a + m^i\,\kappa^a\,)\,\mathrm{d}x \tag{13.5}$$

Let $\boldsymbol{x}$ be the solution of the compatibility equations. By superposition the final results are due to the action of the redundants $\boldsymbol{x}$ and to the applied strain:

$$\mathbf{n}^{\text{init}}(x) = \sum_i x_i\,\boldsymbol{n}^i(x) \qquad \text{(Recall } \boldsymbol{n}^o = 0\text{)} \tag{13.6}$$

$$\varepsilon^{\text{init}}(x) = \sum_i x_i\,\varepsilon^i(x) + \varepsilon^o(x) \qquad \text{(Note } \varepsilon^o = \varepsilon^a\text{)} \tag{13.7}$$

Such initial stresses and strains, which are present prior to any loading (external forces) must be added to the stresses resulting from the external loading, and when ignored, can be source of much trouble. On the other hand, when carefully tuned, initial stresses are often introduced on purpose in structures to reduce the amplitudes of the final internal forces, strains and/or displacements. In these instances we speak of (beneficial) ‘prestresses’.

13.5 Illustrated Examples

13.5.1 Cantilever Propped on Two Supports

The $R = 2$ beam in Fig. 13.7a of flexural stiffness EI and length $2L$ is subjected to an applied strain in the form of a constant curvature $\kappa^a(x) = \bar{\kappa}$ (extension at upper fiber) along its span. As shown in Fig. 13.7a1 this applied curvature can originate from a differential temperature variation along the height of the cross-sections or

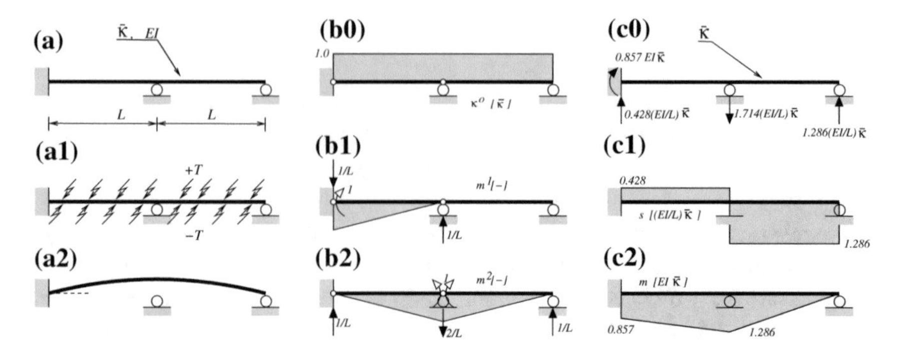

Fig. 13.7 Propped cantilever: analysis

it can be due to a manufacturing imperfection, where the beam arrived curved at the yard (Fig. 13.7a2). A couple of sturdy workmen forced the beam into place by having it built in horizontally in the wall to the left, and by making contact with the two roller supports. This structure is now obviously stressed, even without external loads, and our task is to quantify the initial internal forces in the structure.

This being a beam, we are interested in the bending moments $m(x)$ and related shear forces. The structure is released by two hinges set inside the beam, one at the root and the other one above the middle support. What we have now are two simply supported (and unrelated) beams one next to the other which happen to share a common support.

We subject the released (determinate) structure in turn to the applied strains,[2] and to pairs of unit couples at the releases, on both sides of the hinges (Fig. 13.7b0–b2).

Since the released structure is statically determinate, there are no internal forces due to applied strains, and therefore the strains are the applied strains $\kappa^o(x) = \bar{\kappa}$ as in Fig. 13.7b0. Note that the strain distribution diagram follows the same rule as the moment diagram; it is drawn on the side of the fibers in extension. The moment distributions due to the unit couples in Fig. 13.7b1, b2, and their respective reactions, should by now be classical stuff.

The flexibility coefficients and the loading terms of the compatibility equations, $F_{ij} = \int_x m^i(m^j/EI)\mathrm{d}x$ and $\delta_i^o = \int_x m^i \bar{\kappa}\mathrm{d}x$ give, after integration along the span of the beam, the compatibility equations

$$\frac{1}{6}\frac{L}{EI}\begin{bmatrix} 2 & 1 \\ 1 & 4 \end{bmatrix}\begin{Bmatrix} x_1 \\ x_2 \end{Bmatrix} + \frac{1}{2}\bar{\kappa}L\begin{Bmatrix} -1 \\ -2 \end{Bmatrix} = \begin{Bmatrix} 0 \\ 0 \end{Bmatrix}$$

yielding $x_1 = 0.857\,EI\bar{\kappa}$ and $x_2 = 1.286\,EI\bar{\kappa}$. Note, the units of the redundants $(fl^{-2})l^4 l^{-1} = fl$ correspond indeed to couples. We can now draw the structure with

[2]It will be seen that the applied strain will cause the beam to assume the shape of two circular segments each of length L.

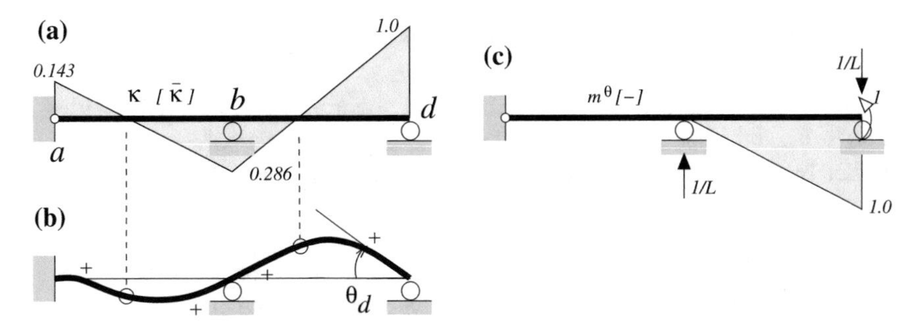

Fig. 13.8 Propped cantilever: curvature and displacements

its reactions in Fig. 13.7c0, and the initial shear s and moment distributions m in Fig. 13.7c1, c2.

When it comes to the drawing the displaced shape or to computing a given displacement or rotation we need the curvatures $\kappa^{\text{init}}(x)$ of the structure. We noted in Sect. 13.4.3 that these are the sum of the mechanical strains m/EI and the applied curvature $\kappa^a = \bar{\kappa}$. The mechanical curvature distribution is the same as the distribution in Fig. 13.7c2 in units $[EI\bar{\kappa}/EI] = [\bar{\kappa}]$. The final curvatures can be seen in Fig. 13.8a.

If we want an estimate of *the equilibrium shape* it is the final curvature distribution κ^{init} which should be consulted. In broad outline, we follow the sign of the curvature and make allowances for the boundary conditions at the supports. The drawing is composed of three regions. Starting from the left we have first a $\pm$ region, that is, extension along the top fibers and compression along the bottom ones, in other words, a concave curve. Next, we meet a $\mp$ region which means a convex curve, and then again a $\pm$ region or a concave curve.

We then try to draw the simplest smooth curve taking into account the boundary conditions (how the beam is supported) and the curvature of the beam. The beam curve should pass through zero at a, b and d, it should have a zero slope at a and it should obey the sequence concave, convex, concave as depicted in Fig. 13.8b.

If *the slope* θ_d at the right extremity is required, you apply a unit couple at d and look for a simple equilibrium system (moment distribution, for beams). Figure 13.8c is one example, m^θ (moment used to compute θ) and the unit-load method gives $\theta_d = \int_b^d m^\theta \, \kappa \, \mathrm{d}s = -0.286\,\bar{\kappa}L$ (radians). The negative sign indicates that the slope is in the opposite sense of the unit couple.

Such stresses (m) and strains (κ) which appear in the structure without any apparent applied forces are *initial stresses* and *initial strains*. They are uninvited guests and if overlooked they could cause havoc, because they superimpose on the mechanical stresses when external loads are added.

In other cases they may be introduced by the designer to improve the structural response. In these cases they are better referred to as *prestresses* and *prestrains*.

13.5.2 Determinate and Redundant Trusses

Consider the determinate and $R = 2$ redundant trusses in Fig. 13.9a, b respectively. In both structures bar ac, that is, bar number 1, instead to being of length $L_1 = L$ as dictated by the geometry of the truss, was initially $L_1 = L(1+\bar{\varepsilon})$. We say that the bar has applied extensional strain $\varepsilon_1^a = \bar{\varepsilon}$ and no applied strain elsewhere, see Table 13.1 (bars 2 and 8 are, of course, missing in the determinate design.) This strain may be due to a uniform temperature increase $\bar{\varepsilon} = \alpha\, T$ in that bar, or the bar could have been manufactured longer than the nominal L. (Whatever the source of the applied strain the analysis procedure is the same.)

In the *determinate* truss (Fig. 13.9a) the applied strain should not cause any problems. When connecting bars bc and ac we will barely notice that bar bc is slightly rotated clockwise about b to accommodate the faulty length of ac. The rest of the truss will also be slightly inclined but nothing more. There will certainly not be any (initial) stresses locked in the determinate truss as a result of the applied strain.

Using a formal approach, the structure has $M = 8$ unknown forces (the axial loads in the M bars) and $N = 8$ nodal degrees of freedom (two at each of the four free nodes), and the truss is thus determinate ($N = M = 8$). Without external forces the nodal equilibrium equation are $\boldsymbol{Q}\,\boldsymbol{n} = 0$ which with the $N \times M$ matrix $\boldsymbol{Q}$ being square gives $\boldsymbol{n} = 0$ and no amount of heating or cooling will change that simple fact. For a statically determinate structure the rule is plain: no external forces means no internal forces.

Since there are no internal forces there are also no mechanical strains ($\epsilon_j = n_j/EA$). The final strains are thus simply the applied strains, that is, $\varepsilon_1 = \bar{\varepsilon}$ in bar 1 and zero strain in all the other bars (Table 13.2.)

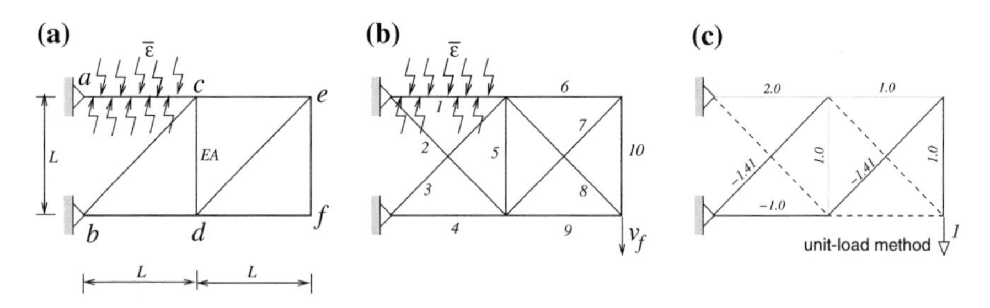

Fig. 13.9 Heated determinate and redundant truss

Table 13.1 Applied strain in truss elements

Bar (m)	1	2	3	4	5	6	7	8	9	10
$\varepsilon_m^a[\bar{\varepsilon}]$	1.0	0	0	0	0	0	0	0	0	0

Table 13.2 Internal forces and elongations: determinate truss

Bar (m)	1		3	4	5	6	7		9	10
$n^o_m[EA\bar{\varepsilon}]$	0		0	0	0	0	0		0	0
$\varepsilon^o_m[\bar{\varepsilon}]$	1.0		0	0	0	0	0		0	0

Table 13.3 Internal forces and elongations: redundant truss

Bar (m)	1	2	3	4	5	6	7	8	9	10
$n^1_m[-]$	−0.707	1.0	1.0	−0.707	−0.707	0	0	0	0	0
$n^2_m[-]$	0	0	0	0	−0.707	−0.707	1.0	1.0	−0.707	−0.707
n^{init}_m $[EA\bar{\varepsilon}]$	−0.117	0.166	0.166	−0.117	−0.105	0.012	−0.017	−0.017	0.012	0.012
$\varepsilon^{\text{init}}_m[\bar{\varepsilon}]$	0.883	0.166	0.166	−0.117	−0.105	0.012	−0.017	−0.017	0.012	0.012

The *redundant* case (Fig. 13.9b) takes a little longer to analyze and is also more interesting. To release the truss we disconnect bars 2 (redundant x_1) and 7 (redundant x_2) from node d. Applying unit forces at the cuts yields respectively the internal forces $\boldsymbol{n}^1$ and $\boldsymbol{n}^2$ shown in rows 1 and 2 of Table 13.3.

These produce the coefficients of the compatibility matrix

$$F_{ij} = \sum_{m=1}^{10} n^i_m n^j_m L_m/(EA)_m$$

and the loading term $\delta^o_i = \sum_{m=1}^{10} n^i_m \varepsilon^o_m L_m$, where ε^o is the strain due to the applied strain, and since the released truss is statically determinate, we have $\varepsilon^o = \varepsilon^a$ which leads to the following compatibility equations:

$$\frac{L}{EA}\begin{bmatrix} 4.33 & 0.5 \\ 0.5 & 4.83 \end{bmatrix}\begin{Bmatrix} x_1 \\ x_2 \end{Bmatrix} + \bar{\varepsilon}L \begin{Bmatrix} -0.707 \\ 0 \end{Bmatrix} = \begin{Bmatrix} 0 \\ 0 \end{Bmatrix},$$

yielding $x_1 = 0.165\, EA\bar{\varepsilon}$ and $x_2 = -0.0171\, EA\bar{\varepsilon}$. The initial stresses (internal forces prior to any loading) are by superposition $\boldsymbol{n}^{\text{init}} = x_1\, \boldsymbol{n}^1 + x_2\, \boldsymbol{n}^2$, and the initial strains are $\boldsymbol{\varepsilon}^{\text{init}} = (\boldsymbol{EA})^{-1}\boldsymbol{n}^{\text{init}} + \boldsymbol{\varepsilon}^a$ (see Table 13.3).

Under all these initial strains *the truss deforms*. This is an initial deformation prior to any loading. We can, for instance, determine the vertical displacement at node f.

We apply a unit force at f in the direction of the displacement we are trying to find, and determine a possible configuration of forces which are in equilibrium with the unit force (in Fig. 13.9c we show one possibility, $\boldsymbol{n}^{v_f}$). The unit force method gives $v_f = \boldsymbol{e}^T_{\text{init}}\, \boldsymbol{n}^{v_f}$ or $v_f = 1.59\, \bar{\varepsilon}L$.

Note, $\boldsymbol{e}_{\text{init}}$ is the vector of initial elongations of the bars. (The elongation of a bar is the product of the strain by the length, for instance, is $e_m = \varepsilon_m L_m$.)

13.5.3 Fixed Elements

The following example helps in clarifying the concepts of applied and initial strains and initial stress. Consider again the beam in the desert on an early morning but we have it now fixed at both ends. The top fibers are heated directly by the sun, the bottom ones are in the shade and perhaps kept cool by the cold sand, and as a result a temperature gradient which we will assume to be linear is established along the width of the beam. Ask any novice student of structures to draw the displaced configuration and chances are that he will come up with the bell-shaped curve in Fig. 13.10a. The reasoning is apparently sound. The top fibers expand more than the bottom ones. This corresponds to a concave curve. But the walls demand a horizontal slope at the boundaries, so we get a compromise shape: a concave curve flattened at the extremities.

Here intuition is a poor advisor. This is not what will happen. It is easy to show that the beam will simply remain straight as in Fig. 13.10b.

Indeed, consider the beam under applied curvature $\bar{\kappa}$. It will try to assume the form of a concave circular segment ($\kappa = -\bar{\kappa}$), or constant curvature line. But the walls will react by applying equal and opposite moments (this is all they can do). The result is a constant curvature convex curve (circular segment) of curvature $\kappa = m/EI$. The sum of the two is also a curve of constant curvature ($\kappa = -\bar{\kappa} + m/EI$). But because the walls impose zero slopes at the extremities the constant curvature curve is forcibly a straight (zero curvature) line. In other words the initial form remains straight and there are no initial deformations. But there are initial stresses (bending moments) because zero curvature $0 = -\bar{\kappa} + m/EI$ requires $m^{\text{init}} = EI\,\bar{\kappa}$.

Following a formal approach (Fig. 13.11a), we depict a uniform element of length L with stiffness properties EA and EI, fixed at both ends. We have established that a temperature gradient can be represented by an applied uniform extensional strain $\bar{\varepsilon}$ and uniform curvature $\bar{\kappa}$. Since the element has both flexural and extensional applied

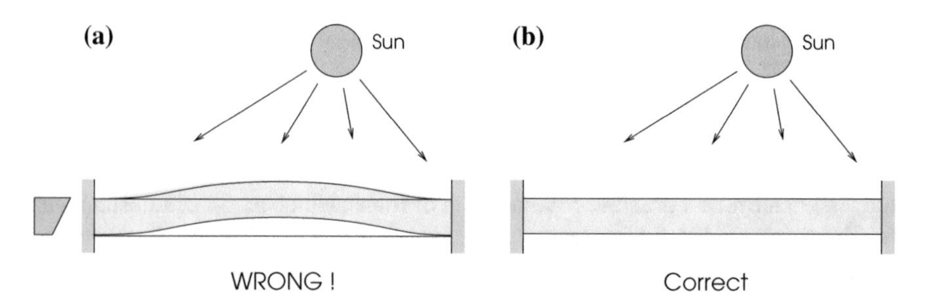

Fig. 13.10 Heated fixed beam

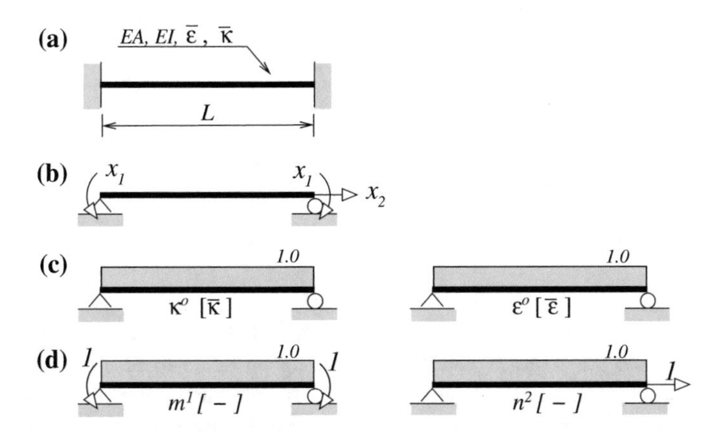

Fig. 13.11 Heated fixed beam: analysis

strain we will perform a frame analysis, that is, the structure is $R = 3$ redundant. In the released structure in Fig. 13.11b the redundants are the end-couples and the axial force in the element. However, the symmetry of the problem requires that the end-moments be equal and opposite so we use this condition from the start. As a result we have only two unknowns: the moments reactions x_1 and the axial reaction x_2.

In Fig. 13.11c we show respectively the flexural κ^o and extensional ε^o strains in the released structure due to the applied strains (they are the same), and in Fig. 13.11d the bending moment and normal force diagrams m^1 and n^2 due to unit couples at 1 and unit force at 2, respectively. Note, $m^2 = n^1 = 0$ because of the orthogonality of the flexural and normal deformation modes. We have two unrelated compatibility equations (one in each mode)

$$F_{11}\, x_1 + \delta_1^o = 0\,, \qquad F_{22}\, x_2 + \delta_2^o = 0,$$

with $F_{11} = \int_0^L (m^1 m^1/EI)\,\mathrm{d}x = L/EI$, $F_{22} = \int_0^L (n^2 n^2/EA)\,\mathrm{d}x = L/EA$, and $\delta_1^o = \int_0^L \kappa^o m^1\,\mathrm{d}x = \bar{\kappa}L$, $\delta_2^o = \int_0^L \varepsilon^o n^2\,\mathrm{d}x = \bar{\varepsilon}L$. This yields $x_1 = -EI\bar{\kappa}$ and $x_2 = -EA\bar{\varepsilon}$.

Consequently, the initial strains are zero

$$\kappa^{\text{init}} = x_1\, \kappa_1^o/EI + \kappa^a = 0 \quad \text{and} \quad \varepsilon^{\text{init}} = x_1\, \varepsilon_1^o/EA + \varepsilon^a = 0$$

but the initial stresses are not!

$$\begin{aligned} \kappa^{\text{init}} &= x_1\, \kappa_1^o/EI + \kappa^a = 0\,, & \varepsilon^{\text{init}} &= x_1\, \varepsilon_1^o/EA + \varepsilon^a = 0\,, \\ m^{\text{init}} &= x_1 = EI\,\bar{\kappa}\,, & n^{\text{init}} &= x_2 = -EA\,\bar{\varepsilon}\,. \end{aligned}$$

Note, m^{init} and n^{init} are respectively positive and negative due to the particular strains created by solar heating ($\kappa^{\text{init}} = -\bar{\kappa},\ \varepsilon^{\text{init}} = \bar{\varepsilon}$).

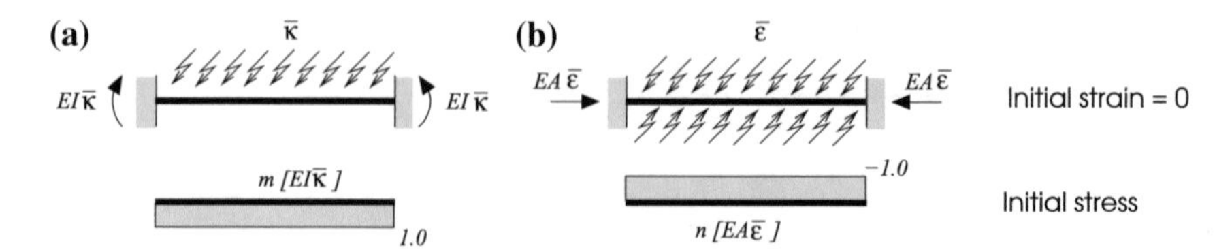

Fig. 13.12 Initial internal forces and deformations in heated fixed beam

It is easy to see why a uniform applied $\bar{\varepsilon}$ as in Fig. 13.12b causes no strains. The walls prevent the bar from expanding by pushing on it. So there are no strains but well a compressive stress. The beam in Fig. 13.12a under an applied $\bar{\kappa}$ does the same. If it were free to expand it would morph into a circular segment of curvature $\bar{\kappa}$. But the walls cannot allow this. They apply reactive couples (no forces). Such couples create a uniform moment distribution and also a uniform curvature. So the final shape of the beam will be of a uniform curvature. But the walls require horizontal tangents at the extremities of this 'curved' beam. So the curvature must be zero.

13.5.4 Partially Heated Beam

Consider the doubly built-in beam in Fig. 13.13a, of length $3L$, of flexural stiffness EI with a hinge at c. The beam has undergone a curving $\bar{\kappa}$ along segment bc either due to a faulty manufacturing or due to differential heating. What are the initial stresses and the initial deformed shape of the structure?

This $R = 1$ structure can be released by a hinge at d. The redundant structure in Fig. 13.13a is the same as the determinate structure in Fig. 13.13b, where x_1 (yet unknown) is the couple applied by the wall on the beam at d. We now analyze a statically determinate structure under applied curvature κ^a and applied loads x_1. The applied curvature is zero but for $-\bar{\kappa}$ along segment bc.

The strains in the released structure are the same as the applied strains, $\kappa^o = \kappa^a$, as shown in Fig. 13.13c. The moment diagram m^1 due to a unit couple at d is simple.

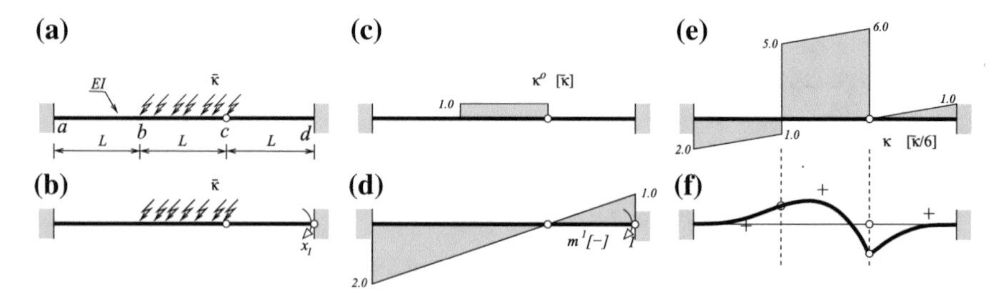

Fig. 13.13 Partially heated beam

It is linear (no distributed loads), it has the values -1.0 at d and zero at c (a hinge), hence the line shown in Fig. 13.13d. Using these basic distributions the compatibility equation yields $x_1 = 0.167\, EI\, \bar{\kappa}$.

The initial moment distribution is thus $m^{\text{init}} = x_1\, m^1$ (the shear is constant, x_1/L). The initial curvature is the sum of the mechanical curvature and the curvature in the released structure due to applied curvature, $\kappa^{\text{init}} = m^{\text{init}}/EI + \kappa^o$ as shown in Fig. 13.13e, from which we can draw an estimate of the deformed shape, Fig. 13.13f.

13.5.5 A Heated Roof

A curvature $\bar{\kappa}$ is applied to element bc of the frame in Fig. 13.14. This can be assumed to be caused by a temperature increase of the exposed surface bc. The structure has a hinged support at a and a horizontal guide (no vertical nor rotational dofs) at d. The flexural stiffness of the frame elements is EI and the axial stiffness is very large $EA \to \infty$. We would like to determine the initial stresses and strains due to the applied strains.

This $R = 1$ structure is made determinate by releasing the bending moment at d, thus transforming the guided support into a horizontal roller.

In Fig. 13.14b, c, we show respectively the curvature distribution due to the applied curvature ($\kappa^o = \kappa^a$) and the moments due to a unit value of the redundant x_1 of the released structure, m^1. The coefficients of the compatibility equation are $F_{11} = \int_a^d m^1 m^1/EI\, \mathrm{d}x = 1.70\, L/EI$ and $\delta_1^o = \int_a^d m^1 \kappa^o\, \mathrm{d}x = \int_b^c m^1 \bar{\kappa}\, \mathrm{d}x = 0.75\, \bar{\kappa} L$. This yields the reactive moment at d, $x_1 = -0.45 EI\, \bar{\kappa}$.

The results are shown in Fig. 13.15 starting with the reactions in Fig. 13.15a. Next the initial internal forces (nsm) and initial κ-distributions are given. Note the discrepancy between m and κ on element bc. The initial displaced shape can now be drawn on the basis of κ (Fig. 13.15b).

Note, for the displaced shape, one uses the κ-distribution and the boundary conditions bearing in mind the well-known assumption of negligible axial extension of flexural members. Consequently, node b moves on an arc of a circle centered at a (along the tangent, in fact).

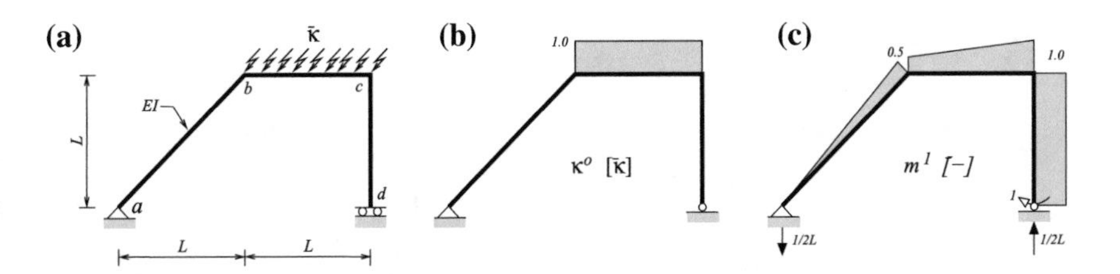

Fig. 13.14 Heated roof: analysis

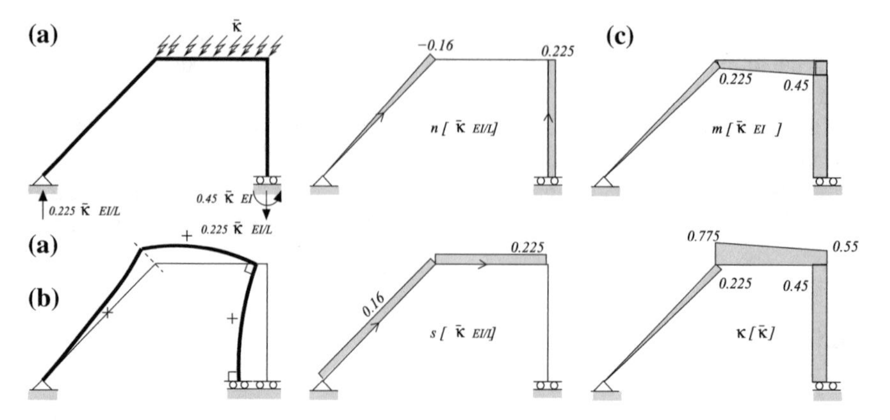

Fig. 13.15 Heated roof: results

13.5.6 Rectangular Box with Applied Curvature on Top

The rectangular shaped frame on simple supports *abde* in Fig. 13.16a is subjected to a uniform applied curvature $\overline{\kappa}$ along the top beam *bd*. The frame is composed of elements of infinite axial stiffness ($EA \to \infty$) and of finite bending stiffness EI, and has redundancy $R = 3$. Indeed a total cut, say at mid-span of *bd* (section c in Fig. 13.16b), will release three redundants and create a statically determinate structure (a tree, albeit without branches, on simple supports). The structure and loading are symmetric with respect to the vertical passing through c. The shear force being intrinsically anti-symmetric must thus be zero at c; $x_3 = 0$. We are left with two static unknowns, the bending moment x_1 and the internal axial force x_2 at c, for which two compatibility equations can be written.

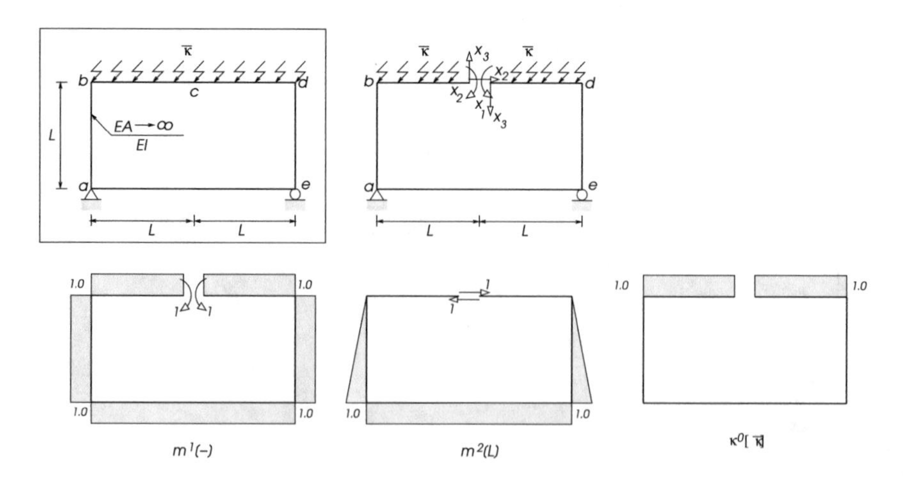

Fig. 13.16 Heated rectangular box: analysis

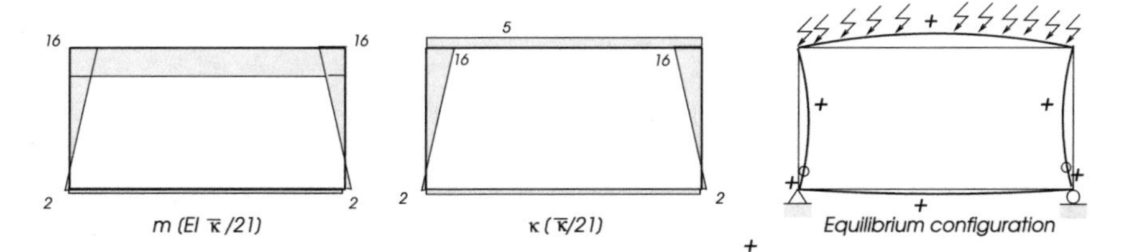

Fig. 13.17 Heated rectangular box: solution

The basic ingredients for the compatibility equations are the bending moments due to unit couples at c ($x_1 = 1$), the bending moments due to unit axial forces at c ($x_2 = 1$) and the curvature of the structure κ^0 due to the applied curvature κ^a, which for a statically determinate structure gives simply $\kappa^0 = \kappa^a$. The three diagrams are given in Fig. 13.16.

Performing the integration for computing the coefficients of the compatibility equations $\boldsymbol{F}\,\boldsymbol{x} + \boldsymbol{\delta}^0 = 0$ gives (without the units)

$$\begin{bmatrix} 6 & 3 \\ 3 & 8/3 \end{bmatrix} \begin{Bmatrix} x_1 \\ x_2 \end{Bmatrix} + \begin{Bmatrix} 2 \\ 0 \end{Bmatrix} = 0 \qquad (13.8)$$

yielding $x_1 = -16/21\, EI\bar{\kappa}$ and $x_2 = 18/21\, EI\bar{\kappa}/L$.

The final bending moment distribution $m = x_1 m^1 + x_2 m^2$ (note $m_0 = 0$), curvature distribution $\kappa = m/EI + \kappa^a$ and displaced shape (equilibrium configuration) are given in Fig. 13.17.

13.6 Summary of the Technique

In the general case, we can have external loads and strain applied to the redundant structure. So, the released structure is subjected to three types of 'loadings':

(a) the redundant forces $\boldsymbol{x}$ (results due to unit applied loads at x_i are denoted with superscript i);
(b) the applied external loads (superscript 'o'); and
(c) the applied strains (*here* superscript 'oo').

The resulting normal forces, shear, bending moments, and curvature distributions are, of course, the sum of the three:

$$n(x) = \sum_i x_i\, n^i(x) + n^o(x) + [n^{oo}(x) = 0]$$

$$s(x) = \sum_i x_i\, s^i(x) + s^o(x) + [s^{oo}(x) = 0]$$

$$m(x) = \sum_i x_i\, m^i(x) + m^o(x) + [m^{oo}(x) = 0]$$

$$\kappa(x) = \sum_i x_i\, \kappa^i(x) + \kappa^o(x) + \kappa^{oo}(x)$$

with $\kappa^o(x) = m^o(x)/EI$. As indicated there are no initial stresses in a statically determinate structure, but there are initial strains, hence the additional term κ^{oo}.

As for the compatibility equations, the flexibility matrix is standard, $F_{ij} = \int_x (m^i m^j/EI)\, dx$. The load vector is $\delta_i^{o+oo} = \int_s m^i\, (\kappa^o + \kappa^{oo})\, dx$.

13.7 Lack-of-Fit

13.7.1 Lack-of-Fit as Applied Gaps

Consider the $R = 1$ beam on three supports in Fig. 13.18a. This unusual assemblage can be two very different things. It is either a curved beam correctly positioned on the three non-collinear supports and it is then free of initial stresses. Or it could be a straight beam forcibly connected to the non-collinear supports, in which case we will have initial stresses and initial strain. The dilemma can be solved by releasing the structure.

In Fig. 13.18c1, we release the structure by disconnecting the beam from the middle support, and notice that it springs back to its original straight condition. The lack of fit at the release, δ^a, tells us that during assembly somebody tampered with the beam and somehow connected it to the three supports (or, for some reason, the outer supports may have settled).

In the example of Fig. 13.18c2 the original structure (Fig. 13.18a) was released by introducing a hinge at the middle support. The elements are again straight and the lack-of-fit is evidenced by the angular discontinuity θ^a that appears at the release.

Lack-of-fit is indeed a close relative of applied deformations. They are different in the sense that with applied deformations $\varepsilon^a(x)$ and $\kappa^a(x)$, we compute the

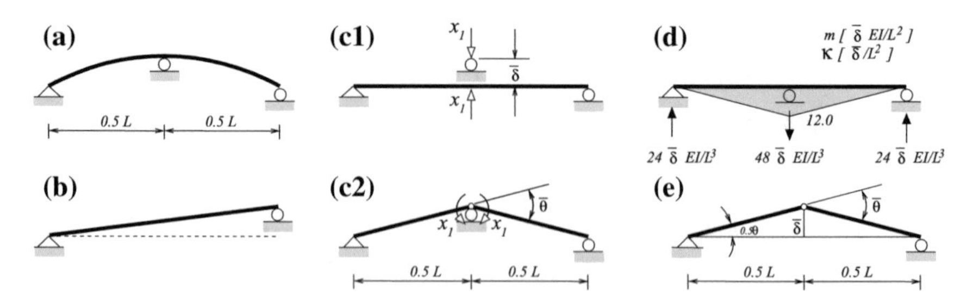

Fig. 13.18 Lack of fit

discontinuities at the releases, whereas here the discontinuity $\boldsymbol{\delta}^a$ are given. The discontinuity $\boldsymbol{\delta}^a$ is as it where an applied gap, hence the 'a' superscript.

Finally, it should be clear that a lack-of-fit (that is, the supports are not where they should be) in a determinate structure as in Fig. 13.18b may not please the customer, but it will not cause any strain or stress in the structure. The beam on two supports will simply not be horizontal.

13.7.2 *The Compatibility Equations*

Consider again the released structure in Fig. 13.18c1. There are no applied strains but only equal and opposite forces x_1 at the release. Note, we have chosen the sense of the redundants in accordance with the definition of the gap $\bar{\delta}$. That is, positive redundant forces cause a widening of the gap.

As usual, the compatibility equation must enforce the closing of the gap. In other words, the widening of the gap due to the redundant plus the applied gap should sum up to a zero gap, $F_{11}\,x_1 + \delta_1^a = 0$.

For the release in Fig. 13.18c1 the applied gap at the cut is $\delta_1^a = \bar{\delta}$, yielding with the flexibility coefficient $F_{11} = L^3/48EI$ the reaction $x_1 = -48\,\bar{\delta}\,EI/L^3$. This produces the bending moment and curvature distributions shown in Fig. 13.18d. Note that in this formulation there is no applied strain. The supports are out of place.

We could of course have chose another way for releasing the structure, such as setting a hinge at the middle support (Fig. 13.18c2). For the elements to remain straight in the released configuration we obtain a gap $\bar{\delta}_1 = \bar{\theta}$. Selecting the sense of the redundant so as to increase the gap we have here $F_{11} = L/3EI$, and the compatibility equation gives the bending moment at the middle support $x_1 = -3\,\bar{\theta}\,EI/L$. From simple trigonometry (see Fig. 13.18e) $\tan(0.5\bar{\theta}) = \bar{\delta}/(0.5\,L)$, and since for small displacements $\tan(0.5\bar{\theta}) = 0.5\bar{\theta}$, we get also $x_1 = -12\,\bar{\delta}\,EI/L^2$, that is, the same result as in Fig. 13.18d (luckily).

13.8 Prestressing

When we have a redundant structure with applied strain and/or lack of fit it will store initial stresses before any external forces are applied. Until now we have considered cases were these applied strains were a nuisance, such as thermal expansion due to the sun's heat or a faulty manufacturing process which resulted in bent or other badly fitting elements. In other words, we would be better off without these initial stresses.

In many cases, however, such initial stresses are introduced on purpose to mitigate the effect of the external loads. In such instances the applied strains or misfits are even tuned to create the desired initial stresses. Here we will call initial stresses which are introduced purposely by the designer of the structure prestresses, and the process

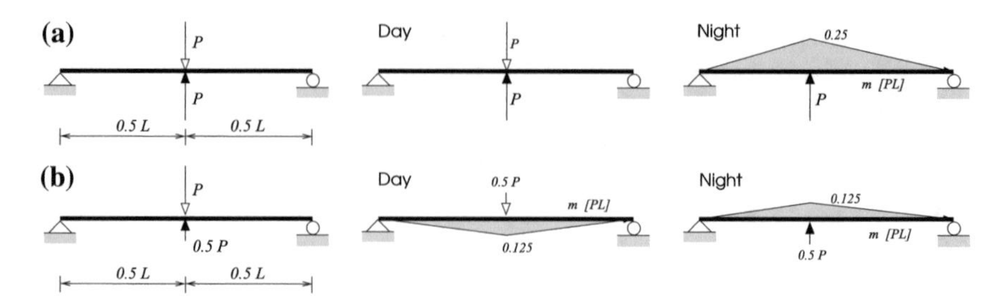

Fig. 13.19 Day and night loadings

of creating these stresses will be named *prestressing* the structure. Only redundant structures can be prestressed.[3]

The analysis should determine what kind of prestresses may be beneficial to the structure, and also indicate the magnitude of the applied strains $\bar{\varepsilon}$, $\bar{\kappa}$ or misfits $\bar{\boldsymbol{\delta}}$ that will produce the desired prestress.

13.8.1 Why Prestressing?

A trivial example can elucidate this question. Consider the simply supported beam subjected to a force P at mid-span in Fig. 13.19a. Let us assume that we are in a position to apply an opposite force so as to reduce the maximum bending moment ($0.25\,PL$) in the beam. What should the magnitude of that force be? The obvious solution is an equal and opposite force which results in a zero bending moment distribution.

Let us now assume that the force P is applied only during the day, and that at night the beam is free of any loads (think of the traffic crossing a bridge) but the opposite force is there day and night. The proposed solution of an equal and opposite force is not valid any more, because we will have zero moment during the day but at night the maximum moment will again reach $0.25\,PL$, albeit of opposite sign. Here we could, for instance, apply an opposite force of half the magnitude of the applied one (Fig. 13.19b) such as to produce a maximum bending moment $0.125\,PL$, both during the day and at night.

Such a permanent reactive force applied to the beam can be the result of prestressing of the structure as will be shown in the following example.

[3]A word of caution: We consider prestressing as a result of deformed elements only. There are other ways to prestress a structure. A simply supported concrete beam can be prestressed (compressed) by embedded cables in tension.

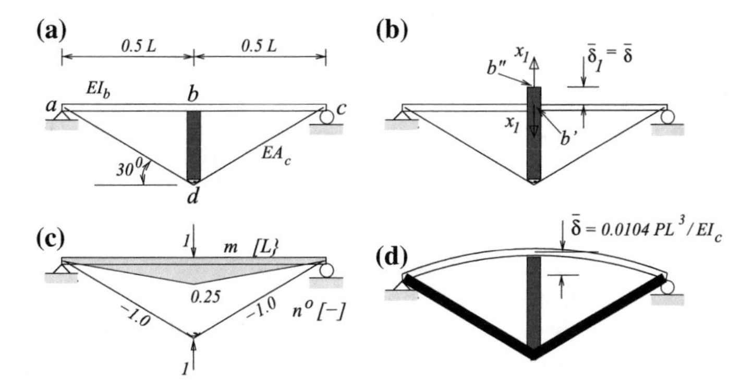

Fig. 13.20 Prestressing beam by underlying cable in tension

13.8.2 Example

In order to reduce the maximum bending moments (and displacements) of a simply supported beam under a central force P the reinforcement in Fig. 13.20a was devised. The additional components are a cable anchored at the beam extremities a and c and a very stiff strut db at mid-span between the beam and the cable.

The interesting feature of the assembly is that the initial length of the strut is longer by an amount $\bar{\delta}$ than its nominal length in Fig. 13.20a. It was wedged in place between b and d, thus applying an upload at b. This causes prestress (internal forces without applied forces), and the purpose of this exercise is to calculate the lack of fit $\bar{\delta}$ necessary to create an internal compressive force $0.50\,P$ in the strut.

The flexural stiffness of the beam is EI_b and the longitudinal stiffness of the cable is EA_c. Since the strut is very rigid we will assume that its axial stiffness is infinite, $EA_{strut} \to \infty$.

We choose to release this $R = 1$ structure by disconnecting the strut from the beam at b, thus producing the determinate arrangement as shown in Fig. 13.20b. The now exposed internal force x_1 is applied at the released ends, that is, b' on the beam and b'' on the strut. In the compatibility equation $F_{11}\, x_1 + \bar{\delta} = 0$, we know the interaction force $x_1 = -0.5P$ and we seek the required discrepancy $\bar{\delta}$. The flexibility coefficient makes allowances for the flexibility of the beam and for the flexibility of the outer bars of the truss, that is, the cable (the inner bar, the strut, is infinitely rigid); $F_{11} = 0.0208\,(L^3/EI_b) + 1.155\,(L/EA_c)$.

Consequently, the discrepancy which will produce the required up-force on the beam is $\bar{\delta} = (0.5P)\, F_{11}$ or

$$\bar{\delta} = \frac{PL^3}{EI_c}\left(0.0104 + \frac{0.5775}{A_c L^2/I_b}\right) \tag{13.9}$$

The solution, depends of course, on the stiffness of the cable (A_c) relative to the stiffness of the beam (I_b).

Clearly, for a given applied discrepancy $\bar{\delta}$, a strong cable will create a larger internal compressive force x_1 than a weak cable. And indeed when $A_c L^2/I_b \rightarrow \infty$, we get the smallest value of the discrepancy, $\bar{\delta} = 0.0104\, PL^3/EI_c$ (see Fig. 13.20d). As the cable becomes more flexible, $\bar{\delta}$ increases and at the limit, when the cable is about to disappear ($A_c L^2/I_b = 0$), we need an infinite discrepancy to produce the force $x_1 = -0.5\, P$, or any other force, for that matter.

As expected, $\bar{\delta} = 0.0104\, PL^3/EI_c$ is the displacement of a simply supported beam of characteristics, L, EI_b, under a central force of magnitude $0.5\, P$.

Note: Instead of wedging a longer strut between the beam and the cable, we can use a strut of nominal length and instead shorten the cable. Noting that due to the 30° slope of the cable, the internal force in the cable is also a tension force of magnitude $0.5\, P$, we can fix the cable at a, run it over the strut, pull the extremity c until a tensile force of $0.5\, P$ is reached, and then fix the cable at c. This is probably what would be done in practice.

13.9 Summing Up

With applied deformations discussed in this chapter and applied loads shown in the previous chapter, we have now concluded the basic topics of the force method. In practice, the two types of loadings occur simultaneously and a combined analysis is warranted. It was not done here because it would complicate things without adding much to the understanding of the technique.

Part IV
Stiffness

Part IV deals with the second major analysis method, the stiffness method, also called the displacement method. In Chap. 14 the method is described using a truss as a didactic example. Next, the concept of an element stiffness matrix is introduced in Chap. 15. Several classic element matrices are computed.

In preparation of the assembly techniques, transformations of the element stiffness matrices due to coordinate modifications are discussed in Chap. 16. Then, in Chap. 17 we describe the equilibrium equations, and show how the system stiffness matrix can be generated by assembly technique.

Part IV concludes with Chap. 18, where we use the fixed-end reactions to allow structures with loads on elements, such as distributed forces or applied strain (temperature variations), to be incorporated in the displacement method.

Chapter 14
Introducing the Stiffness Method

The stiffness or displacement method for analyzing structures is exemplified by means of trusses. This type of structure has been relatively neglected until now because, in the view of the author, truss analysis by the force method is numerically too cumbersome. The stiffness method, on the other hand, is totally independent of the type of structure and applies with equal ease to trusses, beams and frames. We, of course, bear in mind that the computations are all computerized.

14.1 Introduction

The *Displacement or Stiffness method* for analyzing structures comes in several related variants and flavors, from the slope-deflection method to the finite-element method. In fact, the finite-element method, which is now pervasive in applied mathematics, was originally developed by structures engineers for calculating the stresses and displacements in aircraft structures. The common feature of all these techniques is that out of the three basic analysis quantities, the internal forces $\boldsymbol{n}(x)$, the deformations $\epsilon(x)$ and the displacements $\boldsymbol{u}(x)$, what we get first are the displacements. We then compute the deformations by taking the derivatives of the displacements and finally, by virtue of Hooke's law, we obtain what we are most interested in, the internal forces

$$\boldsymbol{u}(x) \rightarrow \epsilon(x) = \mathbf{L}\,\boldsymbol{u}(x) \rightarrow \boldsymbol{n}(x) = \boldsymbol{EA}\,\epsilon(x)$$

We will soon see that this indirect route to the internal forces has its appeal. It is beautifully simple, but it also has shortcomings. The stiffness method generates tedious numerical work of little appeal to us, slow and error-prone humans. This is why this sleeping beauty lay dormant for some time until awakened by the magic kiss of (what else!) the computer. So, whenever we meet long and boring calculations in the sequel always keep in mind that we will not be asked to perform them; the computer will.

M.B. Fuchs, *Structures and Their Analysis*,
DOI 10.1007/978-3-319-31081-7_14

Finally, the stiffness method has mainly to do with numbers (primarily, the displacements at the nodes) rather than functions. And since the theory is linear, we will soon notice that everything is in matrices and vectors (column matrices), but once we get the knack, this should not be a cause of concern.

The stiffness method is very general.

It applies the same technique to all different kinds of structures: trusses, beams, and frames, and if we are willing to include the finite element method, also plates and shells.

Statical redundancy is irrelevant in the stiffness method.

The method does not care whether the structure is statically redundant or determinate. This aspect of the method is often annoying to new students. A simple beam on two supports may require a lengthy analysis with the stiffness method. However, the method solves multi-storey multi-column frames with the same ease.

In this book the stiffness method follows the flexibility method. Consequently the reader will have mustered some experience in understanding the behavior of structures. We will bank on that and assume that the basic aspects of structural analysis are known.

This said, we will start with the analysis of a truss and do it by way of the stiffness method.

14.2 Pin-Jointed Ideal Trusses

Trusses are in some sense a world apart from frames and beams. Broadly speaking, a truss is a structure where the stresses and deformations due to shear and bending are small when compared to those due to the normal forces n, which are constant in every member. On the basis of its triangulated pattern and the application of forces only at the joints, the stresses originate primarily from normal internal forces. So much so that we analyze the structure assuming from the start that $s = m = 0$ in all the members, and restrict the work to calculating only the normal forces in the elements (one force per element), thus lightening the numerical burden.

Clearly, before the analysis, we need some prior knowledge to recognize that a particular frame can behave as a truss. We will give some indications when this can be done, although experience is still the best counselor. We start with the pin-jointed or ideal truss.

An *ideal truss* is a particular type of frame which fits to the following criteria: (1) The members are pin-jointed to one another and to the ground; (2) the structure has a triangulated pattern; (3) the external loads (including the reactions) are forces (no couples allowed) applied to the nodes only; and (4) the members are linear (straight).

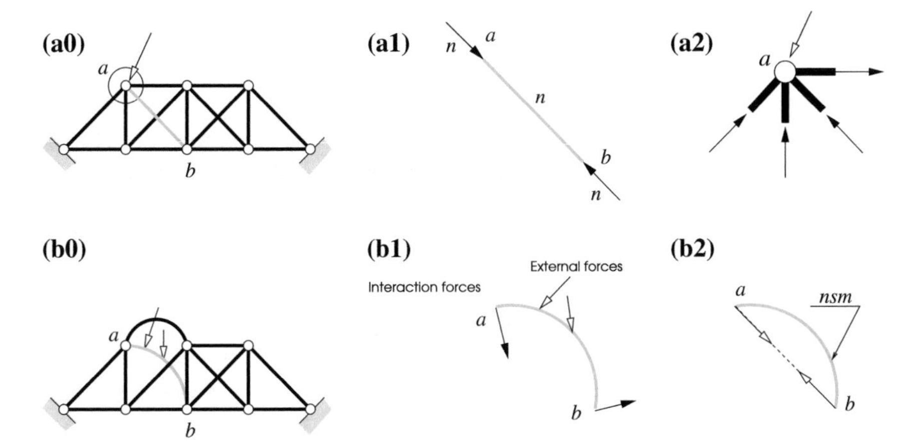

Fig. 14.1 Ideal trusses

The reason for these demands is that for an ideal truss we can guarantee that: (a) The shear and bending moments are zero everywhere ($s = m = 0$); and (b) the only remaining internal forces, the normal force n, is constant in every element.

Indeed, consider the pin-jointed *frame* in Fig. 14.1b0 which obeys to conditions (1) pin-connected and (2) almost triangulated pattern. Due to the presence of hinges at the extremities of all the elements, the end moments of every element are zero. Consequently, only forces can be applied by the surrounding structure to any given element, excluding couples (Fig. 14.1b1). This structure is obviously a frame since there can be shear and moment distributions (and normal forces) in the elements.

Now let us add condition (3), that is, forces (including reactions) are applied at the nodes only. Since an element is subjected to forces at the extremities only, these forces are necessarily equal and opposite (Fig. 14.1b2). We have now only one unknown force per element (the interaction force) but the structure is still a frame (element ab bends.)

Finally, also assuming condition (4) straight elements, we obtain the pin-jointed truss in Fig. 14.1a0. The interaction force is now called n (Fig. 14.1a1) and it is clear that there is no shear and moment in the elements.

As for the redundancy of the pin-jointed *truss*, we note that there is only *one static unknown per element* the normal force n as opposed to two unknowns in beams sm and three unknowns in frames (nsm). Last but not least, there are only *two equilibrium equations per free node*. Indeed, the rotational equilibrium equation is automatically satisfied (see Fig. 14.1a2).

14.3 Rigid-Jointed Trusses

An ideal truss, like anything ideal, is hard to find and indeed most trusses are rigid-jointed, that is, the first design criterium is violated. You will be hard pressed to find even a single hinge in the epitome of truss design, the Eiffel tower in Paris. As a matter of fact, Gustave Eiffel built another memorable albeit hidden truss inside the illustrious Statue of Liberty in New York's harbor. The structure supporting the hand holding the torch, for instance, is a cantilever truss. Both structures are jam-packed with rivets and bolts and still they deserve to be called trusses, that is, they were designed to develop primarily normal internal forces.

How can we know if a frame is a candidate for behaving like a truss? After having ascertained that its members are straight and that the external forces are at the nodes only, we insert hinges at all the nodes of the model. The result is an ideal truss, obviously. If the result is a sound structure and the elements are relatively slender, then usually a truss-type analysis is valid.

Consider, for instance, the frame in Fig. 14.2a0, and let us introduce hinges at the nodes (Fig. 14.2a1). What we have here is by all standards an ideal truss, and with some engineering judgement, we will assume that the original truss will by-and-large behave as its ideal counterpart (small *sm*, significant *n*). The advantage of doing so should not be underestimated. Truss-like frames are structurally very effective since the bending behavior is very much reduced.

There is also a numerical aspect. Both arrangements have 15 members and 6 nodes. But if it behaves as a frame, every member carries three static unknowns, and a node provides three equations; whereas in a truss, each member has only one unknown and a node has two degrees of freedom. Consequently, the frame analysis has $U = 15 \times 3 = 45$ static unknowns, of which $R_{\text{frame}} = 45 - (6 \times 3) = 27$ are redundant. For the truss analysis we have only ($U = 15$) $R_{\text{truss}} = 15 - (6 \times 2) = 3$ redundants. (We have seen that when analyzing a structure by the force method, the difference between a redundancy of 27 and one of 3 is very significant.)

The elementary frames in Fig. 14.2b0, c0 will illustrate the limits of this approach. In frame 14.2b0 both legs are hinged to the ground, and introducing a hinge at the crown (Fig. 14.2b1) produces a stable truss. This allows a truss analysis for the frame.

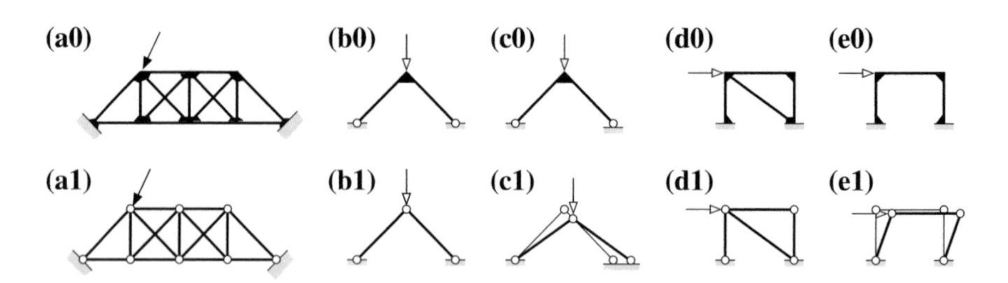

Fig. 14.2 Truss-like frames

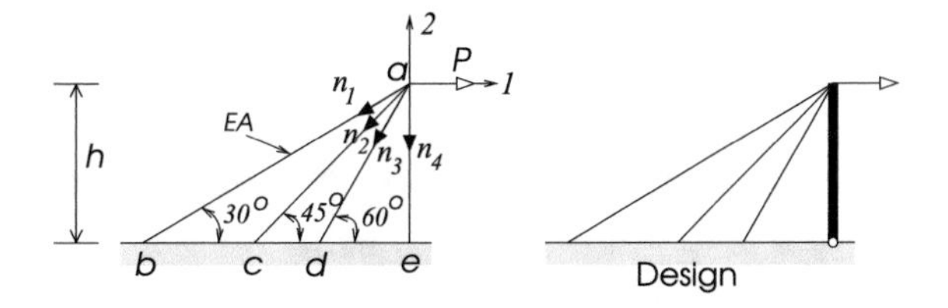

Fig. 14.3 Truss analysis by stiffness method

In frame 14.2c0, on the other hand, the left leg is hinged but the right one has a roller support. (The support can roll on the smooth ground as much as the structural deformations will allow.) Adding a hinge at the apex (Fig. 14.2c1) yields a poor mechanism. As shown, we can displace the structure without deforming any of its members, which is an undeniable indication of instability. Frame 14.2c0 will thus develop primarily flexural deformations and a truss analysis is unacceptable.

Similar cases are shown in Fig. 14.2d, e. In Fig. 14.2d, a truss analysis is possible, whereas trying to make a truss out of the portal frame in Fig. 14.2e is not warranted.

14.4 The Fundamental Analysis Equations for Trusses

Consider the 4-bar truss in Fig. 14.3. The structure has five nodes and four elements. For every node we need the horizontal and vertical components to determine the displacement. Four nodes are fixed at the ground. Their degrees of freedom are closed and their displacement components are known (they are zero.)

The free node *a* has two open degrees of freedom and its displacements components are as yet unknown. We count $N = 2$ open nodal degrees of freedom. They become the components of the $N \times 1$ nodal displacements[1] vector **u**.

At the free node a force may be applied. In the case of the figure, the horizontal component of the applied force is P and the vertical component is zero. These components are included in the 2×1 force vector **p**, using the same numbering sequence as for the displacements.

There are $M = 4$ bars in the truss. Every bar has an unknown internal force and an unknown elongation. These are incorporated in the $M \times 1$ internal forces vector **n** (positive in tension) and elongations vector **e** (positive when extending).

The N equilibrium equations at the free nodes, Hooke's law for each of the M bars and the relations between the elongations of the M bars and the N nodal displacements are respectively

$$\begin{aligned} \mathbf{Qn} &= \mathbf{p} \\ \mathbf{Se} &= \mathbf{n} \\ \mathbf{Ru} &= \mathbf{e} \end{aligned} \tag{14.1}$$

[1]Not to be confused with the 3×1 vector function $\boldsymbol{u}(x)$ of components $u(x)$, $v(x)$ and $\theta(x)$, seen earlier.

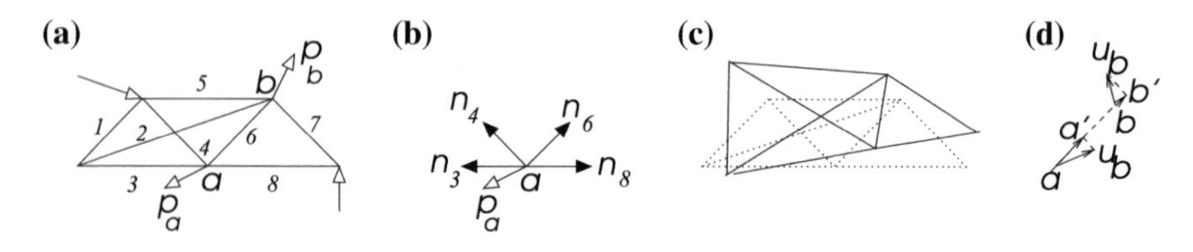

Fig. 14.4 $R = Q^T$

We assume whatever it takes to ensure that these linear relations are physically adequate. The displacements are small enough that the equilibrium equations can be written in the initial shape (rather than in the deformed configuration, where they should have been written) and that $\mathbf{e}$ can be expressed as a linear function of $\mathbf{u}$. Finally the deformations are small enough for the material to be in the range of linear elasticity, that is, the internal forces are linear functions of the deformations.

$\mathbf{Q}$ is called the equilibrium matrix. $\mathbf{S}$ is a diagonal matrix of axial stiffnesses of the elements, $S_{ii} = E_i A_i / L_i$ where the constants in the expression are respectively Young's modulus, the cross-section and the length of element i. Finally, $\mathbf{R}$ is the kinematics matrix. We have here $2M + N$ equations in the $2M + N$ variables $\mathbf{n}(M)$, $\mathbf{e}(M)$ and $\mathbf{u}(N)$.

14.5 Virtual Work, $\mathbf{R} = \mathbf{Q}^T$

As a preamble, we will show that for any truss, $\mathbf{R} = \mathbf{Q}^T$, in other words, the kinematics matrix is the transpose of the equilibrium matrix. This follows from virtual work.[2]

Consider the truss in Fig. 14.4a to which we apply auto-equilibrated forces $\mathbf{p}$, and let $\mathbf{n}$ be internal forces in equilibrium with $\mathbf{p}$. Every node is now in equilibrium under the external and internal forces. For instance, node a is in equilibrium under a force $\mathbf{p}_a$ and forces n_3, n_4, n_6 and n_8 (see Fig. 14.4b). The equilibrium conditions at all the nodes are expressed by $\mathbf{Qn} = \mathbf{p}$, two lines per node (say, equilibrium in x and y directions).

Next we displace all the nodes arbitrarily (Fig. 14.4c). This causes the elements to elongate (contraction is a negative elongation). The elongations as given by $\mathbf{e} = \mathbf{Ru}$. In accordance with $\mathbf{p}$, vector $\mathbf{u}$ carries the components of all the displacements, two components per node. We are reminded that node a is in equilibrium under a force $\mathbf{p}_a$ and forces n_3, n_4, n_6 and n_8. These forces are thus equivalent to zero, and when we displace forces which are equivalent to nothing we obviously do not perform any work. Therefore, displacing a does not produce work. Consequently, displacing all the nodes does not produce work. In other words, the total virtual work of the

[2] See also Chap. 3.

external and internal forces is zero, $evw + ivw = 0$. It is easy to see that evw is $\mathbf{p}^T\mathbf{u}$. We will show that ivw is $-\mathbf{n}^T\mathbf{e}$.

What would be the work of the two internal forces n_6 when we move the nodes? Evidently, we are concerned only with $\mathbf{u}_a$ and $\mathbf{u}_b$ at the end nodes of bar 6 (Fig. 14.4d). Vector $\mathbf{u}_a$ has a lateral component (perpendicular to bar 6) and an axial one, aa'. The work performed by n_6 (the one attached to a) when moving a is therefore $n_6(aa')$. By the same token the work performed by n_6 (the one attached to b) when displacing b is $-n_6(bb')$. It follows that the work done by both forces n_6 is $n_6(aa' - bb')$ or $-n_6(bb' - aa')$. Evidently $bb' - aa'$ is the elongation e_6 of bar ab, and the virtual work done by both forces n_6 is $-n_6e_6$. From here we deduce that the virtual work done by all the internal forces is $ivw = -(n_1e_1 + n_2e_2 + \cdots)$ or $ivw = -\mathbf{n}^T\mathbf{e}$.

Consequently $evw + ivw = 0$ becomes

$$\mathbf{p}^T\mathbf{u} - \mathbf{n}^T\mathbf{e} = 0 \qquad \text{or} \qquad \mathbf{p}^T\mathbf{u} = \mathbf{n}^T\mathbf{e} \tag{14.2}$$

Introducing the first and last equation of (14.1) into the left equation, and noting that the transpose of a product of matrices is the product of the transposed matrices in the opposite sequence $(\mathbf{AB})^T = \mathbf{B}^T\mathbf{A}^T$, we obtain $\mathbf{n}^T\mathbf{Q}^T\mathbf{u} - \mathbf{n}^T\mathbf{R}\mathbf{u} = 0$ or also $\mathbf{n}^T(\mathbf{Q}^T - \mathbf{R})\mathbf{u} = 0$. For this triple product to be zero for any $\mathbf{n}$ and $\mathbf{u}$ the term within the parentheses must be zero, that is,

$$\mathbf{R} = \mathbf{Q}^T \tag{14.3}$$

which is what we set out to establish: the kinematics matrix is the transpose of the equilibrium matrix. The fundamental analysis equations for a truss (14.1) therefore become

$$\mathbf{Qn} = \mathbf{p} \tag{14.4}$$

$$\mathbf{Se} = \mathbf{n} \tag{14.5}$$

$$\mathbf{Q}^T\mathbf{u} = \mathbf{e} \tag{14.6}$$

14.6 Solving the Equations by the Stiffness Method

The method is deceptively simple. We introduce (14.5) into (14.4) by substituting for $\mathbf{n}$. This gives $\mathbf{QSe} = \mathbf{p}$. We now replace $\mathbf{e}$ by its expression in (14.6) to yield

$$\mathbf{Ku} = \mathbf{p} \tag{14.7}$$

with

$$\mathbf{K} \equiv \mathbf{QSQ}^T \tag{14.8}$$

The system of linear equations (14.7) is called the 'equilibrium equations' because they are the same equilibrium equations at the nodes as in (14.4) but in terms of displacements. The square coefficient matrix $\mathbf{K}$ in (14.8) is the system stiffness matrix and it is of the order of the number of open degrees of freedom $N \times N$.

Finding the equations is half the work. The other half is spent computing the displacements and the internal forces. For the displacements we solve (14.7) by any appropriate method. Let us use $\bar{\mathbf{u}}$ to indicate that the displacements are now known. Having $\bar{\mathbf{u}}$ we can compute the elongations $\mathbf{e} = \mathbf{Q}^T\bar{\mathbf{u}}$ using (14.6). Finally, for the internal forces you introduce this result in Hooke's law, $\bar{\mathbf{n}} = (\mathbf{S}\mathbf{Q}^T)\bar{\mathbf{u}}$. Voila!

We could have moved on to the next topic, but there is more to it. It all has to do with getting the stiffness matrix $\mathbf{K}$. Equation (14.8) is not the way to go. We do it differently, by assembling the system (or structure) stiffness matrix from its elements stiffness matrices.

14.7 Element Stiffness Matrices

Consider again $\mathbf{K} = \mathbf{Q}\mathbf{S}\mathbf{Q}^T$ and let $\mathbf{q}_j$ be column j of matrix $\mathbf{Q}$. In the example there are four such columns $\mathbf{q}_1$, $\mathbf{q}_2$, $\mathbf{q}_3$, $\mathbf{q}_4$ where $\mathbf{q}_2$, for instance, is the second column of $\mathbf{Q}$ (see Fig. 14.5). We remind the reader that $\mathbf{S}$ is a diagonal matrix where the diagonal entry $S_{jj} = E_jA_j/L_j$ depends on Young's modulus, the cross-section and the length of element j. We leave it as an exercise to show that the stiffness matrix can also be written as

$$\mathbf{K} = \mathbf{Q}\mathbf{S}\mathbf{Q}^T = \sum_j S_{jj}(\mathbf{q}_j\mathbf{q}_j^T) = \sum_j \mathbf{K}_j \tag{14.9}$$

where

$$\mathbf{K}_j = S_{jj}(\mathbf{q}_j\mathbf{q}_j^T) \tag{14.10}$$

is the element stiffness matrix of bar j.

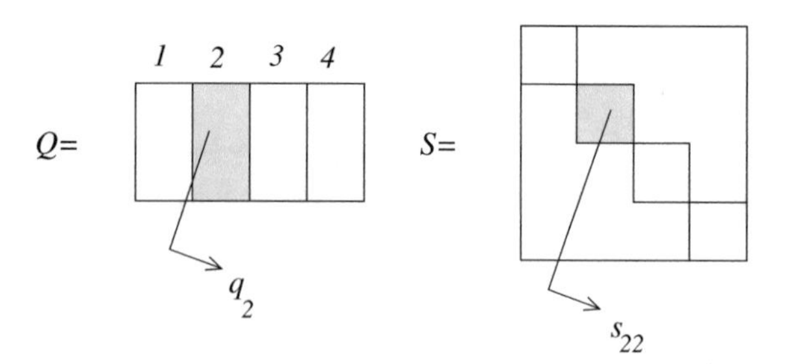

Fig. 14.5 *Q* and *S* matrices

In an equation all terms are of the same type (apples or pears, not both). So everything in (14.9) are $N \times N$ matrices such as $\mathbf{K}$, even $\mathbf{q}_j\mathbf{q}_j^T$. (Note, if $\mathbf{a}$ is an n-vector then $\mathbf{a}^T\mathbf{a}$ is a scalar $(1\times n)(n\times 1) = 1$ and $\mathbf{a}\mathbf{a}^T$ is a matrix $(n\times 1)(1\times n) = n\times n$.) Such matrices are called dyadic matrices and can easily been shown to be singular.

Summing the element stiffness matrices of all the elements of a structure $\sum_j \mathbf{K}_j$ is how the system stiffness matrix is usually assembled. Equation (14.10) is the stiffness matrix of a truss element. We will also learn to develop stiffness matrices of beam elements and when you will graduate to a finite-element course you will be confronted with a menagerie of other elements, membranes, plates, shells and the like. Once we have the stiffness matrices of all the elements of a structure the analysis is practically concluded.

14.8 Solved Example

In the example in Fig. 14.3 there are open degrees of freedom at node e only, for which we can write two equilibrium equations, $\mathbf{Qn} = \mathbf{p}$. The first equation (horizontal equilibrium) is $\cos 30°n_1 + \cos 45°n_2 + \cos 60°n_3 + \cos 90°n_4 = P$, and the second equation (vertical equilibrium) is $\sin 30°n_1 + \sin 45°n_2 + \sin 60°n_3 + \sin 90°n_4 = 0$, or in matrix form

$$\begin{bmatrix} \sqrt{3}/2 & \sqrt{2}/2 & 1/2 & 0 \\ 1/2 & \sqrt{2}/2 & \sqrt{3}/2 & 1 \end{bmatrix} \begin{Bmatrix} n_1 \\ n_2 \\ n_3 \\ n_4 \end{Bmatrix} = \begin{Bmatrix} P \\ 0 \end{Bmatrix} \tag{14.11}$$

The coefficients matrix in the above example is the equilibrium matrix $\mathbf{Q}$. The structure stiffness matrix $\mathbf{K} = \mathbf{QSQ}^T$ is thus

$$\mathbf{K} = \begin{bmatrix} \sqrt{3}/2 & \sqrt{2}/2 & 1/2 & 0 \\ 1/2 & \sqrt{2}/2 & \sqrt{3}/2 & 1 \end{bmatrix} \left(\frac{EA}{h}\right) \begin{bmatrix} 1/2 & 0 & 0 & 0 \\ 0 & \sqrt{2}/2 & 0 & 0 \\ 0 & 0 & \sqrt{3}/2 & 0 \\ 0 & 0 & 0 & 1 \end{bmatrix} \begin{bmatrix} \sqrt{3}/2 & 1/2 \\ \sqrt{2}/2 & \sqrt{2}/2 \\ 1/2 & \sqrt{3}/2 \\ 0 & 1 \end{bmatrix} \tag{14.12}$$

It is understood that (EA/h) belongs to the diagonal stiffnesses matrix $\mathbf{S}$. Executing the triple product yields the stiffness matrix

$$\boldsymbol{K} = \frac{EA}{h} \begin{bmatrix} 0.9451 & 0.9451 \\ 0.9451 & 2.1281 \end{bmatrix} \tag{14.13}$$

Clearly, the stiffness matrix is of the order $(2 \times 4)(4 \times 4)(4 \times 2) = 2 \times 2$.

We noted earlier that the usual method for getting the stiffness matrix is by assembling the matrix using the element stiffness matrices. The element stiffness matrices will here be built on the basis of the columns $\mathbf{q}_j$ of $\boldsymbol{Q}$. For instance,

$$\boldsymbol{K}_1 = s_1\mathbf{q}_1\mathbf{q}_1^T = \frac{1}{2}\frac{EA}{h}\left\{\begin{matrix}\sqrt{3}/2\\1/2\end{matrix}\right\}\left\{\sqrt{3}/2\ 1/2\right\} = (EA/h)\begin{bmatrix}0.375 & 0.2165\\0.2165 & 0.1250\end{bmatrix} \tag{14.14}$$

and similarly for $\mathbf{K}_2$, $\mathbf{K}_3$ and $\mathbf{K}_4$. The stiffness matrix of the entire structure is now 'assembled' by summing up the 4 element stiffness matrices

$$\mathbf{K} = \frac{EA}{h}\left(\begin{bmatrix}0.375 & 0.2165\\0.2165 & 0.1250\end{bmatrix} + \begin{bmatrix}0.3536 & 0.3536\\0.3536 & 0.3536\end{bmatrix} + \begin{bmatrix}0.2165 & 0.3750\\0.3750 & 0.6495\end{bmatrix} + \begin{bmatrix}0 & 0\\0 & 1\end{bmatrix}\right) \tag{14.15}$$

which reproduces the result in (14.13).

Having obtained the stiffness matrix by one method or the another, we can write the equilibrium equations $\mathbf{K}\mathbf{u} = \mathbf{p}$

$$(EA/h)\begin{bmatrix}0.9451 & 0.9451\\0.9451 & 2.1281\end{bmatrix}\left\{\begin{matrix}u_1\\u_2\end{matrix}\right\} = P\left\{\begin{matrix}1\\0\end{matrix}\right\} \tag{14.16}$$

the solution of which is $u_1 = 1.9034$, $u_2 = -0.8453$ in units Ph/EA.

The internal forces are now $\mathbf{n} = (\mathbf{S}\mathbf{Q}^T)\mathbf{u}$ or

$$\begin{Bmatrix}n_1\\n_2\\n_3\\n_4\end{Bmatrix} = \left(\frac{EA}{h}\right)\begin{bmatrix}1/2 & 0 & 0 & 0\\0 & \sqrt{2}/2 & 0 & 0\\0 & 0 & \sqrt{3}/2 & 0\\0 & 0 & 0 & 1\end{bmatrix}\begin{bmatrix}\sqrt{3}/2 & 1/2\\\sqrt{2}/2 & \sqrt{2}/2\\1/2 & \sqrt{3}/2\\0 & 1\end{bmatrix}\left(\frac{Ph}{EA}\right)\begin{Bmatrix}1.9034\\-0.8453\end{Bmatrix}$$
$$= P\begin{Bmatrix}0.6129\\0.5291\\0.1902\\-0.8453\end{Bmatrix} \tag{14.17}$$

which concludes the analysis of this truss by the stiffness (also called displacement) method.

All elements are in tension except for element *ea* which is in compression. A correct design of this structure would require a strut with a suitable moment of inertia along *ea* against possible buckling and slender elements, rods or even cables, for the inclined members (see Fig. 14.3—Design).

The structure being redundant the analysis depends on the relative stiffnesses of the elements, although the change may not be very significant in the present case. If we use EA for the stiffness of the cables, and say, $20EA$ for the strut the normal forces become $\mathbf{n} = \{0.4673\ 0.5291\ 0.4425\ \ -0.9909\}^T$. You can compare this with the previous results (14.17).

14.9 Summing Up

We have chosen to introduce the stiffness method by means of truss analysis. This was an opportunity to discuss this type of structure from a mechanical viewpoint. The emphasis, however, should be on the solved example because it shows the simplicity of the technique once the equations are established.

Chapter 15
Element Stiffness Matrices

In the stiffness or displacement method for analyzing a structure, we start by figuratively dissecting the structure into simple segments which are called elements. If we know the stiffness matrices of these elements the problem is by and large solved (usually by a computer). In this chapter we will mainly show how we can compute the stiffness matrices of the ubiquitous elements of skeletal structures: beams, rods and springs.

15.1 Typical Elements

A structure can be construed as being composed of recurrent constituents which we identify as elements. In Fig. 15.1 we show some classical structures and their typical elements: two-dimensional surface elements such as a shell element of a circular dome or a rectangular plate element, and one-dimensional line elements such a frame element in a plane frame, a beam element in a continuous beam, a truss element in a simple truss and a spring element in a beam on elastic supports.

Shell and plate elements are studied in finite-element-analysis courses. Here we cover structures that can be modeled as being composed of straight segments, to which we impart axial and flexural stiffness properties. The basic principles are the same for all elements and, in fact, there exists no better introduction to the finite-element method for general structures than studying the displacement method for frames and trusses.

In contrast to plates and shells, where elements must be very small,[1] line elements can be long. In fact, we will choose the smallest possible number of elements, although there are exceptions to the rule.

[1]Very small, but not infinitesimally so, hence the term 'finite'.

M.B. Fuchs, *Structures and Their Analysis*,
DOI 10.1007/978-3-319-31081-7_15

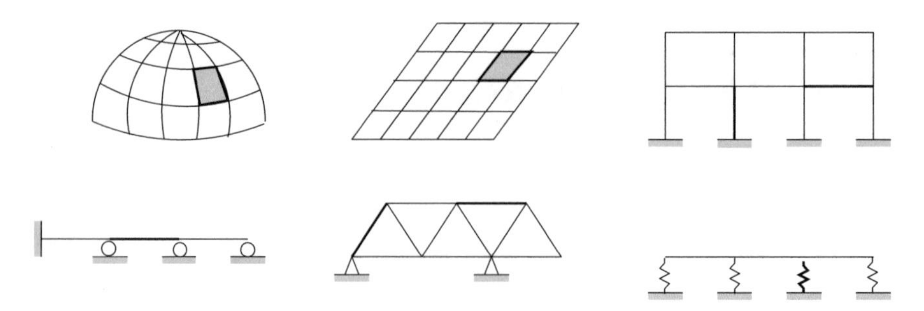

Fig. 15.1 Typical surface and line elements

The modeling is done in the following manner. We position nodes at the junctions, corners and supports of the structure AND at the free extremities. The elements are the lines which connect two consecutive nodes along the structure. A structure is thus the sum of its elements.

For the next step we need to know the stiffness matrix of the elements.

15.2 Element Stiffness Matrix

Consider the element in Fig. 15.2a. Although it is drawn with a geometrically uniform cross-section, the element can be quite general. The cross-section and the material properties can be variable and, for what follows, the axis of the element need not be straight. We now define at each extremity of the element their three degrees of freedom, and we choose to number them as in the sequence shown in Fig. 15.2a. We will see in the sequel that this sequence allocates degrees of freedom 1–4 to the *flexural* mode and degrees of freedom 5–6 to the *extensional* mode, which is helpful for memorizing the stiffness matrix of this type of element. We will have more on this later on.

We define displacements $\boldsymbol{\delta}$ at the six degrees of freedom and corresponding applied forces $\mathbf{f}$. We will now try to answer the following question: Suppose we want to

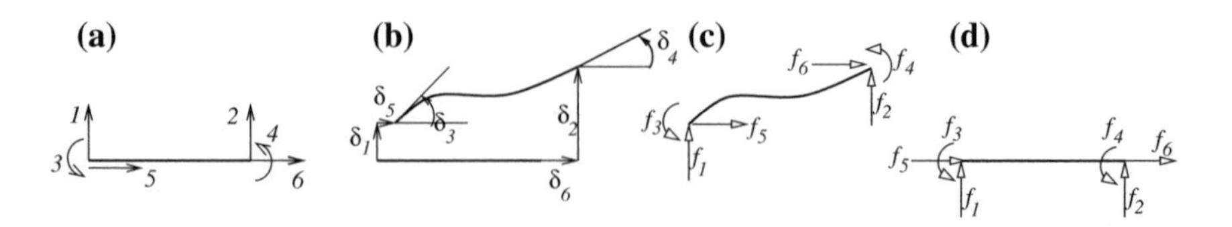

Fig. 15.2 Element degrees of freedom, end-displacements and end-loads

impart displacements $\boldsymbol{\delta}$ at the degrees of freedom, what forces $\mathbf{f}$ must be applied at the extremities in order to produce such displacements?[2]

To visualize the question, imagine a muscular athlete in an itinerant circus holding a steel rod at both ends. A volunteer from the public may ask him to displace the rod horizontally, lift it vertically and rotate it, at the left extremity, and impart some other displacements at the right extremity, as shown in an exaggerated manner in Fig. 15.2b. The question is what forces (Fig. 15.2c) will the athlete apply to produce these displacements. In other words, I give you the displacements $\boldsymbol{\delta}$, you tell me the forces $\mathbf{f}$.

The relation between the forces and the displacements will be assumed to be linear as we are dealing with a linear theory (the displacements must be small enough to warrant linearity). In other words, the relation between $\mathbf{f}$ and $\boldsymbol{\delta}$ can be written as

$$\mathbf{f} = \mathbf{K}\,\boldsymbol{\delta} \quad \text{with} \quad \mathbf{K} = \begin{bmatrix} K_{11} & K_{12} & K_{13} & K_{14} & K_{15} & K_{16} \\ K_{21} & K_{22} & K_{23} & K_{24} & K_{25} & K_{26} \\ K_{31} & K_{32} & K_{33} & K_{34} & K_{35} & K_{36} \\ K_{41} & K_{42} & K_{43} & K_{44} & K_{45} & K_{46} \\ K_{51} & K_{52} & K_{53} & K_{54} & K_{55} & K_{56} \\ K_{61} & K_{62} & K_{63} & K_{64} & K_{65} & K_{66} \end{bmatrix} \tag{15.1}$$

where the 6×6 coefficient matrix is called the stiffness matrix of the element. Note the stiffness relation is not an equation. It must be read from right to left, $\mathbf{f} \leftarrow \mathbf{K}\,\boldsymbol{\delta}$, the product $\mathbf{K}\,\boldsymbol{\delta}$ gives the forces. Consequently, the stiffness matrix of the element harbors the answer regarding the forces to be applied. We pre-multiply the displacements by the element stiffness matrix and get the forces.

Clearly, the element stiffness matrix depends on the properties of the element. The two end-couples needed to curve a long, slender aluminum bar are manifestly smaller than those required to bend a short, bulky steel bar. But before we compute some specific element stiffness matrices, what can be said, in general, regarding element stiffness matrices?

15.3 Some Properties of K

15.3.1 Physical Interpretation of the Coefficients of K

First, we will rewrite $\mathbf{K}\,\boldsymbol{\delta}$ as a sum of vectors. Matrix $\mathbf{K}$ has six columns, each of which can be seen as a vector. We call vector $\mathbf{k}_j$ the j-th column of $\mathbf{K}$. The entries of this vector are thus $K_{1j}\,K_{2j} \ldots K_{6j}$. Now the matrix relation $\mathbf{f} = \mathbf{K}\,\boldsymbol{\delta}$ can also be written as

[2]We are following the very didactic description given in WC Hurty and MF Rubinstein, *Dynamics of Structures*, Englewood Cliffs, NJ: Prentice-Hall, 1964.

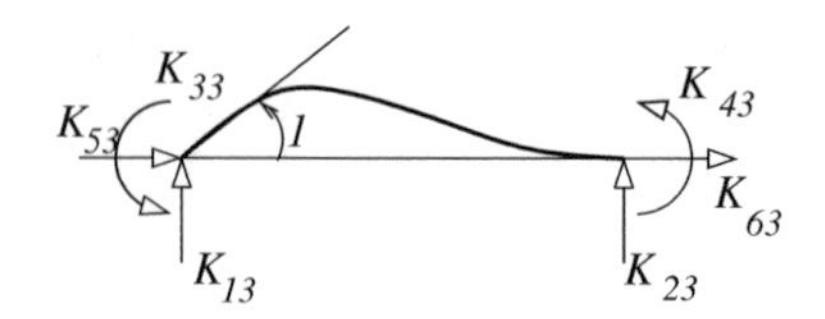

Fig. 15.3 Physical interpretation of column 3 of element stiffness matrix

$$\mathbf{f} = \sum_{i=1}^{6} \delta_i \, \mathbf{k}_i \tag{15.2}$$

This follows from the definition of matrix multiplication.

Now, let us assume that we keep all the end displacements of the element, fixed, except for displacement j which is displaced by a unit value, and we want to know what the forces to be applied are. To do so we set all $\delta_i = 0$ when $i \neq j$ and $\delta_j = 1$, which when introduced in (15.2) gives

$$\mathbf{f} = \mathbf{k}_j \equiv \{K_{1j}\; K_{2j}\; K_{3j}\; K_{4j}\; K_{5j}\; K_{6j}\}^T \tag{15.3}$$

In other words, the coefficients of a column j of $\mathbf{K}$ are the six forces to be applied to the element in order to keep all end-displacements zero, except for displacement j which will have a unit value. (See Fig. 15.3 for a unit rotation at the left extremity, $\delta_3 = 1$, all the remaining degrees of freedom being zero.)

15.3.2 The Stiffness Matrix Is Symmetric

It is easy to show that an element stiffness matrix is symmetric, that is, $K_{ij} = K_{ji}$ for any i, j. We will call $\boldsymbol{\delta}^j$ the vector of end-displacements which are zero everywhere, except for a unit displacement at degree of freedom j. Clearly, $\mathbf{k}_j = \mathbf{K}\,\boldsymbol{\delta}^j$ where we recall that $\mathbf{k}_j$ is the j-th column of $\mathbf{K}$. We have a same equation for index i, $\mathbf{k}_i = \mathbf{K}\,\boldsymbol{\delta}^i$.

Now, the pairs $(\mathbf{k}_j,\ \boldsymbol{\delta}^j)$ and $(\mathbf{k}_i,\ \boldsymbol{\delta}^i)$ are two sets of real forces and corresponding real displacements applied to a same real structure (the element). We can therefore use Betti's theorem (multiplying the forces of one system by the displacements of the second).

$$\mathbf{k}_j^T \,\delta^i = \mathbf{k}_i^T \,\delta^j \qquad \text{This is in fact} \quad K_{ij} = K_{ji} \tag{15.4}$$

which establishes the symmetry of $\mathbf{K}$

$$\mathbf{K} = \mathbf{K}^T \tag{15.5}$$

In Fig. 15.4, we give an example with $j = 2$ and $i = 3$. To the left we have the forces to be applied to the element in order to obtain a unit displacement at 2 and zero elsewhere, and to the right we give the forces for a unit rotation at 3 and zero

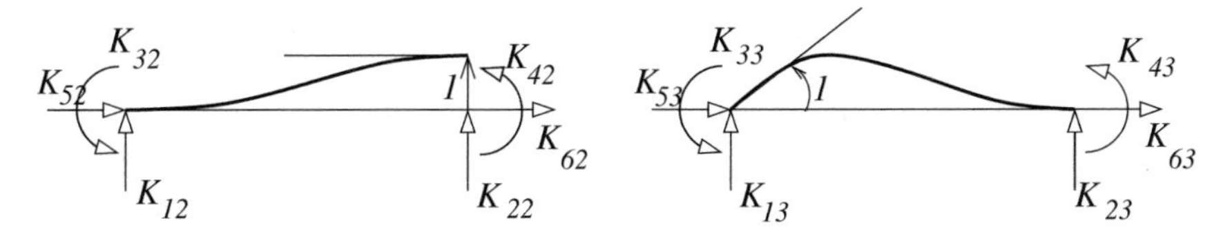

Fig. 15.4 $K_{23} = K_{32}$

displacements at the other degrees of freedom. Betti's theorem gives $K_{12} \times 0 + K_{22} \times 0 + \underline{K_{32} \times 1} + K_{42} \times 0 + K_{52} \times 0 + K_{62} \times 0 = K_{13} \times 0 + \underline{K_{23} \times 1} + K_{33} \times 0 + K_{43} \times 0 + K_{53} \times 0 + K_{63} \times 0$ or $K_{32} = K_{23}$.

Consequently, out of the 36 coefficients of **K** the 15 coefficients below the diagonal are the same as those above the diagonal. We are left with $36 - 15 = 21$ independent coefficients. We will see that there are in fact only four independent coefficients.

15.3.3 Relations Between Rows/Columns

Relations Between Rows

Whatever the displacements imposed to the two extremities, the forces **f** that will be applied to the element will be auto-equilibrated. In other words, we need to have horizontal and vertical translational equilibrium and rotational equilibrium of the element (see Fig. 15.2d)

$$\begin{aligned} f_1 + f_2 &= 0 \\ f_5 + f_6 &= 0 \\ f_3 + f_4 + L f_2 &= 0 \qquad \text{or} \qquad f_3 + f_4 - L f_1 = 0 \end{aligned} \tag{15.6}$$

If $\mathbf{r}_i$ is row i of the element stiffness matrix, then we will note that the load that is applied along degree of freedom i is in matrix notation, $f_i = \mathbf{r}_i^T \boldsymbol{\delta}$. As a result $f_1 + f_2 = 0$ becomes $\mathbf{r}_1^T \boldsymbol{\delta} + \mathbf{r}_2^T \boldsymbol{\delta} = (\mathbf{r}_1^T + \mathbf{r}_2^T)\, \boldsymbol{\delta} = (\mathbf{r}_1 + \mathbf{r}_2)^T \boldsymbol{\delta} = 0$. The three equations in (15.6) can be written

$$\begin{aligned} (\mathbf{r}_1 + \mathbf{r}_2)^T \boldsymbol{\delta} &= 0 \\ (\mathbf{r}_5 + \mathbf{r}_6)^T \boldsymbol{\delta} &= 0 \\ (\mathbf{r}_3 + \mathbf{r}_4 + L\,\mathbf{r}_2)^T \boldsymbol{\delta} &= 0 \end{aligned} \tag{15.7}$$

In every line we have a scalar product of two vectors, the term in parenthesis and $\boldsymbol{\delta}$; and since the latter is an arbitrary vector (any displacements) the terms in parenthesis must be zero. In other words, we have the following three relations between the rows of the element stiffness matrix

$$\mathbf{r}_2 = -\mathbf{r}_1\,, \quad \mathbf{r}_6 = -\mathbf{r}_5 \quad \text{and} \quad \mathbf{r}_2 = -(\mathbf{r}_3 + \mathbf{r}_4)/L \tag{15.8}$$

The first equation, for instance, stipulates that every coefficient in the second row is equal to minus the corresponding coefficient in the first row of the element stiffness matrix.

Relations Between Columns

Due to the symmetry of **K**, we have similar relations between the columns of the matrix.

$$\mathbf{k}_2 = -\mathbf{k}_1\,, \quad \mathbf{k}_6 = -\mathbf{k}_5 \quad \text{and} \quad \mathbf{k}_2 = -(\mathbf{k}_3 + \mathbf{k}_4)/L \tag{15.9}$$

where we remember that k_i is column i of **K**.

Rigid-Body End-Displacements

Based on (15.9) we can show that

rigid-body displacements need no end-forces.

Indeed, let us impose a rigid-body motion to the element, as shown in Fig. 15.5, where u and v are respectively the horizontal and vertical components of an arbitrary translation of the left extremity and θ is an arbitrary rotation. The end displacements of such a rigid-body movement are, of course,

$$\begin{Bmatrix} \delta_1 \\ \delta_2 \\ \delta_3 \\ \delta_4 \\ \delta_5 \\ \delta_6 \end{Bmatrix} = \begin{Bmatrix} v \\ v + L\,\sin\theta \\ \theta \\ \theta \\ u \\ u + L\,(1-\cos\theta) \end{Bmatrix}, \quad \text{or} \quad \boldsymbol{\delta} = \begin{bmatrix} 0 & 1 & 0 \\ 0 & 1 & L \\ 0 & 0 & 1 \\ 0 & 0 & 1 \\ 1 & 0 & 0 \\ 1 & 0 & 0 \end{bmatrix} \begin{Bmatrix} u \\ v \\ \theta \end{Bmatrix} \tag{15.10}$$

In the second matrix equation, we have taken into account small displacements and rotations ($\sin\theta = \theta$, $\cos\theta = 1$).

We now introduce this result into the stiffness relation (15.1)

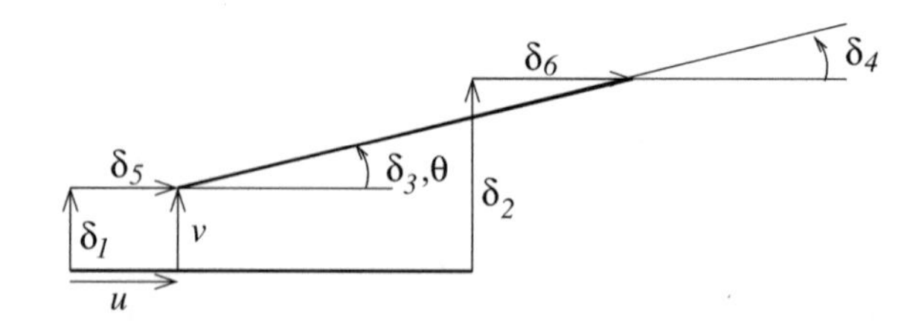

Fig. 15.5 Rigid-body displacement of element

$$\mathbf{f} = \begin{bmatrix} \mathbf{k}_1 & \mathbf{k}_2 & \mathbf{k}_3 & \mathbf{k}_4 & \mathbf{k}_5 & \mathbf{k}_6 \end{bmatrix} \begin{bmatrix} 0 & 1 & 0 \\ 0 & 1 & L \\ 0 & 0 & 1 \\ 0 & 0 & 1 \\ 1 & 0 & 0 \\ 1 & 0 & 0 \end{bmatrix} \begin{Bmatrix} u \\ v \\ \theta \end{Bmatrix} \qquad (15.11)$$

where the first matrix is the element stiffness matrix in terms of its columns $\mathbf{k}_i$ (column i of the matrix). This equation is also $\mathbf{f} = (\mathbf{k}_5 + \mathbf{k}_6)\, u + (\mathbf{k}_1 + \mathbf{k}_2)\, v + (L\,\mathbf{k}_2 + \mathbf{k}_3 + \mathbf{k}_4)\, \theta$ which with (15.9) gives $\mathbf{f} = \mathbf{0}$ for any rigid-body displacement u, v, and θ.

Displacing a free-floating element as a rigid-body indeed requires no forces (remember, the element is weightless). Only when *deforming* something do we need to apply forces. The element stiffness matrix incorporates this property.

15.3.4 An Element Stiffness Matrix Is Singular

We have just seen that there exist three linear relations between the columns of **K** and three more between the rows. These are six excellent reasons for the determinant of an element stiffness matrix $|\mathbf{K}_e|$ to be zero (one relation is enough for a zero determinant). Consequently, the inverse of the element stiffness matrix does not exist, and luckily so. Imagine the havock resulting from an inverse relation $\boldsymbol{\delta} = \mathbf{K}^{-1}\,\mathbf{f}$. If such a relation could be written, it would give us the end displacements of a floating element for any arbitrary loading, including $\mathbf{f} = \{0\ 0\ C\ C\ 0\ 0\}$ or $\mathbf{f} = \{0\ -Q\ 0\ 0\ 0\ 0\}$, for instance. A glance at the two examples in Fig. 15.6 will convince you that when we ask a silly question we are bound to get a silly answer. The body will at best meander like a leaf in the wind. In short, an element stiffness matrix has (luckily) no inverse.

15.3.5 Orthogonality of the Flexural and Extensional Modes

Consider an element in outer space which was deformed by imparting arbitrary displacements at the six degrees of freedom, three at each extremity (see Fig. 15.7). These six displacements are the components of the element nodal displacements vector $\boldsymbol{\delta}$.

Fig. 15.6 An floating element cannot have a flexibility matrix

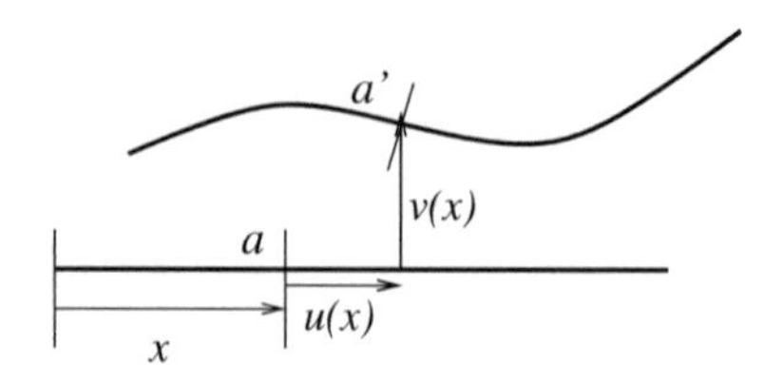

Fig. 15.7 Arbitrary displacement of an element

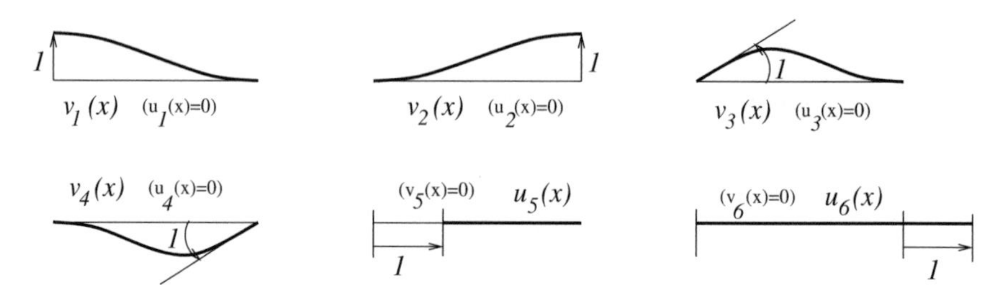

Fig. 15.8 The six elementary displacement modes

The displaced shape can be described by the horizontal components $u(x)$ and the vertical components $v(x)$ of the displacements of points x, where x is a local coordinate along the axis of the element. In Fig. 15.7, we have drawn an exaggerated version of such a displacement (remember, the deformations must remain small). Section a at x, for instance, moved horizontally by $u(x)$, vertically by $v(x)$ and rotated such that the lateral cross-section remains perpendicular to the slope, of the beam.[3]

We are interested in the displaced shapes resulting from six special vectors $\boldsymbol{\delta}^i$ which have zeros everywhere except for a value 1 at position i (see Fig. 15.8). We call $u_i(x)$ and $v_i(x)$ the horizontal and vertical displacement components due to vector $\boldsymbol{\delta}^i$. For nodal displacements $\boldsymbol{\delta}^3 = \{0\,0\,1\,0\,0\,0\}$, for instance, we get the displacement $u_3(x)$ and $v_3(x)$, and for $\boldsymbol{\delta}^5 = \{0\,0\,0\,0\,1\,0\}$, we obtain $u_5(x)$, and $v_5(x)$. We would like to draw your attention to $u_{1-4}(x) = 0$ and $v_{5,6}(x) = 0$. In other words, when we apply nodal values $\boldsymbol{\delta}^1$ to $\boldsymbol{\delta}^4$, we get lateral displacements but no axial displacements.[4] Conversely, for $\boldsymbol{\delta}^5$ and $\boldsymbol{\delta}^6$ there are no lateral displacements, only axial ones. As can clearly be seen in Fig. 15.8, unit displacements at $\boldsymbol{\delta}^1$ to $\boldsymbol{\delta}^4$ induce the flexural mode, whereas $\boldsymbol{\delta}^5$ and $\boldsymbol{\delta}^6$ are related to the extensional mode.

There is an equilibrium counterpart for this. Applying loads along degrees of freedom 1–4 affects vertical and rotational equilibrium, but not axial equilibrium. Similarly, forces acting along degrees of freedom 5 and 6 are ruled by the axial equilibrium equations but not by the lateral and rotational equilibrium equations.

[3]Recall that for an Euler–Bernoulli beam the rotation of a cross-section is equal to the slope of the axis.

[4]When a beam curves it shortens, consequently there are horizontal displacements, but if the flexion is small enough the horizontal displacements can be neglected.

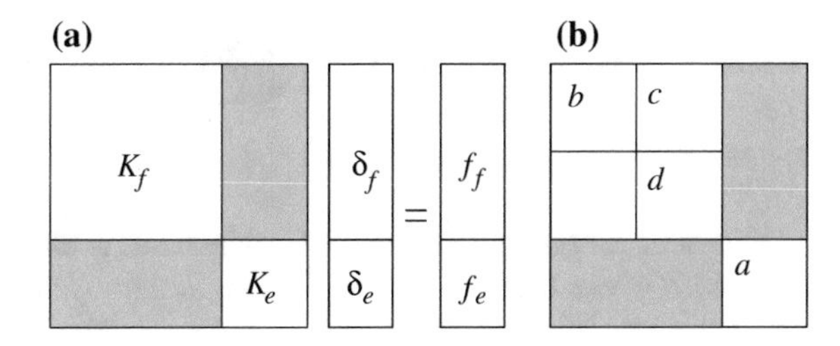

Fig. 15.9 'f'lexion and 'e'xtension submatrices

The concept of *orthogonality* summarizes all this. In the same way as two orthogonal vectors have a zero dot-product, the flexural and extensional modes are oblivious of one another.

Shortening an element does not bend it; bending an element does not shorten it.

This has far reaching consequences in structural theory. Foremost, the element stiffness matrix decomposes into a 4×4 flexural matrix $\mathbf{K}_f$ and a 2×2 extensional matrix $\mathbf{K}_e$ as shown in Fig. 15.9a. The displacements and forces divide likewise into the displacements and forces $\boldsymbol{\delta}_f$ and $\mathbf{f}_f$ related to flexion degrees of freedom 1, 2, 3 and 4, and the displacements and forces $\boldsymbol{\delta}_e$ and $\mathbf{f}_e$ of the extensional mode 5 and 6.

The first four equations of $\mathbf{K}\boldsymbol{\delta} = \mathbf{f}$ become $\mathbf{K}_f\boldsymbol{\delta}_f + \mathbf{0}^{(4\times2)}\boldsymbol{\delta}_e = \mathbf{f}_f$, and the last two equations are $\mathbf{0}^{(2\times4)}\boldsymbol{\delta}_f + \mathbf{K}_e\boldsymbol{\delta}_f = \mathbf{f}_e$, where the zero matrices have the indicated dimensions. As a consequence, $\mathbf{K}\boldsymbol{\delta} = \mathbf{f}$ decomposes into two independent relations: four equations for the flexural mode $\mathbf{K}_f\boldsymbol{\delta}_f = \mathbf{f}_f$ and two equations for the extensional mode $\mathbf{K}_e\boldsymbol{\delta}_e = \mathbf{f}_e$ or

$$[\mathbf{K}_f] \begin{Bmatrix} \delta_1 \\ \delta_2 \\ \delta_3 \\ \delta_4 \end{Bmatrix} = \begin{Bmatrix} f_1 \\ f_2 \\ f_3 \\ f_4 \end{Bmatrix} \quad \text{and} \quad [\mathbf{K}_e] \begin{Bmatrix} \delta_5 \\ \delta_6 \end{Bmatrix} = \begin{Bmatrix} f_5 \\ f_6 \end{Bmatrix} \tag{15.12}$$

where the 4×4 matrix $\mathbf{K}_f$ is the flexural element stiffness matrix and the 2×2 matrix $\mathbf{K}_e$ is the extensional element stiffness matrix. These matrices are also known as the element stiffness matrices in the flexural and extensional modes, respectively.

15.3.6 How Many Independent Coefficients Are There?

Out of the 36 coefficients of **K** how many are independent? The orthogonality of the flexural and extensional modes has reduced their number to $16 + 4 = 20$. The symmetry of the stiffness matrix entails $\mathbf{K}_f^T = \mathbf{K}_f$ and $\mathbf{K}_e^T = \mathbf{K}_e$, and the number of independent coefficients drops to $10 + 3 = 13$. This can be further reduced.

We can show that there are in total only four independent coefficients in **K**: one for the extensional matrix and three for the flexural matrix.

Fig. 15.10 Various extensional elements

Consider the *extensional* matrix $\mathbf{K}_e$ which has coefficients K_{55}, K_{56}, K_{65} and K_{66}. Suppose $K_{55} = a$ (see Fig. 15.9b). Since $\mathbf{r}_6 = -\mathbf{r}_5$ (this was an equilibrium requirement: the coefficients of the 6-th row are minus the corresponding coefficients of the 5-th row), we have $K_{65} = -K_{55} = -a$. Due to symmetry $K_{56} = K_{65} = -a$. And finally from $\mathbf{r}_5 = -\mathbf{r}_6$ again, we find $K_{66} = -K_{56} = a$. The sequence is shown below

$$\begin{bmatrix} a & \\ & \end{bmatrix} \rightarrow \begin{bmatrix} a & \\ -a & \end{bmatrix} \rightarrow \begin{bmatrix} a & -a \\ -a & \end{bmatrix} \rightarrow \begin{bmatrix} a & -a \\ -a & a \end{bmatrix}$$

Consequently, the axial matrix depends on one (stiffness) coefficient, whatever the element may look like or may be made of (Fig. 15.10)

$$\mathbf{K}_e = a \begin{bmatrix} 1 & -1 \\ -1 & 1 \end{bmatrix} \tag{15.13}$$

This stiffness coefficient depends on the shape and material of the element. We will calculate a for a uniform element further on.

For the *flexural* matrix we have two equilibrium conditions for the rows $\mathbf{r}_1 + \mathbf{r}_2 = 0$ and $\mathbf{r}_3 + \mathbf{r}_4 + L\,\mathbf{r}_2 = 0$, where L is the element length, in addition to symmetry. We leave it as an exercise for the reader to establish that, on the basis of $K_{11} = b$, $K_{13} = c$, and $K_{33} = d$, we can fill the flexural matrix. The flexural element stiffness matrix depends therefore on three constants.

Consequently, out the 36 coefficients of *any* (straight) element stiffness matrix we need to compute only $3 + 1 = 4$ constants.

15.4 Formula for a Stiffness Coefficient

We have seen earlier (15.4) that a stiffness coefficient can be expressed as $K_{ij} = \mathbf{k}_j^T \boldsymbol{\delta}^i$. This is almost a tautology. For instance,

$$K_{23} = \mathbf{k}_3^T \boldsymbol{\delta}^2 = \{K_{13}\, K_{23}\, K_{33}\, K_{43}\}^T \{0\ 1\ 0\ 0\}$$

Now let $\varepsilon_i(x)$ and $\kappa_i(x)$ be the deformations and $n_i(x)$ and $m_i(x)$ the internal forces in an element to which we have applied displacements $\boldsymbol{\delta}^i$ at the extremities. We will remember that the internal forces and deformations are related at any section x through Hooke's law, $n_i = EA\,\varepsilon_i$ and $m_i = EI\,\kappa_i$.

Clearly, the external forces $\mathbf{k}_i$ (at the extremities) and the internal forces $n_i(x)$ and $m_i(x)$ constitute an equilibrium system, and the end displacements $\boldsymbol{\delta}^i$ and the deformations $\varepsilon_i(x)$ and $\kappa_i(x)$ are a compatible system. We can repeat all this with end displacements $\boldsymbol{\delta}^j$, and generate the equilibrium system $\mathbf{k}_j$ (at the extremities) with $n_j(x)$ and $m_j(x)$, and the compatible system $\boldsymbol{\delta}^j$ with $\varepsilon_j(x)$ and $\kappa_j(x)$.

From the principle of virtual work we get $\mathbf{k}_j^T \boldsymbol{\delta}^i = \int_x (n_j\varepsilon_i + m_j\kappa_i)\,\mathrm{d}x$ which when using Hooke's law gives

$$K_{ij} = \int_x (\,EA\,\varepsilon_i\,\varepsilon_j + EI\,\kappa_i\,\kappa_j\,)\,\mathrm{d}x \tag{15.14}$$

where we have also replaced $\mathbf{k}_j^T \boldsymbol{\delta}^i$ with K_{ij}. We recall that the deformations $\varepsilon_i\ \kappa_i$ and $\varepsilon_j\ \kappa_j$ are due to unit displacements $\boldsymbol{\delta} = \boldsymbol{\delta}^i$ and $\boldsymbol{\delta} = \boldsymbol{\delta}^j$ at the nodes, respectively.

Note: This formula for a stiffness coefficient of an element is reminiscent of the formula for the flexibility coefficient of a structure

$$F_{ij} = \int_x \left(\frac{n_i\,n_j}{EA} + \frac{m_i\,m_j}{EI} \right) \mathrm{d}x$$

but it is very different. In the flexibility coefficient, $n_i\,n_j$ and $m_i\,m_j$ are the internal forces due to *unit external forces* applied at the nodes of the structure, whereas in (15.14) we have *unit displacements*.

15.5 Computing the Stiffness Matrix of Uniform Elements

We have seen in the previous section that the element stiffness matrix decomposes into two unconnected submatrices: a (4×4) bending element stiffness matrix with degrees of freedom 1–4 and a (2×2) extensional element stiffness matrix for degrees of freedom 5 and 6. In other words, two independent components are hidden inside a longitudinal element of a structure (frame): a flexural or beam element which carries bending stiffness (*EI*), and an extensional or rod element which has axial stiffness (*EA*) as shown in Fig. 15.11.

We will start with the beam element.

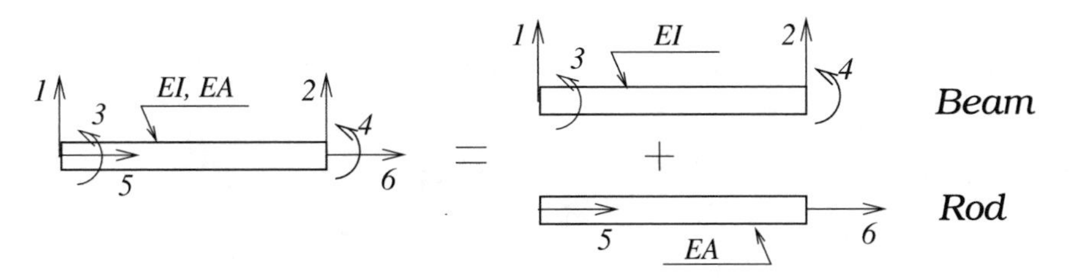

Fig. 15.11 A beam and rod element are hidden inside a frame element

15.5.1 Uniform Beam (Flexural) Element Matrix

We compute the stiffness matrix column by column, and as shown, columns 1 and 3 suffice for the bending matrix since they include constants b, c and d. Here is how we suggest to do it. The entries of column 3 of the stiffness matrix are the forces that must be applied at the degrees of freedom in order to obtain the end displacements $\delta = \delta^3$, that is $\{0\,0\,1\,0\}^T$.

Alternatively, consider the uniform beam in Fig. 15.12a simply supported at the left extremity, built in at the right end, and subjected to a couple C at the left support. It can be construed as an element with loads at the degrees of freedom, an applied couple at 3, and reactions at 1, 2 and 4. The element has thus end-loads $\{R_1\,R_2\,C\,R_4\}^T$ with end-displacements $\{0\,0\,\theta\,0\}^T$.

Now, if C was such that the rotation became $\theta = 1$, we would have the required boundary conditions, which means that the end forces become, by definition, the components of $\mathbf{k}_3$ (Fig. 15.12b). This is how we shall proceed for computing column 3 of the stiffness matrix.

The structure in Fig. 15.12a has redundancy $R = 1$ and can easily be analyzed by the force method. Its reactions and moment distribution are indicated in Fig. 15.12c. (It helps to remember that in the configuration of the figure the carry-over moment, that is, the moment at the wall is always half the applied moment at the hinged support, with an opposite sign.)

To compute the rotation θ at the left support, we apply there a unit couple and look for an equilibrium configuration. Here we conveniently use the exact equilibrium solution, normalized with C, since the couple is unitary. Consequently, $\theta = \int_0^L \kappa\, m^\theta \mathrm{d}x = \int_0^L (m/EI)(m/C)\,\mathrm{d}x = CL/4EI$. Setting $\theta = CL/4EI = 1$ gives $C = 4EI/L$, which is in effect K_{33}. The other reactions follow suit. Therefore, column (and also row) 3 of the stiffness matrix is as in the first matrix in (15.15) (see also Fig. 15.12d)

$$\frac{EI}{L^3}\begin{bmatrix} & & 6L & \\ & & -6L & \\ 6L & -6L & 4L^2 & 2L^2 \\ & & 2L^2 & \end{bmatrix} \qquad \frac{EI}{L^3}\begin{bmatrix} 12 & -12 & 6L & 6L \\ -12 & & -6L & \\ 6L & -6L & 4L^2 & 2L^2 \\ 6L & & 2L^2 & \end{bmatrix} \tag{15.15}$$

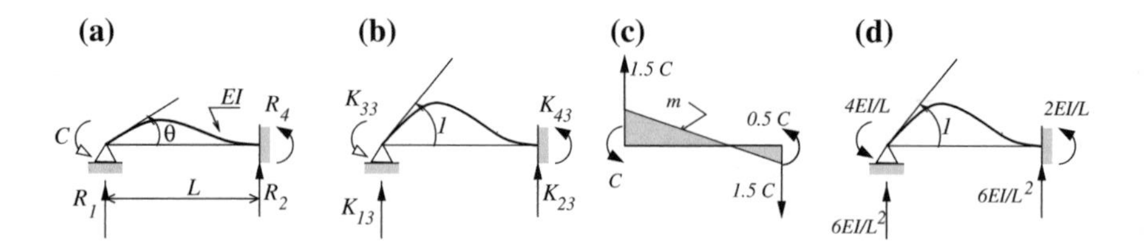

Fig. 15.12 Computing column 3 of beam stiffness matrix

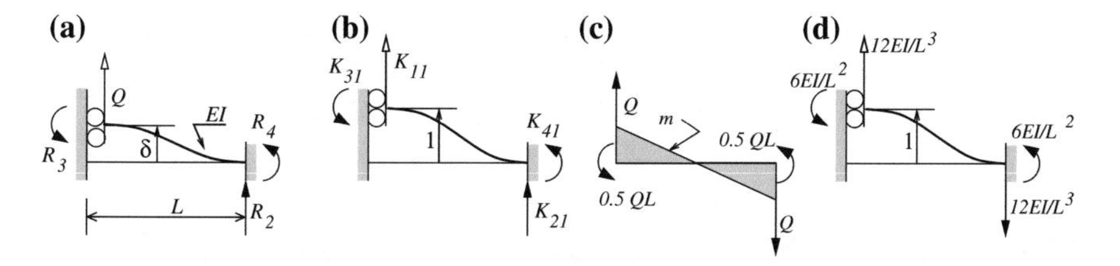

Fig. 15.13 Computing column 1 of beam stiffness matrix

It is instructive to note that the displaced shape in Fig. 15.12b, d is the function that was defined as $v_3(x)$, and its curvature and moment distribution are respectively $\kappa_3(x) = \mathrm{d}^2 v_3/\mathrm{d}x^2$ and $EI\,\kappa_3(x)$.

We can repeat the exercise for column 1. The structure in Fig. 15.13a has now a shear release at the left wall to allow for a vertical displacement at degree of freedom 1, but nothing more. When we apply a force Q at 1 the left end moves up by an amount δ, and we get reactions R_2–R_4 as shown in Fig. 15.13a. If the force Q is such that $\delta = 1$ then by definition $Q = K_{11}$ and the other reactions are K_{21}, K_{31} and K_{41} (Fig. 15.13b). To compute these coefficients, let us find the force Q which causes a unit displacement.

The analysis of the structure under a force Q at degree of freedom 1 is simple. The vertical reaction (at 2) is forcibly a down load Q. Now, the clockwise couple QL of the pair of opposite vertical forces is equilibrated by the moments at the walls. From the viewpoint of the beam, nothing distinguishes the left wall from the right wall. The beam could not care less whether the left support rolls up or whether the right wall is moving down. Consequently, the reactive couples must be equal and each take half of QL (anti-clockwise). The force and reactions at the degrees of freedom and the moment distribution m are indicated in Fig. 15.13c.

We next apply a unit force at degree of freedom 1 and look for an equilibrium configuration. As previously, we conveniently use the exact equilibrium solution, normalized with Q, since now the force is a unit force. Consequently, $\delta = \int_0^L \kappa\, m^\delta\, \mathrm{d}x = \int_0^L (m/EI)(m/Q)\, \mathrm{d}x = QL^3/12EI$. Setting $\delta = QL^3/12EI = 1$ gives $Q = 12EI/L^3$, which is in fact K_{11}.

The final drawing, Fig. 15.13d, shows that the end-forces are the components of row/column 1 of the stiffness matrix as indicated in the second matrix in (15.15).

To fill in the missing coefficients, we remember that $\mathbf{k}_2 = -\mathbf{k}_1$ (and similarly for the rows). Finally, to comply with rotational equilibrium, we have $K_{44} = -(K_{34} + L\,K_{24})$. This completes the computation of $\mathbf{K}_f$

$$\mathbf{K}_f = \frac{EI}{L^3}\begin{bmatrix} 12 & -12 & 6L & 6L \\ -12 & 12 & -6L & -6L \\ 6L & -6L & 4L^2 & 2L^2 \\ 6L & -6L & 2L^2 & 4L^2 \end{bmatrix} \tag{15.16}$$

This is the element stiffness matrix of a uniform straight beam of length L and flexural stiffness EI. We will encounter several other elements with different stiffness matrices. But the rest is for the reader to develop or to find in catalogues.

15.5.2 Uniform Rod (Extensional) Element Matrix

After the flexural matrix, the stiffness matrix of a uniform straight extensional element (15.13) can be obtained in no time. All we need to determine is one constant, whatever the shape of the element, and for a uniform element it is simple.

When we return to basic principles, we find that when $\delta_5 = 0$ and $\delta_6 = 1$, the forces to be applied at the degrees of freedom are, by definition,

$$\begin{Bmatrix} f_5 \\ f_6 \end{Bmatrix} = \begin{Bmatrix} K_{56} \\ K_{66} \end{Bmatrix} \tag{15.17}$$

as shown in Fig. 15.14a. For a uniform element of axial stiffness EA and of length L, when we apply a force P at 6 (see Fig. 15.14b) the element elongates by $\delta = PL/EA$, in other words, $P = (EA/L)\,\delta$. For a unit displacement ($\delta = 1$) the force is called K_{66} and we find $K_{66} = EA/L$. The uniform extensional stiffness matrix is therefore

$$\mathbf{K}_e = \frac{EA}{L}\begin{bmatrix} 1 & -1 \\ -1 & 1 \end{bmatrix} \tag{15.18}$$

Voila!

15.5.3 Extensional and Rotational Spring Element Matrices

A natural extension of the rod element is the linear spring element of stiffness k. Hooke's law for a rod element (see Fig. 15.15a) is $P = (EA/L)\,\delta$, where P is the applied force at the extremities and δ is the total elongation. For a linear spring element (see Fig. 15.15b) we have similarly $P = k\,\delta$, where k is the stiffness

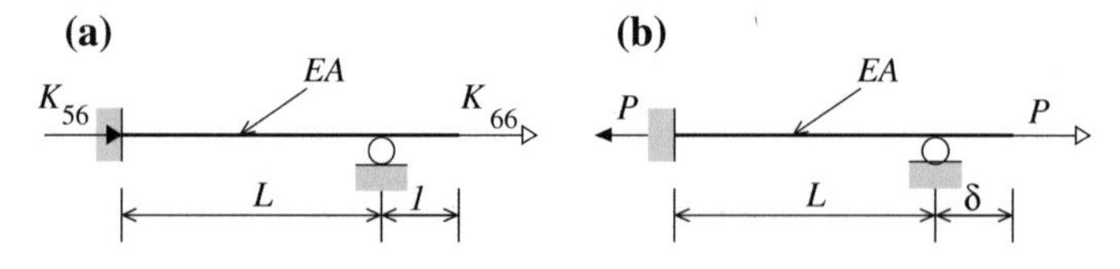

Fig. 15.14 Computing the axial stiffness

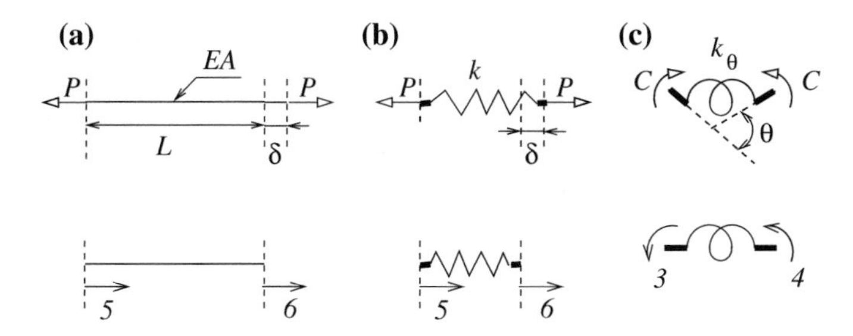

Fig. 15.15 Longitudinal and rotational spring elements

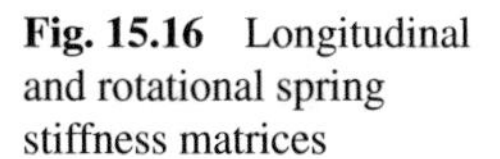

Fig. 15.16 Longitudinal and rotational spring stiffness matrices

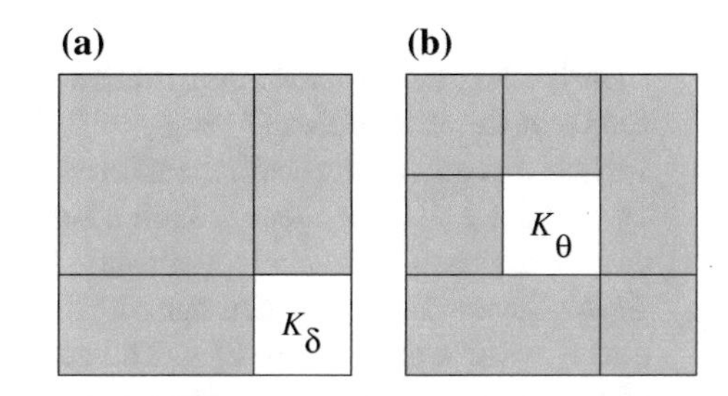

coefficient of the spring. Consequently, the linear spring stiffness matrix is the first expression in (15.19).

$$\mathbf{K}_\delta = k \begin{bmatrix} 1 & -1 \\ -1 & 1 \end{bmatrix} \qquad \mathbf{K}_\theta = k_\theta \begin{bmatrix} 1 & -1 \\ -1 & 1 \end{bmatrix} \tag{15.19}$$

Similarly, for the rotational spring $C = k_\theta \theta$ in Fig. 15.15c) of rotational stiffness k_θ the stiffness matrix is the second expression in (15.19).

Note, such elements are 'point' elements and can be understood to take up no or little space. Also, they have stiffness in their own mode but are infinitely flexible in other modes. You will have noticed that the axial spring element is connected to element coordinates 5 and 6, whereas the rotational spring element is connected to 3 and 4. Having no stiffness beyond these modes, their nominal element stiffness matrices (three *dofs* at each end) are shown in Fig. 15.16, where the shaded regions represent zero sub-matrices.

15.6 Illustrated Examples

Now that we have the stiffness matrix of a uniform flexural element, we have most probably forgotten why we set out to compute it. We recall correctly that the element stiffness matrix gives us the forces $\mathbf{f}$ to apply to the extremities of the element for a given set of end-displacements $\boldsymbol{\delta}$ using $\mathbf{K}\,\boldsymbol{\delta} \rightarrow \mathbf{f}$.

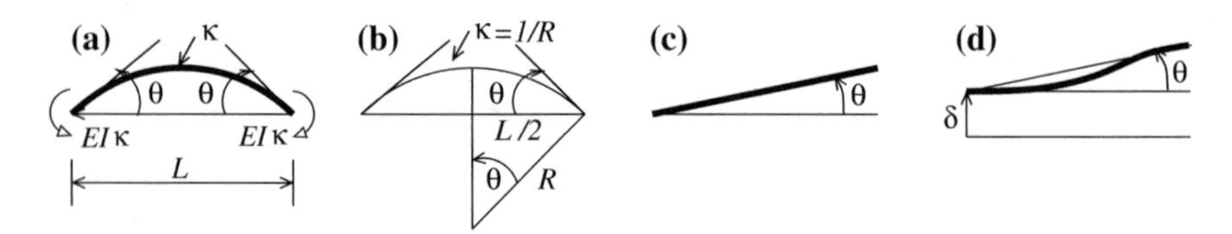

Fig. 15.17 Some trivial beam displacement patterns

1. We start with a case for which we should know the answer, such as what are the end-loads corresponding to applied equal and opposite rotations θ, that is, displacements $\boldsymbol{\delta} = \{0\,0\,\theta\; -\theta\}$ as in Fig. 15.17a?
   Clearly, the solution must be symmetric. Therefore, there cannot be any shear forces at the extremities ($f_1 = f_2 = 0$) because, if there were, we would have vertical disequilibrium (both shear forces must point in the same direction because of symmetry). Consequently, such a segment can have only equal and opposite end-moments. This leads to a constant bending moment along the element, say $EI\,\kappa$, where κ is the constant curvature of the element. The end couples are thus also $f_3 = EI\,\kappa$ and $f_4 = -EI\,\kappa$. The curvature can easily be computed in terms of θ. In Fig. 15.17b we have $\sin\theta = (L/2)/R$ where R is the radius (also radius of curvature) of the circular segment, and since θ is small ($\sin\theta \cong \theta$), then $\theta = (L/2)/R$ or $\kappa = 2\theta/L$. As a result the end moments are $f_3 = -f_4 = (2EI/L)\,\theta$.
   Now let us see if our element stiffness matrix provides the same answer. The basic equation $\mathbf{f} = \mathbf{K}\,\boldsymbol{\delta}$ yields for the present displacements $\mathbf{f} = 0\,\mathbf{k}_1 + 0\,\mathbf{k}_2 + \theta\,\mathbf{k}_3 - \theta\,\mathbf{k}_4$ or

$$\begin{Bmatrix} f_1 \\ f_2 \\ f_3 \\ f_4 \end{Bmatrix} = \frac{EI}{L^3}\left(\theta \begin{Bmatrix} 6L \\ -6L \\ 4L^2 \\ 2L^2 \end{Bmatrix} - \theta \begin{Bmatrix} 6L \\ -6L \\ 2L^2 \\ 4L^2 \end{Bmatrix}\right) = \theta\,\frac{EI}{L}\begin{Bmatrix} 0 \\ 0 \\ 2 \\ -2 \end{Bmatrix}$$

   which is indeed the expected result.
2. The second example also has an obvious answer. What are the end loads corresponding to a rigid body rotation as in Fig. 15.17c?
   Obviously, zero (no deformation, *ergo* no forces). The end-displacements corresponding to the rigid-body rotation are $\boldsymbol{\delta} = \{0\;\; L\theta\;\; \theta\;\; \theta\}^T$. Indeed, the vertical displacement at the right extremity δ_2 is $L\,\sin\theta \cong L\,\theta$. With these values the element equilibrium equations $\mathbf{f} = L\,\theta\mathbf{k}_2 + \theta\,\mathbf{k}_3 + \theta\,\mathbf{k}_4$ become

$$\begin{Bmatrix} f_1 \\ f_2 \\ f_3 \\ f_4 \end{Bmatrix} = \frac{EI}{L^3}\left(L\,\theta \begin{Bmatrix} -12 \\ 12 \\ -6L \\ -6L \end{Bmatrix} + \theta \begin{Bmatrix} 6L \\ -6L \\ 4L^2 \\ 2L^2 \end{Bmatrix} + \theta \begin{Bmatrix} 6L \\ -6L \\ 2L^2 \\ 4L^2 \end{Bmatrix}\right) = \begin{Bmatrix} 0 \\ 0 \\ 0 \\ 0 \end{Bmatrix}$$

3. In the last example, consider displacements $\boldsymbol{\delta} = \{\delta\ (\delta + L\theta)\ 0\ \theta\}$ shown in Fig. 15.17d. Pre-multiplying these displacements by the element stiffness matrix gives $\mathbf{f} = \delta\,\mathbf{k}_1 + (\delta + L\theta)\,\mathbf{k}_2 + \theta\,\mathbf{k}_4$ and also $\mathbf{f} = \delta\,(\mathbf{k}_1 + \mathbf{k}_2) + \theta\,(L\,\mathbf{k}_2 + \mathbf{k}_4)$. Noting that the term in the first parenthesis is zero we find

$$\begin{Bmatrix} f_1 \\ f_2 \\ f_3 \\ f_4 \end{Bmatrix} = \frac{EI}{L^3}\theta \begin{Bmatrix} -12L + 6L \\ 12L - 6L \\ -6L^2 + 2L^2 \\ -6L^2 + 4L^2 \end{Bmatrix} = \frac{EI}{L^3}\theta \begin{Bmatrix} -6L \\ 6L \\ -4L^2 \\ -2L^2 \end{Bmatrix}$$

The nodal forces turn out to be $\mathbf{f} = -\theta\,\mathbf{k}_3$. Now why is that?
With hindsight, we notice that the nodal displacements reduce to a rigid-body translation and rotation (which require no forces) plus a rotation $-\theta$ at degree of freedom 3, for which, by definition, we need $-\theta$ times the components of column 3 of the stiffness matrix.

15.7 Element Stiffness Matrices by Shape Functions

We can raise objections regarding the displacement method, since it seems to require prior knowledge of the force method in order to compute element stiffness matrices. Granted, this may cast a shadow over the stiffness method. But once the stiffness matrices of all the elements composing the structure are known, the displacement method takes off in such a spectacular fashion, that it outweighs the small disadvantage concerning the use of the force method in determining the element stiffness matrices.

In truth, the standard way for producing stiffness matrices does not require the analysis of redundant structures by the flexibility method. Instead, the technique rests on the shape functions or interpolation functions shown in Fig. 15.8.

We will now describe this method for determining the element stiffness matrices, starting with the flexural matrix $\mathbf{K}_f$.

15.7.1 The Flexural Matrix

The four *shape* functions, also called *interpolation* functions, related to the flexural matrix are $v_i(x), i = 1, 2, 3, 4$, where $v_i(x)$ is the displaced shape of the beam element when the displacement along degree of freedom i is given a unit value, $\delta_i = 1$ and the remaining three degrees of freedom are kept zero, $\delta_{j \neq i} = 0$. Following from superposition, the displaced shape of the beam $v(x)$ under arbitrary end-displacements is thus

$$v(x) = \sum_{i=1}^{4} \delta_i \, v_i(x) = \boldsymbol{\delta}^T \mathbf{v}(x) \tag{15.20}$$

where $\mathbf{v}(x)$ is the 4×1 vector function of components $v_i(x)$.

We will compute these interpolation functions for a uniform element of constant EI and length L. We return to the differential equilibrium equations of a infinitesimal lamella $\mathrm{d}x$ of an element: $m''(x) + q(x) = 0$. Hooke's generalized law is $m(x) = EI(x)\,\kappa(x)$, and the curvature for small displacements is $\kappa(x) = v''(x)$, consequently the equilibrium equation in terms of the displacement is $[EI\, v''(x)]'' + q(x) = 0$.

Since the uniform element has no distributed loads $q(x) = 0$ and the flexural stiffness $EI(x)$ is constant, we get the differential equation

$$v''''(x) = 0 \tag{15.21}$$

the general solution of which is $v(x) = c_3\, x^3 + c_2\, x^2 + c_1\, x + c_0$. Noting that $v'(x) = 3\, c_3\, x^2 + 2\, c_2\, x + c_1$ we have upon introduction of the boundary conditions

$$\begin{Bmatrix} \delta_1 \\ \delta_3 \\ \delta_2 \\ \delta_4 \end{Bmatrix} = \begin{Bmatrix} v(0) \\ v'(0) \\ v(L) \\ v'(L) \end{Bmatrix} \quad \text{or} \quad \begin{Bmatrix} \delta_1 \\ \delta_3 \\ \delta_2 \\ \delta_4 \end{Bmatrix} = \begin{bmatrix} 1 & 0 & 0 & 0 \\ 0 & 1 & 0 & 0 \\ 0 & 0 & L^2 & L^3 \\ 0 & 0 & 2L & 3L^2 \end{bmatrix} \begin{Bmatrix} c_0 \\ c_1 \\ c_2 \\ c_3 \end{Bmatrix} \tag{15.22}$$

The solution of this system yields the integration constants which, when introduced in the general solution, produces the equation of the displaced beam

$$v(x) = [(\delta_4+\delta_3)L-2(\delta_2-\delta_1)]\,\bar{x}^3+[3(\delta_2-\delta_1)-(2\delta_3+\Delta_4)L]\,\bar{x}^2+\delta_3\,\bar{x}+\delta_1 \tag{15.23}$$

where

$$\bar{x} = x/L\,, \quad 0 \le \bar{x} \le 1$$

is a non-dimensional coordinate along the beam axis. Finally, we can rearrange this equation and group the terms by nodal displacements

$$v(x) = \boldsymbol{\delta}^T \mathbf{v}(x) = \boldsymbol{\delta}^T \boldsymbol{\phi}(x) = \begin{Bmatrix} \delta_1 \\ \delta_2 \\ \delta_3 \\ \delta_4 \end{Bmatrix}^T \begin{Bmatrix} 1 - 3\bar{x}^2 + 2\bar{x}^3 \\ 3\bar{x}^2 - 2\bar{x}^3 \\ (\bar{x} - 2\bar{x}^2 + \bar{x}^3)L \\ (-\bar{x}^2 + \bar{x}^3)L \end{Bmatrix} \tag{15.24}$$

We have now the shape functions of the beam $v_i(x)$ under unit nodal displacements. These functions are also called interpolation functions, and are sometimes referred to as $\phi_i(x) = v_i(x)$, $i = 1, 2, \ldots, 4$. These are simple polynomials and can easily be differentiated to get the curvature $\kappa = \boldsymbol{\delta}^T\, \mathbf{v}''(x) = \boldsymbol{\delta}^T\, \boldsymbol{\kappa}(x)$ where

$$\boldsymbol{\kappa}(x) = \begin{Bmatrix} \kappa_1(x) \\ \kappa_2(x) \\ \kappa_3(x) \\ \kappa_4(x) \end{Bmatrix} = \begin{Bmatrix} (-6+12\bar{x})/L^2 \\ (6-12\bar{x})/L^2 \\ (-4+6\bar{x})/L \\ (-2+6\bar{x})/L \end{Bmatrix} \tag{15.25}$$

are the curvatures under unitary nodal displacements. Note that the curvatures are linear functions of $\bar{x}$ and also of x.

We can now compute the stiffness coefficients (15.14) with the formula

$$K_{ij} = EI \int_0^1 \kappa_i(\bar{x})\, \kappa_j(\bar{x})\, L\mathrm{d}\bar{x}\ , \quad i,j = 1,2,\ldots,4 \tag{15.26}$$

For instance,

$$K_{23} = EI \int_0^1 \left(\frac{6-12\bar{x}}{L^2}\right)\left(\frac{-4+6\bar{x}}{L}\right) L\mathrm{d}\bar{x} = -6\,\frac{EI}{L^2} = \frac{EI}{L^3}\,(-6L)$$

$$K_{44} = EI \int_0^1 \left(\frac{-2+6\bar{x}}{L}\right)^2 L\mathrm{d}\bar{x} = 4\,\frac{EI}{L} = \frac{EI}{L^3}\,(4L^2)$$

and so on. In this manner we can generate the entire $\mathbf{K}_f$ matrix, (15.15).

15.7.2 The Extensional Matrix

Under arbitrary end-displacements the equilibrium configuration of the extensional element can be written

$$u(x) = \sum_{i=5}^{6} \delta_i\, u_i(x) = \boldsymbol{\delta}^T \mathbf{u}(x) \tag{15.27}$$

where $u_i(x)$ is the axial displacement of the element under nodal displacements $\boldsymbol{\delta}^i$, that is, for a unit displacement $\delta_i = 1$ at degree of freedom i and zero displacement at the other degree of freedom. The two functions $u_i(x)$ are the shape or interpolation functions of the extensional element.

To determine the shape functions consider the differential equilibrium equation of a infinitesimal segment dx of the element, $n'(x) + q(x) = 0$, which with Hooke's generalized law $n(x) = EA(x)\,\varepsilon(x)$ and $\varepsilon(x) = u'(x)$, yields the equilibrium equation in terms of the displacement function $[EA\,u'(x)]' + q(x) = 0$. We assume that the element is uniform (EA is constant) and has no distributed loads ($q(x) = 0$). The differential equation reduces to

$$u''(x) = 0 \tag{15.28}$$

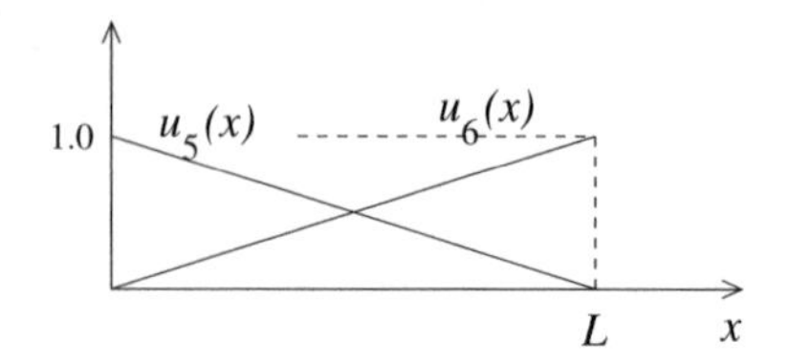

Fig. 15.18 Shape functions of extensional element

the general solution of which is $u(x) = b_1\, x + b_0$. The boundary conditions $u(0) = \delta_5$ and $u(L) = \delta_6$ allow us to compute the constants b_1 and b_0 which yields, after some rearrangement,

$$u(x) = \boldsymbol{\delta}^T \mathbf{u}(x) = \boldsymbol{\delta}^T \boldsymbol{\phi}(x) = \begin{Bmatrix} \delta_5 \\ \delta_6 \end{Bmatrix}^T \begin{Bmatrix} 1 - \bar{x} \\ \bar{x} \end{Bmatrix} \tag{15.29}$$

where $\bar{x} = x/L$ is a non-dimensionalized coordinate along the x axis.

These linear functions (Fig. 15.18) are the shape functions of the extensional element $\phi_i(x) = u_i(x)$, $i = 5, 6$. Note, the displacements are axial; the linear functions are drawn along the lateral ordinate.

A simple differentiation of (15.29) yields the axial strain, $\varepsilon(x) = \mathrm{d}u(x)/\mathrm{d}x$

$$\varepsilon(x) = \boldsymbol{\delta}^T \boldsymbol{\phi}'(x) = \begin{Bmatrix} \delta_5 \\ \delta_6 \end{Bmatrix}^T \begin{Bmatrix} -1 \\ 1 \end{Bmatrix} 1/L \tag{15.30}$$

or $\varepsilon(x) = \boldsymbol{\delta}^T\, \boldsymbol{\varepsilon}(x)$, where $\varepsilon_5, \varepsilon_6$ are the extensional strains under unitary nodal displacements

$$\boldsymbol{\varepsilon} = \begin{Bmatrix} \varepsilon_5 \\ \varepsilon_6 \end{Bmatrix} = \begin{Bmatrix} -1 \\ 1 \end{Bmatrix} 1/L \tag{15.31}$$

The stiffness coefficients follow from (15.14)

$$K_{ij} = EA \int_0^1 \varepsilon_i(\bar{x})\, \varepsilon_j(\bar{x})\, L\mathrm{d}\bar{x}\,, \quad i, j = 5, 6 \tag{15.32}$$

For instance,

$$K_{55} = EA \int_0^1 \left(\frac{-1}{L}\right)\left(\frac{-1}{L}\right) L\mathrm{d}\bar{x} = \frac{EA}{L}$$

We need nothing more (15.18).

15.8 Sample Rod and Beam Element Stiffness Matrices

15.8.1 Stepped Rod Element

The stiffness coefficient k of the stepped axial element shown in Fig. 15.19a is a function of the parameter η. The parameter η is the magnification factor of segment bc of the element compared to the first segment. You will have noticed that the degrees of freedom are numbered 1, 2, instead of 5, 6 used earlier. The numbering is always arbitrary as long as we respect consistency such as not to change the numbers midway in the calculations. The elongation δ of the element under a force P at 2, keeping 1 fixed (Fig. 15.19b), is the sum of the elongations of both halves, $\delta = PL/2EA + PL/2\eta EA = (L/2EA)(1 + 1/\eta)P$ or $P = (EA/L)(2\eta/1 + \eta)\delta$ and when $\delta = 1$ then P is k

$$k = \frac{EA}{L}\frac{2\eta}{1+\eta} \tag{15.33}$$

The solution is sound. For $\eta = 1$ we have a uniform bar of stiffness EA and length L and indeed $k = EA/L$. When η becomes very large, segment bc is in fact a rigid bar (rigid link), the displacement at c is equal to the displacement at b. In practical terms, we now have a bar ab of stiffness EA and length $L/2$, and we expect $k = 2\,(EA/L)$. By dividing the numerator and denominator in (15.33) by η and taking the limit, we obtain indeed the asymptote $k = 2\,(EA/L)$.

We could try to compute the matrix (in fact, the coefficient in front of the matrix) using the *shape functions* (15.29). The problem is that they were developed for uniform elements and ours is a stepped bar. But this is done on a regular basis in finite-element analysis, because usually the exact shape functions of the elements are unknown.

Instead of exact stiffness matrices this will allow us to produce approximate ones (15.32).

$$\begin{aligned}
K_{11} &= \int_0^1 EA(x)\, \varepsilon_i\, \varepsilon_j L\, \mathrm{d}\bar{x} \\
&= EA \int_0^{0.5} \left(\frac{-1}{L}\right)\left(\frac{-1}{L}\right) L\mathrm{d}\bar{x} + \eta\, EA \int_{0.5}^{1} \left(\frac{-1}{L}\right)\left(\frac{-1}{L}\right) L\mathrm{d}\bar{x}
\end{aligned} \tag{15.34}$$

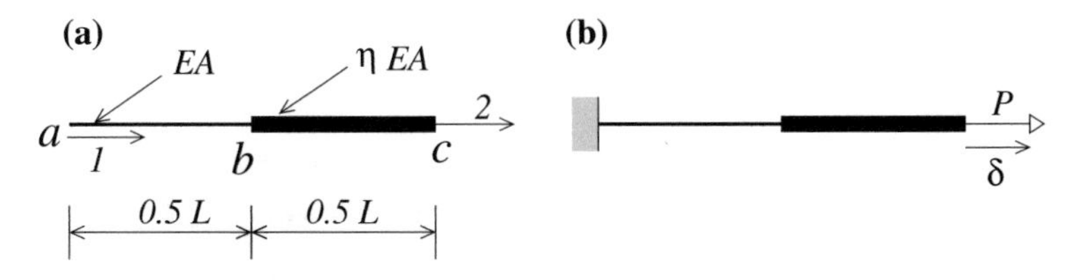

Fig. 15.19 Stepped rod element

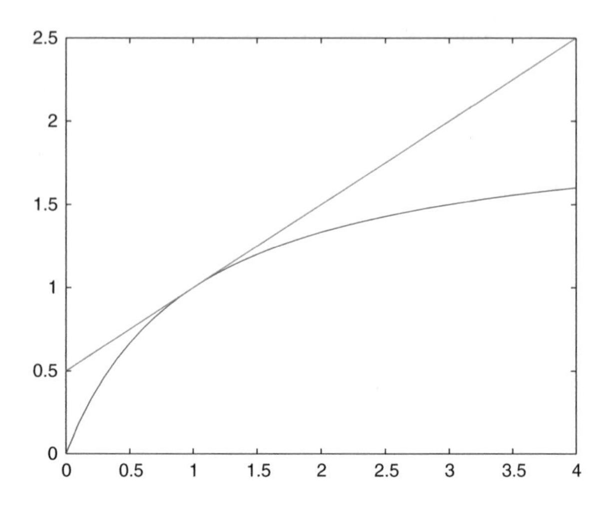

Fig. 15.20 Exact and approximate (linear) stiffness as a function of η

or

$$k_{\text{approx}} = \frac{EA}{L}\frac{1+\eta}{2} \tag{15.35}$$

Note that the slopes of the exact and approximate stiffnesses are respectively $\mathrm{d}k/\mathrm{d}\eta = 2/(1+\eta)^2$ and $\mathrm{d}k_{FE}/\mathrm{d}\eta = 1/2$. Consequently, at $\eta = 1$ both functions have the same value and the same slope.

In Fig. 15.20, we have plotted the exact function $k(x)$ and the linear approximate function in the range $0 \leq \eta \leq 4$. Notice that the approximation is (of course) exact at $\eta = 1$, and that the two curves have indeed a common tangent at that point. More important, however, is the realization that the approximation is of very poor quality, and that it can be used only in the close vicinity of $\eta = 1$. The reason is probably that the shape functions of the displacements are poor approximations of the real displacements.

15.8.2 Tapered Rod Element

Compute the stiffness coefficient k of the tapered axial element shown in Fig. 15.21a. The element length and elastic constant are respectively L and E, and the cross-section varies linearly from A_0–$3A_0$, or $A = A_0(1 + 2x/L)$.

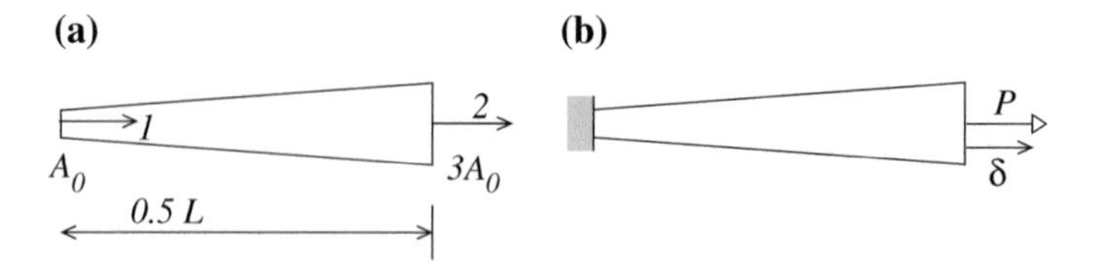

Fig. 15.21 Tapered rod element

We want to compute the exact elongation δ of the element under a force P at 2, keeping 1 fixed (Fig. 15.21b), $\delta = \int_0^L \varepsilon \, \mathrm{d}x$, where the strain at x is $\varepsilon = n/EA$, keeping in mind that the axial force is of course constant along the element $n = P$. Consequently,

$$\delta = \frac{P}{EA_0} \int_0^L \frac{1}{1+2x/L} \, \mathrm{d}x = \frac{PL}{EA_0} \int_0^1 \frac{1}{1+2\bar{x}} \, \mathrm{d}\bar{x}$$
$$= \frac{PL}{EA_0} \left[\frac{\log(1+2\bar{x})}{2} \right]_0^1 = 0.5493 \, \frac{PL}{EA_0} \tag{15.36}$$

and when $\delta = 1$ then P is, by definition, k

$$k = 1.82 \, \frac{EA_0}{L} \tag{15.37}$$

Using the shape functions of the uniform axial element, we get

$$K_{11} = EA_0 \int_0^1 (1+2\bar{x}) \left(\frac{-1}{L} \right) \left(\frac{-1}{L} \right) L \, \mathrm{d}\bar{x} = \frac{EA_0}{L} \, [\bar{x} + \bar{x}^2]_0^1 \tag{15.38}$$

or

$$k_{\text{approx}} = 2 \, \frac{EA_0}{L} \tag{15.39}$$

This is akin to using the mean cross-section $2EA_0$ instead of the tapered one, which is what we would have done if everything else failed. The approximation of the stiffness (15.39) is in this case not far from the exact value (15.37).

15.8.3 Beam Element with Central Hinge

Compute the stiffness matrix $\mathbf{K}_{\text{hinge}}$ of the beam element with a hinge at mid-span, shown in Fig. 15.22a.

We solve the problem by analyzing the beam under a force Q along degree of freedom 1 and a couple C along degree of freedom 3. This will yield columns 1 and 3 of $\mathbf{K}_{\text{hinge}}$ and, from equilibrium considerations and symmetry, we get the rest of the matrix. We note that adding a hinge reduces the structure from redundancy $R = 1$ to statical determinacy, which makes analysis by the force method, easier. (Releasing structures increases the numerical burden when solving by a displacement method.)

Under Q (white arrow in Fig. 15.22b), we get the reactions shown with black arrows and the bending moment m shown in Fig. 15.22c. The diagram is also valid for the curvature κ in a different scale ($[QL/EI]$ instead of $[QL]$).

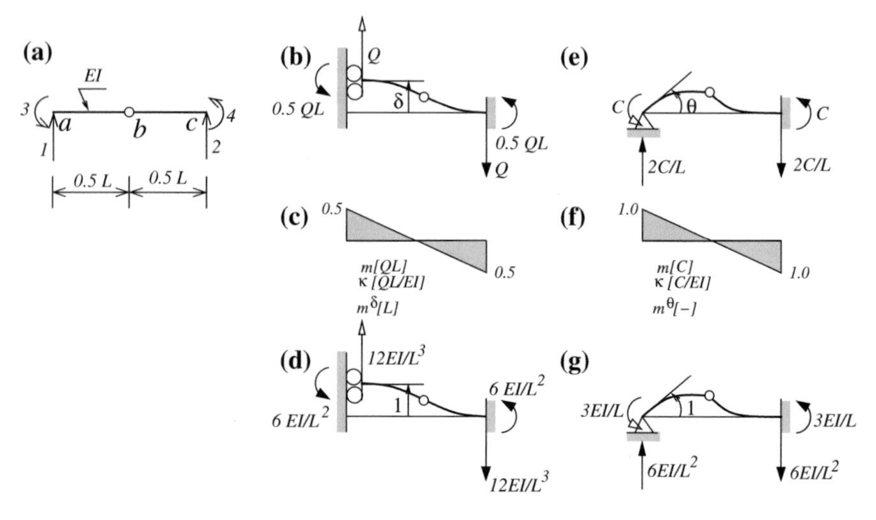

Fig. 15.22 Beam with central hinge element

For computing the displacement δ of force Q, we apply a unit force along degree of freedom 1, and find a moment distribution m^δ in equilibrium with that force. But we just found one for a force Q along the same degree of freedom. So the diagram in Fig. 15.22c is also a valid m^δ after setting $Q = 1$. By virtue of the unit-load method we get $\delta = \int_0^L \kappa m^\delta \mathrm{d}x = (1/12)\, PL^3/EI$. When P is such that it causes $\delta = 1$, we name the force K_{11}. Consequently, $1 = (1/12)\, K_{11}\, L^3/EI$ or $K_{11} = 12\; EI/L^3$. The reactions scale accordingly (see Fig. 15.22d), and constitute the first column (and first row) of the matrix to the left in (15.40).

$$\frac{EI}{L^3}\begin{bmatrix} 12 & -12 & 6L & 6L \\ -12 & & -6L & \\ 6L & -6L & 3L^2 & 3L^2 \\ 6L & & 3L^2 & \end{bmatrix} \qquad \mathbf{K}_{\text{hinge}} = \frac{EI}{L^3}\begin{bmatrix} 12 & -12 & 6L & 6L \\ -12 & 12 & -6L & -6L \\ 6L & -6L & 3L^2 & 3L^2 \\ 6L & -6L & 3L^2 & 3L^2 \end{bmatrix} \tag{15.40}$$

We proceed in a similar manner to compute $\mathbf{k}_3$ (Fig. 15.22e, f). The rotation due to C is $\theta = \int_0^L \kappa m^\theta \mathrm{d}x = (1/3)\, CL/EI$, therefore $K_{33} = 3\, EI/L$ (see Fig. 15.22g and the third column and row in the matrix to the left in (15.40). It is a simple matter to complete matrix $\mathbf{K}_{\text{hinge}}$.

Remark The displaced shape in Fig. 15.22b, e is for information only and is not required for computing the matrix. It is however instructive to note that the deformed shapes of segments *ab* and *bc* are respectively identical in both figures (they have the same curvature), but occupy different positions in the plane due to the boundary conditions. Note also, the slope discontinuity at *b* exists in Fig. 15.22g but not in Fig. 15.22d.

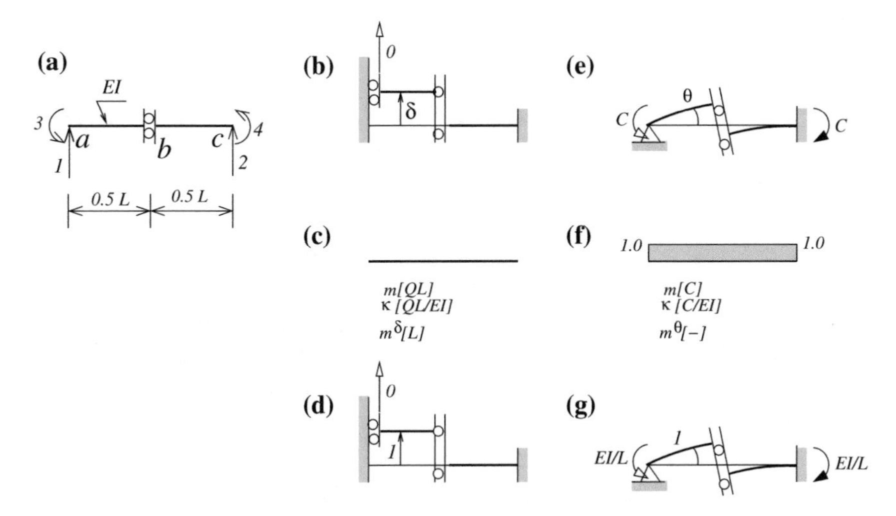

Fig. 15.23 Beam with central guide element

15.8.4 Beam Element with Central Lateral Guide

Compute the stiffness matrix $\mathbf{K}_{\text{guide}}$ of the beam element with a lateral guide at mid-span shown in Fig. 15.23a.

We solve the problem by analyzing the beam under a force Q along degree of freedom 1 and a couple C along degree of freedom 3. This will yield columns 1 and 3 of $\mathbf{K}_{\text{hinge}}$ and, from equilibrium considerations and symmetry, we get the rest of the matrix. Adding a guide also makes the beam statically determinate.

A displacement along degree of freedom 1 with zero displacements at the other degrees of freedom results in rigid-body motion and does not require any forces at the extremities of the element (see Fig. 15.22b–d), hence the first column and row of $\mathbf{K}_{\text{guide}}$ in (15.41) are zero.

$$\mathbf{K}_{\text{guide}} = \frac{EI}{L}\begin{bmatrix} 0 & 0 & 0 & 0 \\ 0 & 0 & 0 & 0 \\ 0 & 0 & 1 & -1 \\ 0 & 0 & -1 & 1 \end{bmatrix} \tag{15.41}$$

We proceed in a similar fashion for computing $\mathbf{k}_3$ (Fig. 15.23e, f). The vertical reactions at a and c are zero, otherwise we would have shear in the beam near the vertical guide which is impossible. For equilibrium we are thus left with equal and opposite couples C. The rotation due to C is $\theta = \int_0^L \kappa m^\theta \mathrm{d}x = CL/EI$, therefore $K_{33} = EI/L$ (see Fig. 15.23g) and the third column and row in $\mathbf{K}_{\text{guide}}$ in (15.41). Next we complete matrix $\mathbf{K}_{\text{guide}}$.

Remark Here also the displaced shape in Fig. 15.22b, e is for information only. Note that the deformed shapes of segments ab and bc are identical circular segments.

15.9 Note

To complete the picture we note that in the general case of six arbitrary displacements $\boldsymbol{\delta}$ one can write (by superposition)

$$u(x) = \sum_{i=5}^{6} \delta_i \, u_i(x) \quad \text{and} \quad v(x) = \sum_{i=1}^{4} \delta_i \, v_i(x) \,. \tag{15.42}$$

The axial component has indexes 5 and 6 whereas the lateral component has indexes 1–4. Taking the first derivative of the first equation, and the second derivative of the second equation, both with respect to x yields respectively the axial strain and the curvature of the element

$$\varepsilon(x) = \sum_{i=5}^{6} \delta_i \, \varepsilon_i(x) \quad \text{and} \quad \kappa(x) = \sum_{i=1}^{4} \delta_i \, \kappa_i(x) \,. \tag{15.43}$$

And since we have come this far, we may as well multiply both sides of the first equation by the axial stiffness $EA(x)$ (yes, it can vary with x) and both sides of the second equation by the flexural stiffness $EI(x)$ to yield the normal force and the bending moment

$$\begin{aligned} n(x) &= \sum_{i=5}^{6} \delta_i \, EA(x) \, \varepsilon_i(x) = \sum_{i=5}^{6} \delta_i \, n_i(x) \\ \text{and} \quad m(x) &= \sum_{i=1}^{4} \delta_i \, EI(x) \, \kappa_i(x) = \sum_{i=1}^{4} \delta_i \, m_i(x) \,. \end{aligned} \tag{15.44}$$

In case you are wondering where these internal forces come from, when we have displaced the end-nodes of the elements (Fig. 15.7), we had to apply loads (of course in equilibrium) to the extremities. These forces and couples cause the internal forces.

15.10 Summing Up

The element stiffness matrices are the building blocks of the stiffness method. We have shown how to compute such matrices, have given their characteristics and have computed the matrices of the two basic elements of skeletal structures: the uniform flexural and the uniform extensional elements. These matrices can now be incorporated in the analysis by the stiffness method.

Chapter 16
Change of Coordinates

Element stiffness matrices are developed in *local* coordinates which are often not the same as the *system* coordinates of the structure. A typical example is when the elements of the structure have different inclinations. This requires a change of coordinates. We show in this chapter how congruent transformations accommodate such cases and several relevant ones.

16.1 The Uniform Frame Element

An element which has both extensional and flexural stiffness is called a frame element. It has six degrees of freedom, and since the extensional and flexural modes are orthogonal, that is, the element deforms in both modes without any interaction between the modes, the stiffness matrix is as given in Fig. 16.1 (gray entries representing zeros).

The upper matrix is the 4×4 flexural (beam) matrix $\mathbf{K}_f$ and the lower one is the 2×2 extensional (rod) one $\mathbf{K}_e$.

This orderly matrix results from a convenient numbering of the degrees of freedom (see Fig. 16.1b). The first four degrees of freedom belong to the flexural mode and the last two are extensional degrees of freedom. It is in a convenient form for memorizing its entries, and we will be using it for small problems that can be solved at the manual or semi-manual level.

For large real-life applications with thousands of degrees of freedom, which will of necessity be computerized, the numbering of the degrees of freedom in Fig. 16.2b is preferred.

As shown in the Fig. 16.2 the degrees of freedom are now grouped by node, first the three degrees of freedom of the left extremity (a) and then the degrees of freedom of the right extremity (b). It is conceptually the same matrix. The components are identical but they appear at different locations and symmetry is maintained. Evidently, this matrix is more difficult to memorize but we recall that the procedure is now algorithmic. If we denote by $\boldsymbol{\delta}_a = \{\delta_1\ \delta_2\ \delta_3\}$ and $\boldsymbol{\delta}_b = \{\delta_4\ \delta_5\ \delta_6\}$ the respective nodal

M.B. Fuchs, *Structures and Their Analysis*,
DOI 10.1007/978-3-319-31081-7_16

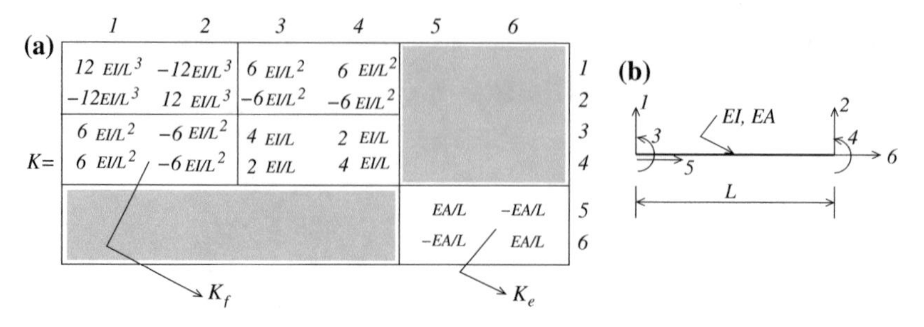

Fig. 16.1 Element stiffness matrix grouped by *beam and rod* degrees of freedom

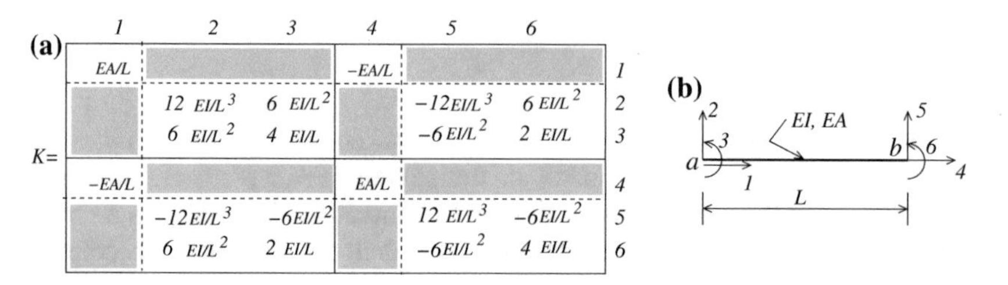

Fig. 16.2 Element stiffness matrix grouped by *nodal* degrees of freedom

displacements and by $\mathbf{f}_a = \{f_1\ f_2\ f_3\}$ and $\mathbf{f}_b = \{f_4\ f_5\ f_6\}$ the corresponding nodal forces (or couples), then the element equilibrium equations can be written in the concise form

$$\begin{Bmatrix} \mathbf{f}_a \\ \mathbf{f}_b \end{Bmatrix} = \begin{bmatrix} \mathbf{K}_{aa} & \mathbf{K}_{ab} \\ \mathbf{K}_{ba} & \mathbf{K}_{bb} \end{bmatrix} \begin{Bmatrix} \delta_a \\ \delta_b \end{Bmatrix} \tag{16.1}$$

where the $\mathbf{K}_{ij}, i, j \leftarrow a, b$ are the 3×3 sub-matrices indicated by solid line partitions in Fig. 16.2. Note, the matrix is now different from its previous form but it remains symmetric.

16.2 Congruent Transformations

Obtaining the stiffness matrix into the new coordinates is simple. For instance, the old degrees of freedom 1 and 4 are now 2 and 6. Consequently, the coefficient $K_{14} = 6EI/L^2$ of the old system is now $K_{26} = 6EI/L^2$ of the new system. All we have to do is to move the 36 coefficients, one by one, to their new locations. This may prove cumbersome for humans, there is, however, an elegant and systematic way of doing this. It involves the concept of transformations of coordinates and it is used extensively when programming the stiffness method.

Let us assume two sets of degrees of freedom which we will call ‘old’ and ‘new’, and to distinguish between them we will assign a ‘hat’ superscript $(\hat{o})$ to all the quantities in the old system. For instance, the force-displacement equations of the frame element in Figs. 16.1 and 16.2 will be respectively

$$\hat{\mathbf{f}} = \hat{\mathbf{K}}\,\hat{\boldsymbol{\delta}} \quad \text{(Old)} \qquad \mathbf{f} = \mathbf{K}\,\boldsymbol{\delta} \quad \text{(New)} \tag{16.2}$$

We would be interested in obtaining the stiffness relations in the new system on the basis of the same relations in the old one. It should be clear that the stiffness of an element is a mechanical property and is independent of any coordinate system.

The point of departure is to express the displacements in the old coordinates in terms of the *same* displacements in the new coordinates by a relation of the type

$$\hat{\boldsymbol{\delta}} = \mathbf{G}\,\boldsymbol{\delta} \tag{16.3}$$

In the present case, for instance, the relation is

$$\begin{Bmatrix} \hat{\delta}_1 \\ \hat{\delta}_2 \\ \hat{\delta}_3 \\ \hat{\delta}_4 \\ \hat{\delta}_5 \\ \hat{\delta}_6 \end{Bmatrix} = \begin{bmatrix} 0&1&0&0&0&0 \\ 0&0&0&0&1&0 \\ 0&0&1&0&0&0 \\ 0&0&0&0&0&1 \\ 1&0&0&0&0&0 \\ 0&0&0&1&0&0 \end{bmatrix} \begin{Bmatrix} \delta_1 \\ \delta_2 \\ \delta_3 \\ \delta_4 \\ \delta_5 \\ \delta_6 \end{Bmatrix} \tag{16.4}$$

Matrix $\mathbf{G}$ must be provided. It is easy to construct. Clearly, when comparing the coordinate numbering in Figs. 16.1 and 16.2, we notice $\hat{\delta}_1 = \delta_2$, $\hat{\delta}_2 = \delta_5$, etc.

Now, an element has displacements $u(x)$ and $v(x)$ and in particular displacements at its extremities. It has loads at its extremities. Since it deforms it has internal deformations $\varepsilon(x)$ and $\kappa(x)$, and by Hooke's law, corresponding internal forces $n(x)$ and $m(x)$. We are reminded that everything is real here: a real element with real loads and real deformations.

Clearly, the displacements and the deformations represent a *compatible system* and the loads and internal forces are an *equilibrium system*, for which we can write the virtual work relation, $evw = -ivw$.

For the internal virtual work we find $-ivw = \int_{\text{elem}} (n\varepsilon + m\kappa)\,\mathrm{d}x$. The external virtual work, depending on the order in which the vectors are arranged (old or new), is $\hat{\boldsymbol{\delta}}^T \hat{\mathbf{f}}$ or $\boldsymbol{\delta}^T \mathbf{f}$. Writing the virtual work principle twice we obtain

$$\begin{aligned} \int_a^b (n\varepsilon + m\kappa)\,\mathrm{d}x &= \hat{\boldsymbol{\delta}}^T \hat{\mathbf{f}} \\ \int_a^b (n\varepsilon + m\kappa)\,\mathrm{d}x &= \boldsymbol{\delta}^T \mathbf{f} \end{aligned} \tag{16.5}$$

or

$$\hat{\boldsymbol{\delta}}^T \hat{\mathbf{f}} = \boldsymbol{\delta}^T \mathbf{f} \tag{16.6}$$

Substituting for $\hat{\delta}$ in (16.3) yields $\boldsymbol{\delta}^T \mathbf{G}^T \mathbf{f} = \boldsymbol{\delta}^T \hat{\mathbf{f}}$ or $\boldsymbol{\delta}^T (\mathbf{f} - \mathbf{G}^T \hat{\mathbf{f}}) = 0$, and since $\boldsymbol{\delta}$ is arbitrary, the term within the parenthesis is zero; therefore, the forces in the new coordinates are related to the old ones by $\mathbf{f} = \mathbf{G}^T \hat{\mathbf{f}}$.

Similarly, upon introducing (16.2) into (16.6), we obtain $\hat{\boldsymbol{\delta}}^T \hat{\mathbf{K}} \hat{\boldsymbol{\delta}} = \boldsymbol{\delta}^T \mathbf{K} \boldsymbol{\delta}$, and when substituting for $\hat{\boldsymbol{\delta}}$ we have $\boldsymbol{\delta}^T (\mathbf{G}^T \hat{\mathbf{K}} \mathbf{G}) \boldsymbol{\delta} = \boldsymbol{\delta}^T \mathbf{K} \boldsymbol{\delta}$ or $\boldsymbol{\delta}^T (\mathbf{K} - \mathbf{G}^T \hat{\mathbf{K}} \mathbf{G}) \boldsymbol{\delta} = 0$. Since this expression is null for any value of $\boldsymbol{\delta}$, forcibly $\mathbf{K} = \mathbf{G}^T \hat{\mathbf{K}} \mathbf{G}$.

In conclusion, the congruent transformation is

$$\boxed{\text{if } \hat{\boldsymbol{\delta}} = \mathbf{G}\,\boldsymbol{\delta} \text{ then } \mathbf{K} = \mathbf{G}^T \hat{\mathbf{K}} \mathbf{G} \text{ and } \mathbf{f} = \mathbf{G}^T \hat{\mathbf{f}}}$$

16.2.1 Permutations

This congruent transformation is valid for any transformation matrix $\mathbf{G}$. In particular for the transformation matrix in (16.4), and indeed, starting from the element stiffness matrix in Fig. 16.1, it produces the element stiffness matrix given in Fig. 16.2.

A matrix of the form in (16.4) is called a *permutation* matrix. Indeed, it permutes the components of a vector from one sequence (as in $\boldsymbol{\delta}$) into another sequence ($\hat{\boldsymbol{\delta}}$).

16.2.2 Stiffness Matrix of an Oblique Element

An important application of congruent transformations of element stiffness matrices is for oblique elements. Consider the element in Fig. 16.3a. It is drawn with an oblique angle α, positive counterclockwise. In the local coordinates (here with 'hat' superscript), which being attached to the element are also oblique, the stiffness matrix in $\hat{\mathbf{f}} = \hat{\mathbf{K}} \hat{\boldsymbol{\delta}}$ is, of course, the nominal element stiffness matrix, as in Fig. 16.2b for a uniform frame element. What would the stiffness matrix be in the coordinate system shown in Fig. 16.3b?

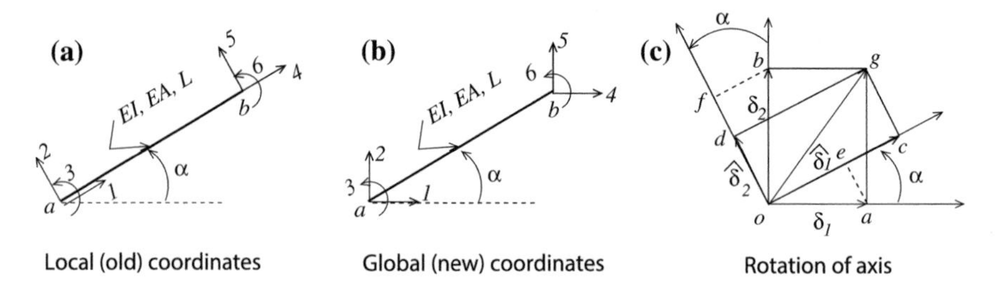

Fig. 16.3 Element rotation. **a** Local (old) coordinates. **b** Global (new) coordinates. **c** Rotation of axis

It should be clear that when you impose displacements to the ends a and b of the element, the forces that must be applied at the ends are the same independently on whether the axis is drawn as in Fig. 16.3a or in Fig. 16.3b. We are here dealing with the coordinates of these displacements and forces. The question that we have posed is: knowing the relationship between the force and displacement coordinates in the (local) system of Fig. 16.3a, can we find the relationship between the force and displacement coordinates in the (global) system of Fig. 16.3b? With the aid of the congruent transformation, the answer is simple to derive.

Let $\hat{\boldsymbol{\delta}}_a$ and $\boldsymbol{\delta}_a$ be respectively the displacement coordinates of extremity a in the local and global systems. Clearly, the following equation holds

$$\begin{Bmatrix} \hat{\delta}_1 \\ \hat{\delta}_2 \\ \hat{\delta}_3 \end{Bmatrix}_a = \begin{bmatrix} c & s & 0 \\ -s & c & 0 \\ 0 & 0 & 1 \end{bmatrix} \begin{Bmatrix} \delta_1 \\ \delta_2 \\ \delta_3 \end{Bmatrix}_a \quad \text{or} \quad \hat{\boldsymbol{\delta}}_a = \mathbf{R}_a\, \boldsymbol{\delta}_a \tag{16.7}$$

where $c = \cos\alpha$, $s = \sin\alpha$. Indeed, what is the component of the displacement in the direction of the axis of the element? In the local coordinates, simply $\hat{\delta}_1$. In the global coordinates it is the sum of the projections of δ_1 and δ_2 on the axis, that is, $\cos\alpha_1\, \delta_1 + \sin\alpha\, \delta_2$, hence the first line in (16.7). The second line of this equation follows suit. The third line is self-evident, $\hat{\delta}_3 = \delta_3$.

The coefficients matrix in (16.7) is called the rotation matrix. It expresses the rotation of the coordinate system from $\hat{\boldsymbol{\delta}}_a$ to $\boldsymbol{\delta}_a$. Now, a similar relation is valid at the other extremity $\hat{\boldsymbol{\delta}}_b = \mathbf{R}_b\, \boldsymbol{\delta}_b$. We can join the two equations

$$\begin{Bmatrix} \hat{\boldsymbol{\delta}}_a \\ \hat{\boldsymbol{\delta}}_b \end{Bmatrix} = \begin{bmatrix} \mathbf{R}_a & 0 \\ 0 & \mathbf{R}_b \end{bmatrix} \begin{Bmatrix} \boldsymbol{\delta}_a \\ \boldsymbol{\delta}_b \end{Bmatrix} \quad \text{or} \quad \hat{\boldsymbol{\delta}} = \mathbf{R}\, \boldsymbol{\delta} \tag{16.8}$$

The rest is standard. The congruent transformation yields $\mathbf{f} = \mathbf{K}\, \boldsymbol{\delta}$ with the inclined element stiffness matrix

$$\mathbf{K} = \mathbf{R}^T \hat{\mathbf{K}}\, \mathbf{R} \tag{16.9}$$

Orthogonal Matrix

Incidently, $\mathbf{R}$ (also $\mathbf{R}_a$ or $\mathbf{R}_b$) is an orthogonal matrix, that is, its inverse equals its transposed

$$\mathbf{R}^{-1} = \mathbf{R}^T. \tag{16.10}$$

Indeed, the congruent transformation gives $\mathbf{f} = \mathbf{R}^T \hat{\mathbf{f}}$. But, the forces and displacements in the local and global coordinates transform in the same way, therefore we have also $\hat{\mathbf{f}} = \mathbf{R}\,\mathbf{f}$, or $\mathbf{f} = \mathbf{R}^{-1}\, \hat{\mathbf{f}}$, hence (16.10).

We are not going to perform the double matrix multiplication (16.9) in its general form but for the case of truss elements. Although this is computer territory the stiffness matrix of an oblique truss element appears in almost every respectable book on structures, and we are not going to be the exception.

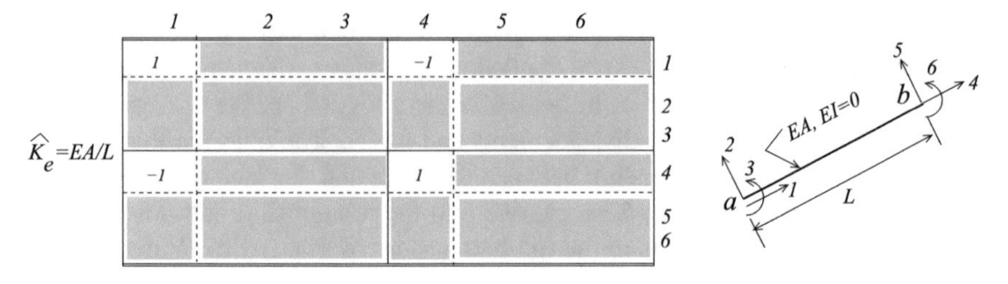

Fig. 16.4 Extensional element matrix in *local* coordinates

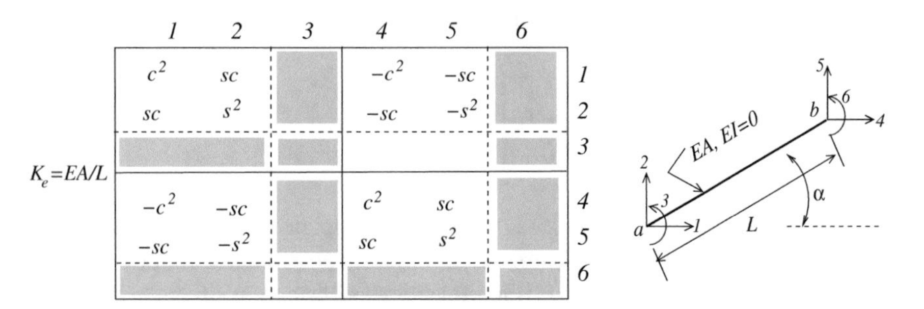

Fig. 16.5 Extensional element matrix in *system* coordinates

Stiffness Matrix of an Oblique Extensional (Truss) Element.

Let us assume that the element in Fig. 16.3 is a truss element. By definition such an element has negligible stiffness in bending but is perfectly sound in tension/compression. Its stiffness matrix $\hat{\mathbf{K}}$ in the 'old' (local) coordinates, that is, coordinates which are attached to the element, is given in Fig. 16.4. Now you can either free your afternoon, and execute the triple product (16.9), or let a symbolic package do the job. Either way you should get the matrix in Fig. 16.5. Note, this matrix is a function of α; sc, for instance, stands for $\sin\alpha\,\cos\alpha$.

In the trivial case of a vertical element ($\alpha = \pi/2$), the element is connected to degrees of freedom 2 and 5, and the stiffness matrix is simply

$$\mathbf{K}_{truss}(\alpha = \pi/2) = \frac{EA}{L} \begin{bmatrix} 0 & 0 & 0 & 0 & 0 & 0 \\ 0 & 1 & 0 & 0 & -1 & 0 \\ 0 & 0 & 0 & 0 & 0 & 0 \\ 0 & 0 & 0 & 0 & 0 & 0 \\ 0 & -1 & 0 & 0 & 1 & 0 \\ 0 & 0 & 0 & 0 & 0 & 0 \end{bmatrix} \tag{16.11}$$

Diagonal Transformation Matrices.

One other particular case that matters in hand calculations concerns a transformation where the matrix is diagonal $\mathbf{G} = \mathbf{D}$ with $D_{ij} = 0$ for $i \neq j$. If we perform the triple product $\mathbf{D}^T\bar{\mathbf{K}}\mathbf{D}$ in two steps, we find that $\bar{\mathbf{K}}\mathbf{D}$ results in a matrix similar to $\bar{\mathbf{K}}$ but where all the columns i have been multiplied by D_{ii}. Pre-multiplying this intermediate matrix by $\mathbf{D}^T$ multiplies all the rows j by D_{jj}. Consequently, $\mathbf{K} = \mathbf{D}^T\bar{\mathbf{K}}\mathbf{D}$ is essentially matrix $\bar{\mathbf{K}}$ where all columns i and rows j are multiplied respectively by D_{ii} and D_{jj}. Note, the diagonal entries $\bar{K}_{ii}$ are positive in the old system and remain so in the new one $K_{ii} = \bar{K}_{ii}\, D_{ii}^2$.

16.3 Summing Up

Element stiffness matrices must be written in coordinates compatible with those of the system stiffness matrix. This chapter has shown how we can perform the necessary coordinate transformation.

Chapter 17
Assembling the System Stiffness Matrix

Having decomposed the structures into nodes and elements, it is clear that knowing the nodal displacements will give us the internal loads in the elements. The displacements at the closed degrees of freedom are known, but we must still compute the displacements at the open degrees of freedom. This is done by writing and solving the equilibrium equations at the open degrees of freedom. We show, in this chapter, that instead of writing the equilibrium equations from first principles, we can instead assemble the system stiffness matrix by a relatively simple algorithmic procedure.

17.1 Modeling the Structure as the Sum of Its Elements

17.1.1 Elements and Nodes

The first step in the displacement method is to decompose the structure into its elements.

An *element* is a line-segment of the structure. To apply the displacement method we need to know the stiffness matrices of all its elements. If there are one or more elements for which the stiffness matrix is not known, the displacement method is not applicable. The most typical element is a straight segment of uniform stiffness.

The *nodes* of a structure are the points where the extremities of the elements are attached (partially or fully) to one another, or to the ground, or to nothing, such as the free end of a cantilever beam.

Both extremities of all the elements are thus attached to nodes. Consider for instance the frame in Fig. 17.1a. This structure was divided into four elements, ab, bc, db, ec, connected at five nodes, a, b, c, d, e. Note the node at the free end a.

The beam in Fig. 17.1b has two elements, ab, bc, and three nodes, a, b, c. The semi-circular arch in Fig. 17.2a can be decomposed into circular segments. If we do not know the stiffness matrix of a circular element we can geometrically approximate

M.B. Fuchs, *Structures and Their Analysis*,
DOI 10.1007/978-3-319-31081-7_17

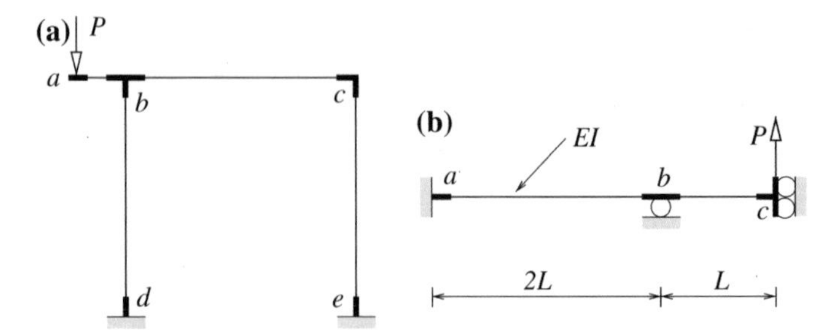

Fig. 17.1 Frame and beam elements and nodes

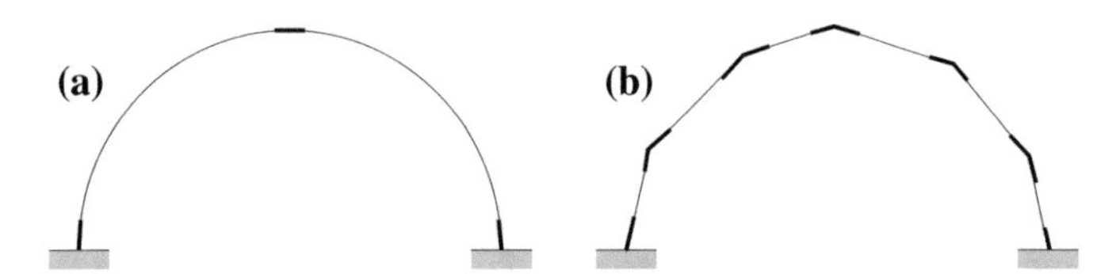

Fig. 17.2 Modeling an arch with frame elements

the arch by straight beam elements (six elements and seven nodes in Fig. 17.2b). The result will then be an approximation of the exact analysis.

The size of the nodes are exaggerated in the figures. In fact they are as small as we can possibly imagine. The nodes only indicate how the elements are connected, but geometrically the nodes do not occupy space. The structure is effectively the sum of its elements.

17.1.2 Nodal or Global Coordinates, Degrees of Freedom

The *degrees of freedom* (dofs) are the sum of all possible motions of the nodes of the structure.

Ours, being plane figures, every node has nominally three displacement components in a frame (horizontal and vertical translation, and rotation), two in a beam (lateral translation and rotation) and two in a truss (horizontal, vertical). The frame in Fig. 17.1a has in total of $5 \times 3 = 15$ nominal dofs, the uniform beam depicted in Fig. 17.1b has $3 \times 2 = 6$ nominal dofs, and the arch-like frame in Fig. 17.2 has $7 \times 3 = 21$ dofs, nominally.

The nominal dofs subdivide into *given dofs* and *'free'* or *open dofs*. Unless specified to the contrary, given dofs are fixed, that is, they are usually *closed*. The free degrees of freedom were numbered sequentially in Fig. 17.3 with zeros indicating fixed dofs. The sequence for assigning the numbers is arbitrary, but it helps to follow a standard method. For frames (see Fig. 17.3a, c) we use the sequence horizontal translation, vertical translation, rotation, and for beams (Fig. 17.3b) lateral translation,

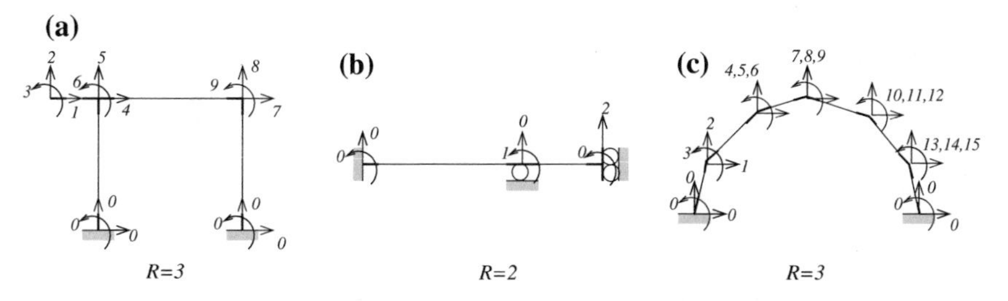

Fig. 17.3 Nodes and degrees of freedom

rotation (remember that for beams we assume that there are no axial displacements nor axial forces).

The given dofs are known from the outset and are determined by the boundary conditions, or they may be imposed by the problem much as the forces are imposed. The free dofs, however, are unknown. In fact they are THE unknowns of the analysis and are often referred to simply as the degrees of freedom in contrast to the given degrees of freedom.

The number of free dofs is, as we will see, very important because it is equal to the size of the *system stiffness matrix*, in other words, to the number of linear equations to be solved in the analysis. For the frame in Fig. 17.3a we have nine free dofs, or simply nine dofs. The analysis of this frame will require the solution of nine linear equations. The uniform beam depicted in Fig. 17.3b has two dofs only and will be solved by two linear equations. Note that the vertical dof at the roller node is zero since the roller prevents movements in that direction, but the rotation is free, while at the right extremity the beam can move upwards but the rotation is fixed. Finally, the arch-like frame in Fig. 17.3c will require no less than 15 equations.

We recall that in the force method the number of equations to be solved is equal to the degree of statical redundancy. We find respectively $R = 3$ and $R = 2$ for the frame and beam in Fig. 17.3 and $R = 3$ for the arch-like frame. The three compatibility equations of the force method for the arch seem to compare very favorably with the 15 equilibrium equations of the displacement method. However this is not the entire story. Obtaining the compatibility equations may be a tedious and error prone endeavor, whereas, as we shall see shortly, the equilibrium equations can be generated in a simple and systematic manner, especially when properly computerized.

17.1.3 Element or Local Coordinates

You will have noticed that there is a difference between the *coordinates* attached to the nodes, the global or system coordinates, and the coordinates of the end-displacements and end-loads of the elements, the local or element coordinates, which are attached to the elements. The displacements of and the loads applied to node *b* of the frame in Fig. 17.1 are given in the system coordinates 4, 5 and 6, while the end-displacements

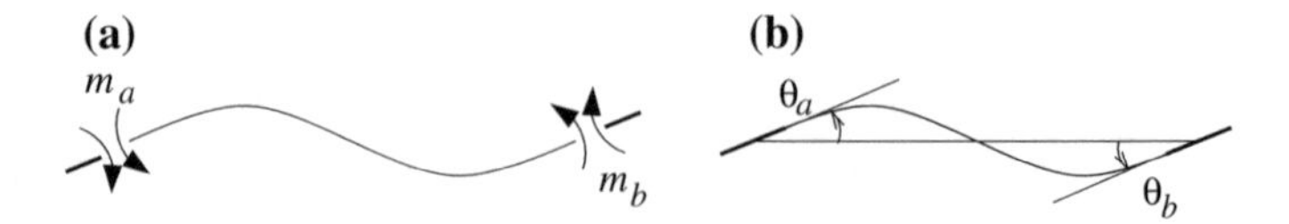

Fig. 17.4 Element end-forces and end-displacements

and end-forces of extremities b of elements ab, db and cb (see Fig. 17.1a) are each given in their respective local coordinates.

Before proceeding, we draw your attention to a fundamental difference between the relationship between element end-loads and nodal loads applied to the element by the node, and element end-displacements and nodal displacements.

The element end-loads are equal *but* opposite to those the element applies to the node. The bending moment m_a applied by the node to the element is, of course, equal and opposite to the moment applied by the element to the node (Fig. 17.4a).

The element end-displacements, on the other hand, are equal to the displacements of the node to which it is attached (Fig. 17.4b). They are usually measured in different coordinate systems (global, local) but they are physically identical. In Fig. 17.4b, we observe that θ_a is common to the node and the element.

The important corollary for our purposes is that, by knowing all the nodal displacements of the structure we know all the end-displacements of the elements, which is the same as saying that we are in possession of everything an analysis can provide. Archimedes is credited for the saying, "Give me a place to stand on, and I will move the Earth." We can paraphrase Archimedes and state, "Give us the nodal displacements and we will give you the structure." There is still a little detail to settle. How do we get the nodal displacements? We use the nodal equilibrium equations.

17.2 The Nodal Equilibrium Equations

Consider again the uniform beam depicted in Fig. 17.1b. The structure has two elements ab, bc, three nodes a, b, c, where out of the $3 \times 2 = 6$ nominal dofs four are fixed. (The lateral displacement and rotation at a, the lateral displacement at b and the rotation at c are equal to zero.)

We are left with only two unknown nodal displacements; the rotation at b which we will call Δ_1, and the lateral displacement at c, that is, Δ_2. Finally, a vertical force P is applied to dof 2.

17.2.1 If the End-Displacements Are Known …

With a leap of imagination, let us assume that we are given the displacements at the free dofs:

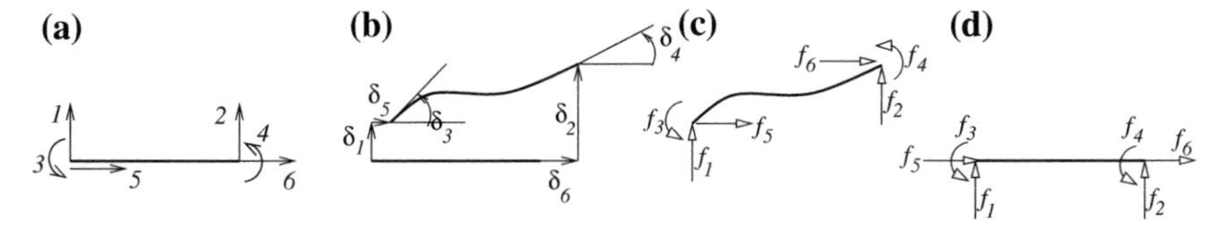

Fig. 17.5 Degrees of freedom of frame element in local coordinates

$$\Delta_1 = \frac{1}{6}\frac{PL^2}{EI} \quad \text{and} \quad \Delta_2 = \frac{1}{6}\frac{PL^3}{EI} \tag{17.1}$$

(Never mind the common 1/6 factor. This is fortuitous.) Note that Δ_1 is non-dimensional (rotations are given in radians) and Δ_2 has the dimension of a length (translation). It is easy to show that the problem is now by and large solved. Recall the sequence for the end-displacements of a beam element in local coordinates (Fig. 17.5).

The displacements at the extremities of every element are now all known (see Fig. 17.6a). For element *ab*, three end-displacements were fixed (local coordinates $\delta_1^{ab} = \delta_2^{ab} = \delta_3^{ab} = 0$) from the outset, and we are given the fourth end-displacement, the counter-clockwise rotation of *b* which is in fact the rotation of node *b*, $\delta_4^{ab} = \Delta_1$.

The same is valid for element *bc*. Two dofs are fixed (the local coordinates $\delta_1^{bc} = \delta_4^{bc} = 0$) and two are now known, $\delta_2^{bc} = \Delta_2$, $\delta_3^{bc} = \Delta_1$.

Keep in mind that we have NOT solved the problem. We are only trying to show that when all the nodal displacements are known the problem is as good as solved.

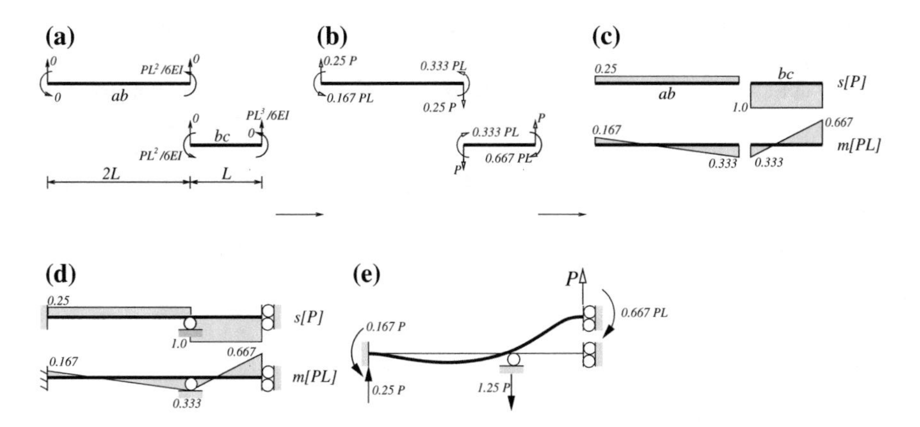

Fig. 17.6 Beam with two open degrees of freedom. **a** End-displacements. **b** End-forces. **c** Shear and moments on elements. **d** Shear an moment diagrams. **e** Reactions and equilibrium configuration

Pre-multiplying the end-displacements of each element by their respective element stiffness matrix gives the element end-loads. For elements ab and bc we get respectively the end-loads $\mathbf{f}^{ab}$ and $\mathbf{f}^{bc}$

$$\begin{Bmatrix} f_1^{ab} \\ f_2^{ab} \\ f_3^{ab} \\ f_4^{ab} \end{Bmatrix} \equiv \begin{Bmatrix} 0.25\,P \\ -0.25\,P \\ 0.167\,PL \\ 0.333\,PL \end{Bmatrix} \leftarrow \frac{EI}{8L^3} \begin{bmatrix} 12 & -12 & 12L & 12L \\ -12 & 12 & -12L & -12L \\ 12L & -12L & 16L^2 & 8L^2 \\ 12L & -12L & 8L^2 & 16L^2 \end{bmatrix} \begin{Bmatrix} 0 \\ 0 \\ 0 \\ PL^2/6EI \end{Bmatrix}$$

and

$$\begin{Bmatrix} f_1^{bc} \\ f_2^{bc} \\ f_3^{bc} \\ f_4^{bc} \end{Bmatrix} \equiv \begin{Bmatrix} -P \\ P \\ -0.333\,PL \\ -0.667\,PL \end{Bmatrix} \leftarrow \frac{EI}{L^3} \begin{bmatrix} 12 & -12 & 6L & 6L \\ -12 & 12 & -6L & -6L \\ 6L & -6L & 4L^2 & 2L^2 \\ 6L & -6L & 2L^2 & 4L^2 \end{bmatrix} \begin{Bmatrix} 0 \\ PL^3/6EI \\ PL^2/6EI \\ 0 \end{Bmatrix}$$

which can also be seen in Fig. 17.6b. The first two entries in the vectors to the left are lateral forces acting upon the extremities (positive up), and the following two are couples (positive anti-clockwise).

For *every* element we readily obtain the shear and moment distributions: the first column in Fig. 17.6c relates to element ab and the second to bc. Since the elements have no distributed forces (distributed forces and temperature variations are dealt with later in the text), the shear distribution is constant and the bending moment is linear.

It is instructive that the diagrams are built on an element by element basis. Juxtaposing the diagrams in Fig. 17.6c produces the shear and moment diagrams for the structure (Fig. 17.6d). Every discontinuity in the shear diagram signifies a corresponding vertical force, and similarly from discontinuities in the moment diagram we can deduce the point moment or reaction as depicted in Fig. 17.6e. For instance, if the diagrams start at a with values $s = 0.25P$ and $m = 0.167PL$ (in the proper sense), it can only be due to reactions applied by the wall to the beam, that is, by a lateral force $0.25\,P$ and a counter-clockwise couple of magnitude 0.167 that the wall exerts on the beam.

The displaced shape in Fig. 17.6e was drawn on the basis of the curvature $\kappa = m/EI$ and the nodal displacements.

17.2.2 *If the End-Displacements Are Wrong …*

What would happen if the nodal displacements are wrong? Assume for instance that we are given a slope or rotation at b, half the value it should be: $\Delta_1 = (0.5)\,PL^2/6EI$ and $\Delta_2 = PL^3/6EI$. Could we detect the mistake? If we pre-multiply the end-displacements by the respective element stiffness matrices, we will obtain the following end-loads at the elements

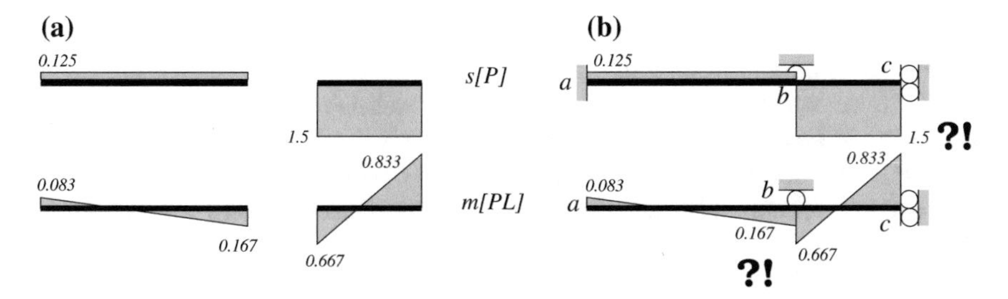

Fig. 17.7 Wrong displacements

$$\mathbf{f}^{ab} = \left\{ 0.125P \; -0.125P \; 0.083PL \; 0.167PL \right\}^T \tag{17.2}$$

$$\mathbf{f}^{bc} = \left\{ -1.5P \; 1.5P \; -0.667PL \; -0.833PL \right\}^T \tag{17.3}$$

leading to the shear and moment distributions shown in Fig. 17.7a.

We recall that these diagrams are created element by element, and observed separately, they reveal nothing unusual. The elements are in equilibrium, indeed, whatever the end-displacements, the elements will always be in equilibrium. It is only when the diagrams are juxtaposed as in Fig. 17.7b will we note discrepancies.

Starting with the bending moments it is obvious that something is very wrong at node b. Indeed, we have there a discontinuity in the bending moment although there cannot be a concentrated couple there (the only possible reaction of a simple support is a force). A similar situation, although less obvious, occurs with the shear diagram at node c. We expect a discontinuity of magnitude P because of the applied external force, but instead we have a drop of the shear force from $1.5\,P$ to zero. Conclusion: the displacements given at the start of this subsection must be wrong.

You may have noted that the discrepancies occurred at the free dofs 1 and 2 and, indeed, an easier way to determine the correctness of the displacements is to check directly the equilibrium of the nodes along the free dofs 1 and 2, as follows. We have indicated the end-loads (17.2) applied to ab in Fig. 17.8, and let us consider in particular the counterclockwise couple $0.083\,PL$. Where does this couple come from? Obviously from node b. Consequently, element ab applies an opposite couple to node b along dof 1. Similarly, the end-loads of element bc at extremity b are equal and opposite to the loads applied by the element on node b. Now, when we look at

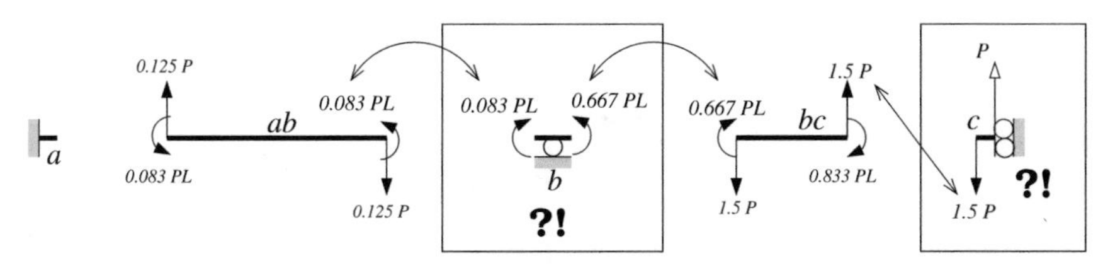

Fig. 17.8 Wrong displacements → no equilibrium at open degrees of freedom

degree of freedom 1 of node b the couples do not cancel each other out, and obviously we do not have rotational equilibrium.

When we repeat the procedure for the vertical translation of node c, we will note that element bc applies a down-force of magnitude $1.5P$ which with the applied force P leaves dof 2 unbalanced. Left as is, node c will crash to the ground. We do not have equilibrium, therefore, the end-loads (17.2) must be wrong, in other words, the displacements at the free dofs are wrong.

Try as we may, no pair of displacements, Δ_1, Δ_2, but the correct ones (17.1) will satisfy all the equilibrium requirements at the free dofs. It all has to do with the uniqueness of solution. So why not determine the two unknowns Δ_1, Δ_2 by imposing equilibrium at degrees of freedom 1 and 2? This is exactly what is done in the stiffness method.

17.2.3 Equilibrium at the Unknown Degrees of Freedom

We will now repeat the procedure for writing the equilibrium equations at the free dofs, but this time the nodal displacements $\mathbf{\Delta} = \{\Delta_1\ \Delta_2\}^T$ (positive counterclockwise for a rotation and up for a translation) will be left as unknowns.

In Fig. 17.9 we show the exploded view of the beam for displacements only. We proceed by determining the dofs at the nodes; a zero for a fixed dof and sequential numbering at the free dofs. This then determines the end-displacements of the elements (the SAME translation and rotation as the node to which the extremity is attached). Note that all the translations are positive in the y-direction and the rotations are positive counterclockwise.

In Fig. 17.10 we show the same exploded view, but now for forces. The positive nodal forces (applied forces) along the dofs are $\mathbf{p} = \{p_1\ p_2\}^T$ (same sense as $\mathbf{\Delta}$), where

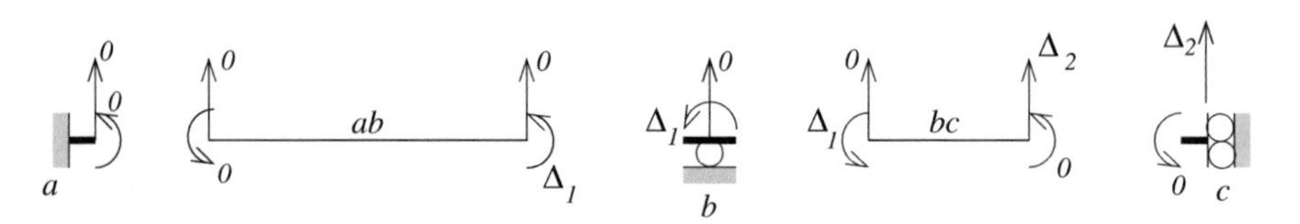

Fig. 17.9 Known displacements at the open degrees of freedom

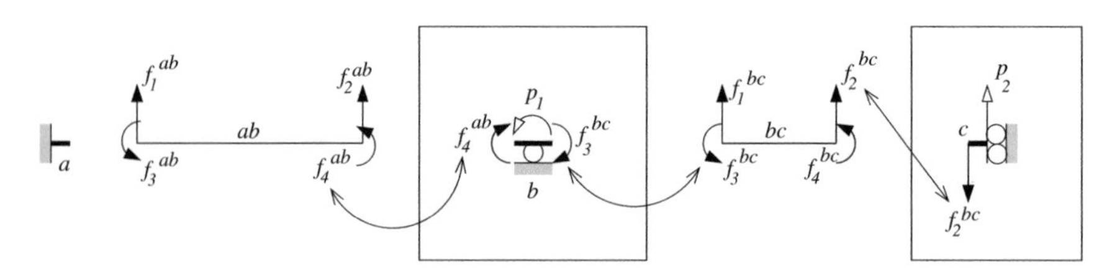

Fig. 17.10 Equilibrium at the open degrees of freedom

in our example $p_1 = 0$ and $p_2 = P$. Here, the sequence is first the element end-loads and then the loads applied to the nodes. For the element end-loads we multiply the element stiffness matrix by the element end-displacements. $\mathbf{f}^{ab} \leftarrow [\mathbf{K}^{ab}]\{0\,0\,0\,\Delta_1\}^T$ and $\mathbf{f}^{bc} \leftarrow [\mathbf{K}^{bc}]\{0\,\Delta_2\,\Delta_1\,0\}^T$.

This yields

$$\begin{Bmatrix} f_1^{ab} \\ f_2^{ab} \\ f_3^{ab} \\ f_4^{ab} \end{Bmatrix} \equiv \begin{Bmatrix} (1.5EI/L^2)\,\Delta_1 \\ (-1.5EI/L^2)\,\Delta_1 \\ (EI/L)\Delta_1 \\ (2EI/L)\,\Delta_1 \end{Bmatrix} \leftarrow \frac{EI}{8L^3}\begin{bmatrix} 12 & -12 & 12L & 12L \\ -12 & 12 & -12L & -12L \\ 12L & -12L & 16L^2 & 8L^2 \\ 12L & -12L & 8L^2 & 16L^2 \end{bmatrix}\begin{Bmatrix} 0 \\ 0 \\ 0 \\ \Delta_1 \end{Bmatrix} \tag{17.4}$$

$$\begin{Bmatrix} f_1^{ab} \\ f_2^{ab} \\ f_3^{ab} \\ f_4^{ab} \end{Bmatrix} \equiv \begin{Bmatrix} 1.5\,x_1 \\ -1.5\,x_1 \\ x_1 \\ 2\,x_1 \end{Bmatrix} \leftarrow \begin{bmatrix} 1.5 & -1.5 & 1.5 & 1.5 \\ -1.5 & 1.5 & -1.5 & -1.5 \\ 1.5 & -1.5 & 2 & 1 \\ 1.5 & -1.5 & 1 & 2 \end{bmatrix}\begin{Bmatrix} 0 \\ 0 \\ 0 \\ x_1 \end{Bmatrix} \tag{17.5}$$

and

$$\begin{Bmatrix} f_1^{bc} \\ f_2^{bc} \\ f_3^{bc} \\ f_4^{bc} \end{Bmatrix} \equiv$$

$$\begin{Bmatrix} -(12EI/L^3)\,\Delta_2 + (6EI/L^2)\,\Delta_1 \\ (12EI/L^3)\,\Delta_2 - (6EI/L^2)\,\Delta_1 \\ -(6EI/L^2)\,\Delta_2 + (4EI/L)\,\Delta_1 \\ -(6EI/L^2)\,\Delta_2 + (2EI/L)\,\Delta_1 \end{Bmatrix} \leftarrow \frac{EI}{L^3}\begin{bmatrix} 12 & -12 & 6L & 6L \\ -12 & 12 & -6L & -6L \\ 6L & -6L & 4L^2 & 2L^2 \\ 6L & -6L & 2L^2 & 4L^2 \end{bmatrix}\begin{Bmatrix} 0 \\ \Delta_2 \\ \Delta_1 \\ 0 \end{Bmatrix} \tag{17.6}$$

$$\begin{Bmatrix} f_1^{bc} \\ f_2^{bc} \\ f_3^{bc} \\ f_4^{bc} \end{Bmatrix} \equiv \begin{Bmatrix} -12\,x_2 + 6\,x_1 \\ 12\,x_2 - 6\,x_1 \\ -6\,x_2 + 4\,x_1 \\ -6\,x_2 + 2\,x_1 \end{Bmatrix} \leftarrow \begin{bmatrix} 12 & -12 & 6 & 6 \\ -12 & 12 & -6 & -6 \\ 6 & -6 & 4 & 2 \\ 6 & -6 & 2 & 4 \end{bmatrix}\begin{Bmatrix} 0 \\ x_2 \\ x_1 \\ 0 \end{Bmatrix} \tag{17.7}$$

where x_1 and x_2 are the non-dimensional nodal displacements.

Having determined the end-loads of the elements, we notice that f_4^{ab} is also the force that element *ab* applies to the free dof of node *b*, and likewise for f_3^{bc}. (Note the opposite senses.) A similar procedure is valid for f_2^{bc} applied by element *bc* to node *c*.

Equilibrium at the free dofs requires

$$\begin{aligned} f_4^{ab} + f_3^{bc} &= p_1 \\ f_2^{bc} &= p_2 \end{aligned} \tag{17.8}$$

Entering the expressions of f_4^{ab}, f_2^{bc} and f_3^{bc} (see (17.5) and (17.7)) in the equilibrium requirements (17.8), and recalling that in the present example $\mathbf{p} = \{0\,P\}^T$, produces the system of linear equations

$$\begin{bmatrix} 6 & -6 \\ -6 & 12 \end{bmatrix} \begin{Bmatrix} x_1 \\ x_2 \end{Bmatrix} = \begin{Bmatrix} 0 \\ 1 \end{Bmatrix} \tag{17.9}$$

which, when solved for the unknowns, yields $x_1 = 1/6$ and $x_2 = 1/6$. After reconciling the units we get indeed the displacements in (17.1).

These equations are an example of the *equilibrium equations* of the stiffness method

$$\mathbf{K}\,\boldsymbol{\Delta} = \mathbf{p} \tag{17.10}$$

In this example we have two unknowns. If we have 5,000, 100,000 unknown dofs or more, it is dealt with in a similar manner, but then we need to fully computerize the procedure. This is why the displacement method is established for an algorithmic solution.

The coefficient matrix in (17.9) is the stiffness matrix of the structure, or the reduced stiffness matrix or the system stiffness matrix, depending on whom you talk to. We will have plenty of opportunities to confuse you with nomenclature in the sequel.

But this is not all. In fact we have perhaps omitted the most efficient part of the stiffness method, because there is a systematic way of producing the equilibrium equations without any need for hard reasoning. We will first show the technique, and then we will try to substantiate the method.

17.3 Assembling the System Stiffness Matrix

The stiffness method can be reduced to the following steps:

1. Model the structure into nodes, known and unknown degrees of freedom, elements and loads (with at present concentrated forces and couples only at nodes);
2. *Write the equilibrium equations for the unknown degrees of freedom* $\mathbf{K}\,\boldsymbol{\Delta} = \mathbf{p}$;
3. Solve the equilibrium equations;
4. Get the internal *nsm* distributions (this is the post-processing stage).

Step 2 lies the heart of the method, and in the example shown it took quite some time and effort. We will replace step 2 by

2. *'Assemble' the stiffness matrix* **K** *and the load vector* **p**.

We will show that assembling **K** is a simple procedure which does not involve any heavy thinking, and is conducive to algorithmic implementation. In essence, we add the local stiffness matrices of all the elements to their corresponding locations in the system matrix.

Let us see how it works for the beam of the previous section.

17.3.1 An Example for Assembling K

The beam in Fig. 17.1b has two elements *ab bc* and three nodes *a b c*, six dofs (two per node), two of which are unknown: the rotation at *b* and the deflection at *c* and four dofs are known (they are fixed).

Starting from node *a*, every fixed dof is tagged 0 and the (free) dofs are numbered 1 2 … in sequence as we encounter them (see Fig. 17.11a).

We now write the element stiffness matrices for all the elements of the beam, in our case, $\bar{\mathbf{K}}^{ab}$ and $\bar{\mathbf{K}}^{bc}$ (the numbers bordering the matrix are the row and column indexes in local coordinates). We also remove all the units L, EI, …, because this is also how a computerized implementation works.

$$\bar{\mathbf{K}}^{ab} = \begin{array}{|rrrr|c} 1 & 2 & 3 & 4 & \\ \hline 1.5 & -1.5 & 1.5 & 1.5 & 1 \\ -1.5 & 1.5 & -1.5 & -1.5 & 2 \\ 1.5 & -1.5 & 2 & 1 & 3 \\ 1.5 & -1.5 & 1 & 2 & 4 \\ \hline \end{array} \qquad \bar{\mathbf{K}}^{bc} = \begin{array}{|rrrr|c} 1 & 2 & 3 & 4 & \\ \hline 12 & -12 & 6 & 6 & 1 \\ -12 & 12 & -6 & -6 & 2 \\ 6 & -6 & 4 & 2 & 3 \\ 6 & -6 & 2 & 4 & 4 \\ \hline \end{array}$$

These matrices must be known. If we do not know the stiffness matrices of all the elements, the stiffness method is not applicable.

The local coordinates of a beam are in the usual sequence 1234 (see Fig. 17.5a) and we add a 'bar' to the element stiffness matrices to indicate local coordinates. We next set up the correspondence between the local coordinates of the element and the nodal dofs in global coordinates (see Fig. 17.12) of the nodes to which the element is attached. For beam *ab* for instance, the end displacements 123 have global

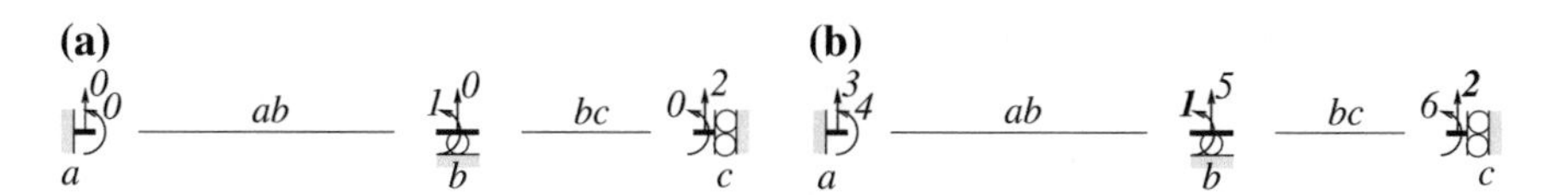

Fig. 17.11 **a** Numbering the degrees of freedom. **b** Numbering the coordinates

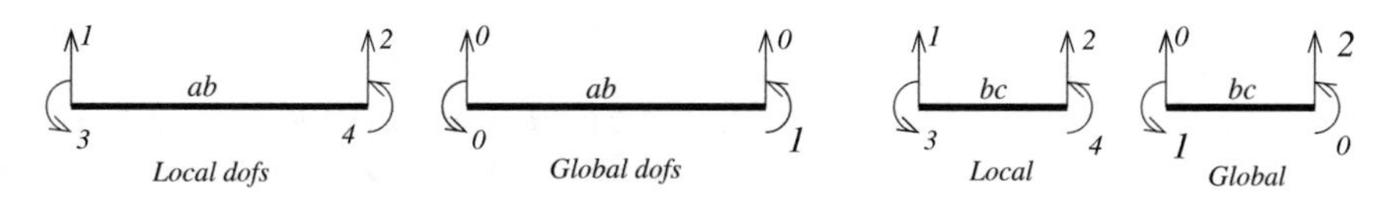

Fig. 17.12 Local and global degrees of freedom

coordinates 000 (fixed) and end-displacement 4 corresponds to global 1. For beam *bc*, local 1234 is global 0210.

We can now proceed with the assembly of the system stiffness matrix.

Two free dofs entail a 2×2 structure stiffness matrix $\mathbf{K}$ which we initially fill with zeros.

$$\mathbf{K} = \begin{array}{|cc|c} \scriptstyle 1 & \scriptstyle 2 & \\ \hline 0 & 0 & \scriptstyle 1 \\ 0 & 0 & \scriptstyle 2 \\ \hline \end{array}$$

We assemble the stiffness matrix by locating in the element matrices components which have an address in $\mathbf{K}$ and adding that component to $\mathbf{K}$. We start with $\bar{\mathbf{K}}^{ab}$.

$$\bar{\mathbf{K}}^{ab} = \begin{array}{|cccc|c} \scriptstyle 0 & \scriptstyle 0 & \scriptstyle 0 & \scriptstyle 1 & \\ \hline 1.5 & -1.5 & 1.5 & 1.5 & \scriptstyle 0 \\ -1.5 & 1.5 & -1.5 & -1.5 & \scriptstyle 0 \\ 1.5 & -1.5 & 2 & 1 & \scriptstyle 0 \\ 1.5 & -1.5 & 1 & \mathbf{2} & \scriptstyle 1 \\ \hline \end{array} \rightarrow \mathbf{K}^{ab} = \begin{array}{|cc|c} \scriptstyle 1 & \scriptstyle 2 & \\ \hline 2 & 0 & \scriptstyle 1 \\ 0 & 0 & \scriptstyle 2 \\ \hline \end{array}$$

In the system matrix $\mathbf{K}$ we have addresses 11, 12, 21, 22. Every component of $\bar{\mathbf{K}}^{ab}$ which has a corresponding address in $\mathbf{K}$ is added to that address. In matrix $\bar{\mathbf{K}}^{ab}$, only address 11 exists in the system matrix. All other addresses, 00, 10, 01 do not appear in $\mathbf{K}$. The contribution of element *ab* to the system stiffness matrix is therefore $\mathbf{K}^{ab}$ (17.11). This is the element stiffness matrix in system coordinates.

$$\bar{\mathbf{K}}^{bc} = \begin{array}{|cccc|c} \scriptstyle 0 & \scriptstyle 2 & \scriptstyle 1 & \scriptstyle 0 & \\ \hline 12 & -12 & 6 & 6 & \scriptstyle 0 \\ -12 & \mathbf{12} & \mathbf{-6} & -6 & \scriptstyle 2 \\ 6 & \mathbf{-6} & \mathbf{4} & 2 & \scriptstyle 1 \\ 6 & -6 & 2 & 4 & \scriptstyle 0 \\ \hline \end{array} \rightarrow \mathbf{K}^{bc} = \begin{array}{|cc|c} \scriptstyle 1 & \scriptstyle 2 & \\ \hline 4 & -6 & \scriptstyle 1 \\ -6 & 12 & \scriptstyle 2 \\ \hline \end{array}$$

Similarly, in $\bar{\mathbf{K}}^{bc}$ we find four entries with a corresponding address in $\mathbf{K}$. The contribution of *bc* to the system stiffness matrix is therefore $\mathbf{K}^{bc}$ (17.11). The numbers bordering the matrix are here the row and column indexes in global coordinates.

The assembly of the stiffness matrix is simply

$$\mathbf{K} = \mathbf{K}^{ab} + \mathbf{K}^{bc} \tag{17.11}$$

Please check that this sum indeed gives the matrix in (17.9).

For the load vector it is even easier. Pick up the external forces at the free degrees of freedom, and copy them to their correct address in $\mathbf{p}$. Clearly, $p(1) = 0$ and $p(2) = P$, hence the load vector in (17.9). Obtaining the nodal loads if we have distributed loads or temperature effects is another affair. We will dedicate an entire chapter to this subject.

In practice, the assembly procedure uses up less space. The element stiffness matrices in global coordinates are not really required. Instead, their components are directly added to the system stiffness matrix. Here we have used $\bar{\mathbf{K}}^{ab}$ and $\bar{\mathbf{K}}^{bc}$ for didactic purposes only.

17.3.2 Equilibrium and Assembly Are Equivalent

We will now show that assembling the system stiffness matrix in the sense of (17.11) is equivalent to establishing equilibrium at the nodes. We will use the frame in Fig. 17.13a to illustrate the concept. Note, the structure is without supports and applied loads. We define at every node their degrees of freedom,three per node, as for node b. These are global or structure or system coordinates. The same coordinate system is chosen at all the nodes.

We now define at every node i, the displacements $\mathbf{\Delta}_i$ and loads $\mathbf{p}_i$ along the degrees of freedom. For instance, vector $\mathbf{\Delta}_b$ is the horizontal and vertical displacement and rotation of node b, as shown in the figure, and $\mathbf{p}_b$ is a horizontal and vertical force and couple applied to node b. This allows us to build the global displacement and force vectors, in the case of the figure of size $5 \times 3 = 15$.

$$\bar{\mathbf{\Delta}} = \{\mathbf{\Delta}_a \, \mathbf{\Delta}_b \, \mathbf{\Delta}_c \, \mathbf{\Delta}_d \, \mathbf{\Delta}_e\}^T \,, \quad \bar{\mathbf{p}} = \{\mathbf{p}_a \, \mathbf{p}_b \, \mathbf{p}_c \, \mathbf{p}_d \, \mathbf{p}_e\}^T \tag{17.12}$$

The 'bar' superscript indicates that the vectors are in augmented coordinates (they include the fixed dofs) to distinguish them from the vectors $\mathbf{\Delta}$ and $\mathbf{p}$ along open dofs of the actual structure.

The Augmented Stiffness Matrix

You may have noticed that what we have in Fig. 17.13a is a floating structure, not unlike the elements which we have discussed previously. And indeed we can look at this floating frame as some super element which, instead of being a line with two extremities, is something more intricate, where we have located nodes at the junctions and the (several) extremities.

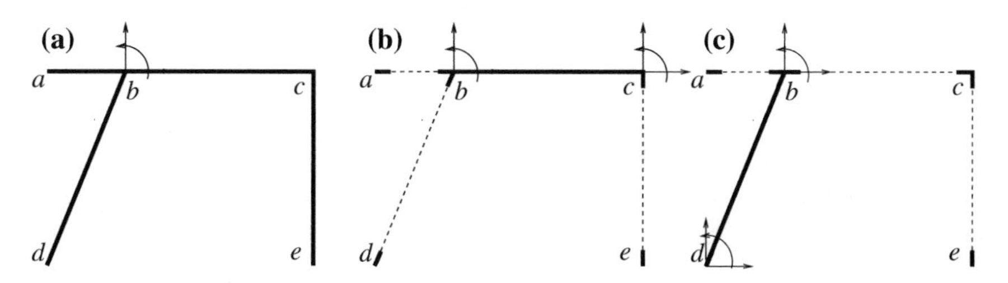

Fig. 17.13 Global degrees of freedom in frame example

Since this super element is floating, we cannot apply any $\bar{\mathbf{p}}$ to the nodes. These forces must be auto-equilibrated. But we can ask the question that was posed with elements: What would be the nodal forces applied to the nodes of the floating structure if the nodes were displaced by $\bar{\boldsymbol{\Delta}}$?

The relationship is of course linear and of the form

$$\bar{\mathbf{p}} = \bar{\mathbf{K}}\bar{\boldsymbol{\Delta}} \tag{17.13}$$

where the coefficient matrix is called the system stiffness matrix in augmented coordinates. This matrix indeed measures stiffness. As for elements, matrix $\bar{\mathbf{K}}$ is singular (why?), symmetric and so on.

We will show that (17.13) represents the equilibrium equations at the nodes of the structure. It will also become clear that, instead of writing the equations and determining $\bar{\mathbf{K}}$ from scratch, we can get this matrix by assembling it on the basis of the element stiffness matrices. (Note, the element stiffness matrix of *ab*, *bc*, *ce* is standard. Element *de* which has a slope with respect to the axes requires a prior coordinates transformation.)

The reasoning is as follows. When we move nodes *b* and *c*, for instance, the forces to be exerted by nodes *b* and *c* on the extremities of element *bc* are

$$\begin{Bmatrix} \mathbf{f}_b^{bc} \\ \mathbf{f}_c^{bc} \end{Bmatrix} = \begin{bmatrix} \mathbf{K}_{bb}^{(bc)} & \mathbf{K}_{bc}^{(bc)} \\ \mathbf{K}_{cb}^{(bc)} & \mathbf{K}_{cc}^{(bc)} \end{bmatrix} \begin{Bmatrix} \boldsymbol{\Delta}_b \\ \boldsymbol{\Delta}_c \end{Bmatrix} \quad \text{or} \quad \mathbf{f}^{bc} = \mathbf{K}^{bc}\,\boldsymbol{\Delta}^{bc} \tag{17.14}$$

Here $\mathbf{f}_b^{(bc)}$ and $\mathbf{f}_c^{(bc)}$ are the two 3×1 force components that must be applied respectively by nodes *b* and *c*, to produce displacements $\boldsymbol{\Delta}_b$ and $\boldsymbol{\Delta}_c$. The 6×6 element stiffness matrix $\mathbf{K}^{(bc)}$ has conveniently been subdivided into four 3×3 submatrices.

The point of interest is that this relation can also be written in the augmented coordinates as follows

$$\begin{Bmatrix} 0 \\ \bar{\mathbf{f}}_b^{bc} \\ \bar{\mathbf{f}}_c^{bc} \\ 0 \\ 0 \end{Bmatrix} = \begin{bmatrix} 0\;0 & 0 & 0\;0 \\ 0\;\mathbf{K}_{bb}^{(bc)} & \mathbf{K}_{bc}^{(bc)} & 0\;0 \\ 0\;\mathbf{K}_{cb}^{(bc)} & \mathbf{K}_{cc}^{(bc)} & 0\;0 \\ 0\;0 & 0 & 0\;0 \\ 0\;0 & 0 & 0\;0 \end{bmatrix} \begin{Bmatrix} \boldsymbol{\Delta}_a \\ \boldsymbol{\Delta}_b \\ \boldsymbol{\Delta}_c \\ \boldsymbol{\Delta}_d \\ \boldsymbol{\Delta}_e \end{Bmatrix} \quad \text{or} \quad \bar{\mathbf{f}}^{bc} = \bar{\mathbf{K}}^{bc}\,\bar{\boldsymbol{\Delta}} \tag{17.15}$$

The second and third equation of (17.15) are the same as (17.14). The remaining equations are simple 'zero equals zero' identities.

This augmented form of the element stiffness relation can also be construed as follows: Consider a structure composed of element *bc*, with all the other elements having zero stiffness (Fig. 17.13b), and let us impose displacements to all the nodes. What forces must the nodes exert? The answer lies in (17.15).

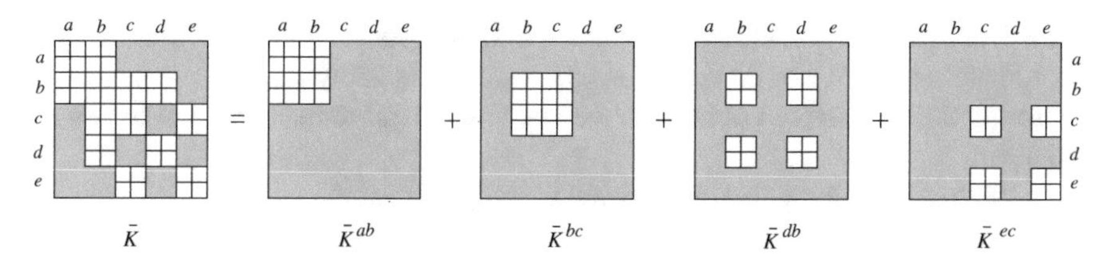

Fig. 17.14 Assembly of system stiffness matrix

The stiffness matrix of element *db* in augmented coordinates is

$$\begin{Bmatrix} 0 \\ \bar{\mathbf{f}}^{db}_b \\ 0 \\ \bar{\mathbf{f}}^{db}_d \\ 0 \end{Bmatrix} = \begin{bmatrix} 0 & 0 & 0 & 0 & 0 \\ 0 & \mathbf{K}^{(db)}_{bb} & 0 & \mathbf{K}^{(db)}_{bd} & 0 \\ 0 & 0 & 0 & 0 & 0 \\ 0 & \mathbf{K}^{(db)}_{db} & 0 & \mathbf{K}^{(db)}_{dd} & 0 \\ 0 & 0 & 0 & 0 & 0 \end{bmatrix} \begin{Bmatrix} \mathbf{\Delta}_a \\ \mathbf{\Delta}_b \\ \mathbf{\Delta}_c \\ \mathbf{\Delta}_d \\ \mathbf{\Delta}_e \end{Bmatrix} \quad \text{or} \quad \bar{\mathbf{f}}^{db} = \bar{\mathbf{K}}^{db}\,\bar{\mathbf{\Delta}} \tag{17.16}$$

All matrices can be seen symbolically in Fig. 17.14.

Equilibrium at the Augmented Coordinates

We can write equations of the type (17.15) or (17.16) for all the elements. Note that $\bar{\mathbf{f}}^{db}$ are the forces applied to element *db* if we impart to the nodes displacements $\bar{\mathbf{\Delta}}$. Now, where do these forces come from? From the nodes, of course. There is no other agent available. Consequently element *db* exerts on the nodes equal and opposite forces.

Now, for the nodes *to be in equilibrium* we need $-(\bar{\mathbf{f}}^{ab} + \bar{\mathbf{f}}^{bc} + \bar{\mathbf{f}}^{bd} + \bar{\mathbf{f}}^{ec}) + \bar{\mathbf{p}} = \mathbf{0}$, that is,

$$\bar{\mathbf{f}}^{ab} + \bar{\mathbf{f}}^{bc} + \bar{\mathbf{f}}^{bd} + \bar{\mathbf{f}}^{ec} = \bar{\mathbf{p}} \tag{17.17}$$

Replacing $\bar{\mathbf{f}}^{ab}$ with $\bar{\mathbf{K}}^{ab}\bar{\mathbf{\Delta}}$, and similarly for the other forces, and replacing $\bar{\mathbf{p}}$ with $\bar{\mathbf{K}}\bar{\mathbf{\Delta}}$ yields $(\bar{\mathbf{K}}^{ab} + \bar{\mathbf{K}}^{ab} + \bar{\mathbf{K}}^{ab} + \bar{\mathbf{K}}^{ab})\bar{\mathbf{\Delta}} = \bar{\mathbf{K}}\bar{\mathbf{\Delta}}$ or $(\bar{\mathbf{K}}^{ab} + \bar{\mathbf{K}}^{ab} + \bar{\mathbf{K}}^{ab} + \bar{\mathbf{K}}^{ab} - \bar{\mathbf{K}})\bar{\mathbf{\Delta}} = 0$, and since the displacements $\bar{\mathbf{\Delta}}$ are arbitrary, the term within the parenthesis must be zero, that is,

$$\bar{\mathbf{K}} = \bar{\mathbf{K}}^{ab} + \bar{\mathbf{K}}^{ab} + \bar{\mathbf{K}}^{ab} + \bar{\mathbf{K}}^{ab} \tag{17.18}$$

which is the *essence of assembly*. In other words, assembly is an expression of nodal equilibrium, which is what we started out to prove.

Equilibrium at the Open Dofs

We can now return to the original structure which is fixed at nodes $d\,e$ and has open degrees of freedom at $a\,b\,c$. When we consider the displacement and force vectors of the floating structure (or super-element), we note an interesting duality.

We will conveniently divide the coordinates into two groups. Subscript '0' is for *open* coordinates at which we can apply forces but we do not know the displacements, and subscript '1' is for closed dofs were we know the displacements but the forces are unknown.

In the frame example this subdivision is straightforward.

$$\begin{aligned} \bar{\mathbf{\Delta}}_0 &= \{\bar{\mathbf{\Delta}}_a\, \bar{\mathbf{\Delta}}_b\, \bar{\mathbf{\Delta}}_c\}^T & \bar{\mathbf{p}}_0 &= \{\bar{\mathbf{p}}_a\, \bar{\mathbf{p}}_b\, \bar{\mathbf{p}}_c\}^T \\ \bar{\mathbf{\Delta}}_1 &= \{\bar{\mathbf{\Delta}}_d\, \bar{\mathbf{\Delta}}_e\}^T & \bar{\mathbf{p}}_1 &= \{\bar{\mathbf{p}}_d\, \bar{\mathbf{p}}_e\}^T \end{aligned} \tag{17.19}$$

At nodes $a\,b\,c$ we know the external forces (a vertical force at a and zero forces elsewhere), but we do not know the displacements. At nodes $d\,e$ on the other hand, we know the displacements (they are all zero) but the applied forces (the reactions, yes) are as yet unknown. With this nomenclature the equilibrium equations of the floating structure can be presented as (see also Fig. 17.15)

$$\begin{bmatrix} \bar{\mathbf{K}}_{00} & \bar{\mathbf{K}}_{01} \\ \bar{\mathbf{K}}_{10} & \bar{\mathbf{K}}_{11} \end{bmatrix} \begin{Bmatrix} \bar{\mathbf{\Delta}}_0 \\ \bar{\mathbf{\Delta}}_1 \end{Bmatrix} = \begin{Bmatrix} \bar{\mathbf{p}}_0 \\ \bar{\mathbf{p}}_1 \end{Bmatrix} \tag{17.20}$$

Bear in mind that the unknowns are the displacements $\bar{\mathbf{\Delta}}_0$ and the reactions $\bar{\mathbf{p}}_1$, and that $\bar{\mathbf{K}}_{00}$ is a square matrix.

Moving the constant term in the first of equations (17.20) to the right-hand side yields

$$\bar{\mathbf{K}}_{00}\bar{\mathbf{\Delta}}_0 = \bar{\mathbf{p}}_0 - \bar{\mathbf{K}}_{01}\bar{\mathbf{\Delta}}_1$$

which can be solved for $\bar{\mathbf{\Delta}}_0$. This concludes the computation of the displacements.

If desired we can proceed to the computation of the reactions by employing the now known displacements in the second of equations (17.20)

$$\bar{\mathbf{p}}_1 \leftarrow \bar{\mathbf{K}}_{10}\bar{\mathbf{\Delta}}_0 + \bar{\mathbf{K}}_{11}\bar{\mathbf{\Delta}}_1$$

Returning to the frame example in Fig. 17.13a, the known displacements are zero, and if when we use the initial nomenclature, $\bar{\mathbf{K}}_{00} = \mathbf{K}$, $\bar{\mathbf{\Delta}}_0 = \mathbf{\Delta}$, $\bar{\mathbf{\Delta}}_1 = \mathbf{0}$ and $\bar{\mathbf{p}}_1 = \mathbf{r}$, where $\mathbf{r}$ are the reactions, the first and second equations in (17.20) become respectively

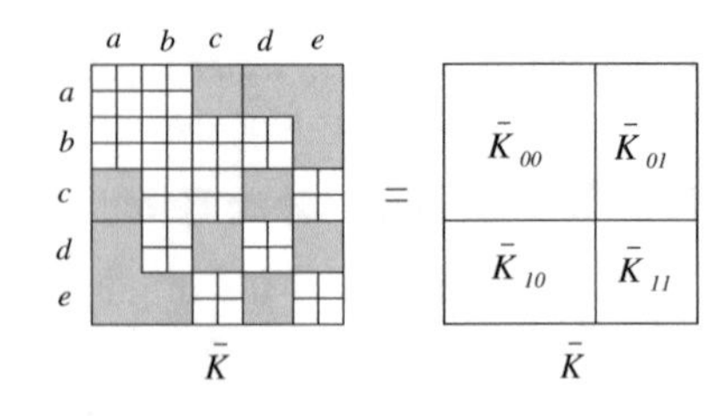

Fig. 17.15 Subdivision of system stiffness matrix

$$\begin{aligned} \mathbf{K}\,\boldsymbol{\Delta} &= \mathbf{p} \\ \mathbf{r} &\leftarrow \bar{\mathbf{K}}_{10}\,\boldsymbol{\Delta} \end{aligned} \tag{17.21}$$

So if we care to assemble also the '**reactions matrix**' $\bar{\mathbf{K}}_{10}$ we obtain the reactions as a by-product of the displacements.

Finally, for the 2-degrees of freedom beam in example Fig. 17.11a, we could have numbered all the coordinates, free and fixed, as in Fig. 17.11b (the numbering sequence is immaterial; for large-scale structures it may affect the efficiency of the computations). This allows to assemble $\bar{\mathbf{K}}_{10}$ as well.

$$\bar{\mathbf{K}}^{ab} = \begin{array}{|cccc|c} 3 & 5 & 4 & 1 & \\ \hline 1.5 & -1.5 & 1.5 & 1.5 & 3 \\ -1.5 & 1.5 & -1.5 & -1.5 & 5 \\ 1.5 & -1.5 & 2 & 1 & 4 \\ 1.5 & -1.5 & 1 & 2 & 1 \\ \hline \end{array} \rightarrow \bar{\mathbf{K}}^{ab}_{10} = \begin{array}{|cc|c} 1 & 2 & \\ \hline 1.5 & 0 & 3 \\ 1 & 0 & 4 \\ -1.5 & 0 & 5 \\ 0 & 0 & 6 \\ \hline \end{array}$$

$$\bar{\mathbf{K}}^{bc} = \begin{array}{|cccc|c} 5 & 2 & 1 & 6 & \\ \hline 12 & -12 & 6 & 6 & 5 \\ -12 & 12 & -6 & -6 & 2 \\ 6 & -6 & 4 & 2 & 1 \\ 6 & -6 & 2 & 4 & 6 \\ \hline \end{array} \rightarrow \bar{\mathbf{K}}^{bc}_{10} = \begin{array}{|cc|c} 1 & 2 & \\ \hline 0 & 0 & 3 \\ 0 & 0 & 4 \\ 6 & -12 & 5 \\ 2 & -6 & 6 \\ \hline \end{array}$$

Summing up the contributions of all the elements $\bar{\mathbf{K}}_{10} = \bar{\mathbf{K}}^{ab}_{10} + \bar{\mathbf{K}}^{bc}_{10}$, and using the displacements at the degrees of freedom (17.1) we readily obtain the reactions

$$\begin{Bmatrix} r_3 \\ r_4 \\ r_5 \\ r_6 \end{Bmatrix} = \begin{Bmatrix} 0.25\,P \\ 0.167\,PL \\ -1.25\,P \\ -0.667\,PL \end{Bmatrix} \leftarrow \begin{bmatrix} 1.5 & 0 \\ 1 & 0 \\ 4.5 & -12 \\ 2 & -6 \end{bmatrix} \begin{Bmatrix} 1/6 \\ 1/6 \end{Bmatrix} \tag{17.22}$$

These are of course the same reactions that we previously deduced from the moment and shear diagrams (Fig. 17.6e). Notice that we have reintroduced the dimensions in this final result (forces in P and couples in PL).

17.4 More on Degrees of Freedom

The main steps of the stiffness method were laid out in the previous sections. By and large this is the whole story. But there are some subtleties which can be introduced here and there to bring the model closer to the physical structure or to reduce the size of a problem. One such issue is the number of degrees of freedom.

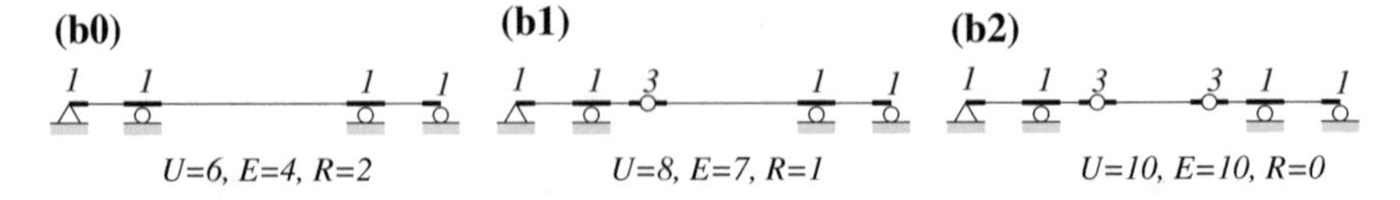

Fig. 17.16 Counting U and E in beams

17.4.1 Degrees of Freedom in the Force and Displacement Methods

We may remember that the issue of degrees of freedom was rather central in the force method. It was one of the ingredients which determined the degree of statical redundancy of a structure. In the displacement method the number of degrees of freedom sets the size of the stiffness matrix **K** or the number of linear equations that we must eventually solve. So whatever we have learned on how to establish degrees of freedom in the force method is also valid here. There is however one little difference. Whereas in the force method degrees of freedom were beneficial since they reduce the redundancy, in the displacement method degrees of freedom are detrimental; they increase the number of simultaneous equations to be solved.

Indeed, let us reexamine some of the examples that were used to illustrate the computation of the redundancy in the force method (Fig. 17.16). Note that the number of equations that can be written at the nodes (E) in the force method is in fact the size of the stiffness matrix. And if in the force method the redundancy of the beams in Fig. 17.16 decreases from left to right, the size of the stiffness matrix increases from 4×4 to 10×10 in the statically determinate case in Fig. 17.16b2.

This is typical. If in the force method we prefer structures with many open degrees of freedom, fixed coordinates are music to the ears of the stiffness method.

The number of elements is of lesser importance in the displacement method. It does not affect the size of the system of equations but we (that is, the computer) may have to invest more time in assembling **K** and in computing the internal distributions of all the elements (post-processing).

17.4.2 Singular Stiffness Matrices

Mechanisms can be detected in the force method when we have more degrees of freedom than statical unknowns, that is, negative redundancy ($R < 0$). But sometimes we encountered structures with $R \geq 0$ which were unstable. So $R \geq 0$ was a necessary but not a sufficient condition for stability. With the stiffness method the test for stability lies with the stiffness matrix. Structure with an inherent instability will have

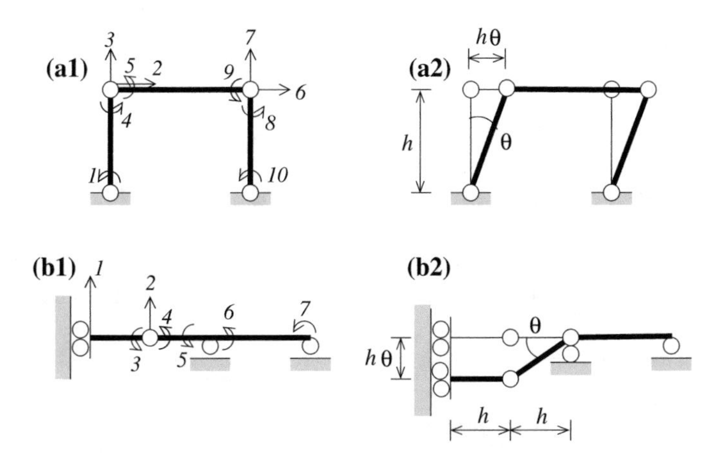

Fig. 17.17 Mechanisms

a singular stiffness matrix. A test for instability is therefore a zero determinant of the stiffness matrix

$$\det(\mathbf{K}) = 0 \qquad \text{Structure has a hidden mechanism} \tag{17.23}$$

Consider for instance the frame in Fig. 17.17a1. The displacement pattern in Fig. 17.17a2 is called a rigid-body displacement since no element deforms. And since no element deforms, such a displacement is very easy to achieve. In fact, it necessitates zero applied forces (forces are there only to deform the structure). With the numbering in Fig. 17.17a1 the rigid body displacements are

$$\mathbf{\Delta} = \left\{ -\theta \; h\theta \; 0 \; -\theta \; 0 \; h\theta \; 0 \; -\theta \; 0 \; -\theta \right\}^T \tag{17.24}$$

and since for such displacements we need zero forces we have $\mathbf{K\Delta} = \mathbf{0}$. Now, for such a relation to hold, necessarily the determinant of $\mathbf{K}$ is zero and the matrix is singular. This is a generalization of a scalar relation of the type $k\delta = 0$ which requires $k = 0$ if $\delta \neq 0$.

Another example of a structure which has an embedded mechanism is the beam in Fig. 17.17b1. For the rigid body displacements

$$\mathbf{\Delta} = \left\{ -h\theta \; -h\theta \; 0 \; \theta \; \theta \; 0 \; 0 \right\}^T \tag{17.25}$$

we need not apply any forces, and forcibly $\det(\mathbf{K})$ is zero.

17.5 Frames

17.5.1 Inextensible Elements

Frame elements are designed to resist bending and are accordingly relatively bulky. As a by-product they are therefore very rigid in the axial direction. So much so that in many analyses of frames, we simply assume that the components of the frame are inextensible. This assumption will be adopted in this text, in other words,

> Frame elements have *finite* bending stiffness *EI* but *infinite* axial stiffness $EA \to \infty$.

This reduces the size of the stiffness matrix but in the process we lose all information regarding the normal forces in the elements. Very often these can be reconstructed on the basis of the shear forces but it involves careful investigation of the translational equilibrium at the nodes. The following example will illustrate the procedure. Consider the six degrees of freedom of frame *abcd* in Fig. 17.18a, where all elements have a common flexural *EI* and axial *EA* stiffness. A straightforward analysis of this structure involves a 6×6 matrix in coordinates $\mathbf{\Delta}_b$ and $\mathbf{\Delta}_c$. Now chances are that the results of the analysis will show that Δ_2 and Δ_5 are very small because frame elements are very stiff axially. Therefore, *ba* and *cd* will not elongate in any relevant manner. Likewise since *bc* will not elongate, Δ_1 and Δ_4 will be almost equal. So if we are willing to apply some engineering judgment we could know right from the start that for all practical purposes

$$\Delta_2 = 0 \qquad \Delta_5 = 0 \qquad \Delta_4 = \Delta_1 \tag{17.26}$$

That is, there are only three independent degrees of freedom as indicated in Fig. 17.18b. We expect therefore to solve a reduced system of equations.

17.5.2 Congruent Transformation

The algorithmic way to reduce the size of the stiffness matrix is by a congruent transformation. We first assemble the equilibrium equations $\hat{\mathbf{K}}\hat{\mathbf{\Delta}} = \hat{\mathbf{p}}$ in the 6×6

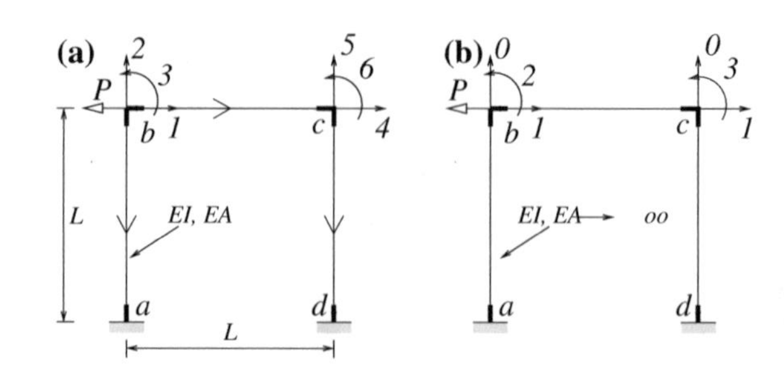

Fig. 17.18 Portal frame $EA \to \infty$ **a** 'old' degrees of freedom, **b** 'New' degrees of freedom

(old) system. In the present example the three elements are identical, so we can use the same (standard) element stiffness matrix for the assembly.

$$
\begin{array}{c|cccccc|ccc}
 & 4 & 0 & 6 & 0 & 5 & 0 & & & cd \\
 & 2 & 5 & 3 & 6 & 1 & 4 & & bc & \\
 & 1 & 0 & 3 & 0 & 2 & 0 & ba & & \\
\hline
 & 12 & -12 & 6L & 6L & 0 & 0 & 1 & 2 & 4 \\
\frac{EI}{L^3} & -12 & 12 & -6L & -6L & 0 & 0 & 0 & 5 & 0 \\
 & 6L & -6L & 4L^2 & 2L^2 & 0 & 0 & 3 & 3 & 6 \\
 & 6L & -6L & 2L^2 & 4L^2 & 0 & 0 & 0 & 6 & 0 \\
 & 0 & 0 & 0 & 0 & \beta & -\beta & 2 & 1 & 5 \\
 & 0 & 0 & 0 & 0 & -\beta & \beta & 0 & 4 & 0 \\
\hline
\end{array}
\qquad (17.27)
$$

Note that $\beta = AL^2/I$ is a measure of the axial stiffness relative to the flexural stiffness. Consider element ab. When we look at the element from the right, node a is at the left end of the element, and the local axis runs from a to b (left to right) as we chose to do here. The numbering of the nodes in the global sequence (left shear, right shear, left moment ...) is then (1 0 3 0 2 0).[1]

Repeating this for the other elements bc (from b to c) and dc (from d to c), the contribution of the three elements to the global stiffness matrix is

$$
\bar{\mathbf{K}}^{ab} = \frac{EI}{L^3}\begin{bmatrix} 12 & 0 & 6L & 0 & 0 & 0 \\ 0 & \beta & 0 & 0 & 0 & 0 \\ 6L & 0 & 4L^2 & 0 & 0 & 0 \\ 0 & 0 & 0 & 0 & 0 & 0 \\ 0 & 0 & 0 & 0 & 0 & 0 \\ 0 & 0 & 0 & 0 & 0 & 0 \end{bmatrix}
$$

$$
\bar{\mathbf{K}}^{bc} = \frac{EI}{L^3}\begin{bmatrix} \beta & 0 & 0 & -\beta & 0 & 0 \\ 0 & 12 & 6L & 0 & -12 & 6L \\ 0 & 6L & 4L^2 & 0 & -6L & 2L^2 \\ -\beta & 0 & 0 & \beta & 0 & 0 \\ 0 & -12 & -6L & 0 & 12 & -6L \\ 0 & 6L & 2L^2 & 0 & -6L & 4L^2 \end{bmatrix}
$$

and

$$
\bar{\mathbf{K}}^{dc} = \frac{EI}{L^3}\begin{bmatrix} 0 & 0 & 0 & 0 & 0 & 0 \\ 0 & 0 & 0 & 0 & 0 & 0 \\ 0 & 0 & 0 & 0 & 0 & 0 \\ 0 & 0 & 0 & 12 & 0 & 6L \\ 0 & 0 & 0 & 0 & \beta & 0 \\ 0 & 0 & 0 & 6L & 0 & 4L^2 \end{bmatrix}
$$

[1] The axial stiffness sub-matrix is the same whether the dofs at the extremities point both right or both left.

giving the equilibrium equations $\hat{\mathbf{K}}\hat{\mathbf{\Delta}} = \hat{\mathbf{p}}$

$$\frac{EI}{L^3}\begin{bmatrix} 12+\beta & 0 & 6L & -\beta & 0 & 0 \\ 0 & 12+\beta & 6L & 0 & -12 & 6L \\ 6L & 6L & 8L^2 & 0 & -6L & 2L^2 \\ -\beta & 0 & 0 & 12+\beta & 0 & 6L \\ 0 & -12 & -6L & 0 & 12+\beta & -6L \\ 0 & 6L & 2L^2 & 6L & -6L & 4L^2 \end{bmatrix} \begin{Bmatrix} \hat{\Delta}_1 \\ \hat{\Delta}_2 \\ \hat{\Delta}_3 \\ \hat{\Delta}_4 \\ \hat{\Delta}_5 \\ \hat{\Delta}_6 \end{Bmatrix} = \begin{Bmatrix} -P \\ 0 \\ 0 \\ 0 \\ 0 \\ 0 \end{Bmatrix}$$

We employ the 'hat' superscripts because this system will be replaced by a new one. We recall that using elements which are axially infinitely stiff ($EA \rightarrow \infty$) imposes three constraints on the displacements (17.26). An alternative way to write the constraints is to express the six 'old' displacements $\hat{\mathbf{\Delta}}$ in terms of the three 'new' ones $\mathbf{\Delta}$ as follows:

$$\begin{Bmatrix} \hat{\Delta}_1 \\ \hat{\Delta}_2 \\ \hat{\Delta}_3 \\ \hat{\Delta}_4 \\ \hat{\Delta}_5 \\ \hat{\Delta}_6 \end{Bmatrix} = \begin{bmatrix} 1 & 0 & 0 \\ 0 & 0 & 0 \\ 0 & 1 & 0 \\ 1 & 0 & 0 \\ 0 & 0 & 0 \\ 0 & 0 & 1 \end{bmatrix} \begin{Bmatrix} \Delta_1 \\ \Delta_2 \\ \Delta_3 \end{Bmatrix} \tag{17.28}$$

We have here an expression between displacements of the form $\hat{\mathbf{\Delta}} = \mathbf{G\Delta}$. This leads to the reduced system of equations $\mathbf{K}\,\mathbf{\Delta} = \mathbf{p}$ where the stiffness matrix is given by the triple product $\mathbf{K} = \mathbf{G}^T\hat{\mathbf{K}}\mathbf{G}$ and $\mathbf{p} = \mathbf{G}^T\hat{\mathbf{p}}$.

It goes without saying that this way of proceeding, even for this very simple portal frame, is not suitable for hand calculations, let alone when thousands of coordinates are involved.

For our purposes we will assemble directly the reduced equilibrium equations.

17.5.3 Direct Assembly

For hand calculations or for didactic reasons a shortcut is preferable. It consists of direct assembly of the reduced equations in terms of the actual degrees of freedom. The other coordinates are set to zero as in Fig. 17.18b.

The element stiffness matrices in global coordinates are as given in (17.27) The contribution of the three elements to the system stiffness matrix (degrees of freedom 1, 2, 3 or addresses 11, 12, 13, 21, 22, 23, 31, 32, 33) is

$$\bar{\mathbf{K}}^{ba} = \frac{EI}{L^3}\begin{bmatrix} 12 & 6L & 0 \\ 6L & 4L^2 & 0 \\ 0 & 0 & 0 \end{bmatrix} \qquad \bar{\mathbf{K}}^{bc} = \frac{EI}{L^3}\begin{bmatrix} 0 & 0 & 0 \\ 0 & 4L^2 & 2L^2 \\ 0 & 2L^2 & 4L^2 \end{bmatrix}$$

and

$$\bar{\mathbf{K}}^{cd} = \frac{EI}{L^3}\begin{bmatrix} 12 & 0 & 6L \\ 0 & 0 & 0 \\ 6L & 0 & 2L^2 \end{bmatrix}$$

Note, β does not appear in any of these matrices. Indeed, $\mathbf{K}^{bc}(1,1)$ receives $\beta - \beta + \beta - \beta$ and remains zero. Similarly, for the force vector $\mathbf{p}(1)$ there is once $-P$ and once zero, consequently $\mathbf{p}(1) = -P$. The equilibrium equations at the degrees of freedom are thus

$$\frac{EI}{L^3}\begin{bmatrix} 24 & 6L & 6L \\ 6L & 8L^2 & 2L^2 \\ 6L & 2L^2 & 8L^2 \end{bmatrix}\begin{Bmatrix} \Delta_1 \\ \Delta_2 \\ \Delta_3 \end{Bmatrix} = \begin{Bmatrix} -P \\ 0 \\ 0 \end{Bmatrix} \tag{17.29}$$

For the solution of these equations just remove the symbols EI/L^3, L, L^2, P, and solve the equations as a regular system of linear equations $\mathbf{Ax} = \mathbf{b}$

$$\begin{bmatrix} 24 & 6 & 6 \\ 6 & 8 & 2 \\ 6 & 2 & 8 \end{bmatrix}\begin{Bmatrix} x_1 \\ x_2 \\ x_3 \end{Bmatrix} = \begin{Bmatrix} -1 \\ 0 \\ 0 \end{Bmatrix} \tag{17.30}$$

the solution of which is

$$\mathbf{x} = \{-0.0595\ 0.0357\ 0.0357\}$$

or with the proper units

$$\{\Delta_1\ \Delta_2\ \Delta_3\} = \frac{PL^2}{EI}\{-0.0595L\ 0.0357\ 0.0357\} \tag{17.31}$$

We can now compute the element end-forces by multiplying the element stiffness matrices by their end-displacements in local coordinates, that is,

$$\begin{aligned} \bar{\mathbf{f}}^{ba} &\leftarrow \bar{\mathbf{K}}^{ba}\bar{\boldsymbol{\Delta}}^{ba} \\ \bar{\mathbf{f}}^{bc} &\leftarrow \bar{\mathbf{K}}^{bc}\bar{\boldsymbol{\Delta}}^{bc} \\ \bar{\mathbf{f}}^{cd} &\leftarrow \bar{\mathbf{K}}^{cd}\bar{\boldsymbol{\Delta}}^{cd} \end{aligned}$$

and since the element stiffness are in this case the same standard matrix we find $[\bar{\mathbf{f}}^{ba}\ \bar{\mathbf{f}}^{ba}\ \bar{\mathbf{f}}^{ba}] \leftarrow \bar{\mathbf{K}}[\bar{\boldsymbol{\Delta}}^{ba}\ \bar{\boldsymbol{\Delta}}^{ba}\ \bar{\boldsymbol{\Delta}}^{ba}]$ or (having removed the symbols)

$$\begin{bmatrix} -0.5 & 0.428 & -0.5 \\ 0.5 & -0.428 & 0.5 \\ -0.214 & 0.214 & -0.214 \\ -0.286 & 0.214 & -0.286 \\ 0 & 0 & 0 \\ 0 & 0 & 0 \end{bmatrix}$$

$$\leftarrow \begin{bmatrix} 12 & -12 & 6 & 6 & 0 & 0 \\ -12 & 12 & -6 & -6 & 0 & 0 \\ 6 & -6 & 4 & 2 & 0 & 0 \\ 6 & -6 & 2 & 4 & 0 & 0 \\ 0 & 0 & 0 & 0 & \beta & -\beta \\ 0 & 0 & 0 & 0 & -\beta & \beta \end{bmatrix} \begin{bmatrix} -0.0595 & 0 & -0.0595 \\ 0 & 0 & 0 \\ 0.0357 & 0.0357 & 0.0357 \\ 0 & 0.0357 & 0 \\ 0 & 0 & 0 \\ 0 & 0 & 0 \end{bmatrix}$$

This produces the shear (positive sense of the element axis for the shear diagram is bottom–top for the columns and left–right for the beam) and moment distributions shown in Fig. 17.19. It is easy to see that forces are in units P and bending moments in PL.

This procedure gives *zero* normal forces for all elements (see last two entries in each force vector). This does not make much sense. Clearly nodes b, c have no translational equilibrium without normal loads as is clear form Fig. 17.19a. We must surmise a horizontal force acting to the left of intensity $0.5P$ and an up-force of $0.428P$ (see emphasized vectors in the inset of Fig. 17.19a). These forces are applied by elements bc and ba respectively. Therefore, the node applies equal and opposite forces to the elements and consequently $n_{ba} = -0.428P$ and $n_{bc} = -0.5P$. By a similar argument we find $n_{cd} = +0.428P$, hence the normal diagram in Fig. 17.20. In the same figure we also have the displaced shape (based on the curvature distribution, in this case also the bending moment diagram) and the external force and reactions. We note that for rotational equilibrium we have an anti-clockwise couple PL (external force and horizontal reactions), which is balanced by the couples applied by the ground on each column and the couple produced by the vertical reactions $(0.286 + 0.286 + 0.428)PL$.

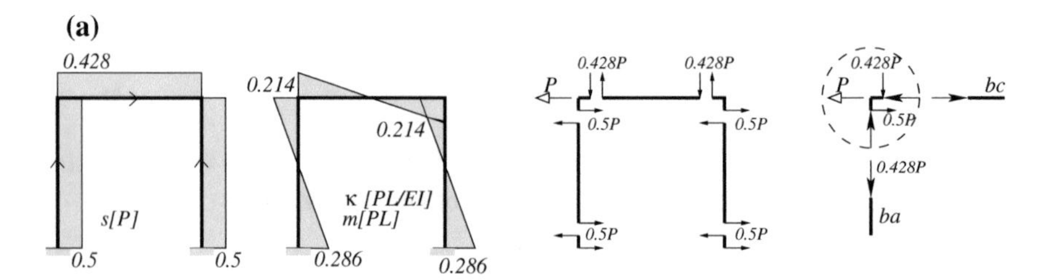

Fig. 17.19 Portal frame *sm*-diagrams. **a** Shear forces on elements and nodes

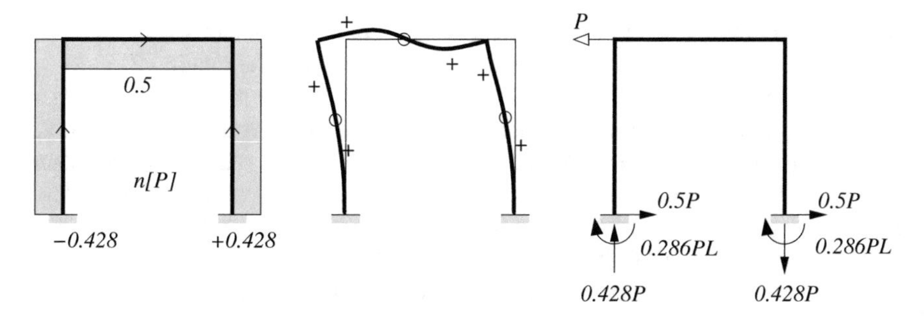

Fig. 17.20 Portal frame n-diagram

In summary, by assuming that elements are infinitely stiff in the axial direction, we reduce the system of equations, but we also lose all information regarding the normal forces. It rests with the engineer to complement the results from statics, as much as possible.

17.5.4 Axial Elements Combined with Frame Elements

The barbershop emblem in Fig. 17.21a was originally hanging from a cantilever beam emblazoned with traditional multicolored spiral stripes. The local municipality deemed the emblem not safe enough and ordered it to be strengthened by a diagonal cable as indicated in the figure. We are asked to analyze the structure, that is, to compute its internal forces. The emblem of weight $2P$ is secured to the beam by two straps. The left one, which carries half the weight of the emblem is for all practical purposes taken by the wall at a. The other half, P, is suspended at b (Fig. 17.21b).

In view of a displacement analysis we indicate in Fig. 17.21c the coordinates at the nodes, and as shown there are only two degrees of freedom. Indeed, the beam and rod are fixed at the wall and the only node that can move is b. With the assumption

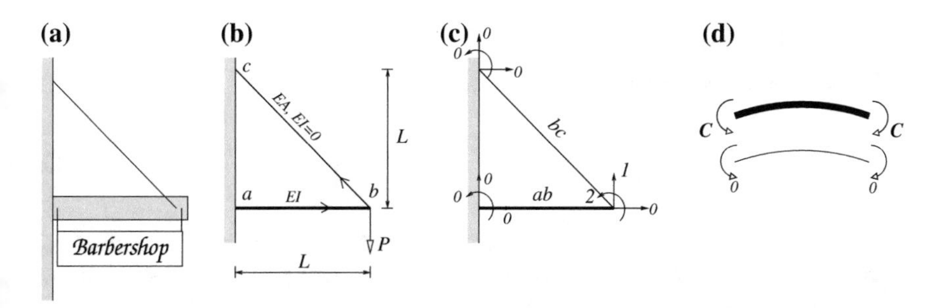

Fig. 17.21 Barbershop board. **a** Actual structure. **b** Structural model. **c** Degrees of freedom. **d** Flexing a beam and a rod

of axially infinitely stiff frame elements, $EA \to \infty$, the horizontal displacement is closed, and we are left with the vertical displacement 1 and the rotation 2.

What makes this very simple structure special, is that we have an element (cb) with no flexural rigidity attached to a node that translates and rotates. Well, the node will need to apply a force to translate (because it stretches or contracts the rod) but, as far as the rod is concerned, node b can rotate as much as it wishes. There is little flexural rigidity in the element (think of a chain) and almost zero end-couples will be found at the extremities (Fig. 17.21d).

The stiffness matrix of the beam element ab in system coordinates is

$$\frac{EI}{L^3}\left|\begin{array}{cccc|c} 0 & 1 & 0 & 2 & ab \\ \hline 12 & -12 & 6L & 6L & 0 \\ -12 & \mathbf{12} & -6L & \mathbf{-6L} & 1 \\ 6L & -6L & 4L^2 & 2L^2 & 0 \\ 6L & \mathbf{-6L} & 2L^2 & \mathbf{4L^2} & 2 \\ \hline \end{array}\right. \tag{17.32}$$

The rod element is a truss element inclined by 45°. If we consider the element as bc, that is, the axis is positive from b to c, the inclination is $\alpha = 135°$. Since the element length is $1.41L$ we find with $\cos(135°) = -0.707$ and $\sin(135°) = 0.707$ that the rod stiffness matrix in system coordinates is

$$\frac{EA}{1.41L}\left|\begin{array}{cccccc|c} 0 & 0 & 0 & 0 & 1 & 2 & bc \\ \hline 0.5 & -0.5 & 0 & -0.5 & 0.5 & 0 & 0 \\ -0.5 & 0.5 & 0 & 0.5 & -0.5 & 0 & 0 \\ 0 & -0 & 0 & 0 & 0 & 0 & 0 \\ -0.5 & 0.5 & 0 & 0.5 & -0.5 & 0 & 0 \\ 0.5 & -0.5 & 0 & -0.5 & \mathbf{0.5} & 0 & 1 \\ 0 & -0 & 0 & 0 & 0 & 0 & 2 \end{array}\right. \tag{17.33}$$

The assembled equilibrium equations are now

$$\frac{EI}{L^3}\begin{bmatrix} 12 + 0.355\beta & -6L \\ -6L & 4L^2 \end{bmatrix}\begin{Bmatrix} \Delta_1 \\ \Delta_2 \end{Bmatrix} = \begin{Bmatrix} -P \\ 0 \end{Bmatrix} \tag{17.34}$$

where $\beta = A_{\mathrm{rod}}L^2/I_{\mathrm{beam}}$.

For the numerics we solve the algebraic system

$$\begin{bmatrix} 12 + 0.355\beta - 6 \\ -6 \quad\quad\quad\quad 4 \end{bmatrix}\begin{Bmatrix} x_1 \\ x_2 \end{Bmatrix} = \begin{Bmatrix} -1 \\ 0 \end{Bmatrix} \tag{17.35}$$

The result, as usual with redundant structures, depends on the relative stiffness of the rod (stretching) and the beam (curving). Originally, with $\beta = 0$, we get the simple cantilever deflection $\Delta_1 = -0.333PL^3/EI$. With the introduction of a cable, with $\beta = 4$ for instance, the deflection is already 2/3 of what it was $\Delta_1 = -0.2262PL^3/EI$ (the slope is reduced from $\Delta_2 = -0.5$ to $-0.339PL^2/EI$).

Assuming a beam and wire with circular cross-sections let the radius of the beam be $R_{\text{beam}} = 0.1L$. Introducing $I_{\text{beam}} = 0.25\pi R_{\text{beam}}^4$ and $A_{\text{rod}} = \pi R_{\text{rod}}^2$ into $\beta = 4$ yields $R_{\text{rod}}/R_{\text{beam}} = 0.1$. You can judge for yourself that we do not need much wire to achieve this stiffening. This is the ratio of the diameters shown in the elements of Fig. 17.21d.

To close this exercise we find for the rod (note $\beta = 4$)

$$\begin{Bmatrix} f_1 \\ f_2 \\ f_3 \\ f_4 \\ f_5 \\ f_6 \end{Bmatrix} \leftarrow P \begin{Bmatrix} -0.0802 \\ 0.0802 \\ 0 \\ 0.0802 \\ -0.0802 \\ 0 \end{Bmatrix}$$

$$\leftarrow \frac{EA}{1.41L} \begin{bmatrix} 0.5 & -0.5 & 0 & -0.5 & 0.5 & 0 \\ -0.5 & 0.5 & 0 & 0.5 & -0.5 & 0 \\ 0 & -0 & 0 & 0 & 0 & 0 \\ -0.5 & 0.5 & 0 & 0.5 & -0.5 & 0 \\ 0.5 & -0.5 & 0 & -0.5 & 0.5 & 0 \\ 0 & -0 & 0 & 0 & 0 & 0 \end{bmatrix} \frac{PL^2}{EI} \begin{Bmatrix} 0 \\ 0 \\ 0 \\ 0 \\ -0.226L \\ -0.339 \end{Bmatrix}$$

This amounts to a tensile force in the wire $n_{\text{rod}} = \sqrt{(f_4^2 + f_5^2)} = 0.113P$.

17.5.5 Trusses

Unless fully computerized, analyzing trusses by the displacement method is a none starter. There is simply too much tedious work involved. We will indicate the main steps in a very simple case but we will leave the actual implementation for large structures to algorithmic processes.

Consider the 3-bar truss in Fig. 17.22a. Nodes $a\,b\,c$ are fixed and node d is free. Since truss elements are oblivious of end-rotations, we simply ignore them in truss models. Consequently, this structure has only translational degrees of freedom Δ_1 and Δ_2 shown in Fig. 17.22b. The elements have been determined as $da\ db\ dc$ and, consequently, their slope α is respectively 135° 90° 45°. The element stiffness matrices in global coordinates and the equilibrium equations are respectively

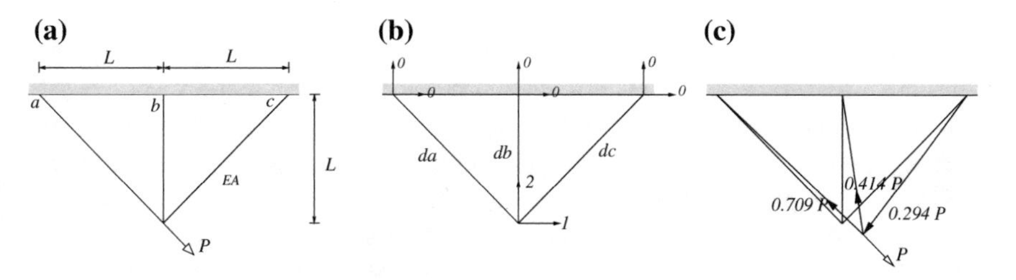

Fig. 17.22 Three-bar truss **a** Structural model. **b** Degrees of freedom. **c** Equilibrium configuration

$$\frac{EA}{1.41L}\begin{array}{|rrrr|c}
1 & 2 & 0 & 0 & da \\ \hline
0.5 & -0.5 & -0.5 & 0.5 & 1 \\
-0.5 & 0.5 & 0.5 & -0.5 & 2 \\
-0.5 & 0.5 & 0.5 & -0.5 & 0 \\
0.5 & -0.5 & -0.5 & 0.5 & 0
\end{array}
\quad
\frac{EA}{L}\begin{array}{|rrrr|c}
1 & 2 & 0 & 0 & db \\ \hline
0 & 0 & 0 & 0 & 1 \\
0 & 1 & 0 & 1 & 2 \\
0 & 0 & 0 & -0 & 0 \\
0 & -1 & 0 & 1 & 0
\end{array}
\tag{17.36}$$

$$\frac{EA}{1.41L}\begin{array}{|rrrr|c}
1 & 2 & 0 & 0 & dc \\ \hline
0.5 & 0.5 & -0.5 & -0.5 & 1 \\
0.5 & 0.5 & -0.5 & -0.5 & 2 \\
-0.5 & -0.5 & 0.5 & 0.5 & 0 \\
-0.5 & -0.5 & 0.5 & 0.5 & 0
\end{array}$$

$$\frac{EA}{L}\begin{bmatrix} 0.707 & 0 \\ 0 & 1.707 \end{bmatrix}\begin{Bmatrix} \Delta_1 \\ \Delta_2 \end{Bmatrix} = \begin{Bmatrix} 0.707P \\ -0.707P \end{Bmatrix}$$

which yields $\Delta_1 = PL/EA$ and $\Delta_2 = -0.414PL/EA$. The end-forces for the three elements are thus

$$\begin{Bmatrix} f_1 \\ f_2 \\ f_4 \\ f_5 \end{Bmatrix}^{da} = P\begin{Bmatrix} 0.5014 \\ -0.5014 \\ -0.5014 \\ 0.5014 \end{Bmatrix} \leftarrow \frac{EA}{1.41L}\begin{bmatrix} 0.5 & -0.5 \\ -0.5 & 0.5 \\ -0.5 & 0.5 \\ 0.5 & -0.5 \end{bmatrix}\frac{PL}{EA}\begin{Bmatrix} 1.0 \\ -0.414 \end{Bmatrix}$$

$$\begin{Bmatrix} f_1 \\ f_2 \\ f_4 \\ f_5 \end{Bmatrix}^{db} = P\begin{Bmatrix} 0 \\ -0.414 \\ 0 \\ 0.414 \end{Bmatrix} \leftarrow \frac{EA}{L}\begin{bmatrix} 0 & 0 \\ 0 & 1 \\ 0 & 0 \\ 0 & -1 \end{bmatrix}\frac{PL}{EA}\begin{Bmatrix} 1.0 \\ -0.414 \end{Bmatrix}$$

and

$$\begin{Bmatrix} f_1 \\ f_2 \\ f_4 \\ f_5 \end{Bmatrix}^{dc} = P\begin{Bmatrix} 0.2078 \\ 0.2078 \\ -0.2078 \\ -0.2078 \end{Bmatrix} \leftarrow \frac{EA}{1.41L}\begin{bmatrix} 0.5 & 0.5 \\ 0.5 & 0.5 \\ -0.5 & -0.5 \\ -0.5 & -0.5 \end{bmatrix}\frac{PL}{EA}\begin{Bmatrix} 1.0 \\ -0.414 \end{Bmatrix}$$

These are the loads applied by the nodes to the elements. We easily find $n_{da} = 0.709P$, $n_{db} = 0.414P$ and $n_{dc} = -0.294P$. And indeed, these internal forces balance the applied load as shown in Fig. 17.22c.

Before proceeding we will draw the attention to two features. Try any other displacements and compute the ensuing normal forces and you will never have equilibrium at the degrees of freedom. There is only one solution to a structural problem.

Also, the normal forces were computed under the assumption that the bars had 135° 90° 45° slopes respectively. But at equilibrium these slopes have changed, so the assumption is wrong, and so is presumably the solution. This is indeed the case. The question is how wrong are we? Well, within the framework of (very) small displacements the error is (very) small, and compared to the uncertainties regarding the forces, supports, manufacturing tolerances and the like, what we miss by writing the equilibrium equations in the original configuration is negligible.

17.6 Illustrated Examples

17.6.1 *Fixed Beam with Elastic Supports*

Consider the fixed beam of bending stiffness EI on two additional elastic supports of stiffness k, loaded at b by a point force Q, depicted in Fig. 17.23a. In addition we are given the value of the spring stiffness in terms of EI and L

$$k = 5EI/L^3.$$

Assembling K and Solving the Equilibrium Equations

The beam will be modeled with 6 nodes $abcdef$, three beam elements ab, bc, cd and two spring elements be, cf. At every node we have nominally two dofs (vertical translation and rotation—we assume that in beams nothing happens in the axial direction). Consequently, we have only four unknown nodal displacements, and the system stiffness matrix **K** will thus be of size 4×4.

$$\frac{EI}{L^3}\begin{array}{|cccc|ccc} 2 & 0 & 4 & 0 & & & dc \\ 1 & 2 & 3 & 4 & & bc & \\ 0 & 1 & 0 & 3 & ab & & \\ \hline 12 & -12 & 6L & 6L & 0 & 1 & 2 \\ -12 & 12 & -6L & -6L & 1 & 2 & 0 \\ 6L & -6L & 4L^2 & 2L^2 & 0 & 3 & 4 \\ 6L & -6L & 2L^2 & 4L^2 & 3 & 4 & 0 \\ \hline \end{array} \qquad \frac{EI}{L^3}\begin{array}{|cc|cc} 0 & 2 & & cf \\ 0 & 1 & be & \\ \hline 5 & -5 & 0 & 0 \\ -5 & 5 & 1 & 2 \\ \hline \end{array} \qquad (17.37)$$

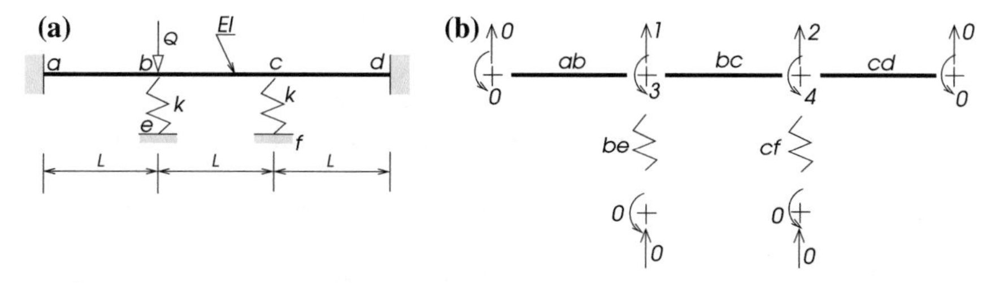

Fig. 17.23 Fixed beam on two springs

The fixed dofs will be labeled '0' as shown in the exploded view in Fig. 17.23b. We first relate the local dofs of the three beam elements to the global ones (17.37 left), and next the local dofs of the two spring elements to the global ones (17.37 right).

For assembly of the system stiffness matrix, we prepare a zero 4×4 matrix, and add into it all the components of the five matrices in (17.37) at their address in the global matrix ($i, j = 1..4$).

$$\frac{EI}{L^3}\begin{bmatrix} 29 & -12 & 0 & 6L \\ -12 & 29 & -6L & 0 \\ 0 & -6L & 8L^2 & 2L^2 \\ 6L & 0 & 2L^2 & 8L^2 \end{bmatrix}\begin{Bmatrix} \Delta_1 \\ \Delta_2 \\ \Theta_3 \\ \Theta_4 \end{Bmatrix} = \begin{Bmatrix} -P \\ 0 \\ 0 \\ 0 \end{Bmatrix} \tag{17.38}$$

or after removing the units gives

$$\begin{bmatrix} 29 & -12 & 0 & 6 \\ -12 & 29 & -6 & 0 \\ 0 & -6 & 8 & 2 \\ 6 & 0 & 2 & 8 \end{bmatrix}\begin{Bmatrix} x_1 \\ x_2 \\ x_3 \\ x_4 \end{Bmatrix} = \begin{Bmatrix} -1 \\ 0 \\ 0 \\ 0 \end{Bmatrix} \tag{17.39}$$

the solution of which is

$$\mathbf{x} = \{-0.0588 \quad -0.0321 \quad -0.0374 \quad 0.0532\}^T$$

Computing the End-Loads and Drawing the *sm*-diagrams

Multiplying the element stiffness matrices by the end-displacement gives the loads applied to the extremities of the elements. For example, pre-multiplying the end-displacements of beam element *ab* by its element stiffness matrix yields the loads applied to the element

$$\begin{Bmatrix} 0.4813 \\ -0.4813 \\ 0.2781 \\ 0.2032 \end{Bmatrix} \leftarrow \begin{bmatrix} 12 & -12 & 6 & 6 \\ -12 & 12 & -6 & -6 \\ 6 & -6 & 4 & 2 \\ 6 & -6 & 2 & 4 \end{bmatrix}\begin{Bmatrix} 0 \\ -0.0588 \\ 0 \\ -0.0374 \end{Bmatrix}$$

where the first two components are the shear forces respectively at a and b (measured in units P), and the following two are the bending moments at a and b respectively (measured in units PL).

Repeating the procedure for elements bc and cd produces the beam element end-loads, shown in Fig. 17.24a.

For springs eb and fc, we respectively multiply the spring element stiffness matrices by their end-displacements to obtain the forces in both springs

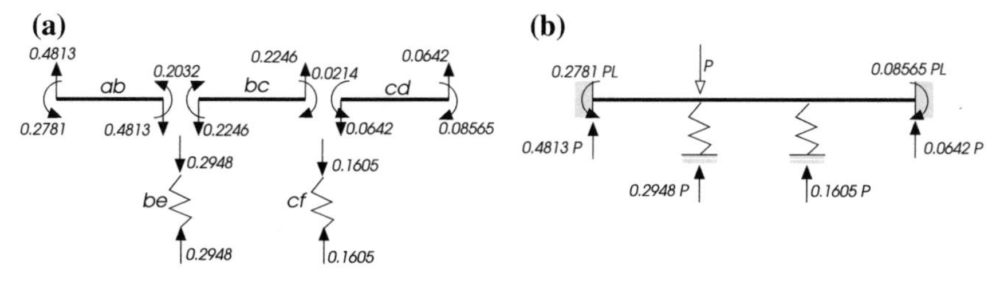

Fig. 17.24 Fixed beam on two springs: end-loads

$$\begin{Bmatrix} -0.2948 \\ 0.2948 \end{Bmatrix} \leftarrow \begin{bmatrix} 5 & -5 \\ -5 & 5 \end{bmatrix} \begin{Bmatrix} 0 \\ -0.0588 \end{Bmatrix}$$

and

$$\begin{Bmatrix} -0.1605 \\ 0.1605 \end{Bmatrix} \leftarrow \begin{bmatrix} 5 & -5 \\ -5 & 5 \end{bmatrix} \begin{Bmatrix} 0 \\ -0.0321 \end{Bmatrix}$$

as also shown in Fig. 17.24a. The reactions follow immediately (Fig. 17.24b).

Noting that the shear is constant on every element and the bending moment linear (no distributed forces), we can now draw the shear and bending moment diagrams element by element (see Fig. 17.25). We will leave the exploded view of the *sm*-diagrams (a) in the left part, for didactic reasons, with the standard diagrams (b) to the right.

The diagrams are now completed with the equilibrium configuration (bottom of Fig. 17.25b), where we have denoted the sides of the fibers in tension with a + sign and the positions of zero-curvature (change of sign of the curvature) by small circles. These points correspond to sections of zero bending moments (no applied curvature).

We will note that the bending moments on both sides of node b, calculated separately, are equal and opposite ($0.2032PL$). So are the bending moments on both sides of node c. These nodes are thus in rotational equilibrium. The same is valid for translational equilibrium at these nodes (see the drawing at the bottom of Fig. 17.25a) for translational equilibrium of node b.

17.6.2 Beam with Guides and a Roller

The beam of uniform stiffness EI in Fig. 17.26a has vertical guides at the extremities a and c, a simple support at $1/3$ span and is subjected to a point force P at a. (A vertical guide allows for vertical translation but prevents rotation of the node.)

Assembling K and Solving the Equilibrium Equations

The analysis model consists of two uniform elements ab and bc and three nodes a, b and c (Fig. 17.26b). Fixed dofs are denoted by zero and the unknown (free)

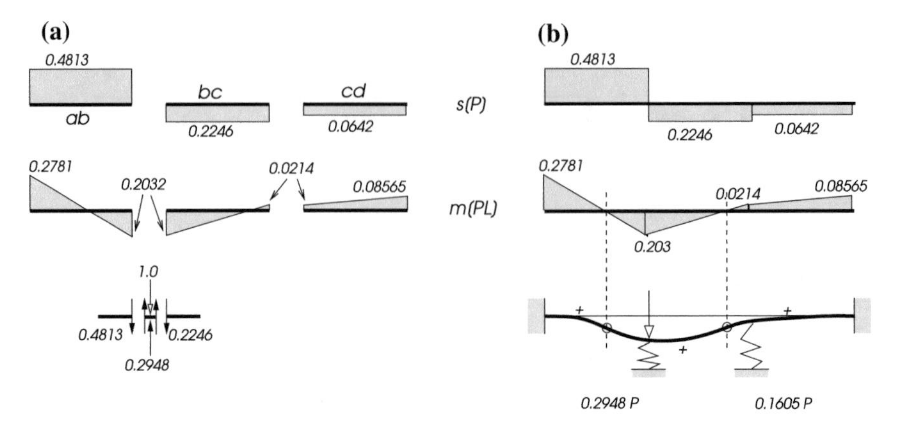

Fig. 17.25 Fixed beam on two springs: *sm*-diagrams

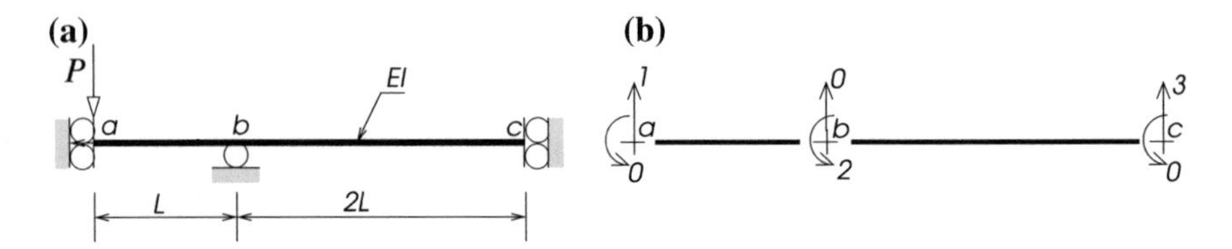

Fig. 17.26 Beam with two guides and roller

dofs are numbered from 1 to 3. The sequence of numbering is immaterial at this point although it is always good engineering practice to follow a systematic way of numbering.

Elements *ab* and *bc* of lengths L and $2L$ respectively have the following element stiffness matrices, where we have indicated in the upper and right margins the correspondence between the local and global dofs.

1	0	0	2	*ab*
12	−12	6	**6**	1
−12	12	−6	−6	0
6	−6	4	2	0
6	−6	2	**4**	2

and

0	3	2	0	*bc*
1.5	−1.5	1.5	1.5	0
−1.5	**1.5**	**−1.5**	−1.5	3
1.5	**−1.5**	**2**	1	2
1.5	−1.5	1	2	0

(17.40)

Assembling the boldface components of these matrices in the corresponding addresses in the system stiffness matrix

1	2	3	
12	6	0	1
6	4 + 2	−1.5	2
0	−1.5	1.5	3

yields the equilibrium equations

$$\begin{bmatrix} 12 & 6 & 0 \\ 6 & 6 & -1.5 \\ 0 & -1.5 & 1.5 \end{bmatrix} \begin{Bmatrix} x_1 \\ x_2 \\ x_3 \end{Bmatrix} = \begin{Bmatrix} -1 \\ 0 \\ 0 \end{Bmatrix} \tag{17.41}$$

the solution of which is

$$\mathbf{x} = \{-1/4 \quad 1/3 \quad 1/3\}^T$$

where displacements are given in PL^3/EI units and rotations in PL^2/EI.

Computing the End-Loads and Drawing the *sm*-diagrams

Post-multiplying the element stiffness matrices by their corresponding above end-displacements produces the end-loads applied to the elements. For element *ab* we have

$$\begin{Bmatrix} -1 \\ 1 \\ -5/6 \\ -1/6 \end{Bmatrix} \leftarrow \begin{bmatrix} 12 & -12 & 6 & 6 \\ -12 & 12 & -6 & -6 \\ 6 & -6 & 4 & 2 \\ 6 & -6 & 2 & 4 \end{bmatrix} \begin{Bmatrix} -1/4 \\ 0 \\ 0 \\ 1/3 \end{Bmatrix}$$

where the first two components are the shear forces respectively at *a* and *b* (measured in units *P*) and the following two are the bending moments at *a* and *b* respectively, measured in units *PL*.

For *bc* we obtain similarly

$$\begin{Bmatrix} 0 \\ 0 \\ 1/6 \\ -1/6 \end{Bmatrix} \leftarrow \begin{bmatrix} 1.5 & -1.5 & 1.5 & 1.5 \\ -1.5 & 1.5 & -1.5 & -1.5 \\ 1.5 & -1.5 & 2 & 1 \\ 1.5 & -1.5 & 1 & 2 \end{bmatrix} \begin{Bmatrix} 0 \\ 1/3 \\ 1/3 \\ 0 \end{Bmatrix}$$

The first two components are the zero shear forces at the extremities and the last two are bending moments (*PL*) applied by the nodes to the extremities (see Fig. 17.27a, upper drawing). The shear and bending moments, given in Fig. 17.27b follow suit, and are drawn element-by-element. (We have used different shading to emphasize the point.)

We also note that the moment diagram is identical to the curvature distribution κ in PL/EI units, from which we can infer the displaced shape (Fig. 17.27b, bottom). Element *bc* deforms as a circular arc of radius $1/\kappa$. The point of zero curvature, or inversion of the curvature sign is indicated by a 'virtual' hinge. Finally, the reactions are shown in Fig. 17.27a (bottom).

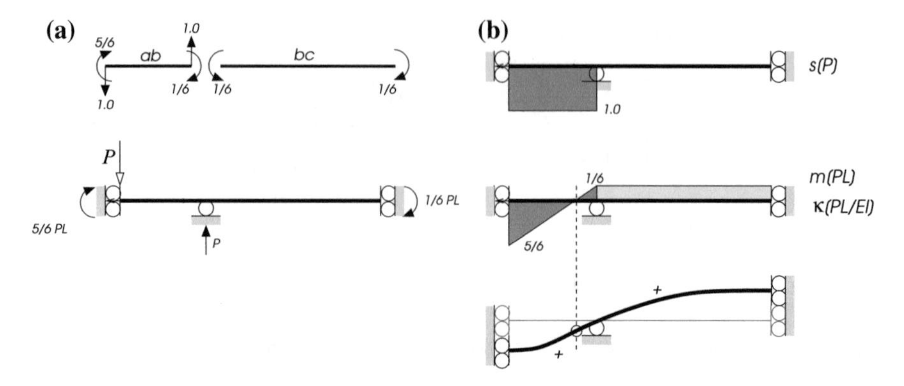

Fig. 17.27 Beam with two guides and roller: solution

17.7 Summing Up

This chapter on the assembly of the system stiffness matrix, in conjunction with the chapter on element stiffness matrices, constitute the core of the stiffness or displacement method of analysis. When we know the stiffness matrices of all the elements of the structure we can easily incorporate them in the system stiffness matrix which is identical to writing the equilibrium equations at the open degrees of freedom. After solving these equations we have the displacements of the extremities of all the elements, in other words, we know the internal forces in the elements. All this, of course, should be computerized.

Bear in mind that the method is only valid for structures loaded by point forces and couples at the nodes. The next chapter will show us a way to adapt the technique to structures with distributed forces along its members.

Chapter 18
Loads on Elements

At this juncture we have the major ingredients of the stiffness method. There is, however, one important aspect that has still not been addressed. You will remember that, all that has been done relates to structures with point forces and point couples applied to the open degrees of freedom. But, more often than not loads are applied to elements, be they concentrated loads or distributed ones.

In this chapter we will see how such element loads can be replaced by equivalent loads applied at the nodes.

18.1 Element Loads

Ignoring force P in Fig. 18.1a, the frame can be modeled as a structure with five nodes and four elements. What happens when a load is applied to an element, such as force P at mid-span of element bc? This could in principle be dealt with by adding a node f at the application point of force P. The model has now five elements compared to the original four; the number of nodes has increased from five to six and there are now 12 unknown degrees of freedom instead nine (three at every node along *abfc*). This amounts to an increase of $1/3$ in the number of unknowns and equations. We can live with that.

But what about Fig. 18.1b? Shall we add six nodes (12 degrees of freedom) and as many elements? That is, instead of two elements and two unknown degrees of freedom, we now have eight elements and 14 unknowns; an increase of 700 % in the number of equations! This does not seem to be a sensible way to tackle this type of problem. Moreover, with outright distributed forces and temperature loading, we would have absolutely no idea how to proceed (see Fig. 18.1c), where the structure is subjected to sun and wind.

The technique for dealing with such cases (in fact, by far the majority of cases) hinges on the concept of equivalent nodal loads. That is, we are going to replace the

M.B. Fuchs, *Structures and Their Analysis*,
DOI 10.1007/978-3-319-31081-7_18

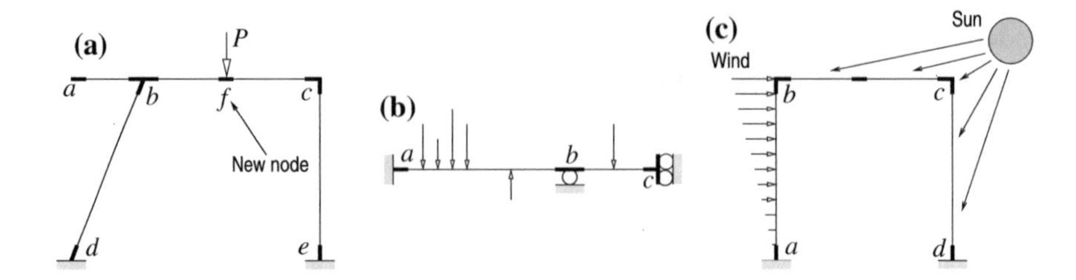

Fig. 18.1 Forces on elements

loads on the elements with equivalent loads applied to the nodes; and we already know how to deal with structures with loads at nodes. Such equivalent nodal loads are computed by way of the fixed-end reactions or in short **f.e.r**.

18.2 Fixed-End Reactions – f.e.r.

Consider the frame *abcd* in Fig. 18.2a loaded by a vertical force P at mid-span of the beam bc. Clearly, the entire structure will deform and will assume the equilibrium shape approximately as indicated. At a and d we will have reactions, and there will be bending moments and shear forces along the entire structure.

When we now build rigid walls at b' and c' (Fig. 18.2b), you will agree that parts ab' and $c'd$ have no means of knowing what is going on along span $b'c'$. These walls factually isolate $b'c'$ from the rest of the structure. Consequently, ab' and $c'd$ remain in their pristine condition while span $b'c'$ will behave as a fixed beam subjected to whatever loading there may be there (P in this case). Along $b'c'$ we will have bending moments, shear forces and often also normal forces (although not in this case), and at the walls b' and c' we will have reactions. These are, for obvious reasons, called fixed-end reactions or **f.e.r**. But for the rest of the structure ab' and $c'd$, there are no internal forces let alone reactions at a and d.

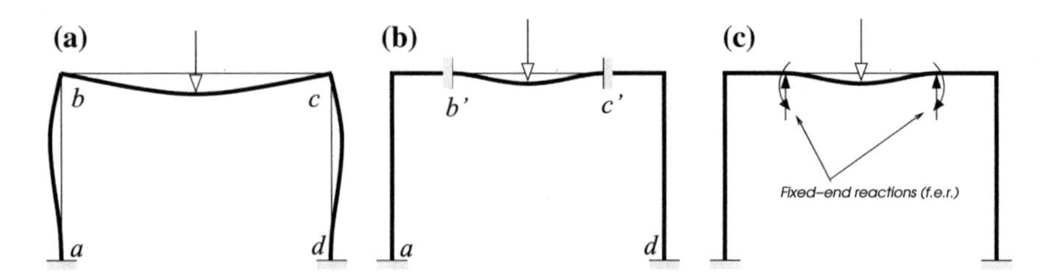

Fig. 18.2 Fixed end reactions – f.e.r.

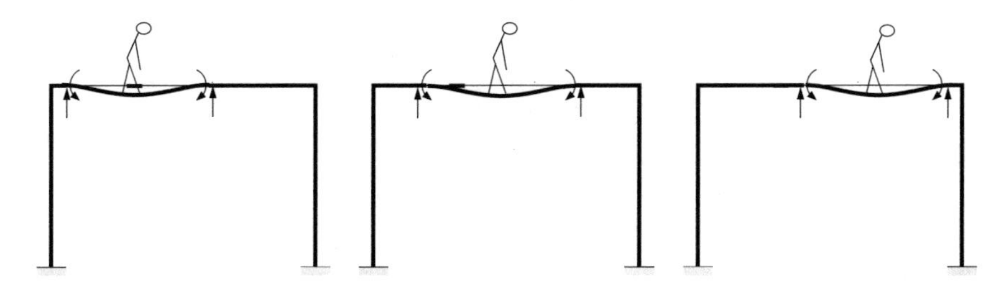

Fig. 18.3 Intruder with f.e.r.

Now, and this is important, we could achieve the same isolation by applying the f.e.r. at b' and c' instead of building the walls, as in Fig. 18.2c. Whatever was said concerning Fig. 18.2b applies to Fig. 18.2c. A knowledgable intruder could walk unnoticed in a building by applying the appropriate f.e.r. in front and behind himself. As visualized in Fig. 18.3 the rest of the structure will remain undisturbed.

18.2.1 Typical f.e.r.

There exist published solutions for f.e.r. of uniform fixed beams under all sorts of loadings. We will be assuming that you have access to these solutions. If you are unlucky enough to have a loading for which you do not find published results, simply solve the doubly built-in beam by the force method (the degree of statical redundancy is 2 or 3 depending on the loading) or by any other means.

For our purposes, we will give here the results of a uniform beam under a central point load Q (Fig. 18.4a), a uniform distributed force of magnitude q (b) and (c) under applied curvature $\bar{\kappa}$ such as a temperature gradient (heated on top and cooled at the bottom), which causes a free beam to bend with curvature $\bar{\kappa}$ (upper fibers extended).

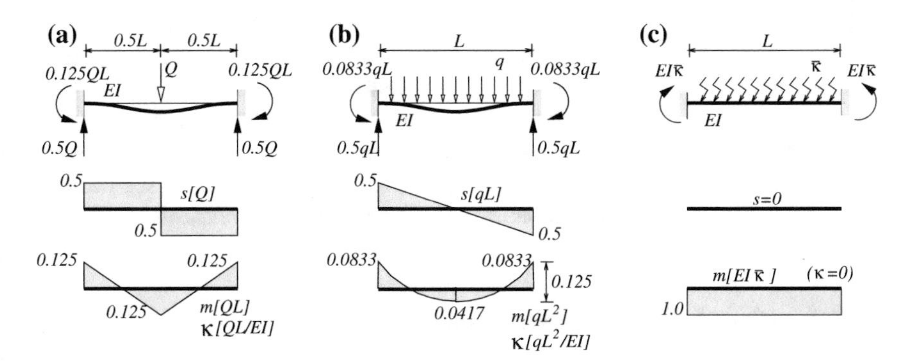

Fig. 18.4 Typical f.e.r.'s of a uniform beam. **a** Mid-span force. **b** Uniform distributed force. **c** Uniform applied curvature

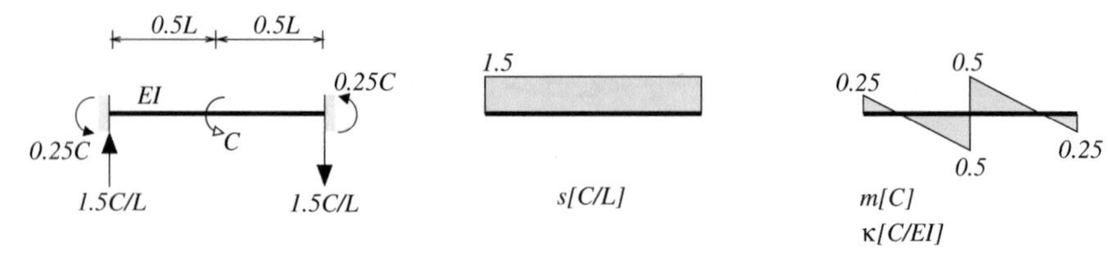

Fig. 18.5 f.e.r. of a couple

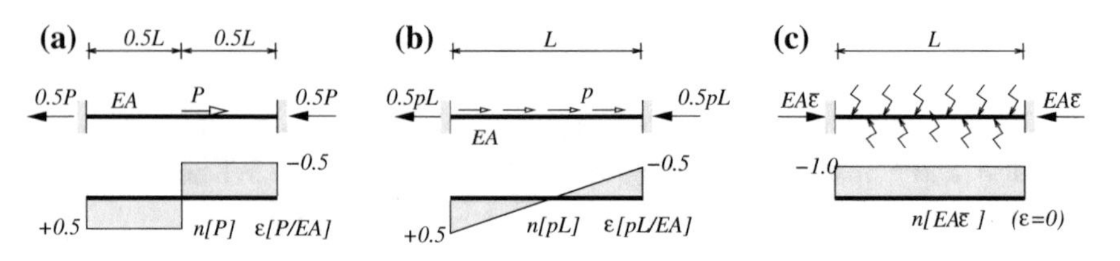

Fig. 18.6 Typical f.e.r.'s of a uniform rod. **a** Mid-span axial force. **b** Uniform distributed axial load. **c** Uniform temperature

The latter can also be considered as a deformation due to a botched manufacturing or introduced on purpose to create prestressing.

The curvature distributions in cases (a) and (b) are m/EI, and note that under the differential temperature the curvature is zero, although there are bending moments. It so happens that the curvature due to the bending moments exactly balances the imposed curvature, and the beam remains straight. (We have drawn the equilibrium configurations in row 1 for all three cases.)

An additional example is the f.e.r. of a central couple on a uniform beam (Fig. 18.5). Note that the walls add each a couple $0.25C$ *to* the one applied, and all these couples are equilibrated by the couple $(1.5C/L \times L)$ created by the equal and opposite vertical reactions. Also, the shear force ignores the moment discontinuity at mid-span, and remains constant along the span, as is expected from a beam without distributed or applied forces along the span.

For completeness, we have added the f.e.r. and the related normal forces and extension distributions for a uniform rod of axial stiffness EA and length L due to an axial mid-force between walls (Fig. 18.6a), a uniform axial distributed loading (b) and (c) a uniform applied extension $\bar{\varepsilon}$. The results in Fig. 18.6 are intuitive even for Fig. 18.6c. Although there are no extensional strains, there is a compressive internal force $EA\bar{\varepsilon}$. This should help in visualizing the corresponding case in the beam of (Fig. 18.4c), where there is no curvature but there are bending moments.

18.2.2 The Method in Three Steps

We will outline the few simple steps for dealing with loads on elements, including distributed loads, and we will use the example in Fig. 18.7a to illustrate. We analyze

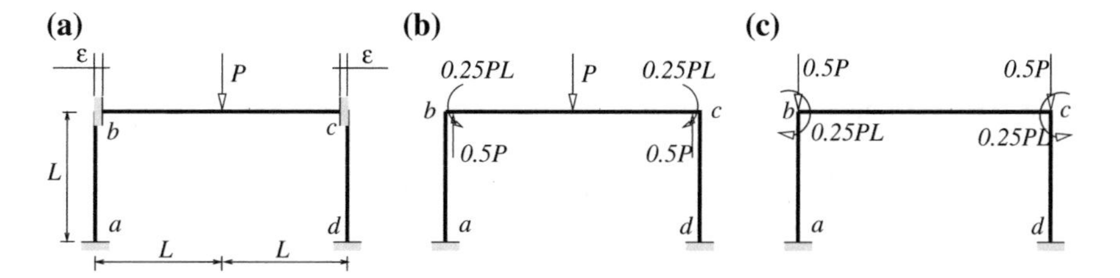

Fig. 18.7 $P = (P + \text{f.e.r.}) + (-\text{ f.e.r.})$. **b** Problem II (P + f.e.r.). **c** Problem I (−f.e.r.)

the structure under two loading systems: once under the *external forces* + *the f.e.r.* (as we shall see in a short while, this is very easy), and once under − *f.e.r.* (here we have loads at the nodes, thus we can solve this). The final analysis is of course the sum of the two.

1. **Analyze the structure under *P* + f.e.r. (Problem II)**
   In a figurative sense, we are building in the loaded element as close as we wish to its extremities, and we determine the f.e.r. (the reactions applied by the walls) and the *nsm*-distributions of the element. These reactions and distributions are in fact copied from known solutions.
   In Fig. 18.7a, the walls were set at a distance $\varepsilon \to 0$ from the nodes. In the case of the example, the f.e.r. (the reactions at the walls) are vertical up-forces $0.5P$ and equal and opposite couples $0.25PL$, as shown in Fig. 18.7b, where we have replaced the wall by their reactions.
   Note, if more than one element is loaded we compute all the f.e.r. of all the loaded elements.
   We have thus analyzed the structure under the external loads (symbol '*P*') plus the f.e.r. We already know that the structure, except the loaded element, is unaffected by such a loading. Excluding the loaded element, nothing moves or deforms, there are no reactions (at *a* and *d*), no internal forces ($nsm = 0$), and the degrees of freedom (*b* and *c*) have no displacements nor rotations.
   As for the loaded element(s), we are assuming that the analysis is given. We simply read off the *nsm* and, if required, also the κ-diagrams from textbooks, formulas, books or electronic media. I know that this does not go down well with purists, because it assumes that somebody has solved such built-in elements with, forcibly, the force method. It follows that the displacement method relies in part on the force method, but there is nothing wrong with that.
2. **Analyze the structure under – f.e.r. (Problem I)**
   Now we analyze the structure under the − f.e.r. forces (Fig. 18.7c). To make things clear, we apply the − f.e.r. flush on the nodes. Consequently this System I is a classical problem with applied loads at the nodes, which we can solve by the standard displacement method.

3. **Sum up steps 1 and 2 (Problem I + II)**
   Finally, we sum up the results of steps (1) and (2). Indeed, by superposition the sum of Systems II and I is the solution of the original problem.

$$(P + \text{ f.e.r.}) + (-\text{f.e.r.}) = P \tag{18.1}$$

In conclusion, we replace the original problem with loaded elements which we cannot solve by the two types of problems: problem II the solution of which can be found in published records, and problem I which is a regular problem with loads at the nodes that can be solved.

18.2.3 Equivalent Nodal Loads

When we apply the − f.e.r. loads to the nodes of the structure, we obtain at all the nodes the same nodal displacements as under the initial external loads P. These − f.e.r. are therefore also called equivalent nodal loads, that is,

the − f.e.r. and the external loads are equivalent for producing the nodal displacements

but they are not structurally identical. Indeed, the moment distributions under P and − f.e.r. are different. We must add the solution of System II to the − f.e.r. solution to get the solution under P.

18.3 Illustrated Examples

18.3.1 Fixed Beam with Intermediate Support

Consider the uniform beam *abc* in Fig. 18.8a of bending stiffness EI, fixed at the extremities, with a roller support at b and subjected to a uniform distributed force $q = Q/L$ along its entire span. Segments *ab* and *bc* are of lengths L and $2L$ respectively.

The only unknown degree of freedom of the structure is the rotation at support b (positive clockwise, see Fig. 18.8c.) We replace loading (q) by loading II : (q+f.e.r.) and loading I : (−f.e.r.). The solution of the problem is the sum of solutions II and I.

Problem II (q + f.e.r.): Extremity a of ab and the vertical displacement at b are fixed. To fix the rotation at b of beam ab, we apply the missing f.e.r.: $-(Q/L)L^2/12 = -QL/12$ (clockwise). For extremity b of bc, we need $(Q/L)(2L)^2/12 = 4QL/12$ (counterclockwise). Consequently the f.e.r. are a couple $4QL/12 - QL/12 = 3QL/12$) at b (counterclockwise). This is equivalent to fixing node b or building a wall at b. We have now two juxtaposed fixed beams, and the corresponding shear and moment (and curvature) diagrams are the juxtapositions of the solutions as given in Fig. 18.8b (diagrams s^{II}, m^{II} and κ^{II}).

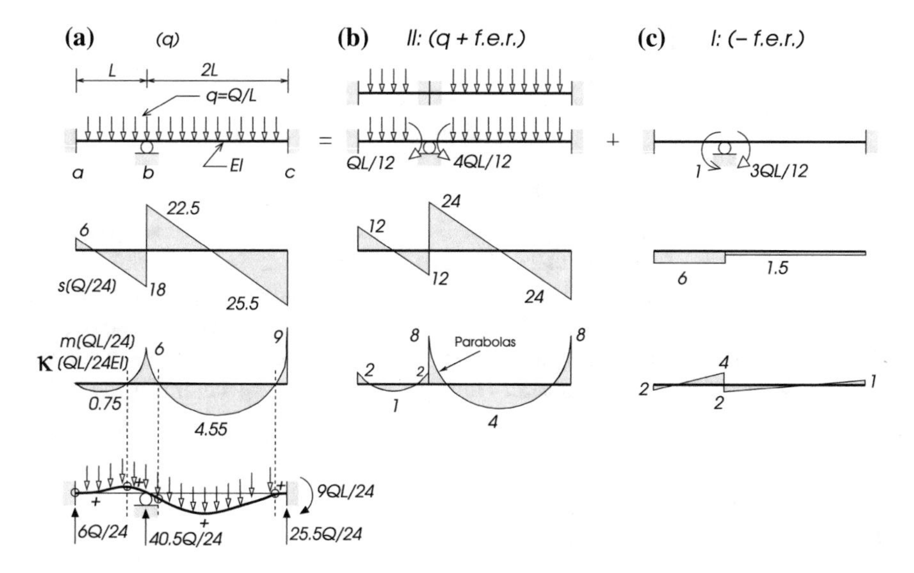

Fig. 18.8 Fixed beam with intermediate support

Problem I (– f.e.r.): The problem to solve is the original structure with – f.e.r., that is, a clockwise couple $3QL/12$) at b (see Fig. 18.8c). This is a structure with a loading along a dof and can be solved by the standard displacement method.

$$K^{ab} = \begin{array}{|rrrr|l}
0 & 0 & 0 & 1 & \\ \hline
12 & -12 & 6 & 6 & 0 \\
-12 & 12 & -6 & -6 & 0 \\
6 & -6 & 4 & 2 & 0 \\
6 & -6 & 2 & \mathbf{4} & 1 \\ \hline
\end{array} \qquad K^{bc} = \begin{array}{|rrrr|l}
0 & 0 & 1 & 0 & \\ \hline
1.5 & -1.5 & 1.5 & 1.5 & 0 \\
-1.5 & 1.5 & -1.5 & -1.5 & 0 \\
1.5 & -1.5 & \mathbf{2} & 1 & 1 \\
1.5 & -1.5 & 1 & 2 & 0 \\ \hline
\end{array} \qquad (18.2)$$

The system stiffness matrix is the scalar $4+2=6$ (assembled from K^{ab} and K^{bc}), yielding the equilibrium equations (in numerical form) $6\,x_1 = -3/12$. The rotation at b is thus $\theta_1 = -(1/24)QL^2/EI$, where we have re-established the correct dimensions. Indeed, the stiffness coefficient is expressed in EI/L and the couple at b in QL.

We can now compute the end-loads on both elements.

$$\mathbf{f}^{ab} = \frac{1}{24}\begin{Bmatrix} -6 \\ 6 \\ -2 \\ -4 \end{Bmatrix} \leftarrow \begin{bmatrix} 12 & -12 & 6 & 6 \\ -12 & 12 & -6 & -6 \\ 6 & -6 & 4 & 2 \\ 6 & -6 & 2 & 4 \end{bmatrix} \begin{Bmatrix} 0 \\ 0 \\ 0 \\ -1/24 \end{Bmatrix}$$

$$\mathbf{f}^{bc} = \frac{1}{24}\begin{Bmatrix} -1.5 \\ 1.5 \\ -2 \\ -1 \end{Bmatrix} \leftarrow \begin{bmatrix} 1.5 & -1.5 & 1.5 & 1.5 \\ -1.5 & 1.5 & -1.5 & -1.5 \\ 1.5 & -1.5 & 2 & 1 \\ 1.5 & -1.5 & 1 & 2 \end{bmatrix} \begin{Bmatrix} 0 \\ 0 \\ -1/24 \\ 0 \end{Bmatrix}$$

from where we can draw the s^I, m^I and κ^I diagrams.

The final distributions are the sum of the I and II solutions: $sm\kappa = (sm\kappa)^I + (sm\kappa)^{II}$ (Fig. 18.8a). The reactions follow suit (bottom sketch in Fig. 18.8c).

Note: The bending moment is zero at the built-in extremity a. This is very unusual because restraining walls usually apply moments. However, this being the case, the equilibrium configuration has a virtual hinge at the root. Consequently, the displaced shape of segment ab is above line ac. The sagging of bc is such that it lifts the entire span ab.

18.3.2 Frame with Rigid Element

The frame bordered by a rectangle in Fig. 18.9 is composed of elements of infinite axial stiffness ($EA \to \infty$), as usual. This implies that we will have a reduced number of degrees of freedom at the nodes (which is good), but as a corollary the stiffness method will not produce the internal axial forces.

In addition, in the present example the vertical element bc is very rigid, both extensionally and in flexure. This rigid element will be treated with both $EA \to \infty$ and $EI \to \infty$. As we shall see, the number of degrees of freedom at the nodes will be further reduced, but we will not receive any information regarding *nsm* in element bc. We will try to generate the missing data from equilibrium considerations.

Due to the presence of external forces on the elements, the original problem (column 1 in Fig. 18.9) is divided into problem II (Q + f.e.r.) (column 2 in the figure) and problem I (−f.e.r.) shown in column 3. The solution is the sum of I and II.

Problem II: (Q + f.e.r.)
This solution corresponds to the structure where all the nodes are fixed. The moment and shear distributions can be copied from tables (also given earlier in this text), and are shown in the second columns of Fig. 18.9.

Problem I: (–f.e.r.)
In this problem, we apply the minus nodal forces which were added to Problem II to lock the nodes. Or also, we apply the minus nodal forces of Problem II to the degrees of freedom of Problem I.

The degrees of freedom shown in the third column are almost standard, and result directly from the boundary conditions and the axial rigidity of the elements. For instance, denoting by u, v, θ the horizontal, vertical and rotational displacements of a node, we find that if $v_b = \Delta_2$, so is $v_c = \Delta_2$, because element bc does not extend. It is however noteworthy that the bending rigidity of bc entails $\theta_b = \theta_c = 0$. Indeed

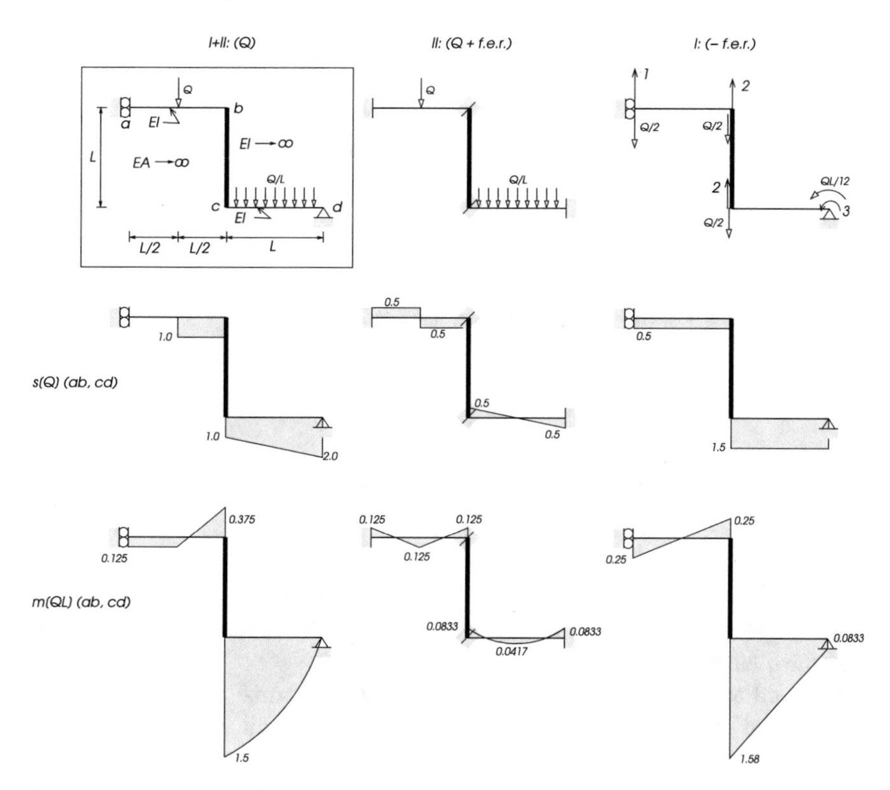

Fig. 18.9 Frame with rigid element

any rigid rotation of *bc* would change the lengths of either *ab* or *cd* or both, which contradicts ($EA \rightarrow \infty$).

$$
\begin{array}{|rrrrrr|ccc}
2 & 0 & 0 & 3 & 0 & 0 & & & dc \\
0 & 0 & 0 & 0 & 2 & 2 & & bc & \\
1 & 2 & 0 & 0 & 0 & 0 & ab & & \\
\hline
12 & -12 & 6 & 6 & 0 & 0 & 1 & 0 & 2 \\
-12 & 12 & -6 & -6 & 0 & 0 & 2 & 0 & 0 \\
6 & -6 & 4 & 2 & 0 & 0 & 0 & 0 & 0 \\
6 & -6 & 2 & 4 & 0 & 0 & 0 & 0 & 3 \\
0 & 0 & 0 & 0 & 1 & -1 & 0 & 2 & 0 \\
0 & 0 & 0 & 0 & -1 & 1 & 0 & 2 & 0 \\
\hline
\end{array}
\qquad (18.3)
$$

The element stiffness matrices are given in array (18.3), where the flexural and extensional submatrices are, in principle, multiplied by their respective *EI* and *EA*. Note that the axial components of elements *ab* and *cd* do not enter the system stiffness matrix $\boldsymbol{K}$, and that element *bc* does not assemble at all in $\boldsymbol{K}$. It is connected to global degree of freedom Δ_2 which cancels out at assembly $(1 - 1 - 1 + 1 = 0)$.

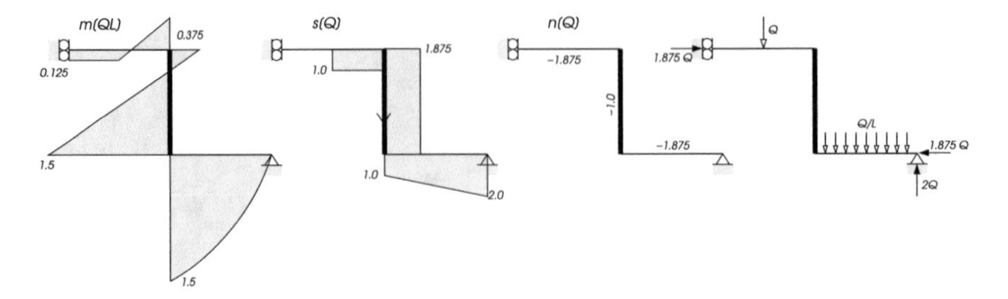

Fig. 18.10 Inferring the *nsm*-distributions of the rigid element

The equilibrium equations are

$$\begin{bmatrix} 12 & -12 & 0 \\ -12 & 24 & 6 \\ 0 & 6 & 4 \end{bmatrix} \begin{Bmatrix} x_1 \\ x_2 \\ x_3 \end{Bmatrix} = \begin{Bmatrix} -0.5 \\ -1.0 \\ 0.0833 \end{Bmatrix} \tag{18.4}$$

the solution of which is $x_1 = -0.583$, $x_2 = -0.542$ and $x_3 = 0.833$. This produces the *sm* diagrams of Problem I (third column in Fig. 18.9). The sum of solutions *I* and *II* are the final diagrams (first column).

As noted at the start, we are missing the internal normal forces in the entire structure and all the internal forces in element *bc*. We will generate them from equilibrium considerations. The bending moments on *bc* are easy to determine. It is a straight line joining the bending moments at the extremities. From the bending moments we get the shear force on *bc*. Having the shear everywhere, we can deduce the normal forces from equilibrium at the nodes. The final *msn* diagrams as well as the reactions are shown in Fig. 18.10.

18.3.3 Fixed Haunched Beam

A haunched beam is a beam whose moment of inertia increases significantly towards the supports. To understand the influence of such a thickening in a fixed beam we will study a stylized case. Consider the uniform beam with infinite bending stiffness at the extremities subjected to a point force Q at mid-span shown in Fig. 18.11b. To follow the trend, we have added two limiting cases: a beam with a haunch of almost zero length in Fig. 18.11a and a haunch which covers almost half the span (on both sides) in Fig. 18.11c.

Note that the shear diagram is the same in the three cases. Indeed, the structure being symmetric the reactions at a and a' are $Q/2$, hence the shear diagrams. (We recall that drawing a shear diagram is an exercise in equilibrium and does not depend on the stiffnesses of the structure.)

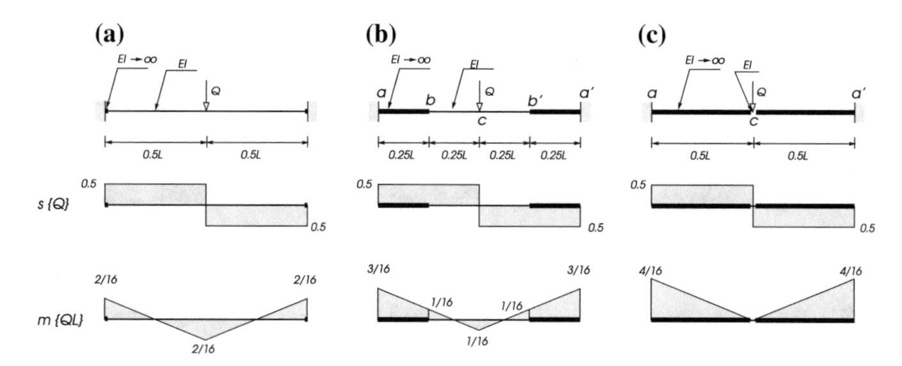

Fig. 18.11 Haunched beam

The moment distributions of the limiting cases are easy. Without haunches (Fig. 18.11a) the structure is a uniform clamped beam with the well-known bending diagram. (We use $2/16$ instead of $1/8$ for comparison.)

With haunches which cover *almost* the entire span (there is always a small length of beam with a finite stiffness EI) the moment diagram is as in Fig. 18.11c. We have here in fact two cantilevers of length $L/2$, each taking half the load ($Q/2$).

The trend is clear. Adding haunches seems to translate the bending diagram up, as is. The case in Fig. 18.11b should be somewhere in between the limiting cases. We note that segment bb' is a clamped beam of length $L/2$. Indeed, segments ab and $b'a'$ are rigid and fixed in the wall; extremities b and b' can therefore neither translate nor rotate. The moment distribution on segment bb' has thus the classical distribution shown in Fig. 18.11b, with extreme values $Q(L/2)/8 = QL/16$. The shear being constant, the slope of the moment distribution is also constant, and the moment distribution on ab, $b'a'$ is simply the continuation of the straight line. As expected, the diagram is in between the diagrams of the limiting cases.

18.3.4 Fixed Beam with Central 'Haunch'

An otherwise uniform clamped beam (bending stiffness EI) with a 'haunch' centrally located is subjected to a point-force Q at mid-span. To study the influence of such a haunch on the bending moment distribution, we will assume a central segment with infinite bending stiffness $EI \to \infty$ shown in Fig. 18.12b.

The limit cases are a haunch along (almost) the entire length (Fig. 18.12a) and a haunch over a segment of zero length shown in (Fig. 18.12c). The shear distribution is here also trivial.

The beam in Fig. 18.12a with a zero bending stiffness at the supports (EI is to ∞ like zero to one) is equivalent to a beam with hinges at the extremities. We have thus

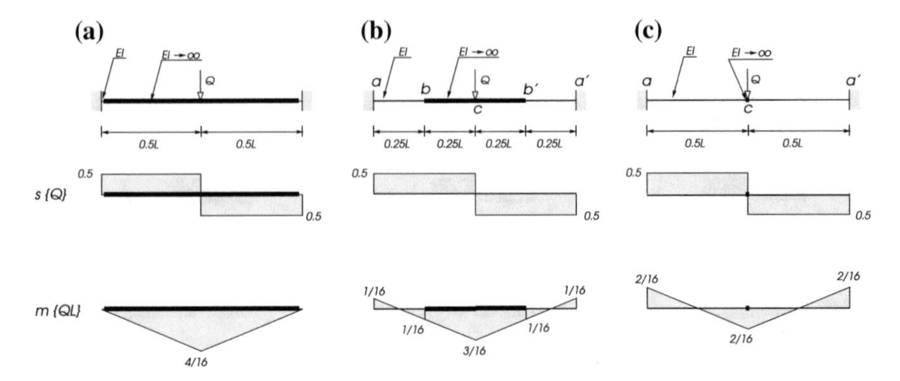

Fig. 18.12 Fixed beam with central segment of infinite rigidity

the classical bending moment distribution of a beam on two simple supports. The other limit case (Fig. 18.12c) is a doubly fixed beam.

A clue for drawing the distribution in the intermediate case is to note that segment bb' must deflect horizontally. Consequently, we get an S-shape deformation for segment ab (and $b'a'$), where the couple produced by the shear forces ($Q/2 \times L/4$) is balanced equally by the bending moment at the extremities of the segment, hence the values $QL/16$. We keep in mind that the bending distribution is piecewise linear.

Note: The main importance of these last two exercises is probably the notion that the internal forces in a redundant structure very much depend on the distribution of stiffnesses. This can be used as a design tool when planning structures.

18.3.5 Applied Strain Yielding Zero Equivalent Loads

The 2-span beam in the first line of Fig. 18.13a is fixed at a, has a roller support at b and a vertical guide at c. The spans are both of length L and have uniform bending stiffness EI. As shown in Fig. 18.13c, the structure has two unknown nodal degrees of freedom.

Since the beam is subjected to uniform applied curvature of magnitude $\bar{\kappa}$ (extension along the upper fibers), we replace loading $\bar{\kappa}$ (the original problem) by Problem II, where the beam is subjected to ($\bar{\kappa}$ + f.e.r.) plus Problem I where the applied loads are − f.e.r.

In Problem II (Fig. 18.13b), the degrees of freedom are all fixed. This is achieved by complementing the loading with what the walls would have applied if all the degrees of freedom were really fixed. As shown in Fig. 18.13b, to lock everything at the extremities of ab we need to apply a counterclockwise couple $EI\bar{\kappa}$ at b; and to lock the extremities of bc requires the same couples (clockwise) at b only.

(These values are taken from the f.e.r. of a built-in beam under applied strain. There are forcibly no vertical reactions in this case.) Consequently, there are no applied

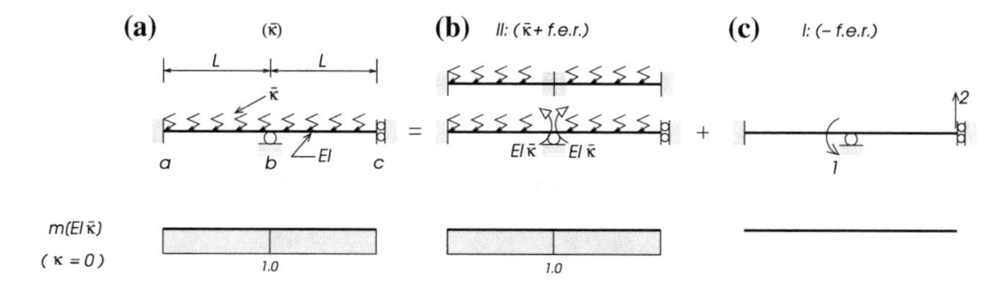

Fig. 18.13 A case of applied curvature without equivalent nodal loads

f.e.r. and Problem I has no applied loads at the degrees of freedom (Fig. 18.13c). The solution (Fig. 18.13a) is Problem II (Fig. 18.13b).

18.3.6 Continuous Beam with Element Loads

The continuous beam in Fig. 18.14 has a concentrated force in the middle of the first span and uniform distributed forces along the following two spans. The beam is fixed at a, has simple supports at b and c and a vertical guide at d. The amplitude of the distributed force is measured in units P/L to facilitate the drawing of the shear diagram (units P) and the bending moment diagram (units PL). Since we have loads on the elements, we are solving the problem in two step by means of the f.e.r. The figure is divided into three columns. The first column (0) is the original problem; the second column (II) is the solution for loads (P + f.e.r.) and the third column is the solution for − f.e.r. Clearly, the solution of (0) is the sum of the solutions (II) and (I).

Problem II

The loading of Problem II is the span loads plus the fixed-end reactions. Applying fixed-end reactions to the extremities of each of the three elements or building physical walls at the extremities are one and the same thing. These walls isolate each element from its surroundings and the analysis of such a structure is simply the juxtaposition of Fig. 18.4a for span ab (with $Q = P$) and Fig. 18.4b for spans bc and cd (with $q = P/L$) as can be seen in Fig. 18.14II. So Problem II is assumed solved, since all we have to do is to look up the solution for the particular span loadings.

Problem I

Case I is a structure with loads − f.e.r. applied at the nodes, and here we can use the classical displacement method. The problem has three unknown degrees of freedom, as indicated in the first row of Fig. 18.14I: the rotations of nodes b and c and the vertical displacement of the guide at d.

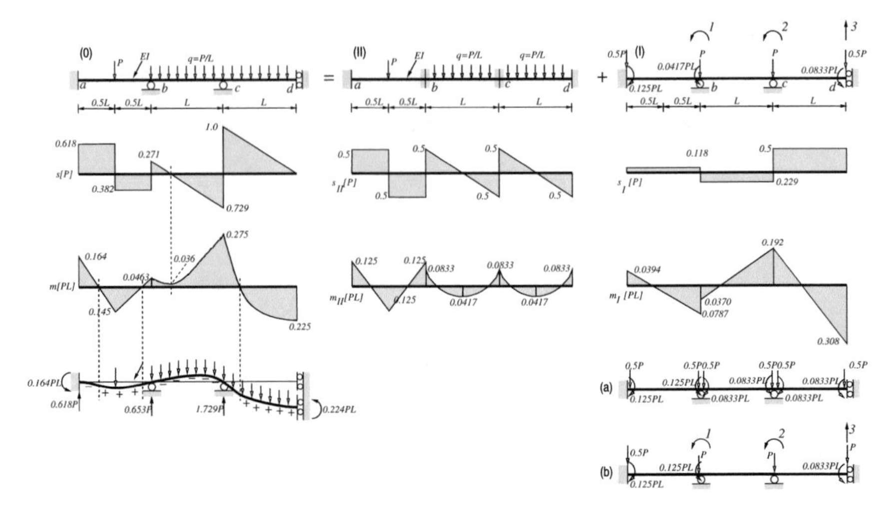

Fig. 18.14 Three-span continuous beam with element loads

The moment and shear diagrams are shown in Fig. 18.14I. For the interested reader we give some intermediate results. The structure has three unknown dofs, and the stiffness matrix of the three elements has the standard form of a uniform beam of length L and bending stiffness EI.

A word regarding the nodal loads. Strictly speaking, when we apply the − f.e.r. on every element, very very close to the extremities, we obtain the loads as given in Fig. 18.14a (at the bottom of the third column). We may as well put them flush on the supports. Adding the forces and moments at every support, we obtain the picture in Fig. 18.14b.

These loads divide into two types:

- Forces and couples applied directly on a support, such as the couple at a or the force P at b: We must realize that such loads do not enter the structure. They disappear immediately into the ground, where they are reacted by something infinitely rigid, as opposed to the flexible structure. We discard these reactions, and we will deduce the proper reactions from the final shear and moment diagrams.
- Loads at degrees of freedom: These are taken by the structure, and they constitute the equivalent nodal loads (equivalent in the sense of producing the exact nodal displacements).

With these loads the equilibrium equations are (without dimensions)

$$\begin{bmatrix} 8 & 2 & 0 \\ 2 & 8 & -6 \\ 0 & -6 & 12 \end{bmatrix} \begin{Bmatrix} x_1 \\ x_2 \\ x_3 \end{Bmatrix} = \begin{Bmatrix} 0.0417 \\ 0 \\ -0.5 \end{Bmatrix} \tag{18.5}$$

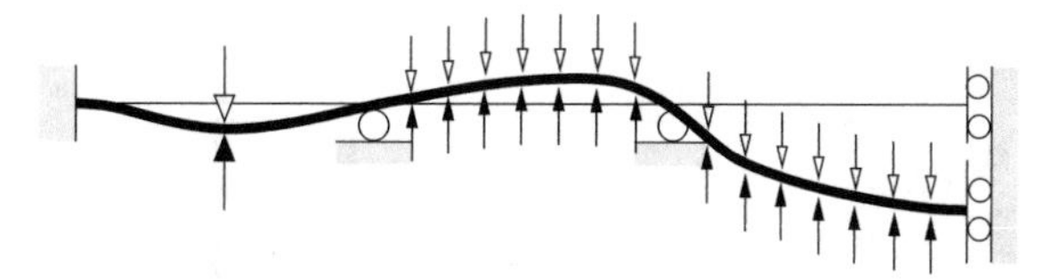

Fig. 18.15 Three-span continuous beam: equilibrium configuration

with the solution

$$\{x_1\ x_2\ x_3\} = \{0.0197\ -0.0579\ -0.0706\} \tag{18.6}$$

This produces the following end-loads at the three elements respectively

$$\left[\mathbf{f}^{ab}\ \mathbf{f}^{bc}\ \mathbf{f}^{cd}\right] = \begin{bmatrix} 0.118 & -0.229 & 0.5 \\ -0.118 & 0.229 & -0.5 \\ 0.0394 & -0.037 & 0.192 \\ 0.0787 & -0.1922 & 0.308 \end{bmatrix}$$

$$\leftarrow \begin{bmatrix} 12 & -12 & 6 & 6 \\ -12 & 12 & -6 & -6 \\ 6 & -6 & 4 & 2 \\ 6 & -6 & 2 & 4 \end{bmatrix} \begin{bmatrix} 0 & 0 & 0 \\ 0 & 0 & x_3 \\ 0 & x_1 & x_2 \\ x_1 & x_2 & 0 \end{bmatrix}$$

where the shear forces are given in P and the bending moments in PL. (We read the above relation from right to left.)

Problem (II + I)

Adding up the distributions $s = s_{II} + s_I$ and $m = m_{II} + m_I$ yields the solutions of the original problem as given in the first column of Fig. 18.14, where we also show the equilibrium configuration.

Note the importance of the equilibrium shape. Only in the equilibrium configuration can the structure apply loads equal and opposite to the applied ones. With this shape, in the present example (Fig. 18.15), the structure applies an equal and opposite force P and equal and opposite distributed forces P/L, thus effectively annulling the external loads and keeping the structure at rest.

18.3.7 Frame with Element Loads

The frame in Fig. 18.160 (upper left drawing) has element loads on two of its components. The bending stiffness of all its elements is EI and we assume, as usual, infinite axial stiffness $EA \to \infty$.

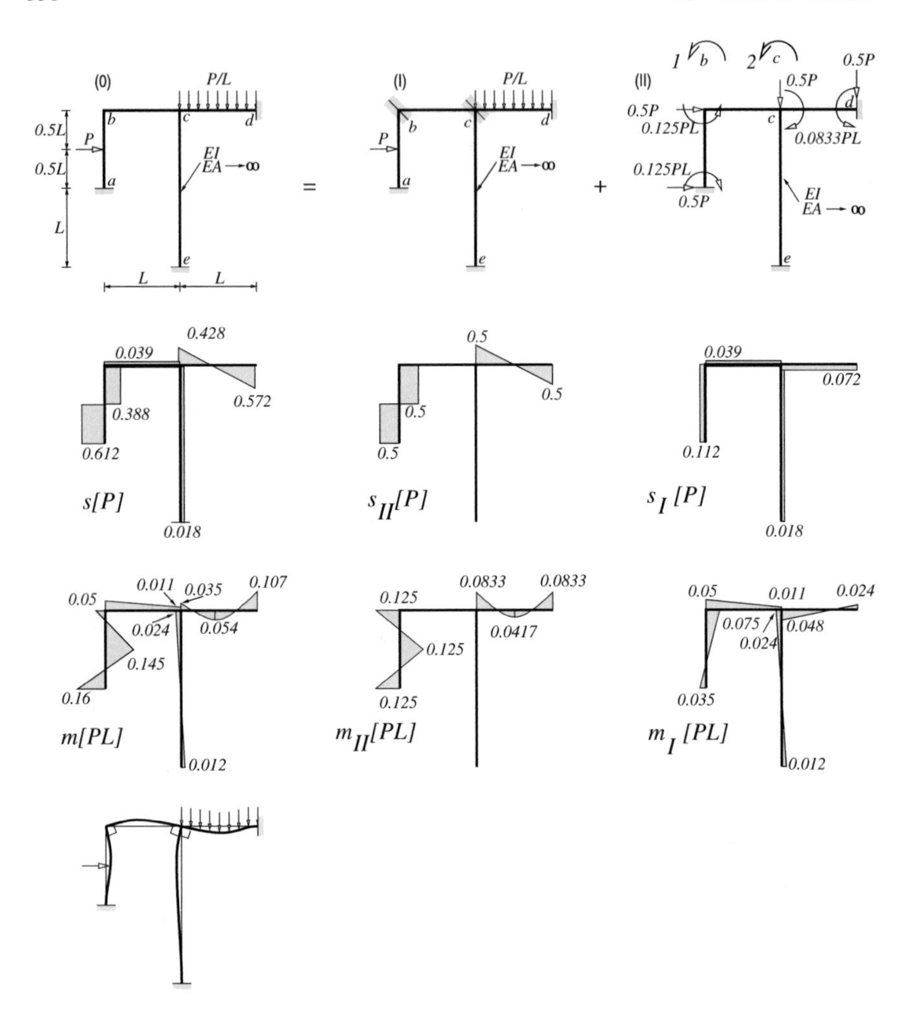

Fig. 18.16 Frame with element loads

Problem II

We block all the nodal displacements along the degrees of freedom as in Fig. 18.16II. Every element is now isolated from the rest of the world, and all it recognizes are its own element loads. The analysis of the frame in case II is the juxtaposition of four independent element analyses. For the loaded elements *ab* and *cd* we copy the moment and shear distributions from suitable references (they can also be found in this book). Note that for the elements which are free of loading (*bc* and *ec*) we get zero distributions, $s_{II}^{bc} = m_{II}^{bc} = s_{II}^{ec} = m_{II}^{ec} = 0$. The loading of problem II is thus '*P*' + f.e.r.

Problem I

We now apply – f.e.r. to the relevant nodes. Since we have only nodal loads, we use the classical displacement method for case I. The moment and shear diagrams are shown in the third column of Fig. 18.16. Having assumed infinite axial stiffness, the displacement method will not provide any information related to the n-distributions. The latter will have to be reconstituted as will be shown in the sequel.

The solution of Problem I requires numerical effort; we will summarize the main steps.

In a first instance we note that nodes b and c cannot displace vertically, because this would imply an extension or contraction of elements ab, ec. These nodes can similarly not translate horizontally, lest elements bc, cd change lengths, which would not be commensurate with assumption $EA \to \infty$. There are consequently only two unknown dofs: the rotations at nodes b and c, Θ_1 and Θ_2 respectively.

From the f.e.r. we retain only those nodal loads which are applied to free dofs, that is,

$$p = PL \begin{Bmatrix} 0.125 \\ -0.0833 \end{Bmatrix} \tag{18.7}$$

when assembling the 2×2 system stiffness matrix, we note that elements ab, bc and cd have the standard element stiffness matrix (without units in the following)

$$\begin{array}{|rrrr|ccc}
0 & 0 & 2 & 0 & & & cd \\
0 & 0 & 1 & 2 & & bc & \\
0 & 0 & 0 & 1 & ab & & \\
\hline
12 & -12 & 6 & 6 & 0 & 0 & 0 \\
-12 & 12 & -6 & -6 & 0 & 0 & 0 \\
6 & -6 & 4 & 2 & 0 & 1 & 2 \\
6 & -6 & 2 & 4 & 1 & 2 & 0 \\
\hline
\end{array} \tag{18.8}$$

Element ec has a length $2L$ and consequently

$$\begin{array}{|rrrr|c}
0 & 0 & 0 & 2 & ec \\
\hline
1.5 & -1.5 & 1.5 & 1.5 & 0 \\
-1.5 & 1.5 & -1.5 & -1.5 & 0 \\
1.5 & -1.5 & 2 & 1 & 0 \\
1.5 & -1.5 & 1 & \mathbf{2} & 2 \\
\hline
\end{array} \tag{18.9}$$

By assembly we get the following dimensionless equilibrium equations

$$\begin{bmatrix} 8 & 2 \\ 2 & 10 \end{bmatrix} \begin{Bmatrix} x_1 \\ x_2 \end{Bmatrix} = \begin{Bmatrix} 0.125 \\ -0.0833 \end{Bmatrix} \tag{18.10}$$

the solution of which gives, after restitution of the dimensions, the node rotations

$$\{\Theta_1\ \Theta_2\} = \frac{PL^2}{EI}\{0.0186\ -0.0121\} \tag{18.11}$$

with these displacements we can now compute the internal forces $\left[\mathbf{f}^{ab}\ \mathbf{f}^{bc}\ \mathbf{f}^{cd}\right]$

$$\begin{bmatrix} 0.112 & 0.039 & -0.072 \\ -0.112 & -0.039 & 0.072 \\ 0.037 & 0.050 & -0.048 \\ 0.075 & -0.011 & -0.024 \end{bmatrix} \leftarrow \begin{bmatrix} 12 & -12 & 6 & 6 \\ -12 & 12 & -6 & -6 \\ 6 & -6 & 4 & 2 \\ 6 & -6 & 2 & 4 \end{bmatrix} \begin{bmatrix} 0 & 0 & 0 \\ 0 & 0 & 0 \\ 0 & x_1 & x_2 \\ x_1 & x_2 & 0 \end{bmatrix}$$

and

$$\mathbf{f}^{ec} = \begin{Bmatrix} -0.018 \\ 0.018 \\ -0.012 \\ -0.024 \end{Bmatrix} \leftarrow \begin{bmatrix} 1.5 & -1.5 & 1.5 & 1.5 \\ -1.5 & 1.5 & -1.5 & -1.5 \\ 1.5 & -1.5 & 2 & 1 \\ 1.5 & -1.5 & 1 & 2 \end{bmatrix} \begin{Bmatrix} 0 \\ 0 \\ 0 \\ x_2 \end{Bmatrix}$$

where shear forces are given in P and bending moments in PL. The corresponding s_I and m_I distributions are depicted in the third column of Fig. 18.16.

Solution (II + I)

The solution of the original problem is the sum of intermediate solutions II and I as shown in the first column of Fig. 18.16, where we have also added a cartoon of the displaced shape.

This is however only part of the picture. We are missing the normal distribution and the corresponding reactions. The reason lies in the assumption of axially infinitely rigid elements. This has allowed us to reduce the number of dofs (and equilibrium equations to be solved) from the nominal six (three at each node b and c) to two. The price is the loss of direct information regarding the axial forces. But one can try to reconstitute the normal distributions on the basis of the sm-distributions.

It should be clear that when we have strong enough computational power to analyze the structure, unless you have a dynamic problem or a design problem of substance, we will seldom trouble ourself with reducing the number of dofs. We just let the machine do the job with the nominal number of dofs.

Distribution of Normal Forces *n*

In a first step we replace the distributed load along cd by the statically equivalent central force P. (Bear in mind that the distributed and concentrated forces are not one and same thing. They are equivalent for static equilibrium purposes only.)

Next, in Fig. 18.17c we indicate the force reactions at supports $a\,e\,d$ which can be read off the shear diagram in Fig. 18.16. The horizontal force $0.406P$ at d can be added in order to maintain horizontal equilibrium of the entire structure. Now, equilibrium of node c (Fig. 18.17b) requires a compressive force $n_{ec} = -0.389P$ in the central column, and cascading to node b we need $n_{ec} = -0.039P$ for the vertical

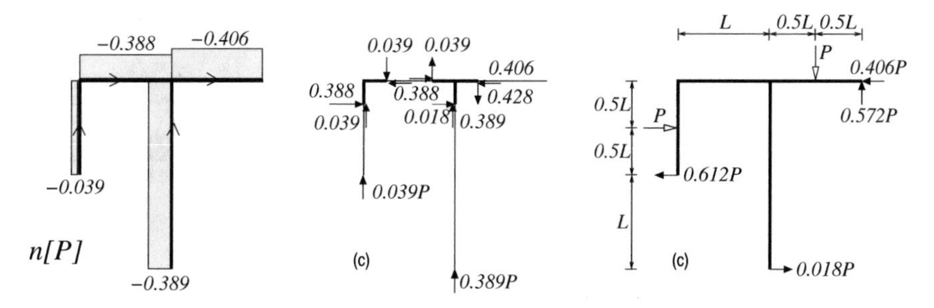

Fig. 18.17 Frame with element loads: inferring n-distribution

equilibrium of that node; hence the vertical reactions at a and e. We have everything needed to draw the n-distribution, where you will notice that since all the normal forces are compressive they have a negative sign and are drawn above the line.

This essentially concludes the analysis. We may check our results by verifying that the resultant and the resultant moment (with respect to any point) of all forces and couples (external and reactions) applied to the structure are zero. For instance, the sum of the vertical forces (starting from the left) is $(0.039 + 0.389 - 1.0 + 0.572)P = 0$.

In order not to lose sight of the overall picture, let us recall that the normal stresses are the sum of the bending stresses (tension and compression) and the stresses originating from the axial force (here in compression). We have no means of computing the stresses since we do not know the shape of the cross-sections of the beam. The purpose of structural analysis is to obtain the *nsm* distributions and the equilibrium configuration (for this EAI of the cross-sections is enough).

Stresses are extremely important but they are not part of structural analysis.

They are in the realm of Elasticity, Solid Mechanics, Strength of Materials and the like, and are thus not included in this and similar texts.

18.3.8 Fixed Beam with Central Guide

Consider the uniform fixed beam *abc* of bending stiffness *EI* in Fig. 18.18a with a vertical guide at b and a point force P in the middle of span *ab*. Analyze the beam and use congruent transformation to get the equilibrium equations for the fixed uniform beam in Fig. 18.18b.

Case (a): Fixed Beam with Guide

The beam in Fig. 18.18, modeled with two elements *ab* and *bc*, is fixed at a and c and has three unknown degrees of freedom at b: the vertical translations 1 and 3 on both sides of the guide and the *common* rotation 2 (the rotation of the guide).

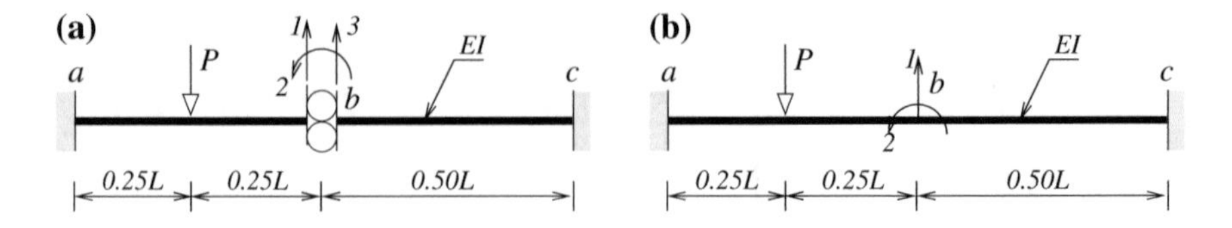

Fig. 18.18 Fixed beam with central guide

The equilibrium equations of problem *I* are

$$\begin{bmatrix} 96 & -24 & 0 \\ -24 & 16 & 24 \\ 0 & 24 & 96 \end{bmatrix} \begin{Bmatrix} x_1 \\ x_2 \\ x_3 \end{Bmatrix} = \begin{Bmatrix} -1/2 \\ 1/16 \\ 0 \end{Bmatrix} \tag{18.12}$$

the solution of which is

$$\mathbf{x} = \{ -0.0091 \; -0.0156 \; 0.0039 \}$$

where the displacements are given in PL^3/EI and the rotation in PL^2/EI.

We will note that the assembly of the system stiffness matrix was based on the element stiffness matrices of uniform beams of length $L/2$

3	0	2	0		*bc*
0	1	0	2	*ab*	
96	−96	24	24	0	3
−96	96	−24	−24	1	0
24	−24	8	4	0	2
24	−24	4	8	2	0

(18.13)

and on the f.e.r. of a fixed beam of length $L/2$ with a central point force P ($P/2$ and $PL/16$ for the force and couple reactions respectively).

Adding the solutions of *I* and *II* yields the shear and bending distributions shown in Fig. 18.19a, where we have added the equilibrium configuration.

Case (b): Removing the Guide

In case (b) we solve a uniform beam built in at the extremities with a force P at one-quarter span. For the solution of Problem I, instead of assembling the system stiffness matrix from scratch, we will be using the congruent transformation method. The old displacements $\hat{\Delta}$ in Fig. 18.18a have three components whereas the new ones Δ in Fig. 18.18b have two components.

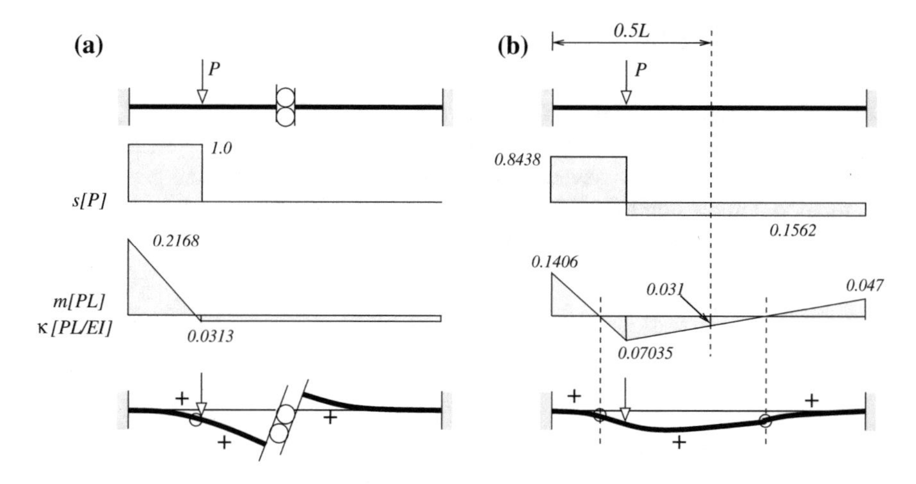

Fig. 18.19 Fixed beam with central guide: congruent transformation

The relation between them $\hat{\Delta} = \boldsymbol{G}\Delta$ is

$$\begin{Bmatrix} \hat{\Delta}_1 \\ \hat{\Delta}_2 \\ \hat{\Delta}_3 \end{Bmatrix} = \begin{bmatrix} 1 & 0 \\ 0 & 1 \\ 1 & 0 \end{bmatrix} \begin{Bmatrix} \Delta_1 \\ \Delta_2 \end{Bmatrix} \tag{18.14}$$

Employing $\boldsymbol{K} = \boldsymbol{G}^T \hat{\boldsymbol{K}} \boldsymbol{G}$ and $\boldsymbol{p} = \boldsymbol{G}^T \hat{\boldsymbol{p}}$ with $\hat{\boldsymbol{K}}$ and $\hat{\boldsymbol{p}}$ given earlier in the text, yields the equilibrium equations of Problem I in the new coordinate system $\boldsymbol{K}\Delta = \boldsymbol{p}$ (Fig. 18.18b).

The equilibrium equations of problem I are

$$\begin{bmatrix} 196 & 0 \\ 0 & 16 \end{bmatrix} \begin{Bmatrix} \Delta_1 \\ \Delta_2 \end{Bmatrix} = \begin{Bmatrix} -1/2 \\ 1/16 \end{Bmatrix} \tag{18.15}$$

The solution is now

$$\boldsymbol{x} = \{-0.0026 \; 0.0039\}$$

where the displacements are again given in PL^3/EI and the rotation in PL^2/EI.

Proceeding with the solution of Problem I and adding the solution of Problem II gives the shear and moment distributions shown in Fig. 18.19b. The figure is completed with the corresponding equilibrium configuration.

18.4 Summing Up

The stiffness method is now complete. We have limited the presentation to loads applied mostly to uniform line elements. These are in fact the only elements used in this text. As mentioned several times it does not retract from the generality of the method. On the contrary. What was explained for the uniform element is valid, almost *verbatim*, for any other type of element.

Part V
Four Additional Topics

In Part V we present four short chapters related to the design of optimal structures which the author developed over the years. Although of an advanced nature, the subjects are presented in a relatively simple form and should be within grasp of readers with a basic understanding of structures and structural analysis, in particular, readers who have mastered the main topics of this book.

Chapter 19, shows that a simple support in a beam is optimally positioned along the span, from a stiffness aspect, if the slope of the beam on the support is zero (is horizontal).

Next, Chap. 20 describes a class of trusses which are both fully-stressed (efficient use of material) and redundant (safe). It is shown that it is an exceptional design. As a rule redundant trusses cannot be fully stressed if the bars are made of a same material.

Chapter 21 shows how a frame can be viewed as a generalized truss. The (unimodal) bars of the classical truss are in a frame element, unimodal normal, shear and moment components. Many concepts related to trusses can now be transposed, as is, to frames, including the explicit expression of the next chapter.

Finally, Chap. 22 presents a formula which gives the internal forces in structures composed of unimodal elements (bars, springs, unimodal beam elements), explicitly in terms of the stiffnesses of the components. This unique analytical expression partially due to Cauchy and Binet is at this stage difficult, if not impossible, to implement due to the 'curse of dimensionality'. Its construct is however beautiful.

Chapter 19
A Strain Energy Theorem—Moving Supports

Consider an elastic beam with a set of given supports, subjected to external loads, such as the cantilever beam in Fig. 19.1a which is subjected to distributed and point forces. We now want to move the simple support to maximize the 'stiffness' of the beam with respect to that loading. As is customary in this sort of problem, we will seek to position the support such as to minimize the 'strain energy' U of the loaded beam, which is a measure of its 'flexibility'.

We have not discussed the notion of strain energy in a structure until now so we will start with defining that quantity.

19.1 Strain Energy

Consider the *displacement* function of the beam under the applied *loads* and let $\kappa(\mathrm{x})$ be the curvature of the displacement function.[1]

Clearly the *displacement* and the curvature κ are a compatible set and the applied *loads* and the bending moments m are an equilibrium set. We can write a 'virtual' work equation[2] in the form $evw = ivw$

$$(loads) \times (displacements) = \int_0^L m\kappa \, \mathrm{dx} \tag{19.1}$$

Multiplying both sides of the equation by $1/2$ we obtain in the left-hand-side the strain energy U stored in the structure. Noting that by Hooke's law $m = EI\kappa$, (19.1) produces the classical expression of the strain energy of linear elastic structures

$$U = \frac{1}{2} \int_0^L \frac{m^2}{EI} \, \mathrm{dx} \tag{19.2}$$

[1] We use temporarily x to differentiate from the support coordinate x.

[2] Everything is in fact real in this case.

M.B. Fuchs, *Structures and Their Analysis*,
DOI 10.1007/978-3-319-31081-7_19

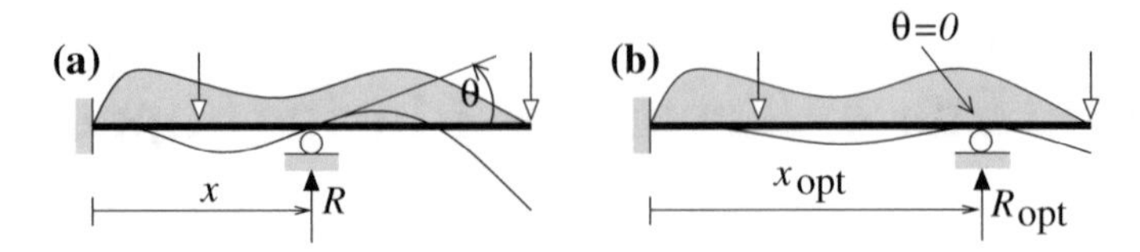

Fig. 19.1 Moving a simple support

The strain energy stored in a structure is often chosen as representing, in some sense, its flexibility. Indeed from (19.1) we note that the larger the displacements under a set of loads the larger the strain energy is. Consequently, minimizing the strain energy as akin to stiffening the structure.

19.2 Minimum Strain Energy

Let x be the distance of the support from the left end of the beam in Fig. 19.1a. The strain energy U is thus a function of x. A necessary condition for the extremum of the strain energy is that its derivative with respect to the position x of the support is zero

$$\frac{\mathrm{d}U(x)}{\mathrm{d}x} = 0 \tag{19.3}$$

It was shown[3] that the derivative of the strain energy with respect to the position x of a simple support is

$$\boxed{\frac{\mathrm{d}U}{\mathrm{d}x} = R\,\theta}$$

where R is the reaction at the support and θ is the slope of the beam at the support (Fig. 19.1a).

The strain energy has therefore an extremal value when $R\theta = 0$. Two cases can occur:

$R = 0$ The support is at a point where it provides no reaction. This is certainly not a position that will stiffen the structure. In fact, at that location the support can even be removed without altering the stiffness. It is a stationary point of the stiffness.

$\theta = 0$ Here is a minimum point of the strain energy.

> A support with zero slope ($\theta = 0$) is optimally positioned.

[3] Z Mroz and GIN Rozvany, *J. of Opt. Th. and Applications*, Vol. 15, No. 1, 1975.

The beam in Fig. 19.1b has thus the maximum stiffness compared to all other values of $x \neq x_{\text{opt}}$.

Many Variables

Note that the property is valid for multiple design variables. Indeed, let x_i be the position of support i of a set of moving supports and let θ_i be the slope of the beam at that support under a given loading. The beam will be the stiffest for these loads if all θ_i are simultaneously zero.

19.3 Designing for Zero Slope

As noted, at the optimal position of a support the slope of the beam is zero (Fig. 19.2a). Clearly, when we replace the optimally positioned simple support by a clamped support the bending moments at both sides of the clamped support are equal (Fig. 19.2b). When we move the clamped support (Fig. 19.2c), there will be an unbalance of bending moments at the clamp because the support must reestablish zero slope.

It is therefore suggested to replace the search for zero slope by a search for zero moment unbalance:

$$m_R - m_L = 0 \tag{19.4}$$

where m_L is the built-in moment to the left of the clamp and m_R is the built-in moment to the right of the clamp.

A simple support with zero moment unbalance when clamped, is optimally positioned.

This approach to the design problem often simplifies the design as can be seen in the following example.

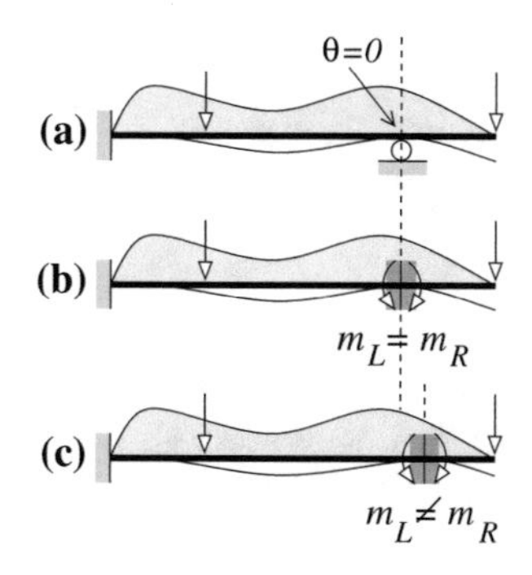

Fig. 19.2 Moving a clamped support

19.4 Illustrated Examples

19.4.1 Example 1

The cantilever beam ac of length L and uniform flexural stiffness in Fig. 19.3 is stiffened by a simple support b a distance x from the built-in left extremity. What would be the optimal position of the intermediate support b under a uniform load q?

We fix support b and compute m_L and m_R at b. Bending moment m_L is the f.e.r. of a uniform beam of length x under uniform loading, $m_L = qx^2/12$ and m_R is the root moment of a cantilever of length $L - x$, $m_R = q(L - x)^2/2$. Equating both moments we get $x_{\text{opt}} = 0.71L$.

19.4.2 Example 2

The propped cantilever beam ac of length L and uniform flexural stiffness in Fig. 19.4 is stiffened by a simple support b a distance x from the built-in left extremity. What would be the optimal position of the intermediate support b under a uniform load q?

We fix support b and compute m_L and m_R at b. Bending moment m_L is the f.e.r. of a clamped beam of length x under uniform loading, $m_L = qx^2/12$, and m_R is the root moment of a propped cantilever of length $L - x$, $m_R = q(L - x)^2/8$.

Equating both moments yields $x_{\text{opt}} = 0.55L$.

19.4.3 Example 3: Several Moving Supports

In this example we seek the optimal layout of five additional intermediate supports to stiffen a simply supported uniform beam under triangular loading (Fig. 19.5a). The design variables are the five components x_i of vector $\boldsymbol{x}$, where x_i is the normalized

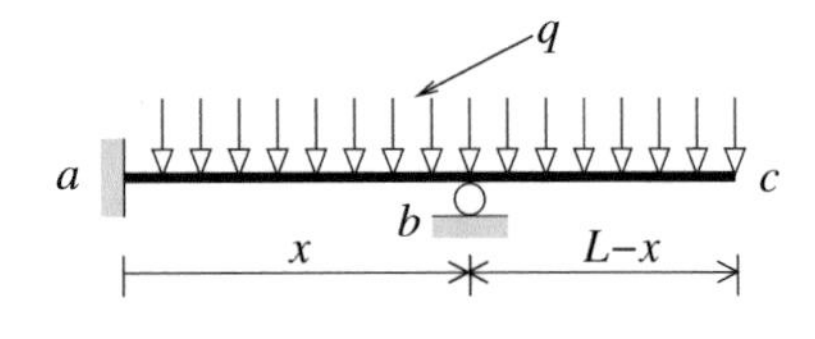

Fig. 19.3 Propped cantilever example

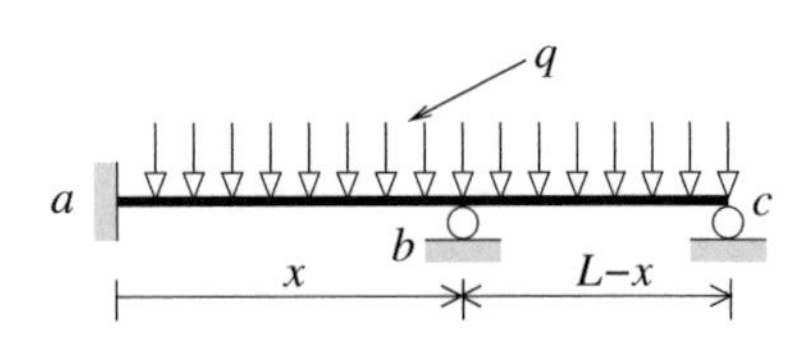

Fig. 19.4 Cantilever with two additional supports

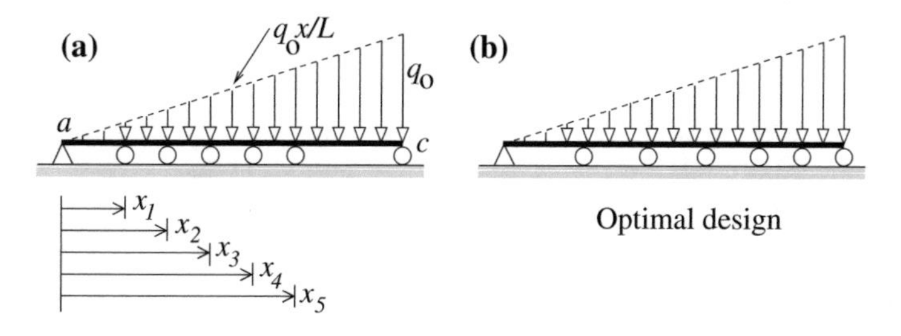

Fig. 19.5 Several moving supports

distance (distance divided by the beam length) of the support from the left extremity. Here we have to simultaneously annul the slopes of the beam over the five additional supports.

In line with the equivalence between the bending moment unbalance at a built-in moving support and the zero-slope requirement, we can find the optimal positions of the supports by simultaneously annulling the moment unbalances over the supports, that is, solve the five non-linear equations $\Delta m_i(\boldsymbol{x}) = 0$ with $i = 1 \ldots 5$. We note that the unbalance at support i is $\Delta m_i = m_i^R - m_i^L$.

We will not discuss methods for solving such equations. One approach would be to minimize the positive function $\sum_{i=1}^{5}(\Delta m_i)^2$, that is, the sum of the squares of the unbalanced moments at the design supports. This function is indeed at its lowest value (zero) when all unbalances are zero. The optimal design variables are[4]

$$\boldsymbol{x} = \{0.234 \;\; 0.424 \;\; 0.598 \;\; 0.752 \;\; 0.888\}^T.$$

19.5 Summing Up

The optimality criterion of zero slope over a roller support of a beam is easy to memorize and relatively easy to implement. We must however bear in mind that the criterion is valid for a given set of loads. If the loading on the beam changes, so does the optimal location of the support.

[4]MB Fuchs and MA Brull, *Computers & Structures*, Vol. 10, 1979.

Chapter 20
Fully-Stressed Trusses

The notion of fully-stressed structures is very appealing in structural design. The basic definition of a fully-stressed structure is a structure where every part is stressed to the maximum permissible stress of the material it is made of. An intuitive premise of most designers is that such a structure, even if not optimal, cannot be that bad.

However, when we try to design such a structure, we note that unless we relax the definition of fully-stressed design (f.s.d.) it is not that easy. We will address the complication in the case of trusses. Although redundant trusses can in general not be fully stressed we present an intriguing exception: the circle-chord truss. That structure under some restrictions is both redundant and fully stressed.

20.1 Fully Stressed Trusses

The stress and strain in bar j of a truss are respectively

$$\sigma_j = \frac{n_j}{A_j} \qquad \epsilon_j = \frac{e_j}{L_j} \qquad \sigma_j = E\,\epsilon_j \tag{20.1}$$

where n_j is the force in bar j, e_j is its elongation, and A_j and L_j are the cross-sectional area and length of the bar. The stress and strain are related by Hooke's law (third relation in (20.1)) where E is the Young modulus of the material. *All bars are assumed to be made of the same material.* This is important because it lies at the heart of whatever follows.

It is easy to show that, except for some very special structures,

Redundant trusses cannot be fully stressed.

In fact, the statement means that redundant trusses cannot be fully strained. In other words, we cannot have the same strain in all the bars of a redundant truss. And

M.B. Fuchs, *Structures and Their Analysis*,
DOI 10.1007/978-3-319-31081-7_20

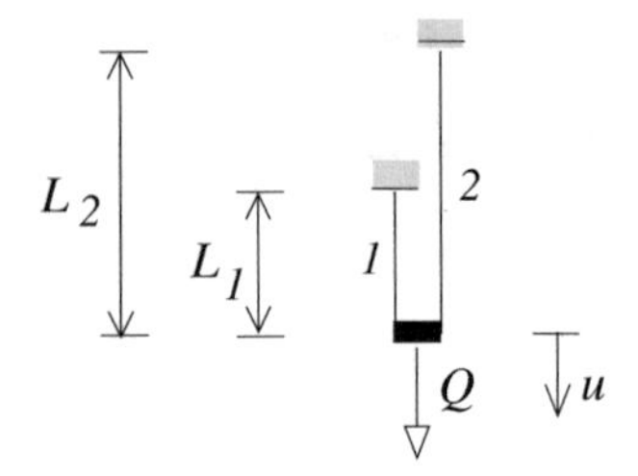

Fig. 20.1 Two-bar redundancy

since E is common to all bars, we also cannot have the same stress in all the bars. A simple example will illustrate this statement.

Consider the uniaxial 2-bar truss in Fig. 20.1. The arrangement has redundancy $R = 1\,(U = 2, E = 1)$. Under a vertical force Q the free node will displace vertically by an amount u. Consequently, the elongations of both bars are equal, $e_1 = e_2 = u$. Elongations are strains multiplied by lengths, $\epsilon_1 L_1 = \epsilon_2 L_2$. Unless $L_1 = L_2$ the strains cannot be equal, and the same is true for the stresses

$$\sigma_1 = \frac{L_2}{L_1}\,\sigma_2 \tag{20.2}$$

Note that this result is independent of the cross-sectional areas.

20.2 General Redundant Truss

In the case of a general redundant truss of N open nodal degrees of freedom and M $(>N)$ bars, we can select a statically determinate stable substructure of N bars, and express the elongations of the R $(= M - N)$ redundant bars $\boldsymbol{e}_1$ in terms of the basic set of elongations $\boldsymbol{e}_0$

$$\boldsymbol{e}_1 = \boldsymbol{G}\,\boldsymbol{e}_0 \tag{20.3}$$

Indeed, the elongations of the basic set depend on the nodal displacements $\boldsymbol{u}$, $\boldsymbol{e}_0 = \boldsymbol{R}_0\boldsymbol{u}$. Likewise the elongations of the redundant bars are $\boldsymbol{e}_1 = \boldsymbol{R}_1\boldsymbol{u}$. Introducing $\boldsymbol{u} = \boldsymbol{R}_0^{-1}\boldsymbol{e}_0$ in the previous equation yields (20.3) with $\boldsymbol{G} = \boldsymbol{R}_1\boldsymbol{R}_0^{-1}$. The $R \times N$ coefficients matrix $\boldsymbol{G}$ depends only on the geometry of the structure.

Let $\boldsymbol{l}_1$ and $\boldsymbol{l}_0$ be respectively the lengths of the redundant bars and the basic bars. We have $\boldsymbol{e}_1 = (\boldsymbol{l}\boldsymbol{\epsilon})_1$ and $\boldsymbol{e}_0 = (\boldsymbol{l}\boldsymbol{\epsilon})_0$, where the components of these vectors are the products of the lengths and the strains of the elements, $e_m = l_m\epsilon_m$. When introduced into (20.3), we have, after multiplying both sides by Young's modulus, $(\boldsymbol{l}\boldsymbol{\sigma})_1 = \boldsymbol{G}\,(\boldsymbol{l}\boldsymbol{\sigma})_0$.

Now, for the truss to be fully stressed all the stresses should equal a same value, say $\bar{\sigma}$. Consequently, for a truss to be fully stressable we need

$$\boldsymbol{l}_1 = \boldsymbol{G}\,\boldsymbol{l}_0 \tag{20.4}$$

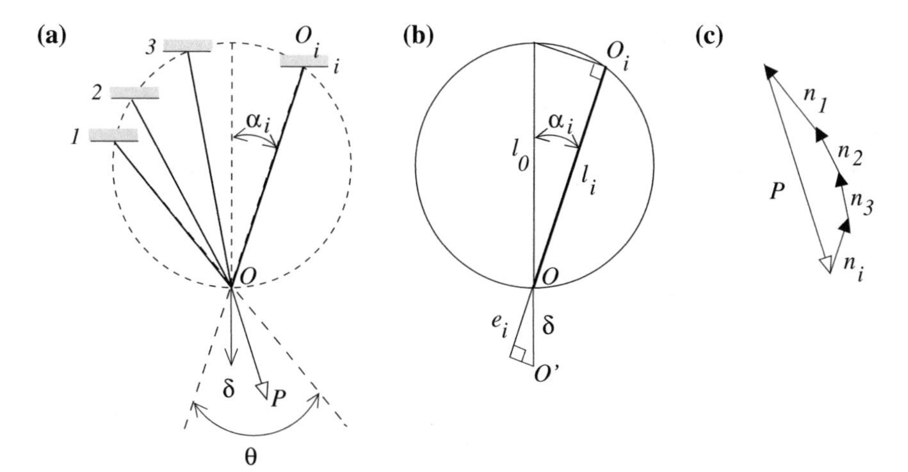

Fig. 20.2 Circle-chord truss

a condition rarely met.

20.3 Circle-Chord Trusses

We will present circle-chord trusses which are a class of trusses which although redundant can be fully stressed.[1]

Consider the truss embedded in a circle in Fig. 20.2a. All nodes are positioned on the perimeter of the circle. All nodes are fixed except for node O which is free to move. All bars i radiate from the common free node O to its fixed node O_i.

As shown in Fig. 20.2a, bar i makes and angle α_i with the diameter passing through O. (For convenience we have drawn the diameter through O vertical.) The two outer bars form an intercept angle θ.

We can show the following properties for the circle-chord truss:

1. If the displacement δ of node O is along the diameter the truss is fully stressed;
2. For a force applied at the free node O and within the intercept angle θ, we can find cross-sections A_i which, for the vertical displacement δ, will be in equilibrium with the applied force.
3. All such fully-stressed trusses will have the same volume of material, independently of the number of bars.

Property 1

Consider bar i in Fig. 20.2b. Let l_0 be the length of the diameter and l_i be the length of bar i. We have

[1]Fuchs and Felton, *AIAA J.* **12**, 11, 1974.

$$l_i = l_0 \cos \alpha_i \tag{20.5}$$

Now, assuming that node O displaces vertically by an amount δ to position O', the elongation of bar i is $e_i = \delta \cos \alpha_i$ or $\epsilon_i l_i = \delta \cos \alpha_i$. Indeed, the extremity O of bar i can reach O' by elongating by e_i and swinging in place about O_i, that is, moving along the tangent.

Introducing (20.5) yields $\epsilon_i l_0 \cos \alpha_i = \delta \cos \alpha_i$ or $\epsilon_i = \delta / l_0$. Multiplying both sides by E gives

$$\sigma_i = \frac{E\,\delta}{l_0} \tag{20.6}$$

in other words, the stresses in all bars are equal if the free node translates along the diagonal.

Property 2

For the circle-chord truss in Fig. 20.2a with a force P within the intercept angle there are many possible equilibrium solutions, one of which is shown in Fig. 20.2c. Clearly, having the applied force within the intercept angle guarantees that the bars will all be in tension. Note, if the truss is fully stressed, the force in any bar is $n_i = \bar{\sigma} A_i$. In other words, if the cross-sections of the bars are proportional to the arrows drawn in Fig. 20.2c the truss is fully-stressed and satisfies equilibrium and also compatibility (we recall that the displacement of node O is vertical).

Property 3

The volume of material of the truss is the sum of the volumes of its elements

$$V = \sum_i A_i l_i \tag{20.7}$$

Introducing (20.5) into the above relation gives $V = l_0 \sum_i A_i \cos \alpha_i$. But $\sum_i A_i \cos \alpha_i = P/\bar{\sigma}$, therefore the volume of a circle-chord truss is

$$V = \frac{P l_0}{\bar{\sigma}} \tag{20.8}$$

That is, the volume of a circle-chord truss is the same as an optimal bar of length l_0. As noted, this result is independent of the actual configuration of the truss.

20.4 Extended Circle-Chord Trusses

The three-bar circle-chord truss in Fig. 20.3a under the force shown is redundant and fully-stressed, so it is both a safe design (more bars than necessary for stability) and an efficient design (the material is fully exploited). However, ceilings are not usually

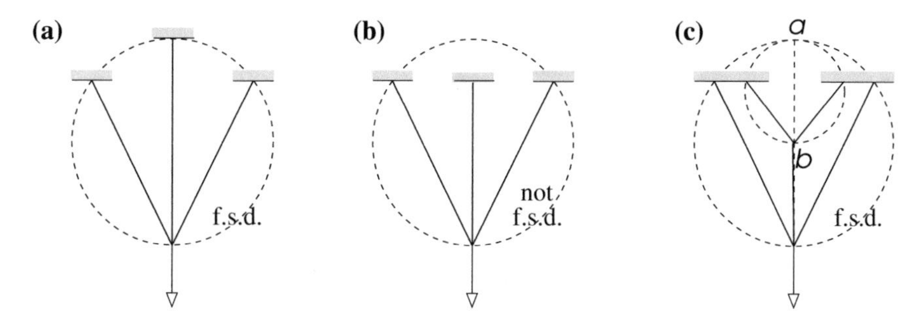

Fig. 20.3 Three-bar f.s.d. truss

circular. Three-bar trusses would look more like the design in Fig. 20.3b, which is unfortunately not fully stressed.

Arguably, the truss in Fig. 20.3c provides an adequate solution. We can show that we can replace segment *ab* of the middle-bar (Fig. 20.3c) by a circle-chord truss of diameter *ab* and maintain the fully-stressed property. The interior 2-bar circle-chord truss was designed such that the anchor-points are along a same line. This last design is now fully stressed, redundant and has all its supports along a line.

20.5 Summing Up

The volume of material of the extended circle-chord truss in Fig. 20.3c is the same as in Fig. 20.3a. Any circle-chord truss whose diameter has the length of the bar it replaces is 'equal' to that bar.

Chapter 21
Frames Viewed as Generalized Trusses

A truss is often considered as a structure relatively different from a frame. The truss is composed of unimodal normal elements and when properly triangulated it opposes the external forces applied at the nodes by normal internal forces, constant in every element.

It is shown that a frame can likewise be considered as being composed of unimodal normal, shear and moment elements and when subjected to nodal loads it behaves in many respects as a generalized truss.

21.1 Element Stiffness Matrices in Modal Coordinates

The extensional (axial) $\mathbf{K}_a$ and flexural (bending) $\mathbf{K}_b$ stiffness matrices of a uniform element of length L, Young modulus E, cross-sectional area A and moment of inertia I are respectively

$$\mathbf{K}_a = k_a \begin{bmatrix} 1 & -1 \\ -1 & 1 \end{bmatrix} \qquad \mathbf{K}_b = k_b \begin{bmatrix} 12 & -12 & 6L & 6L \\ -12 & 12 & -6L & -6L \\ 6L & -6L & 4L^2 & 2L^2 \\ 6L & -6L & 2L^2 & 4L^2 \end{bmatrix} \tag{21.1}$$

with $k_a = EA/L$ and $k_b = EI/L^3$.

Let $\mathbf{K}$ represent symbolically one of these matrices. An eigenvalue λ_i and corresponding eigenvector $\mathbf{u}_i$ of this matrix are such that $\mathbf{K}\,\mathbf{u}_i = \lambda_i \mathbf{u}_i$. Moving the right-hand side to the left produces the *characteristic equation*

$$(\mathbf{K} - \lambda_i \mathbf{I})\mathbf{u}_i = 0 \tag{21.2}$$

where $\mathbf{I}$ is a unit diagonal matrix of appropriate dimensions.

M.B. Fuchs, *Structures and Their Analysis*,
DOI 10.1007/978-3-319-31081-7_21

These equations have of course a trivial solution $\mathbf{u}_i = \mathbf{0}$. For a non-trivial solution to these equations the coefficient matrix must be singular, in other words, its determinant must be zero

$$|\mathbf{K} - \lambda_i \mathbf{I}| = 0 \tag{21.3}$$

This scalar algebraic equation in λ_i is called the *characteristic equation* of matrix $\boldsymbol{K}$. Clearly, the order of the equation equals the order (size) of the matrix and also the number of solutions (roots) λ_i of the characteristic equation.

Omitting subscript i we obtain for both matrices in (21.1), respectively

$$\lambda(\lambda - 2k_a) = 0 \qquad \lambda^2(\lambda - 2k_b)(\lambda - 30k_b) = 0 \tag{21.4}$$

21.1.1 Axial Stiffness Matrix

The eigenvalues of the axial element are

$$\lambda = 0 \qquad \lambda_1 = 2k_a \tag{21.5}$$

The eigenvector which corresponds to the zero eigenvalue is the solution of

$$k_a \begin{bmatrix} 1 & -1 \\ -1 & 1 \end{bmatrix} \begin{Bmatrix} u_a \\ u_b \end{Bmatrix} = \begin{Bmatrix} 0 \\ 0 \end{Bmatrix} \tag{21.6}$$

which gives a rigid-body displacement in the axial direction, $u_L = u_R$ (see Fig. 21.1).

For the eigenvector of the non-zero eigenvalue $\lambda_1 = 2k_a$ we have

$$k_a \begin{bmatrix} -1 & -1 \\ -1 & -1 \end{bmatrix} \begin{Bmatrix} u_a \\ u_b \end{Bmatrix} = \begin{Bmatrix} 0 \\ 0 \end{Bmatrix} \tag{21.7}$$

which yields $u_L = -u_R$. This is a deformation mode. The element either extends or contracts.

We normalize the eigenvectors such that the sum of the absolute values of the components equals 1 (see Fig. 21.1a). The matrix composed of the eigenvectors, that is, the eigen-matrix $\mathbf{U}_a$, is thus[1]

$$\mathbf{U}_a = \begin{bmatrix} 1/2 & 1/2 \\ 1/2 & -1/2 \end{bmatrix} \qquad \mathbf{U}_b = \begin{bmatrix} 1/2 & -1/6 & 0 & -1/3 \\ 1/2 & 1/6 & 0 & 1/3 \\ 0 & 1/3 & 1/2 & 1/6 \\ 0 & 1/3 & -1/2 & 1/6 \end{bmatrix} \tag{21.8}$$

[1] MB Fuchs, Computers & Structures, Vol. 63, No. 10, 1997.

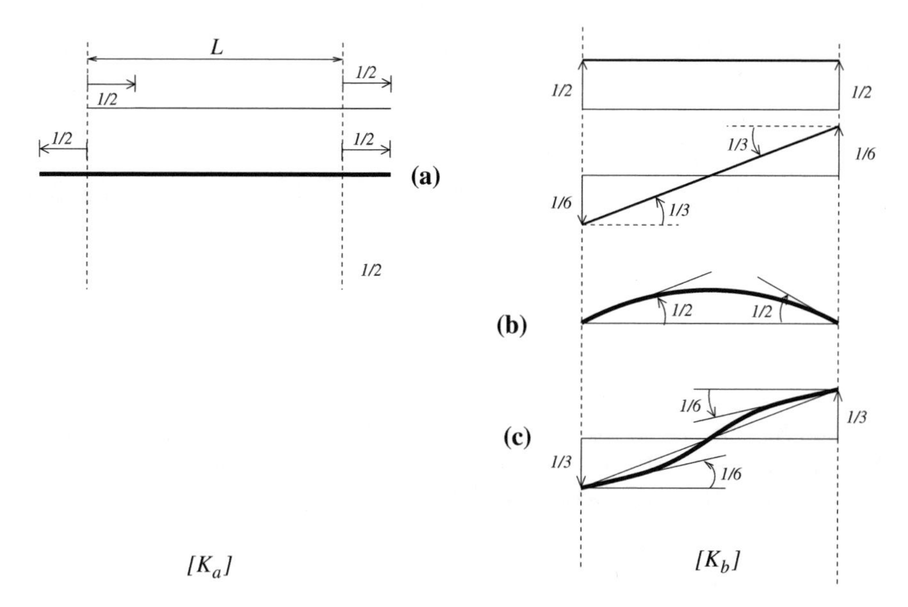

Fig. 21.1 Eigen-vectors

21.1.2 Bending Stiffness Matrix

Repeating the procedure for the bending matrix, we obtain $\mathbf{U}_b$, the four columns of which are the four eigenvectors of the bending matrix (see Fig. 21.1b). Here also, the eigenvectors were normalized as in the axial case. We will recognize that the first two columns are rigid-body modes which correspond to the two zero eigenvalues $\lambda^2 = 0$. These are respectively a rigid-body vertical translation $v_L = v_R$ and a rigid-body rotation $v_L = -v_R$ and $\theta_L = \theta_R = 2v_R/L$. (We recall, for small angles, an angle is equal to its tangent.)

The following two eigenvectors are deformation modes. The third column of $\mathbf{U}_b$ is pure bending and the fourth column represents 'pure' shear or shear bending.

21.2 Coordinate Transformation

Consider the change of variables

$$\mathbf{u} = \mathbf{U}\mathbf{u}' \tag{21.9}$$

where $\mathbf{U}$ is either eigenvalue matrix $\mathbf{U}_a$ or $\mathbf{U}_b$. This in fact saying that the end-displacements of the elements are linear combinations of the modal displacements $\mathbf{u} = \sum_i u'_i \mathbf{u}_i$ where $\mathbf{u}_i$ is column i of $\mathbf{U}$.

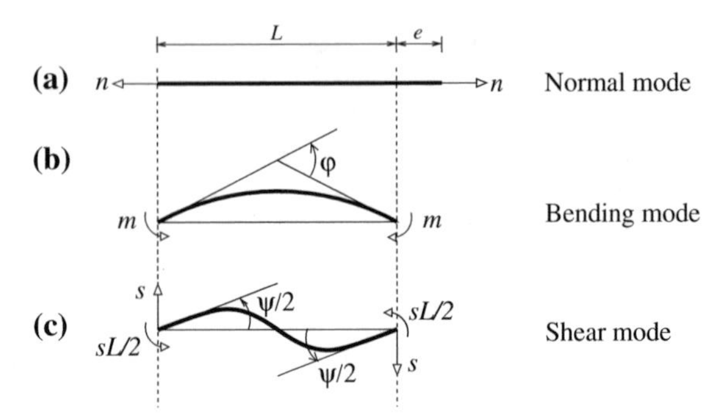

Fig. 21.2 Normal, pure bending and shear bending modes

Following the rule for a congruent transformation, we find in modal coordinates $\mathbf{K}'\mathbf{u}' = \mathbf{p}'$ with $\mathbf{K}' = \mathbf{U}^T\mathbf{K}\mathbf{U}$ and $\mathbf{p}' = \mathbf{U}^T\mathbf{p}$. Performing these matrix multiplications for the axial and bending matrices yields respectively

$$\mathbf{K}_a = \begin{bmatrix} 0 & 0 \\ 0 & k_a \end{bmatrix} \qquad \mathbf{K}_b = \begin{bmatrix} 0 & 0 & 0 & 0 \\ 0 & 0 & 0 & 0 \\ 0 & 0 & k_b & 0 \\ 0 & 0 & 0 & 3k_b \end{bmatrix} \tag{21.10}$$

Calculating the modal forces $\mathbf{p}'$ and modal displacements $\mathbf{u}'$ gives, after discarding the rigid body modes, the element equilibrium equation in modal coordinates

$$\begin{Bmatrix} n \\ m \\ s/2 \end{Bmatrix} = \begin{bmatrix} EA/L & 0 & 0 \\ 0 & EI/L & 0 \\ 0 & 0 & 3EI/L^2 \end{bmatrix} \begin{Bmatrix} e \\ \phi \\ \psi \end{Bmatrix} \quad \text{or} \quad \mathbf{n}_i = \mathbf{S}_i\mathbf{e}_i \tag{21.11}$$

where n is the normal force, m is the average bending moment in the element (recall that the bending moment is linear) and s is the shear force.

The corresponding modal deformations can be seen in Fig. 21.2. The elongation e is the extension of the element in the normal mode, and ϕ and ψ are the angles governing pure bending and the shear-bending deformations respectively, as shown in Fig. 21.2b, c.

In shorthand the relation is $\mathbf{n}_i = \mathbf{S}_i\mathbf{e}_i$ where we have added subscript i to indicate that these are the stiffness relations for a general frame element i.

21.3 Unimodal Normal, Moment and Shear Elements

Note that the element stiffness matrix of the frame element $\mathbf{S}_i$ is diagonal. Consequently the three stiffness relations are uncoupled (not connected). We can therefore imagine (see Fig. 21.3) that

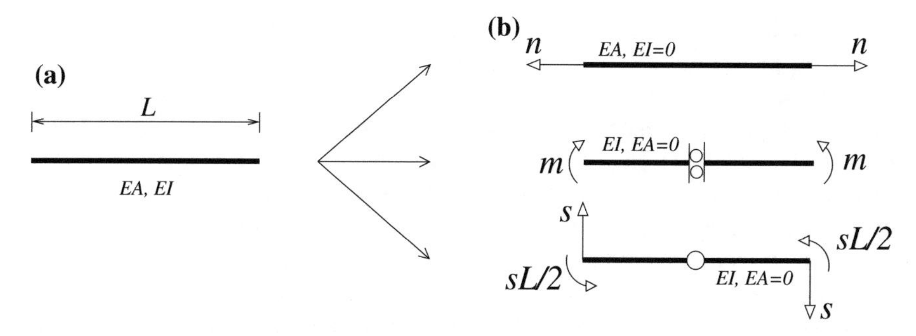

Fig. 21.3 Three unimodal components

a frame element is in fact three unimodal elements mounted in parallel.

A *normal* element that has no bending resistance and that can take only axial forces.
A *moment* element that has zero axial stiffness and presents resistance only to pure bending (two equal and opposite end-couples).
A *shear* element that has also zero axial stiffness and takes only equal and opposite shear forces applied at the extremities along with equilibrating equal end-couples.

These unimodal elements act as *filters*. From a general loading applied to the extremities, the normal element will carry only the normal force and will be oblivious to bending, the shear element will carry only shear forces and related end-couples and the moment element will carry the average bending moment and filter-out the rest. Together, the three unimodal elements behave like a regular frame element.

In Fig. 21.3b, we have shown stylized renderings of such unimodal elements. A slender element with no bending stiffness ($EI = 0$) that can take only axial forces represents the unimodal normal element. The following element assumed to have no axial stiffness ($EA = 0$) represents the moment element. Indeed, the shear release prevents it from carrying shear. Finally, the last element in Fig. 21.3b is the shear bending element. It is also assumed to have no axial stiffness. It is designed to carry the shear force and related end-couples.

Together, the three unimodal elements will carry any loading applied to the extremities of the standard element (Fig. 21.3a).

21.4 Frames Viewed as Trusses

The stiffness matrix in modal coordinates is diagonal, consequently the three deformation modes are uncoupled. The decomposition of a prismatic element into axial and bending elements is widely used. We can now further decompose the bending element into a moment element and a shear element which are mutually orthogonal.

A corollary of all this is that a frame composed of flexural elements with applied loads at the joints (including couples) can be analyzed as a generalized truss. In a truss we have only (unimodal) axial elements. In a frame we have now all three unimodal elements. Interestingly, a beam (by definition, beams do not carry axial loads and do not deform axially) is a generalized truss with moment and shear elements.

Nothing can distinguish the equations for a frame from those of a truss. Consequently, everything we know about trusses can now be applied as is to frames, in particular, the formula for the internal forces explicitly in terms of the stiffnesses which are here the two unimodal bending stiffnesses and the axial stiffness.

We will show an example for constructing the explicit expression for a propped cantilever as if it were a truss.

21.5 Explicit Analysis of a Propped Cantilever

The propped cantilever in Fig. 21.4a is composed of two uniform elements of length L and bending stiffness EI_{ab} and EI_{bc} respectively. It has '$M = 2$' static unknowns (the shear and moment at one end of the beam) and '$N = 1$' open degree of freedom (the slope at c) and the redundancy is thus $R = 1$. This is the classical way of doing.

We will repeat this in view of what we have shown above. Since we need uniform elements and can have loads only at nodes, we add a node at b and replace every segment by two unimodal elements (see Fig. 21.4b: an m-element (elements 1 and 3) and a s-element (elements 2 and 4). We now have $M = 4$ and $N = 3$ (three open degrees of freedom at the nodes). Naturally, the redundancy remains the same: $R = 4 - 3 = 1$.

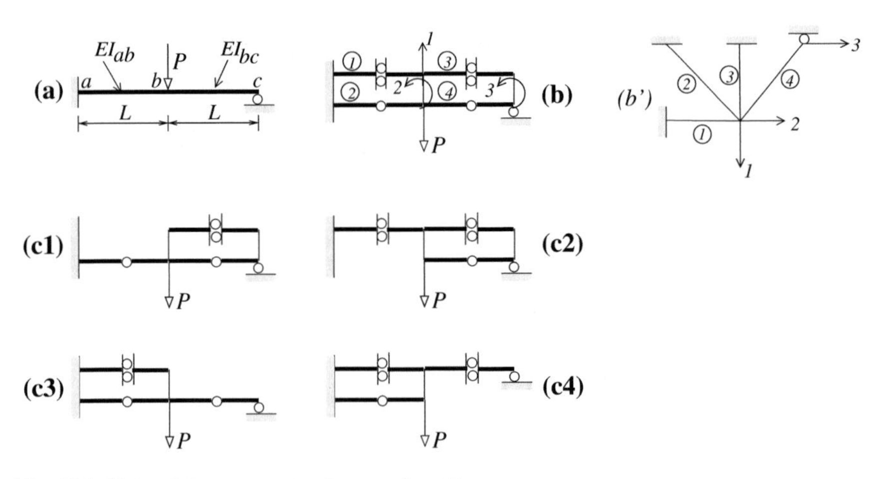

Fig. 21.4 Unimodal components of propped cantilever

Since this model of the propped cantilever is in principle identical to that of the truss in Fig. 21.4b' ($M = 4$ unimodal n-elements and $N = 3$ open degrees of freedom), we can simply repeat the steps which produce the explicit expressions of a truss.[2] What is different with a beam (or a frame) is the equilibrium equations $\mathbf{Q}\,\mathbf{n} = \mathbf{p}$ and the content of the diagonal matrix of the element stiffnesses $\mathbf{S}$.

The equilibrium equations are

$$\begin{bmatrix} 0 & -2 & 0 & 2 \\ 1 & L & -1 & L \\ 0 & 0 & -1 & -L \end{bmatrix} \begin{Bmatrix} m_1 \\ s_2/2 \\ m_3 \\ s_4/2 \end{Bmatrix} = \begin{Bmatrix} -P \\ 0 \\ 0 \end{Bmatrix} \tag{21.12}$$

where m_1 is the average bending moment in element 1, s_2 is the shear in element 2 and so on.

The first equation $-s_2 + s_4 = -P$ is translational equilibrium along dof 1. The second equation $m_1 + s_2L/2 - m_3 + s_4L/2 = 0$ is rotational equilibrium along dof 2, and the third equation $-m_3 - s_4L/2 = 0$ is rotational equilibrium along dof 3.

The diagonal matrix of the stiffnesses is

$$\mathbf{S} = \begin{bmatrix} EI_{ab}/L & 0 & 0 & 0 \\ 0 & 3EI_{ab}/L^2 & 0 & 0 \\ 0 & 0 & EI_{bc}/L & 0 \\ 0 & 0 & 0 & 3EI_{bc}/L^2 \end{bmatrix} \tag{21.13}$$

The structure, being of redundancy $R = 1$, removing one component will render it statically determinate. Removing each time one element, we obtain the four statically determinate substructures shown in Fig. 21.4c1–c4. They all turn out to be stable. Solving in turn the four systems of equations $\mathbf{Q}_j\mathbf{n}_j = \mathbf{p}$, $j = 1 \ldots 4$ produces the respective internal forces

	$\mathbf{n}_1$	$\mathbf{n}_2$	$\mathbf{n}_3$	$\mathbf{n}_4$
	0	$P/3$	$P/3$	$P/3$
	PL	0	$PL/2$	$-P/2$
	$-PL/2$	$P/2$	0	0
	$-PL/2$	$P/2$	0	0
$\lvert Q_j\rvert^2$	$36L^2$	4	$4L^2$	4

We note that internal force i in substructure i (the substructure in which bar i was removed) is, of course, zero.

We also calculate the squares of the determinants of the coefficients matrices $\lvert Q_j\rvert^2$ (see bottom line of table).

[2]This example follows the method detailed in the subsequent chapter.

Finally, the determinants of the matrices of the stiffnesses of the four substructures are

$$
\begin{aligned}
|\mathbf{S}_1| &= (3EI_{ab}/L^2)(EI_{bc}/L)(3EI_{bc}/L^2)\\
|\mathbf{S}_2| &= (EI_{ab}/L)(EI_{bc}/L)(3EI_{bc}/L^2)\\
|\mathbf{S}_3| &= (EI_{ab}/L)(3EI_{ab}/L^2)(3EI_{bc}/L^2)\\
|\mathbf{S}_4| &= (EI_{ab}/L)(3EI_{ab}/L^2)(EI_{bc}/L)
\end{aligned}
\tag{21.14}
$$

Recall, substructure i is missing element i.

Introducing all this into (22.2) with (22.4) and (22.6) yields, for instance, for the shear force in unimodal element 2 (the shear in segment ab in fact)

$$
s_2 = 2P\,\frac{9I_{bc} + 2I_{ab}}{28I_{bc} + 4I_{ab}} \tag{21.15}
$$

A cursory check gives for $I_{bc} = I_{ab}$ (uniform beam) $s_2 = 11P/16$, a known result, and for $I_{bc} = 0$ (P is taken by ab only) $s_2 = P$, which is an obvious result.

21.6 Summing Up

We have shown that frames are not that different from trusses when properly presented. In fact, frames can be seen, in many respects, as generalized trusses. Consequently, many theories developed for trusses can be extended to frames, in particular, the explicit analysis expressions of the next chapter.

Chapter 22
Explicit Analysis

Given a structure, we have to solve equations in order to calculate the distribution of the internal forces in that structure, be it the compatibility equations if one uses the force method, or the equilibrium equations of the displacement method. One way or the other, we solve a system of linear equations. This is what makes structural analysis an implicit affair.

In the realm of design we often need to redesign and hence reanalyze the structure several times before reaching a satisfactory solution. This may prove cumbersome.

There are, however, relations which allow us to write the internal forces of a truss explicitly in terms of the stiffnesses thus bypassing any need for reanalysis.[1] The explicit equations have a major shortcoming but they are theoretically very interesting.

We will present here the explicit expressions of the internal forces as a function of the stiffnesses of the bars of the truss and indicate a way to prove the relations.

22.1 The Explicit Expression

Consider a redundant truss composed of N nodal degrees of freedom and M $(>N)$ bars. The degree of redundancy is thus $R = M - N$.

The $2M+N$ analysis variables are the bar forces $\boldsymbol{n}$ and elongations $\boldsymbol{e}$ and the nodal displacements $\boldsymbol{u}$. The $2M + N$ analysis equations are respectively Compatibility (elongations-displacements) $\boldsymbol{e} = \boldsymbol{R}\boldsymbol{u}$, Elasticity $\boldsymbol{t} = \boldsymbol{S}\boldsymbol{e}$ and Equilibrium $\boldsymbol{Q}\boldsymbol{n} = \boldsymbol{p}$, where the entries of vector $\boldsymbol{p}$ are the externally applied forces along the nodal degrees of freedom. The element stiffnesses are the components $S_{jj} = E_j A_j / L_j$ of the diagonal stiffness matrix $\boldsymbol{S}$. The virtual work principle causes the link $\boldsymbol{R} = \boldsymbol{Q}^T$.

[1] MB Fuchs, *Int J Solids Structures* **29**(*16*) *1992*

M.B. Fuchs, *Structures and Their Analysis*,
DOI 10.1007/978-3-319-31081-7_22

Any truss formed by a subset of N (out of M) bars is statically determinate. There are

$$C_N^M = \frac{M!}{N!R!} \tag{22.1}$$

such substructures although many may not be stable.

Let $|\boldsymbol{K}|$ be the determinant of the structure stiffness matrix, and let $|\boldsymbol{K}_k|$ be the determinant of the stiffness matrix of substructure k.

We can show that the internal forces $\boldsymbol{n}$ of the truss can be expressed as

$$\boldsymbol{n} = \frac{1}{|\boldsymbol{K}|} \sum_k |\boldsymbol{K}_k| \, \boldsymbol{n}_k \tag{22.2}$$

where $\boldsymbol{n}_k$ are the forces in substructure k if it were to support by itself the applied forces $\boldsymbol{p}$.

Note: We assume that substructure k is composed of all the nodes and all the bars with the R 'missing' bars allocated a zero stiffness, and therefore zero internal forces.

When the explicit expression is written as

$$\boxed{\boldsymbol{n} = \sum_k \frac{|\boldsymbol{K}_k|}{|\boldsymbol{K}|} \boldsymbol{n}_k}$$

we find that the internal forces $\boldsymbol{n}$ are a linear combination of the internal forces $\boldsymbol{n}_k$. The contribution of each substructure is proportional to the determinant of its stiffness matrix.

Determinant $|\boldsymbol{K}_k|$

Noting that the stiffness matrix of substructure k is the product of three square matrices $\boldsymbol{K}_k = \boldsymbol{Q}_k \boldsymbol{S}_k \boldsymbol{R}_k$, we find that the determinant of $\boldsymbol{K}_k$ is the product of the determinants of the three matrices

$$|\boldsymbol{K}_k| = |\boldsymbol{Q}_k| \, |\boldsymbol{S}_k| \, |\boldsymbol{R}_k| \tag{22.3}$$

Since $\boldsymbol{R}_k = \boldsymbol{Q}_k^T$, we have $|\boldsymbol{R}_k| = |\boldsymbol{Q}_k|$, and therefore

$$|\boldsymbol{K}_k| = |\boldsymbol{Q}_k|^2 |\boldsymbol{S}_k| \tag{22.4}$$

Determinant $|\boldsymbol{K}|$

The $N \times N$ structure stiffness matrix $\boldsymbol{K} = \boldsymbol{Q}\boldsymbol{S}\boldsymbol{Q}^T$ is in fact the product of a $N \times M$ and a $M \times N$ matrices $(\boldsymbol{Q}\boldsymbol{S}^{1/2})(\boldsymbol{S}^{1/2}\boldsymbol{Q}^T)$, where the diagonal matrix $\boldsymbol{S}^{1/2}$ has $S_{jj}^{1/2}$ on its main diagonal, and zero elsewhere. For the determinant of such a matrix we refer to the Binet–Cauchy formula.

Binet–Cauchy Formula

Binet and Cauchy have shown that the determinant of an $N \times N$ square matrix which is the product of two rectangular matrices $\boldsymbol{A}$ and $\boldsymbol{B}$ is

$$|\boldsymbol{AB}| = \sum_k |\boldsymbol{A}_k \boldsymbol{B}_k| \tag{22.5}$$

where index k runs over all $N \times N$ submatrices of $\boldsymbol{A}$ and $\boldsymbol{B}$.

Consequently, $|\boldsymbol{K}| = \sum_k |\boldsymbol{Q}_k \boldsymbol{S}_k^{1/2}||\boldsymbol{S}_k^{1/2} \boldsymbol{Q}_k^T| = \sum_k |\boldsymbol{Q}_k||\boldsymbol{S}_k^{1/2}||\boldsymbol{S}_k^{1/2}||\boldsymbol{Q}_k^T|$, and since the determinant of a diagonal matrix is the product of its coefficients $|\boldsymbol{K}| = \sum_k |\boldsymbol{Q}_k||\boldsymbol{S}_k||\boldsymbol{Q}_k^T|$ or also

$$|\boldsymbol{K}| = \sum_k |\boldsymbol{Q}_k|^2 |\boldsymbol{S}_k| \tag{22.6}$$

Determinant $|\boldsymbol{S}_k|$

The determinant of a diagonal matrix is the product of its diagonal entries. Let $m, n, o, p, \ldots$ be the bars which belong to substructure k. We call π_k that product, that is

$$|\boldsymbol{S}_k| = \pi_k = s_m s_n s_o s_p \ldots \tag{22.7}$$

and recalling that s_m, for instance, is $s_m = E_m A_m / L_m$, we obtain

$$|\boldsymbol{S}_k| = \pi_k = \frac{E_m E_n E_o E_p \ldots}{L_m L_n L_o L_p \ldots} x_m x_n x_o x_p \ldots \tag{22.8}$$

(We have changed the A's into x's on purpose to emphasize that the cross-sections are in the present context variables in a design process.)

The coefficients of the numerator and denominator are constants such that the fraction in (22.8) can be replaced by a constant B_k.

$$B_k = \frac{E_m E_n E_o E_p ..}{L_m L_n L_o L_p ..} \tag{22.9}$$

We obtain

$$|\boldsymbol{S}_k| = \pi_k = B_k \; x_m x_n x_o x_p .. \tag{22.10}$$

Introducing (22.10), (22.6) and (22.4) into (22.2), yields the internal forces in the truss explicitly in terms of the bar cross-sections $\boldsymbol{x}$

$$\boldsymbol{n} = \frac{\sum_k ....|\boldsymbol{Q}_k|^2 \; B_k \; \boldsymbol{n}_k \; x_m x_n x_o x_p .. \; ....}{\sum_k ....|\boldsymbol{Q}_k|^2 \; B_k \; x_m x_n x_o x_p .. \; ....} \tag{22.11}$$

We will recognize that, except for the explicit components of $\boldsymbol{x}$, all the other symbols in (22.11) are constants. Indeed, $\boldsymbol{Q}_k$ are the coefficient matrices of the equilibrium equations, and depend only on the nodal coordinates and bars connectivity. The constant B_k is the ratio of elasticity constant and bar length. Finally, the internal forces $\boldsymbol{n}_k$ do not depend on the stiffnesses because the substructures are statically determinate.

22.2 Example

The truss in Fig. 22.1a of height L and $N = 2$ open degrees of freedom at node a, has $M = 3$ bars with cross-sections x_1, x_2 and x_3 where Young's modulus E is common to all three bars. It is subjected to an applied force P along degree of freedom 2 at the free node.

We intend to construct the explicit expression (22.11) for the three internal forces n_1, n_2 and n_3.

The degree of redundancy of the truss is $R = M - N = 1$, and the number of statically determinate substructures (22.1) is of course 3. These two-bar trusses can be seen in Fig. 22.1b1–b3.

Building the Solution

The equilibrium equations of the three-bar truss $\mathbf{Qn} = \mathbf{p}$ are

$$\begin{bmatrix} \sqrt{2}/2 & 0 & -\sqrt{2}/2 \\ \sqrt{2}/2 & 1 & \sqrt{2}/2 \end{bmatrix} \begin{Bmatrix} n_1 \\ n_2 \\ n_3 \end{Bmatrix} = \begin{Bmatrix} 0 \\ P \end{Bmatrix} \tag{22.12}$$

which for the three substructures gives the following equilibrium solutions

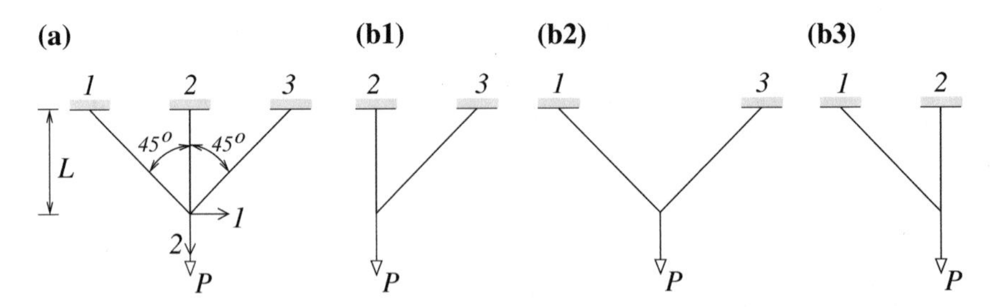

Fig. 22.1 Substructures of three-bar truss

$$\text{Substructure 1} \quad \begin{bmatrix} 0 & -\sqrt{2}/2 \\ 1 & \sqrt{2}/2 \end{bmatrix} \begin{Bmatrix} n_2 \\ n_3 \end{Bmatrix} = \begin{Bmatrix} 0 \\ P \end{Bmatrix}$$
$$\text{Substructure 2} \begin{bmatrix} \sqrt{2}/2 & -\sqrt{2}/2 \\ \sqrt{2}/2 & \sqrt{2}/2 \end{bmatrix} \begin{Bmatrix} n_1 \\ n_3 \end{Bmatrix} = \begin{Bmatrix} 0 \\ P \end{Bmatrix} \tag{22.13}$$
$$\text{Substructure 3} \quad \begin{bmatrix} \sqrt{2}/2 & 0 \\ \sqrt{2}/2 & 1 \end{bmatrix} \begin{Bmatrix} n_1 \\ n_2 \end{Bmatrix} = \begin{Bmatrix} 0 \\ P \end{Bmatrix}$$

the solutions of which are

$$\begin{Bmatrix} n_1 \\ n_2 \\ n_3 \end{Bmatrix}_1 = P \begin{Bmatrix} 0 \\ 1 \\ 0 \end{Bmatrix} \quad \begin{Bmatrix} n_1 \\ n_2 \\ n_3 \end{Bmatrix}_2 = P \begin{Bmatrix} \sqrt{2}/2 \\ 0 \\ \sqrt{2}/2 \end{Bmatrix} \quad \begin{Bmatrix} n_1 \\ n_2 \\ n_3 \end{Bmatrix}_3 = P \begin{Bmatrix} 0 \\ 1 \\ 0 \end{Bmatrix} \tag{22.14}$$

we can write the internal forces in the truss (22.11) explicitly in terms of the cross-sectional areas $\boldsymbol{x}$ of the bars

$$\boldsymbol{n} = \frac{|\boldsymbol{Q}_1|^2 \, B_1 \, \boldsymbol{n}_1 \, x_2 x_3 + |\boldsymbol{Q}_2|^2 \, B_2 \, \boldsymbol{n}_2 \, x_1 x_3 + |\boldsymbol{Q}_3|^2 \, B_3 \, \boldsymbol{n}_3 \, x_1 x_2}{|\boldsymbol{Q}_1|^2 \, B_1 \quad x_2 x_3 + |\boldsymbol{Q}_2|^2 \, B_2 \quad x_1 x_3 + |\boldsymbol{Q}_3|^2 \, B_3 \quad x_1 x_2} \tag{22.15}$$

(The spaces in the above equation were left on purpose to emphasize the structured form of the expression.)

Noting (22.9), (22.13) and (22.14), we have $B_1 = E/\sqrt{2}L$, $B_2 = E/L$, $B_3 = E/\sqrt{2}L$, and $|\boldsymbol{Q}_1| = \sqrt{2}/2$, $|\boldsymbol{Q}_2| = 1$, $|\boldsymbol{Q}_3| = \sqrt{2}/2$.

Equation (22.15) becomes

$$\begin{aligned} n_1 &= \frac{2\,x_1 x_3}{x_2 x_3 + 2\sqrt{2}\,x_1 x_3 + x_1 x_2} \, P \\ n_2 &= \frac{x_2 x_3 + x_1 x_2}{x_2 x_3 + 2\sqrt{2}\,x_1 x_3 + x_1 x_2} \, P \\ n_3 &= \frac{2\,x_1 x_3}{x_2 x_3 + 2\sqrt{2}\,x_1 x_3 + x_1 x_2} \, P \end{aligned} \tag{22.16}$$

Checking the Limits

Let us consider the limits for symmetric designs ($x_3 = x_1$). The internal forces are

$$\begin{aligned} n_1 &= \frac{2\,x_1^2}{2\,x_1 x_2 + 2\sqrt{2}\,x_1^2} \, P \\ n_2 &= \frac{2\,x_1 x_2}{2\,x_1 x_2 + 2\sqrt{2}\,x_1^2} \, P \\ n_3 &= \frac{2\,x_1^2}{2\,x_1 x_2 + 2\sqrt{2}\,x_1^2} \, P \end{aligned} \tag{22.17}$$

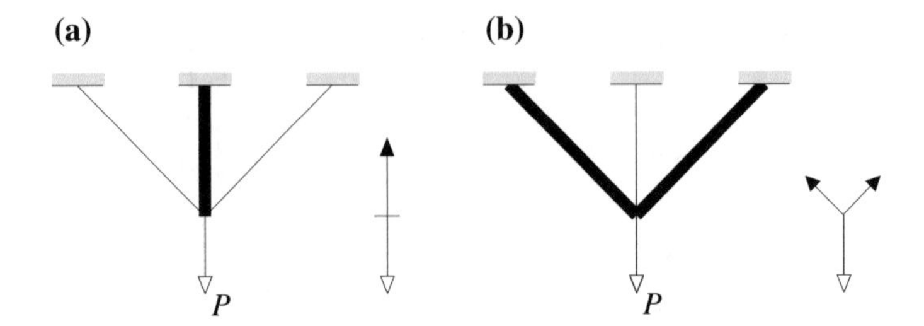

Fig. 22.2 Flow of forces in three-bar truss

For trusses with a very stiff central bar (see Fig. 22.2a), force P will flow through this central bar with little forces in the inclined members. And indeed, dividing the numerators and denominators in (22.17) by x_2 and taking the limit $x_2 \rightarrow \infty$ yields $\mathbf{n} = P\,\{0 \;\; 1 \;\; 0\}^T$.

Conversely, in trusses with very stiff exterior members, as in Fig. 22.2b, we expect the internal forces to flow through the exterior bars with little force in the middle bar. Dividing the numerators and denominators in (22.17) by x_1^2 and taking the limit $(x_1 \rightarrow \infty)$ produces the result, $\mathbf{n} = P\,\{\sqrt{2}/2 \;\; 0 \;\; \sqrt{2}/2\}^T$.

22.3 Summing Up

This nice method for 'constructing' explicit expressions for the internal forces in the bars of a truss is heavily burdened by the combinatoric nature of the number of substructures. Except for very small structures, the number of substructures which can be found in a truss is usually inordinate. This renders the formula almost unusable as a straightforward manner.

The explicit expression, however, with its stiffness matrices, the subdivision of the equilibrium equations into a basic and a redundant set seems to be a perfect epitome of structural analysis by combining seamlessly concepts of the stiffness method and the flexibility method.

Part VI
Epilogue

We have now completed a tour of the principles of the analysis of structures. It is probably appropriate to present a concise overview of what we have learned, starting with the three pillars of structures: equilibrium, deformations and the materials law, also known as constitutive law, in our case elasticity.

The elements of the structure will now be linear springs.

Chapter 23
Plus ça Change

As a young student of structures I was often mystified by the different number of unknowns of a structural problem: R=M-N if solved by the force method or N if solved by the displacement method. Well, in truth, an analysis of a structure solves the structure equations in 2M+N unknowns. The symmetric coefficient matrix of these equations comprises the three pillars of structures: equilibrium, deformations and elasticity. The force and displacement methods are subsets of the structure equations. This is exemplified by a simple assemblage of linear springs.

23.1 The Three Pillars of Structures

23.1.1 Equilibrium

Consider the structure composed of elementary extensional springs shown in Fig. 23.1a. It has $M = 4$ elements and thus $U = 4$ static unknowns, the internal forces (axial of course) in the springs, $\boldsymbol{n} = \{n_1\, n_2\, n_3\, n_4\}^T$. The $N = 2$ open nodes have only one degree of freedom each ($E = 2$). Let $\boldsymbol{p} = \{p_1\, p_2\}^T$ be the external forces applied to the open degrees of freedom. The equilibrium equations at the free dofs are for dof 1: $n_1 - n_2 - n_3 = p_1$ and for dof 2: $n_2 + n_3 + n_4 = p_2$ (see Fig. 23.1b) or in matrix notation

$$\boldsymbol{Q}\,\boldsymbol{n} = \boldsymbol{p} \qquad \text{(Equilibrium)} \qquad (23.1)$$

as given in Fig. 23.1c.

Subdividing the equilibrium matrix column-wise into four vectors $\boldsymbol{q}_j$ as shown in Fig. 23.1c leads to the equilibrium equations written as

$$\sum_j n_j\, \boldsymbol{q}_j = \boldsymbol{p} \qquad (23.2)$$

M.B. Fuchs, *Structures and Their Analysis*,
DOI 10.1007/978-3-319-31081-7_23

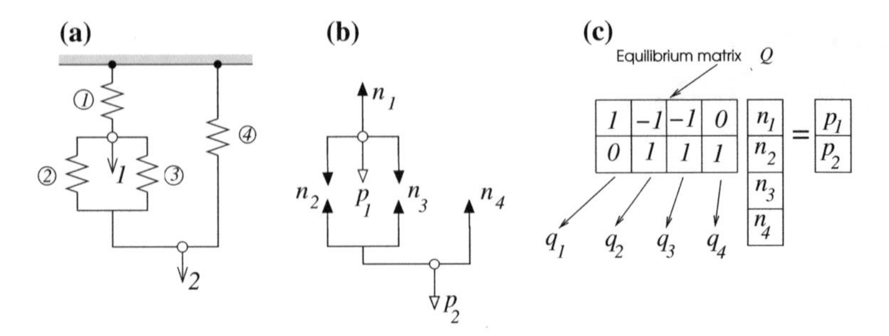

Fig. 23.1 Equilibrium

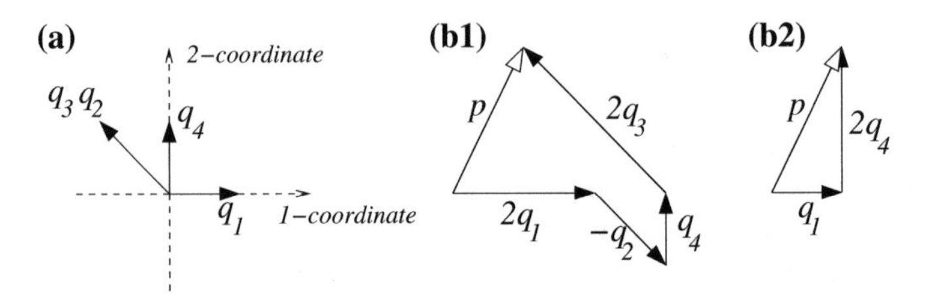

Fig. 23.2 Graphical representation of equilibrium

If we consider the products $n_j\,\boldsymbol{q}_j$ as force vectors (in the sense of $\boldsymbol{p}$), they have the graphical representation shown in Fig. 23.2a.[1]

Equation (23.2) states that equilibrium requires that we can construct any vector $\boldsymbol{p}$ by a linear combination of the columns of $\boldsymbol{Q}$. The present structure is redundant $R = 2$, and there are a multitude of possibilities to equilibrate $\boldsymbol{p}$. In Fig. 23.2b, two solutions for an applied force vector $\boldsymbol{p} = \{1\ 2\}^T$ (horizontal and vertical components) are shown: $\boldsymbol{n} = \{2\ {-1}\ 2\ 1\}^T$ and $\boldsymbol{n} = \{1\ 0\ 0\ 2\}^T$. In the latter solution, for example, the first spring is taking p_1 and the fourth one is holding p_2 while springs 2 and 3 are inactive. There are an infinite number of equilibrium solutions.

Note: A structure has one and only one solution. Here we are toying with the notion that, if there were only the equilibrium equations, how many equilibrium solutions and what kind of solutions can we get?

23.1.2 Deformation (Virtual Work)

Virtual Work

We now assume some displacements $\boldsymbol{u}$ at the $N = 2$ dofs. These displacements are arbitrary, and have no relation to the forces $\boldsymbol{p}$. We define the expression $\boldsymbol{p}^T\boldsymbol{u}$

[1] Note, a vector $\boldsymbol{q}_j$ represents the dimensionless external forces that will give a unit force in element j, that is, $n_j = 1$ and zero elsewhere.

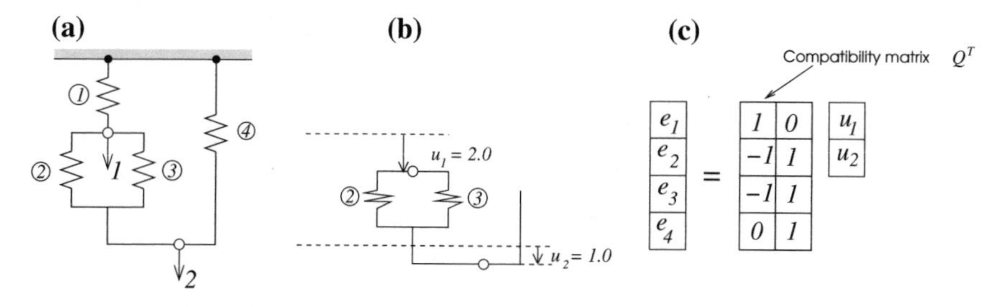

Fig. 23.3 Elongations-displacements

as the virtual work of the forces $\boldsymbol{p}$ over the displacements $\boldsymbol{u}$. It is work because it has the dimension of work (*forces* $\times$ *displacements*), and it is virtual because the displacements are not related to the forces, and are certainly not the result of the forces (the stiffness of the M springs is not even defined). Introducing (23.1) in the virtual work expression yields, after some manipulation,

$$\boldsymbol{p}^T\boldsymbol{u} = (\boldsymbol{Q}\boldsymbol{n})^T\boldsymbol{u} = \boldsymbol{n}^T\boldsymbol{Q}^T\boldsymbol{u} = \boldsymbol{n}^T(\boldsymbol{Q}^T\boldsymbol{u}) \tag{23.3}$$

Next, we call

$$\boldsymbol{e} = \boldsymbol{Q}^T\boldsymbol{u} \qquad \text{(Compatibility)} \tag{23.4}$$

which, after introduction in (23.3), yields the famous virtual work expression

$$\boldsymbol{p}^T\boldsymbol{u} = \boldsymbol{n}^T\boldsymbol{e} \qquad \text{(Virtual work: } evw = ivw) \tag{23.5}$$

Elongations

It is only after we establish the physical meaning of $\boldsymbol{e}$ that the virtual work equation can have meaning and become useful. It is not difficult to notice that the components of $\boldsymbol{e}$ are the M elongations of the springs due to the N nodal displacements $\boldsymbol{u}$.

In Fig. 23.3c we note that for displacements $\boldsymbol{u} = \{2\ 1\}^T$, for instance, the components of $\boldsymbol{e}$ are $\{2\ -1\ -1\ 1\}^T$. These are indeed the elongations of springs 1 to 4 respectively (Fig. 23.3a). In Fig. 23.3b we show graphically the contraction of springs 2 and 3 as a result of $\boldsymbol{u}$.

23.1.3 Elasticity

The third pillar of structural theory (with equilibrium and elongation-displacements) is the constitutive law or material properties, in our case, Elasticity. The elements are made of elastic materials governed by Hooke's Law, $\sigma = E\epsilon$. For a spring i, the linear relation is between the force n_i applied to the spring and its elongation e_i, that

Fig. 23.4 Elasticity

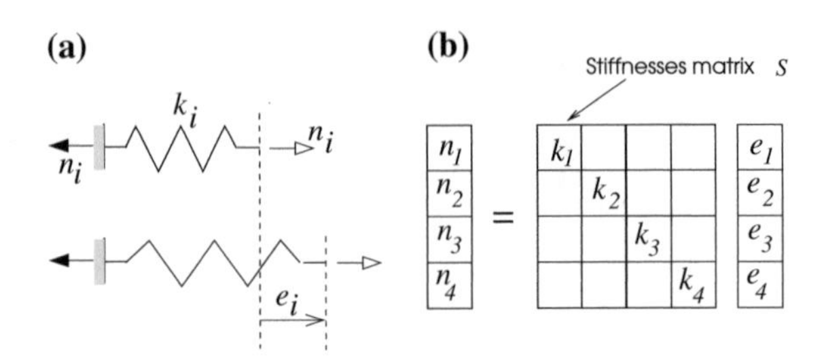

is, $n_i = k_i e_i$ (Fig. 23.4a) where k_i is the spring's stiffness. For a given force, if the stiffness k_i increases the elongation, e_i, is reduced.

We can write this equation for all the springs in matrix form

$$\boldsymbol{n} = \boldsymbol{S}\boldsymbol{e} \qquad \text{(Elasticity)} \tag{23.6}$$

where S is a diagonal matrix with the stiffnesses of the springs along the diagonal (Fig. 23.4b).

23.2 Methods of Analysis

23.2.1 The Basic Method

We will show that the basic method encompasses both classic analysis methods. Indeed, the equilibrium equations (in terms of the N nodal displacements) and the compatibility equations (in terms of R redundant forces) are subsets of the basic equations.

We have N equilibrium equations, M deformation equations and M elasticity equations in N displacements, M internal forces and M deformations.

$$\begin{aligned} \boldsymbol{p} &= \boldsymbol{Q}\,\boldsymbol{n} \text{ Equilibrium} \\ \boldsymbol{e} &= \boldsymbol{Q}^T\boldsymbol{u} \text{ Deformation} \\ \boldsymbol{n} &= \boldsymbol{S}\,\boldsymbol{e} \text{ Elasticity} \end{aligned} \tag{23.7}$$

These $N + 2M$ equations in $N + 2M$ variables can be combined in a single matrix equation with a symmetric coefficient matrix

$$\begin{bmatrix} \mathbf{0} & \boldsymbol{Q} & \mathbf{0} \\ \boldsymbol{Q}^T & \mathbf{0} & -\boldsymbol{I} \\ \mathbf{0} & -\boldsymbol{I} & \boldsymbol{S} \end{bmatrix} \begin{Bmatrix} \mathbf{u} \\ \mathbf{n} \\ \mathbf{e} \end{Bmatrix} = \begin{Bmatrix} \mathbf{p} \\ \mathbf{0} \\ \mathbf{0} \end{Bmatrix} \tag{23.8}$$

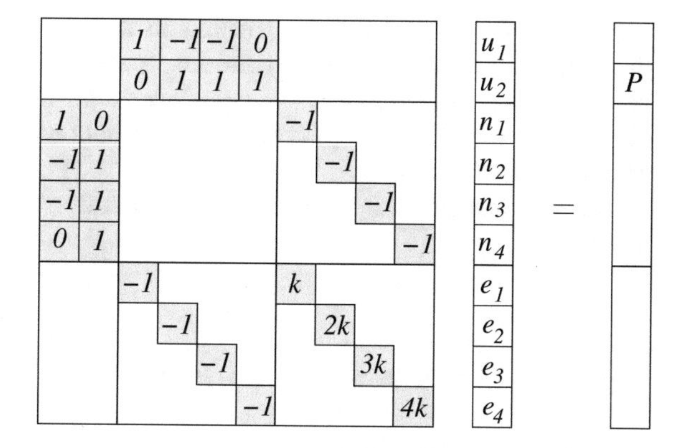

Fig. 23.5 Structure equations for 4 springs example

where **0** stands for zero matrices of appropriate dimensions; $\boldsymbol{I}$ is a $M \times M$ unit diagonal matrix and $\boldsymbol{S}$ is the $M \times M$ diagonal, element stiffnesses matrix. It has the element stiffnesses S_{jj} on its diagonal.

We call these $N+2M$ equations the *structure equations* and their coefficient matrix is accordingly the *structure matrix*. Note the symmetry of the structure matrix.

If we assume springs of stiffness $k_1 = k$, $k_2 = 2k$, $k_3 = 3k$, $k_4 = 4k$, and a force P applied at degree of freedom 2 (see Fig. 23.1), the structure equations take the form given in Fig. 23.5, where blanks represent zeros.

Note that, for the solution of the linear system of equations $\boldsymbol{Ax} = \boldsymbol{b}$, we use only numbers and omit the parameters k and P defining the dimensions; and so the result is $\boldsymbol{x} = \{0.1724\ 0.2069\ 0.1724\ 0.0690\ 0.1034\ 0.8276\ 0.1724\ 0.0345\ 0.0345\ 0.2069\}^T$.

When parsed one can reconstitute the proper dimensions

$$\begin{aligned}\mathbf{u} &= (P/k)\{0.1724\ \ 0.2069\}^T\\ \mathbf{n} &= P\{0.1724\ \ 0.0690\ \ 0.1034\ \ 0.8276\}^T\\ \mathbf{e} &= (P/k)\{0.1724\ \ 0.0345\ \ 0.0345\ \ 0.2069\}^T\end{aligned} \tag{23.9}$$

It is noteworthy that

the tendency for forces is to flow along the stiffer paths.

For instance, force P, which acts along degree of freedom 2, splits into $0.8276P$ for the right branch and $0.1724P$ for the left one. Similarly, the latter splits between springs 2 and 3 following their relative stiffnesses 2/5 and 3/5. However, the elongations e_2 and e_3 are equal as imposed by the geometry.

In practice the structure equations are never used, probably due to the excessive size of the coefficient matrices. Instead, the stiffness method with its $N \times N$ stiffness matrix and the simple method to assemble that matrix is preferred. For small-scale structures with low redundancy we often use the force method with its $R \times R$ flexibility matrix, where R is the degree of static redundancy of the structure.

We will show that the displacement and force analysis methods are subsets of the structure equations (Fig. 23.6). Indeed, when we solve the structure equations in parts we could stumble on the N equations of the displacement method or the R (degree of static redundancy) compatibility equations, although the latter equations are a little convoluted.

The two methods are however used without ever passing through the structure equations.

23.2.2 The Displacement Method

N Equilibrium Equations

In system (23.7) we substitute the elasticity relation for $\boldsymbol{n}$ in the equilibrium equations, $\boldsymbol{p} = \boldsymbol{Q}\boldsymbol{S}\boldsymbol{e}$, and next we replace $\boldsymbol{e}$ by the elongation-displacements relation yielding $\boldsymbol{p} = \boldsymbol{Q}\boldsymbol{S}\boldsymbol{Q}^T\boldsymbol{u}$.

We define the $N \times N$ stiffness matrix

$$\boldsymbol{K} = \boldsymbol{Q}\boldsymbol{S}\boldsymbol{Q}^T \tag{23.10}$$

yielding the N equilibrium equations in terms of nodal displacements

$$\boldsymbol{K}\,\boldsymbol{u} = \boldsymbol{p} \tag{23.11}$$

This is the displacement method. Having solved these equations we get the nodal displacements $\bar{\boldsymbol{u}}$ (see footnote[2]). Pre-multiplying by $\boldsymbol{Q}^T$ yields the deformations $\boldsymbol{e} = \boldsymbol{Q}^T\bar{\boldsymbol{u}}$. Finally the internal forces are $\boldsymbol{n} = \boldsymbol{S}\,\bar{\boldsymbol{e}}$.

Assembly

It is easy to show that (23.10) can also be written

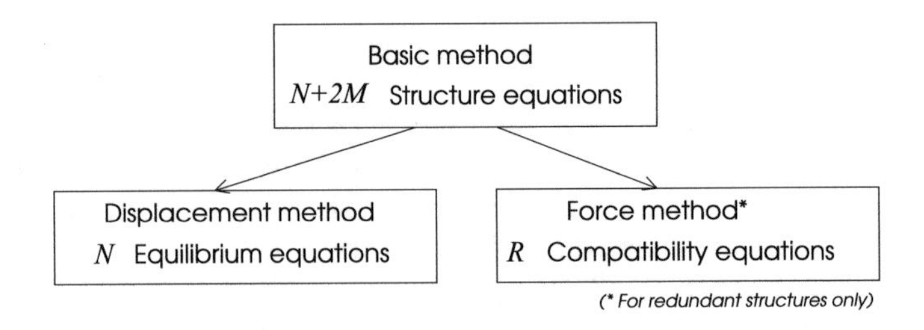

Fig. 23.6 Basic, displacement and force methods

[2]Barred variables indicate calculated values.

$$K = \sum_{j=1}^{M} s_j \, \boldsymbol{q}_j \, \boldsymbol{q}_j^T \tag{23.12}$$

where we recall that $\boldsymbol{q}_j$ is column j of the equilibrium matrix $\boldsymbol{Q}$. Note that term j in the sum is the stiffness matrix of element j in global coordinates,

$$\boldsymbol{K}_j = s_j \, \boldsymbol{q}_j \, \boldsymbol{q}_j^T \tag{23.13}$$

which leads to the assembly method of the stiffness matrix

$$\boldsymbol{K} = \sum_{j=1}^{M} \boldsymbol{K}_j \tag{23.14}$$

This relation is arguably the reason for the success of the stiffness method of analysis.

Note: Matrices of the form $\boldsymbol{a}\boldsymbol{a}^T$ are called dyadic matrices and are singular; and indeed element stiffness matrices are singular.

Example

Using the 'structure' in Fig. 23.1, we find, either by (23.10) or (23.14),

$$\boldsymbol{K} = \begin{bmatrix} 6 & -5 \\ -5 & 9 \end{bmatrix} \tag{23.15}$$

and the solution of $\boldsymbol{K}\,\boldsymbol{u} = \boldsymbol{p}$ is indeed as given in the first line of (23.9). The deformations and the internal forces follow suit.

23.2.3 The Force Method

R Compatibility Equations

In a redundant truss we can remove or disconnect $R = M - N$ carefully chosen bars and still have a stable structure. Not every choice will do. We could remove R bars and end up with an unstable structure. However, we will assume that the substructure is stable. The same is valid for the system of springs. In the following we will be using the truss paradigm when referring to the system of springs.

Let us divide the equilibrium matrix accordingly into a subset *1* of R redundant bars and a subset *0* of N bars composing the determinate structure. For convenience, we denote

$$\boldsymbol{R} = \boldsymbol{Q}^T \tag{23.16}$$

and we will also assume that the redundant bars are numbered last. Consequently, matrix $\boldsymbol{R}$ is partitioned into a square non-singular matrix $\boldsymbol{R}_0$ and matrix $\boldsymbol{R}_1$. We partition $\boldsymbol{Q}$, $\boldsymbol{S}$, $\boldsymbol{e}$ and $\boldsymbol{n}$ properly.

$$\boldsymbol{R} = [\boldsymbol{R}_0\ \boldsymbol{R}_1] \quad \boldsymbol{Q} = [\boldsymbol{Q}_0\ \boldsymbol{Q}_1] \quad \boldsymbol{S} = [\boldsymbol{S}_0\ \boldsymbol{S}_1] \quad \boldsymbol{e} = [\boldsymbol{e}_0\ \boldsymbol{e}_1] \quad \boldsymbol{n} = [\boldsymbol{n}_0\ \boldsymbol{n}_1] \tag{23.17}$$

The non-singularity of $\boldsymbol{R}_0$ (and of $\boldsymbol{Q}_0$) is what makes the substructure stable.

The equilibrium equations, elongation-displacements equations and elasticity equations become

$$\begin{aligned} \boldsymbol{p} &= \boldsymbol{Q}_0\,\boldsymbol{n}_0 + \boldsymbol{Q}_1\,\boldsymbol{n}_1 && \text{Equilibrium} \\ \boldsymbol{e}_0 &= \boldsymbol{R}_0\,\boldsymbol{u} && \text{Elongation-displacements basic bars} \\ \boldsymbol{e}_1 &= \boldsymbol{R}_1\,\boldsymbol{u} && \text{Elongation-displacements redundant bars} \\ \boldsymbol{n}_0 &= \boldsymbol{S}_0\,\boldsymbol{e}_0 && \text{Elasticity basic bars} \\ \boldsymbol{n}_1 &= \boldsymbol{S}_1\,\boldsymbol{e}_1 && \text{Elasticity redundant bars} \end{aligned} \tag{23.18}$$

The elongation-displacements equations (second and third lines in (23.18)) are thus split into two sets. Since $\boldsymbol{R}_0$ is non-singular, we have $\boldsymbol{u} = \boldsymbol{R}_0^{-1}\boldsymbol{e}_0$ which, when introduced in the expression of $\boldsymbol{e}_1$, yields

$$\boldsymbol{e}_1 = \boldsymbol{G}\ \boldsymbol{e}_0 \tag{23.19}$$

where

$$\boldsymbol{G} = \boldsymbol{R}_1\ \boldsymbol{R}_0^{-1} \tag{23.20}$$

is a matrix depending solely on the geometry of the structure. Equation (23.19) is R compatibility equations, which will allow us to compute the R unknown forces in the redundant bars.

We define the diagonal matrices of the flexibilities of the elements

$$\boldsymbol{F}_0 = \boldsymbol{S}_0^{-1} \qquad \boldsymbol{F}_1 = \boldsymbol{S}_1^{-1} \tag{23.21}$$

Introducing the elasticity equations $\boldsymbol{e}_0 = \boldsymbol{F}_0\,\boldsymbol{n}_0$ and $\boldsymbol{e}_1 = \boldsymbol{F}_1\,\boldsymbol{n}_1$ in the compatibility equation (23.19) yields the compatibility equations in terms of internal forces.

$$\boldsymbol{F}_1\boldsymbol{n}_1 = \boldsymbol{G}\ \boldsymbol{F}_0\boldsymbol{n}_0 \tag{23.22}$$

The equilibrium equations (line 1 in (23.18)) can be put in the form $\boldsymbol{Q}_0\,\boldsymbol{n}_0 = \boldsymbol{p} - \boldsymbol{Q}_1\,\boldsymbol{n}_1$. Note that, assuming that the redundant bars were removed or disconnected from the 'released structure' ($\boldsymbol{n}_1 = \boldsymbol{0}$), the forces in the released structure under the applied loads are

$$\boldsymbol{n}_r = \boldsymbol{Q}_0^{-1}\,\boldsymbol{p} \tag{23.23}$$

and consequently the R compatibility equations become, after some manipulation,

$$(\boldsymbol{F}_1 + \boldsymbol{G}\ \boldsymbol{F}_0\,\boldsymbol{G}^T)\,\boldsymbol{n}_1 = \boldsymbol{G}\ \boldsymbol{F}_0\boldsymbol{n}_r \tag{23.24}$$

These are the compatibility equations of the force method.

In Summary

After creating the equilibrium matrix $\boldsymbol{Q}$ and selecting the R redundant bars we solve in sequence the following equations

$$\boldsymbol{Q}_0\, \boldsymbol{n}_r = \boldsymbol{p} \tag{23.25}$$

$$(\boldsymbol{G}\,\boldsymbol{F}_0\,\boldsymbol{G}^T + \boldsymbol{F}_1)\,\boldsymbol{n}_1 = \boldsymbol{G}\,\boldsymbol{F}_0\,\bar{\boldsymbol{n}}_r \tag{23.26}$$

$$\boldsymbol{Q}_0\, \boldsymbol{n}_0 = \boldsymbol{p} - \boldsymbol{Q}_1\, \bar{\boldsymbol{n}}_1 \tag{23.27}$$

We solve N linear equations to find the forces in the released structure under the applied forces. Next, we solve the R compatibility equations for the forces $\boldsymbol{n}_1$ in the redundant bars, and finally we solve N equations for the forces $\boldsymbol{n}_0$ in the basic structure.

Example

Using the 4-springs case, we recall that this is an $R = 2$ redundant structure. We conveniently select elements 3 and 4 as the redundants. Springs 1 and 2 thus constitute the released structure. The basic ingredients are

$$\boldsymbol{Q}_0 = \begin{bmatrix} 1 & -1 \\ 0 & 1 \end{bmatrix} \quad \boldsymbol{Q}_1 = \begin{bmatrix} -1 & 0 \\ 1 & 1 \end{bmatrix}$$

$$\boldsymbol{F}_0 = \begin{bmatrix} 1 & 0 \\ 0 & 1/2 \end{bmatrix} \quad \boldsymbol{F}_1 = \begin{bmatrix} 1/3 & 0 \\ 0 & 1/4 \end{bmatrix} \quad \boldsymbol{p} = \begin{Bmatrix} 0 \\ 1 \end{Bmatrix}$$

where we have omitted the dimension constants from the flexibility and load matrices.

The compatibility equations (23.26) are thus

$$\begin{bmatrix} 0.8333 & 0.5 \\ 0.5 & 1.75 \end{bmatrix} \begin{Bmatrix} n_3 \\ n_4 \end{Bmatrix} = \begin{Bmatrix} 0.5 \\ 1.5 \end{Bmatrix} \tag{23.28}$$

the solution of which is indeed $n_3 = 0.1034$ and $n_4 = 0.8276$.

The forces in the basic elements are obtained by solving (23.27), that is,

$$\begin{bmatrix} 1 & -1 \\ 0 & 1 \end{bmatrix} \begin{Bmatrix} n_1 \\ n_2 \end{Bmatrix} = \begin{Bmatrix} 0.1034 \\ 0.0690 \end{Bmatrix} \tag{23.29}$$

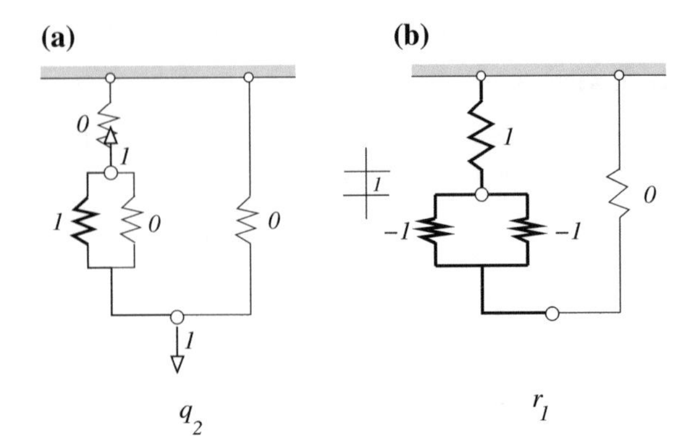

Fig. 23.7 Writing $\boldsymbol{Q}$ and $\boldsymbol{R}$.

23.3 Equivalence

23.3.1 Columns and Rows of Q

Calculating $\mathbf{Q}$ can be done column-wise or row-wise. We will show that the columns of $\boldsymbol{Q}$ represent nodal forces and that the rows of $\boldsymbol{Q}$ (or columns of $\boldsymbol{R}$) are element elongations. This insight is the basis for two methods for obtaining $\boldsymbol{Q}$.

Columns of Q

The equilibrium equations (23.1) can be written as the linear combination

$$\sum_{j=1}^{M} \boldsymbol{q}_j n_j = \boldsymbol{p} \tag{23.30}$$

where $\boldsymbol{q}_j$ is column j of matrix $\boldsymbol{Q}$ and n_j is the force in element j.

Let us assume that $n_m = 1$ and $n_{j \neq m} = 0$. In this case we get $\boldsymbol{q}_m = \boldsymbol{p}$. We note that vector $\boldsymbol{q}$ has the meaning of nodal forces, and in particular

Column $\boldsymbol{q}_m$ is nodal forces in equilibrium with internal forces $n_m = 1, n_{j \neq m} = 0$

For instance, in the 4-springs example, the second column, $\boldsymbol{q}_2 = \{-1\ 1\}^T$ is the nodal forces in equilibrium with internal forces $\boldsymbol{n} = \{0\ 1\ 0\ 0\}^T$. This can be seen in Fig. 23.7a.

This is, by the way, an effective method of creating matrix $\boldsymbol{Q}$ columnwise. For a unit force in element 1 (positive in tension), we need a unit force at dof 1 and nothing at 2, hence the first column of $\boldsymbol{Q}$ is $\boldsymbol{q}_1 = \{1\ 0\}^T$, and so forth.

Rows of Q (Columns of R)

The elongation-displacement equations (23.7) can be written as the linear combination

$$\sum_{j=i}^{N} \boldsymbol{r}_i u_i = \boldsymbol{e} \tag{23.31}$$

where $\boldsymbol{r}_i$ is column i of matrix $\boldsymbol{R}$ (also row i of matrix $\boldsymbol{Q}$) and u_i is the displacement of dof i.

Let us assume that $u_m = 1$ and $u_{j \neq m} = 0$. In this case we get $\boldsymbol{r}_m = \boldsymbol{e}$. We note that vector $\boldsymbol{r}$ has the meaning of element elongations, and in particular

Row $\boldsymbol{r}_m$ is elements elongation due to a unit displacement $u_m = 1$ and $u_{j \neq m} = 0$

For instance, in the 4-springs example, the first row $\boldsymbol{r}_1 = \{1\ -1\ -1\,0\}^T$, is the element elongations compatible (or due to) nodal displacements $\boldsymbol{u} = \{1\,0\}^T$ as is evident in Fig. 23.7b.

23.3.2 On the Equilibrium and Compatibility Equivalence

The stiffness matrix is the assembly of the stiffness matrices of all the springs of the structure. Following the logic of the force method, the bars composing the structure are divided into N determinate bars composing the released structure (subscript 0) and $R = M - N$ redundant bars (subscript 1) similar to what was done in (23.18).

The stiffness matrix $\boldsymbol{K}$ can accordingly be divided into the stiffness matrix $\boldsymbol{K}_0$ resulting from the assembly of the determinate bars and the stiffness matrix $\boldsymbol{K}_1$ resulting from the assembly of the redundant bars.

Clearly, with

$$\boldsymbol{K} = \boldsymbol{K}_0 + \boldsymbol{K}_1 \tag{23.32}$$

the equilibrium equations become

$$(\boldsymbol{K}_0 + \boldsymbol{K}_1)\,\boldsymbol{u} = \boldsymbol{p} \tag{23.33}$$

Pre-multiplying both sides of the above equation by $\boldsymbol{K}_0^{-1}$ yields

$$(\boldsymbol{I} + \boldsymbol{K}_0^{-1}\boldsymbol{K}_1)\,\boldsymbol{u} = \boldsymbol{K}_0^{-1} p \tag{23.34}$$

where I is the $N \times N$ diagonal unit matrix

We note that $\boldsymbol{K}_1 = \boldsymbol{Q}_1\boldsymbol{S}_1\boldsymbol{R}_1$ and $\boldsymbol{K}_0 = \boldsymbol{Q}_0\boldsymbol{S}_0\boldsymbol{R}_0$ were subscript 1 refers to that part of the structure with the redundant bars and 0 corresponds to the determinate substructure.

Pre-multiplying both sides of (23.34) by $\boldsymbol{R}_1$ gives $\boldsymbol{R}_1(\boldsymbol{I} + \boldsymbol{K}_0^{-1}\boldsymbol{K}_1)\,\boldsymbol{u} = \boldsymbol{R}_1\boldsymbol{K}_0^{-1} p$ or, after expansion,

$$\boldsymbol{F}_1\boldsymbol{S}_1\boldsymbol{R}_1\boldsymbol{u} + \boldsymbol{R}_1\boldsymbol{K}_0^{-1}\boldsymbol{K}_1\boldsymbol{u} = \boldsymbol{R}_1\boldsymbol{K}_0^{-1} p \tag{23.35}$$

where we have multiplied the first term by the unit matrix $\boldsymbol{F}_1\boldsymbol{S}_1$ for further use. Noticing that $\boldsymbol{K}_0^{-1} = \boldsymbol{R}_0^{-1}\boldsymbol{F}_0\boldsymbol{Q}_0^{-1}$ and that $\boldsymbol{S}_1\boldsymbol{R}_1\boldsymbol{u}$ is the forces $\boldsymbol{n}_1$ in the redundant bars, (23.35) becomes

$$\boldsymbol{F}_1\boldsymbol{n}_1 + (\boldsymbol{R}_1\boldsymbol{R}_0^{-1})\boldsymbol{F}_0(\boldsymbol{Q}_0^{-1}\boldsymbol{Q}_1)(\boldsymbol{S}_1\boldsymbol{R}_1\boldsymbol{u}) = (\boldsymbol{R}_1\boldsymbol{R}_0^{-1})\boldsymbol{F}_0(\boldsymbol{Q}_0^{-1}p) \qquad (23.36)$$

which is exactly the R compatibility equations (23.24)

$$(\boldsymbol{F}_1 + \boldsymbol{G}\,\boldsymbol{F}_0\,\boldsymbol{G}^T)\,\boldsymbol{n}_1 = \boldsymbol{G}\,\boldsymbol{F}_0\boldsymbol{n}_r$$

where we recall that $\boldsymbol{n}_r$ are the forces in the stand-alone determinate structure under the applied forces.

23.4 Plus ça Change . . .

I am by nature a minimalist and this book is an expression of a minimalistic view of structures. The theory of structures has been developed over the years by many towering scientists. Cremona, Mohr, Navier, Bernoulli, Pasternak, Betti, Hooke, Castigliano, Menabrea, Volterra and more recently Cross, Timoshenko, Argyris, to mention but a few, have all left their enduring mark on structural analysis.

The theory has matured and it was a privilege to present a streamlined theory based on three simple concepts, equilibrium, deformations (virtual work) and elasticity, and show how the analysis of a variety of structures from simple spring arrangements to complex frames are analysed on the basis of these three principles.

Plus ça change, plus c'est la même chose

Printed in the United States
By Bookmasters